# Osiris

A RESEARCH JOURNAL DEVOTED TO THE HISTORY OF SCIENCE AND ITS CULTURAL INFLUENCES



EDITORIAL OFFICE
DEPARTMENT OF SCIENCE AND TECHNOLOGY STUDIES
CORNELL UNIVERSITY
726 UNIVERSITY AVENUE
ITHACA, NEW YORK 14850 USA

## SUGGESTIONS FOR CONTRIBUTORS TO OSIRIS

OSIRIS is devoted to thematic issues, often conceived and compiled by guest editors.

1. Manuscripts should be **typewritten** or processed on a **letter-quality** printer and **double-spaced** throughout, including quotations and notes, on paper of standard size or weight. Margins should be wider than usual to allow space for instructions to the typesetter. The right-hand margin should be left ragged (not justified) to maintain even spacing and readability.

2. Bibliographic information should be given in **footnotes** (not parenthetically in the text), typed separately from the main body of the manuscript, **double-** or even **triple-spaced,** numbered consecutively throughout the article, and keyed to reference numbers typed above the line in the text.

   a. References to **books** should include author's full name; complete title of the book, underlined (italics); place of publication and publisher's name for books published after 1900; date of publication, including the original date when a reprint is being cited; page numbers cited. *Example:*

   [1]Joseph Needham, *Science and Civilisation in China,* 5 vols., Vol. I: *Introductory Orientations* (Cambridge: Cambridge Univ. Press, 1954), p. 7.

   b. References to articles in **periodicals** should include author's name; title of article, in quotes; title of periodical, underlined; year; volume number, Arabic and underlined; number of issue if pagination requires it; page numbers of article; number of particular page cited. Journal titles are spelled out in full on first citation and abbreviated subsequently. *Example:*

   [2]John C. Greene, "Reflections of the Progress of Darwin Studies," *Journal of the History of Biology,* 1975, *8*:243–272, on p. 270; and Dov Ospovat, "God and Natural Selection: The Darwinian Idea of Design," *J. Hist. Biol.,* 1980, *13*:169–174, on p. 171

   c. When first citing a reference, please give the title in full. For succeeding citations, please use an abbreviated version of the title with the author's last name. *Example:*

   [3]Greene, "Reflections" (cit. n. 2), p. 250.

3. Please mark clearly for the typesetter all unusual alphabets, special characters, mathematics, and chemical formulae, and include all diacritical marks.

4. A small number of **figures** may be used to illustrate an article. Line drawings should be directly reproducible; glossy prints should be furnished for all halftone illustrations.

5. Manuscripts should be submitted to OSIRIS with the understanding that upon publication **copyright** will be transferred to the History of Science Society. That understanding precludes OSIRIS from considering material that has been submitted or accepted for publication elsewhere.

---

OSIRIS (SSN 0369-7827) is published once a year.

Subscriptions are $39 (hardcover) and $25 (paperback).

Address subscriptions, single issue orders, claims for missing issues, and advertising inquiries to *Osiris,* The University of Chicago Press, Journals Division, P.O. Box 37005, Chicago, Illinois 60637.

**Postmaster:** Send address changes to *Osiris,* The University of Chicago Press, Journals Division, P.O. Box 37005, Chicago, Illinois 60637.

Osiris is indexed in major scientific and historical indexing services, including *Biological Abstracts, Current Contexts, Historical Abstracts,* and *America: History and Life.*



Hardcover edition, ISBN 0-226-00090-7
Paperback edition, ISBN 0-226-00091-5

# Commemorative Practices in Science: Historical Perspectives on the Politics of Collective Memory

Edited by Pnina G. Abir-Am and Clark A. Elliott

A RESEARCH JOURNAL
DEVOTED TO THE HISTORY OF SCIENCE
AND ITS CULTURAL INFLUENCES
SECOND SERIES VOLUME 14 1999

**Cover:** A montage of images from three scientific commemorations: Arthur Szyk's portrait of Copernicus, commissioned by the Kosciuszko Foundation in preparation for the 1943 quadracentennial of Copernicus's death; Harvard President James B. Conant addressing participants at the university's three-hundreth anniversary celebration in 1936; and guests of honor at a West Berlin ceremony recognizing the 1957 centennial of Max Planck's birth.

# Acknowledgments

We are very grateful for the gracious cooperation of an international community of contributors—hailing from Canada, France, Germany, Israel, and the United States—who have enabled us to produce the first multi-author, English-language volume on the history of scientific commemorations. We are particularly thankful to Professor Charles S. Maier, Director of the Minda de Gunzburg Center for European Studies at Harvard University and a pioneer in the study of commemoration, history, and memory, for finding the time to write the volume's Preface.

Eight out of the fourteen papers in this volume originated in a symposium held on April 10 and 11, 1995, as part of the Boston Colloquium for Philosophy of Science at Boston University (established in 1960 by Robert S. Cohen and Marx Wartofsky). We are indebted to Professor Alfred I. Tauber for including our symposium in his Colloquium and for extending hospitality to our contributors, especially to co-editor Pnina G. Abir-Am, who organized the symposium during a research fellowship at Boston University.

This volume required a large number of referees from inside and outside the United States to whom we are much indebted for helpful advice. In addition, we benefited from the generosity of several colleagues, most notably Barton Bernstein, Cathryn Carson, Joy Harvey, Mike Neufeld, and Alex Roland.

One of us (PGA) wishes to acknowledge the ongoing hospitality and intellectual stimulation provided by the Dibner Institute for the History of Science and Technology at M.I.T. and its Burndy Library; by the Center for European Studies at Harvard University; and by the National Science Foundation, especially through its STS-95-25794 research grant.

Above all, we wish to express our gratitude to the *Osiris* Editor-in-Chief, Professor Margaret W. Rossiter of Cornell University, for her unfailing confidence in our project, to the *Osiris* Editorial Board for accepting our volume proposal, and to the *Osiris* staff for its brave and devoted handling of the production process.

Pnina G. Abir-Am
Clark A. Elliott

# Preface

"SCIENTIFIC COMMEMORATION"—THE NOTION initially sounds fraught with self-contradiction. Commemoration answers the need to celebrate, to look back, often with filial piety. Science is an activity that, at least in the popular mind, restlessly looks forward through its continual refinement, if not abandonment, of old concepts. Hence, what does commemoration have to do with science? And what does commemoration have to teach about the history of science? In fact, as this volume demonstrates, a great deal. The study of commemorations opens a new approach to the history of science, as it has already done for historians of society and politics more generally. Conversely, the history of scientific commemoration has much to teach historians in general about issues of collective memory and representation. As a preeminent cultural activity of the twentieth century, science offers a privileged vantage point from which to learn how commemoration functions. As an enterprise claiming canons of rational procedure, its commemorations reveal not that these claims are false, but that in science too, many agendas, strategic sites, and heroes are possible—and commemorators will often compete by constructing persuasive, yet contrasting, representations of the past.

Like all the other ways to transmit representations of the past, the practice of commemoration reveals itself as increasingly complicated the more we reflect on it. An activity that initially sounds straightforward—answering a need to celebrate and recollect as a group, and thereby to establish a collective identity on the basis of shared descent—emerges, so this volume establishes, as complex, strategic, and self-interested. Thus, the importance of this collection transcends the history of science alone, although that particular history certainly remains central for understanding the century now closing.

The diverse commemorative practices prevailing in science that are analyzed in this volume expose the ambiguities and ambitions that mark rituals of collective remembering in all spheres of human life. Scientists, like others, commemorate for many different reasons. We commemorate to reincorporate a shared past whose meaning seems destined otherwise to fade; yet we commemorate simultaneously to establish that the object of memory is precisely history, past, no longer controlling, and that collective identity persists as a vital force in the present. Commemoration thus honors, but also safely confines to the past, the achievements of an innovative ancestor out of whose shadow a new generation struggles to emerge. This collection, including Pnina Abir-Am's contributions, also remind us that we commemorate for political reasons: to legitimize succession, to delineate the inheritors of authority, and to demonstrate whose claims can be set aside.

Commemoration thus has a dual thrust. By revisiting an earlier achievement, it celebrates community solidarities—genuinely inherited or belatedly organized—with respect to the past. In claiming to strengthen solidarities, to negotiate collec-

 0369-7827/99/1401-0001$02.00

tive identities, to establish the boundaries of groups on the basis of intellectual or institutional legacies, commemorations must remain in implicit tension with analytical and critical history, which always deconstructs alleged solidarities to reveal conflicts and contradiction. This does not prevent some commemorations from being designed as critical narratives in their own right, as in the case of the Smithsonian's abortive *Enola Gay* exhibit, recalled in this volume by Stanley Goldberg. Such a "counter-commemoration" proposed an alternative community constituted by victims abroad and dissenters at home. As such, it revealed that the commemorative effort is exquisitely political, hinging on the power of various constituencies to get their own memories publicly accepted. So, too, of course, were the competitive commemorations of the Planck centennials organized by East and West Germans, and discussed in this volume by Dieter Hoffmann: the GDR seizing on Planck's contributions to enhance its ideological and national claims; the West Germans endeavoring implicitly to claim a subsisting German unity.

By proposing a teleological narrative that celebrates earlier science as a prelude for the work of the commemorators, or marginalizes rival approaches, such commemorative practices remain profoundly political. To shape such a purposeful past, the guardians of memory select scattered events, bring together a particular cohort, and exclude others. And, even if the impulse to commemorate springs from the normal traditions of an academic discipline and avoids a partisan agenda, it still reaffirms a scientific community, reveals boundaries, and makes a claim for self-respect. Thus, such a commemoration often legitimates changes in the direction of an established institution, as in the case of the Pasteur Institute in Paris or the Laboratory at Cold Spring Harbor. Staking out control of the past, through selective representations, has emerged as crucial for influencing the future flow of resources—material and spiritual.

As an act of collective memory, commemoration is, as historians have understood since Maurice Halbwachs' pioneering study, really an act of constructing memory not retrieving it—a construction process keyed to a particular institution or event through which a group passes. Commemorations are sited in time analogously to the way Pierre Nora's famous *lieux de mémoire* are sited in space; indeed, they might be described as *journées de mémoire.* Commemoration is thus subject to all the ambiguities that have now been explored by writers who deal with history and memory. These are terms that we cannot simply oppose to each other: the first as allegedly analytical and "objective"; the second as subjective and cathectic. In fact, history and memory intertwine and nourish each other, even as they necessitate the importance of the other.

As she explains in the volume's Introduction, Pnina G. Abir-Am has organized the examples of scientific commemoration around three poles of orientation. First, that of the "great mind"—often a man's; occasionally a woman's—whose celebration can legitimate new ideas or even whole nations (as in the case of Copernicus for Poland). Second, commemorations of institutions, such as Harvard University or the Lawrence-Berkeley Laboratory, which justify a continuing or a transformed organizational mission. Third, the commemoration of a methodology or a discipline, permitting a more sophisticated claim on scientific custodianship, and even (as in the case of the Niels Bohr commemorations and the zany collective mathematical play that the "Bourbaki" group sustained) exploiting inside humor and a self-mockery to confirm the cleverness and disinterest of the *magistri ludi.*

Without apology, commemorative activity incorporates dimensions of memory as well as analytic history. The tension between history and memory inherent in commemoration is only one of the contested terrains for contemporary historiography. Late twentieth century historical output—monographs, narratives, exhibitions, films and documentaries, commemorations—emerges out of at least three ongoing methodological struggles: one between analysis and narration, a second between the unique and the comparable, and, the one so central to commemoration, between history and memory.

The first tension, that between narrative and analysis, helps shape the form of the historical product, its artfulness and plausibility as an account of events—the skill, so to speak, that makes an argument persuasive. Historians may debate the appropriate degree of "emplotment," to use Hayden White's term, and postmodern writers may even deny any firm boundary between history and fiction in a world of docudramas, filmed history, historical novels, etc. Probably just as corrosive has been the growing license for the intrusions of the participant-historian (a testimony to the importance of ethnographic approaches) and the ordering of the story around her or his role in mediating history.

The second tension is the familiar one between elements of uniqueness and comparability. The definition of history as a "science" of explaining unique developments was the contribution of German Historismus and Weberian ideal-type sociology. Both history and social theory "explained," but the former did so by virtue of sequential ordering of unique events—the vicissitudes of states and other institutions—and the latter by the application of encompassing typologies. Yet, the historian could not renounce ideal types. Hence, this dilemma became all the more acute with the need to explain the movements so characteristic of the nineteenth and twentieth centuries: nationalism, fascism, Nazism, and communism.

Nonetheless, over the past two decades historians have tended to retreat from the application of social-science approaches or modeling. They have fallen back into a neo-Historismus that celebrates narrative and accepts its sufficiency for historical explanation. In any case, the history of commemoration faces this tension no more and no less than historiography in general. Comparative histories of commemoration involve deriving ideal types, while single case studies, as those in this volume testify, rely on the explanatory adequacy of historical narrative, even if they apply insights from ethnography, literary theory, or other disciplines.

The problems raised by this volume's enterprise lie rather in the third field of tension: the uneasy relation of memory to history, which has become troubling especially in the last decade for reasons that Pnina G. Abir-Am summarizes in the volume's Introduction. The anniversaries of World War II, the Holocaust, and the fall of the Communist project have given the watersheds of contemporary history (with their anniversaries) an overpowering presence. We are, in effect, doomed to commemorate—and to render such events meaningful, all the more when they have been tragic and sacrificial.

The successive efforts to interpret these developments (what the Germans call historicization) have continually evoked the claims of memory as well as the art of representation. It is not that all commemoration involves representing the catastrophes or triumphs of the twentieth century, but that historians understand every act of commemoration places at stake claims of meaning and significance.

The historian of scientific commemoration must be all the more alert to the tensions inherent in sorting out history and memory because scientists often make claims to objectivity that their own commemorations (no more than any other commemorations) cannot satisfy. Entangled in memory and the politics of science, scientific commemoration thus provides a privileged ground for reflecting on commemoration and memory in general.

The historian of commemoration sooner or later must ask whether it is possible to organize commemorative activity that does not merely relegate the past to a safe and agreed upon foundational act, vitally important but now safely transcended. Historians will eventually explore the conflictual strands within given commemorations. But can the commemorators themselves carry out their assignments in a context of debate? Can one both commemorate and argue with, so to speak, the object of commemoration—that is, build dialectic into commemoration as it is part of historiographical praxis? Obviously, the common public understanding of commemoration precludes such an adversarial approach.

The history of the Smithsonian's *Enola Gay* exhibit as recounted in this volume provides an instructive lesson about how commemorative expectations collided with an effort at public education. Exhibit organizers aspired to a critical exhibit that challenged a prevalent narrative, and the guardians of memory wanted celebration untroubled by the complexities of history. This is not to argue that the proposed exhibit was more correct or should not have been challenged in the preparatory stages. Indeed, contested interpretations could have been built into the exhibit perhaps without evoking such a furious reaction. Probably it is too much to expect that soldiers who have "lived" a history should be asked to critique it; but this volume suggests that it is also unlikely that scientists will write their own history, although—as the studies of the Los Alamos National Laboratory and CERN suggest—they may discover advantages in having historians discern it for them.

Ultimately, history and commemoration must remain in fundamental tension. Commemoration involves the invocation and nurturing of solidarity; history requires the undermining and questioning of solidarities. At the same time the historical impulse cannot be liberated from the commemorative. It is illusory to expect that modern science, the adventure par excellence of the twentieth century, can escape that tension. Rather by understanding how it is played out in commemorative practice, the historian of science gains a new insight into how science proceeds as a collective enterprise: its strategies, alliances, rewards. The history of commemoration, in science as in war or politics or artistic endeavor, tells us far more than the commemorators ever wanted to disclose.

Charles S. Maier

# Introduction

## *By Pnina G. Abir-Am*

THE COMMEMORATIVE MANIA that swept the world over the past decade—which peaked between 1989 and 1995 with the convergence of the bicentennial of the French revolution and the half-centennial of World War II—highlighted a revival of public interest in rearticulating the relationship between history and memory in postmodern society.[1] Since international society became, in these same years, increasingly decentralized, depolarized, and interactive, with new boundaries constantly replacing the traditional national ones, this interest in history

[1] On the bicentennial of the French Revolution see "La révolution dans l'imaginaire politique Français," *Le Débat,* 1983, *26;* François Furet, *La révolution* (Paris: Hachette, 1988); Pascal Ory, *Une nation pour mémoire:1889, 1939, 1989, trois jubilés révolutionnaires* (Paris: Presse de la Fondation Nationale des sciences politiques, 1992); Patrick Hutton, "The Role of Memory in the Historiography of the French Revolution," *History and Theory,* 1991, *30:*56–69. Hutton's essay, in addition to being more accessible to American readers, is particularly useful in demonstrating the shifting images of the Revolution at each successive half-centennial, thus illustrating the changing impact of presentist agendas upon the constructs of collective memory of the same historical event.

A plethora of books were issued to commemorate the ending of World War II. See, for example, the survey by Samuel Hynes, "Fifty Years of Remembering World War II," *New York Times Book Review,* 13 April 1995, pp. 12, 14. (Incidentally, Hynes pronounced Martin Gilbert's *The Day the War Ended: May 8, 1945—Victory in Europe* (London, Holt: 1995) to be the most readable. For scholarly debates on the problems of commemorating the end of World War II in the United States, see the "Special Section: The Last Act," *Technology and Culture,* July 1998, *39:*457–98, in which several historians comment on four books about the derailed *Enola Gay* exhibition at the Smithsonian Institution.

On the poststructuralist and postmodernist roots of the revival of interest in the relationship between memory and history, see "Memory and History," a special issue of *History and Anthropology,* 1986, *2,* especially the comprehensive and masterful introduction by Nathan Wachtel, pp. 207–24. See also "Memory and Counter-Memory," a special issue of *Representations,* 1989, *26,* especially the incisive introduction by coeditors Natalie Zemon Davis and Randolph Starn, pp. 1–6. See also Pnina G. Abir-Am's, introduction to *La mise en mémoire de la science: Pour une ethnographie historique des rites commemoratifs,* ed. Pnina G. Abir-Am (Paris: Editions des Archives Contemporaines, 1998), pp. 1–14. On the implications of the convergence of commemorative timing with real-time events see *Les lieux de mémoire,* 9 vols., ed. Pierre Nora (Paris: Gallimard, 1984–1993); volume 1 has recently been translated by Arthur Goldhammer for Harvard University Press. This extensive treatment of the problems of history and memory by over one hundred historians, which is also a methodological gold mine, deals with French issues only. For a broader claim that the new preoccupation with memory reflects the bankruptcy of collective aspirations for the future in society, see Charles S. Maier, "A Surfeit of Memory? Reflections on History, Melancholy and Denial," *History and Memory,* 1993, *4:*136–52. On the convergence of multiple commemorative moments in the late 1980s with other real-time events, most notably *glasnost* (which encouraged the rehabilitation of large tracts of formerly repressed public memory in Eastern Europe) in precipitating a turning point in scholarly preoccupation with the nexus of history and memory, see Stephane Courtois, "Archives du communisme: Mort d'une mémoire, naissance d'une histoire," *Le Débat,* 1993, *77:*146–56; Richard S. Esbenshade, "Remembering to Forget: Memory, History, National Identity in Postwar East-Central Europe," *Representations,* 1995, *49:*72–95.

and memory is likely to continue as a major element of both scholarly and public discourse in the twenty-first century.[2]

As befits a world in which science has become a major source of cultural literacy and economic welfare, the large-scale commemorative revival of the late 1980s and early 1990s featured a considerable scientific component. For example, the bicentennial of several scientific institutions created by the French Revolution and the half-centennial of the atomic bombing of Japan were landmarks in an increasingly sophisticated tradition of commemoration related not only to technoscience but also to prominent, even controversial, public events.[3]

These two major commemorations generated much analysis by a variety of commentators, including professional historians, who dramatically expanded the scholarly discourse on commemorations into almost a subdiscipline in its own right.[4] Nevertheless, the comparative potential of this discourse remains largely unexplored since, with minor exceptions, it remains linguistically and thematically segregated, especially with regard to its French and American perspectives. These two cultures are not only among the most substantive contributors to the commemorative literature, they also provide contrasting traditions of commemoration.

By bringing together a critical mass of case studies and exploring its comparative potential, whether crosscultural or transhistorical, this volume fills a methodological gap in the commemorative literature. Furthermore, this volume focuses on the scientific component of commemorative activity (which is often missing from the literature) and clarifies the features it shares with commemorative practices in other spheres of society. For example, in parts I and II, commemorative objects in science that also prevail in the cultural, social, or political realms are explored.

In contrast, part III focuses on commemorative objects that are unique to science: discoveries, theories, disciplines, innovations, techniques, and instruments. Including such a wide range of objects raises the question how the *epistemological status* of a given object has an impact on the commemorative practices that develop around

[2] On nationalism as a key agenda in commemorative activity, see *The Invention of Tradition,* eds. Eric Hobsbawm and Thomas Ranger (Cambridge: Cambridge University Press, 1992); Eric Hobsbawm, "From One Centenary to Another," in his *Echoes of the Marseillaise: Two Centuries Look Back on the French Revolution* (New Brunswick, N.J.: Rutgers University Press, 1990), pp. 67–90; *Commemorations: The Politics of National Identity,* ed. John R. Gillis (Princeton: Princeton University Press, 1994).

[3] On events in France, see for example the symposium published for the bicentennial of the Ecole Polytechnique, *La Formation Polytechnicienne,* eds. Bruno Belhoste, Amy Dalmedico Dahan, and Antoine Picon (Paris: Dunod, 1994), among many other bicentennial publications issued by the Conservatoire National des Arts et Métiers, Museum National d'Histoire Naturelle, Ecole Normale Supérieure (hereafter ENS), Académie de Médecine, and other institutions established or restructured during the French Revolution.

On anniversaries of the atomic bomb, see chapter 8 in this volume of *Osiris.* See also Barton J. Bernstein, "The Struggle Over History: Defining the Hiroshima Narrative" in *Judgement at the Smithsonian,* ed. Philip Nobile (New York:1995), pp. 127–256; Martin Harwit, *An Exhibit Denied: Lobbying the History of the "Enola Gay"* (New York: Copernicus, 1996); *History Wars: The "Enola Gay" and Other Battles for the American Past,* eds. Edward T. Linenthal and Tom Engelhardt (New York: Metropolitan Books, 1996); *Hiroshima's Shadow: Writings on the Denial of History and the Smithsonian Controversy,* eds. Kai Bird and Lawrence Lifschultz (Stony Creek, Conn.: Pamphlateer Press, 1998); see also note 1.

[4] The most convenient entry into this new area is the journal *History and Memory,* established in 1989, which has great international authorship and is particularly strong in comparative studies. For a manifesto-type essay opening its first issue (by an intellectual historian who also worked in history of science), see Amos Funkenstein, "Collective Memory and Historical Consciousness," *History and Memory,* 1989, *1*:5–26.

it. This status ranges from the idiosyncratic subjectivity of great scientific minds to the solid objectivity of discoveries and the quasi objectivity of scientific societies and laboratories. As this volume documents, diverse degrees of linkage may obtain among these three categories of commemorative objects, with some great minds becoming inseparable from certain discoveries or institutions, a situation sometimes acknowledged by conflating two or more commemorative objects in one event.

Other questions that this volume seeks to address pertain to the role of *cultural context* in commemorative practices in science, particularly the French and American traditions, which are explored in over a dozen case studies. Here a major challenge is to evaluate whether the commemorative rites for a great scientist share more features with those held for a compatriot in another field or with those for a comparable scientist from another country. In the same vein, this volume challenges the nationalism that remains prominent in commemorations outside science through examples in which scientists have been commemorated outside of their country of origin.

Another major objective of this study is to shed light on the *historicity* of the commemorative events under scrutiny, especially trends in terms of the timing and frequency of staging commemorations throughout the twentieth century. Was the commemorative mania of the early 1990s a phenomenon unique to this century, or was it part of a longer tradition in science? For example, the late nineteenth century also saw intense commemorative activity, in part resulting from the deliberate efforts of the positivist movement to substitute royalist and religious memories of kings and saints with those of great minds (which often meant scientists), and to establish science as a secular religion.[5] During that time, a technoscientific device, the Eiffel Tower, commissioned for the centennial of the French Revolution (a grand affair during the commemoration-conscious Third Republic) became France's most famous landmark of progress.[6] Similarly, the centennial of the Ecole Polytechnique in 1894 was a commemorative event on an unusually large scale, as befit an elite military school during a critical period of preparation for *revanche* (to regain Alsace and Lorraine, lost to Germany in 1871). It was also a state affair, since the president of the Republic, Sadi-Carnot, was an alumnus as well as the grandson of a famous scientist who contributed to the principle of conservation of energy and participated in the French Revolution.[7]

Though some commemorative activity had persisted in science throughout the nineteenth century, the emphasis at the end of the century was more systematic.[8]

[5] On the practices and outlook of the positivist movement with regard to scientific commemoration, see Annie Petit, "La commémoration de l'héritage scientifique dans le positivisme: Théories et pratiques" in Abir-Am, *La mise en mémoire* (cit. n. 1), pp. 159–90. For further examples see chapter 2 in this volume of *Osiris;* and the section on Lavoisier's centennial by Bernadette Bensaude-Vincent, "Lavoisier par documents et monuments: Reflexions sur deux cents ans de commemoration," in Abir-Am, *La mise en mémoire* (cit. n. 1), pp. 265–88; *idem, Lavoisier: Mémoire d'une revolution* (Paris: Flammarion, 1993).

[6] On state-sponsored commemorations of great minds during the Third Republic as a way to inculcate a new civic memory, see Avner Ben Amos, "The Other World of Memory: State Funerals of the French Third Republic as Rites of Commemoration," *History and Memory,* 1989, *1:*85–108.

[7] See Anousheh Karvar, "Le centenaire de l'Ecole Polytechnique: Rites d'une élite nationale sur fond de crise" in Abir-Am, *La mise en mémoire* (cit. n. 1), pp. 191–206.

[8] See Pascal Duris, "Sous la bannière linnéene: Le culte de Linné en France et à l'étranger au XIXe siècle" in Abir-Am, *La mise en mémoire* (cit. n. 1), pp. 251–64; *idem, Linné et la France, 1780–1950* (Geneva: Droz, 1993).

How did this rich commemorative legacy at the start of the century unfold in science, given that this century had plenty to commemorate with the key role that science played in sustaining and terminating two world wars?[9] Do the peaks of commemorative activity in twentieth century science correspond to the great wars? How do commemorations in science relate to those in society, given the increasingly paradigmatic role of science in the cultural imagination, the obsession with commemorating social and cultural memory, and the century's totalitarian legacy?

For example, the dramatic public responses to cinematic and televised renditions of memories of the fascist and Nazi eras has been attributed to the post-1968 generation's need for a collective memory of crucial historical events of which it has no personal memories.[10] Furthermore, this post-1968 demand was sustained in the 1980s and 1990s by the collapse of grand political utopias that rested on collective aspirations for the future, such as socialism, as well as by the rise of a politics of identity that encourages a preoccupation with the past as part of justifying demands for special political rights in the present.[11] As noted above, the convergence of commemorative and real-time events in the late 1980s and early 1990s has produced a constellation of historicity particularly conducive to intense commemorative activities in both science and society.

It is intriguing to ask whether the intervals between commemorations in science have become any shorter (i.e., partial centennials as opposed to multiple centennials), particularly later in the twentieth century when the commemorative imperative may have become stronger. If so, how is the problem of countermemories of living witnesses, which invariably affects the staging of events with a commemorative span shorter than half a centennial, handled at such events?

Along similar lines, is the mid-century, with its World War II-memory-saturated generation, a time of collective amnesia, with commemorative action during that

[9] For an argument conceptualizing the twentieth century in terms of three postwar reconfigurations following the settlements of 1918, 1945, and 1989, see Charles S. Maier, "1918, 1945, 1989: Three Eras of Post-imperial Stability?" lecture given at the Center for European Studies, March 1995; for an adaptation of this argument to science, especially transdisciplinary biomolecular science, see Pnina G. Abir-Am, "The Molecularization of Biology in the 20th Century: Three Eras of Transdisciplinary Stability," in *Science in the 20th Century,* eds. John Krige and Dominique Pestre (Amsterdam: Harwood Publishers, 1997), pp. 495–520.

[10] See Yosefa Loshitzky, "'Memory of my Memory': Processes of Private and Collective Remembering in Bertolluci's *The Conformist* and the *Spider's Stratagem,*" *History and Memory,* 1991, *3:*87–114; Anton Kaes, "History and Film: Public Memory in the Age of Electronic Dissemination," *History and Memory,* 1990, *2:*111–29; Henri Rousso, *The Vichy Syndrome: History and Memory in France Since 1944* (Cambridge, Mass.: Harvard University Press, 1991; first French edition Paris: Seuil, 1987). These works deal with Italy, Germany, and France, respectively. There is a large literature on issues of commemorating the Holocaust. Kaes, in "History and Film" examines the impact of NBC's television series *Holocaust* and Claude Lanzmann's film *Shoah* on the public memory of the Holocaust, especially in Germany; see also James E. Young, "The Biography of a Memorial Icon," *Representations,* spring 1989, *26:*69–106; Anette Wievorka, "1992: Reflexions sur une commemoration," *Annales Economies, Sociétés, Civilisations,* May–June 1993, 703–14; Yael Zerubavel, "The Death of Memory and the Memory of Death: Massada and the Holocaust as Historical Metaphors," *Representations,* 1994, *45:*72–100; *The Limits of Representation: Studies on the Holocaust,* ed. Saul Friedlander (Los Angeles: University of California Press, 1992); Pierre Vidal-Naquet, *Assassins of Memory* (New York: Columbia University Press, 1992; original French edition, Paris: La Decouverte, 1987).

[11] See Maier, "Surfeit of Memory?" (cit. n. 1); David Hollinger, "How Wide the Circle of 'We'? American Intellectuals and the Problem of the Ethnos since World War II," *American Historical Review,* April 1993, 317–37; *idem,* "Group Preferences, Cultural Diversity, and Social Democracy: Notes toward a Theory of Affirmative Action," *Representations,* 1996, *55:*31–40.

time confined mainly to communist regimes desperate to legitimize themselves through a new cult of ideological leaders? If so, what role, if any, did science play in these "memory wars"? Last, are the serial commemorations of some scientific objects throughout the century reliable indicators of the linkage between presentist agendas, both scientific and social, and public constructs of memory of the same object?

Still another dimension of this volume pertains to *disciplinary variation* in commemorative practices, both with regard to the choice of commemorative objects (i.e., a discovery, an institution, or a great mind) and the frequency or scale of the staged commemorations. Is it plausible that disciplines facing sharp changes in their social utility (whether rising or declining) are more likely to stage commemorative events than fields that are not experiencing rapid change in size or public mission? Is physics more likely to commemorate its founders, institutions, or discoveries than biology or chemistry? Do some disciplines produce commemorative events of wide public interest, or are they all confined to professional elites that will understand the technicalities of their signposts to posterity?

The possible variation in disciplinary practices of commemoration according to changing social utility raises the question of the *political uses* of commemorations. High-ranking political figures were invariably involved in many scientific commemorations in the six countries discussed in this volume. Furthermore, political bodies occasionally intervened directly with the course of scientific commemorations, such as when the French National Assembly voted twice not to commemorate the great eighteenth-century encyclopedist Denis Diderot.[12] A recent, better-known case pertains to the French president's decision, on the occasion of the centennial of the discovery of radium in 1998, that codiscoverer Marie Curie be reburied in the Panthéon, France's collective graveyard for great minds. In the same vein, the *Enola Gay* exhibition at the Smithsonian Institution that was prepared to commemorate the half-centennial of the dropping of the atomic bomb was cancelled after twenty-one members of the United States Congress and several political organizations expressed their opposition to that revisionist exhibition.[13]

This volume also raises a series of *methodological questions* regarding the dual historiographic and ethnographic perspective required of analysts who participated in preparing the scientific commemorations that they later studied. It thus expands upon previous discussions on reflexivity and the postmodern emphasis on writing about one's own participation in commemorative events in a critical, yet empathetic, manner.[14]

In his preface to this volume, Charles S. Maier asks where studies of scientific commemoration fit into the wider historical studies of commemorative practices in

[12] See Roselyne Rey, "Diderot et l'Encyclopédie, commemorations manquées" in Abir-Am, *La mise en mémoire* (cit. n. 1), pp. 15–24. On the commemorative practices related to the public entombment of French heros in the Panthéon see Mona Ozouf, "Le Panthéon: L'Ecole Normale des morts" in Nora, *Les lieux de mémoire* (cit. n. 1), vol. 1, pp. 139–66.

[13] For details see notes 1 and 3, and chapter 8 in this volume of *Osiris.*

[14] These methodological issues have been discussed, at least in a preliminary fashion, in my reply to comments of three respondents: see Pnina G. Abir-Am, "Toward a Historical Ethnography of Scientific Anniversaries: The 50th Anniversary of the First Protein X-ray Photo, 1984, 1934," *Social Epistemology,* 1992, *7*:321–54 and the exchanges of response and reply, 355–81. The role of reflexivity in the production of private and public memory is also discussed in Kaes, "History and Film" (cit. n. 10), and Loshitzky, "'Memory of my own memory'" (cit. n. 10).

society, at a time when the culture-at-large is increasingly imbued with science-derived values, facts, and goods, and when the science fiction of today becomes the scientistic culture of tomorrow. He suggests how scientific commemorations might shed light on the logic of collective memory in the late twentieth century, a time when multiculturalism and postmodernism have not only decentered and fragmented much of traditional historical discourse, but have elevated collective memory into the ultimate political asset.

## PART I: COMMEMORATING GREAT MINDS: SCIENTISTS AS CULTURAL HEROES

Part I, which includes six essays and almost a dozen case studies (most essays link their main case study to predecessor or successor ones) explores the practice of commemorating heroes of science and sometimes of society as well. Chapter 1 examines several commemorations of Nicholas Copernicus (1473–1543)—founder of heliocentric astronomy, distinguished mathematician, Renaissance scholar, indeed an icon of modern science—in Poland, Germany, and in the United States during the last three centuries. The chapter dwells in particular on the 1943 quadricentennial of his death as well as of his masterpiece *De revolutionibus,* and on the quinquecentennial of his birth in 1973.

Chapters 2 through 4 highlight the practice of commemorating great nineteenth-century French scientists, both at the turn of and throughout the twentieth century. Chapter 2 examines the contrasting commemorative practices that developed in France around two great biomedical scientists, Claude Bernard (1813–1878) and Louis Pasteur (1822–1895). Chapter 3 explores the commemoration of a pioneering French woman scientist, Clemence Royer (1930–1902)—best known as the author of the first French translation of Darwin's *On the Origin of Species*—and its linkage to Republican gender politics. Chapter 4 compares two commemorations of yet another French biomedical scientist, Jean-Martin Charcot (1825–1893).

This ensemble sheds light on the specificity of commemorative practices for scientists in France (which become occasions for the glorification of the Republic) and in the biomedical sciences (which emphasize preserving sites of biomedical authenticity), as well as on the changing practices of commemoration in time. Thus, while the modernist post-World War I era turned out to be conducive to the construction of an heroic, success-oriented, unified public memory, the postmodernist 1990s produced a more critical, fragmented, and failure-focused memory.

Chapter 5 examines three commemorations of another major figure in nineteenth-century biomedical science, Robert Koch (1843–1910), a rival of Pasteur, whose seminal discovery of the tubercle bacillus (rather than his own more controversial persona) became the object of a fortieth anniversary in Germany in 1922 and fiftieth anniversaries in several countries in 1932. Chapter 6 deals with the 1958 centennial of the birth of German physicist Max Planck (1858–1947), which became a major bone of contention between East and West Germany since it was held in both East and West Berlin at the height of the cold war.

Taken together, chapters 1, 5, and 6 shed light on the specificity of German commemorative practices for scientists. They also illuminate the political constraints of different commemorative moments in German history (see Table 1). In sum, part I explores the commemoration of great scientific minds in several cultural/national

**Table 1. Distribution of commemorations mentioned in this volume by year, object, historicity, country, and discipline**

| Number/ Author | Year | Commemorative Object | Historicity | Country | Discipline |
|---|---|---|---|---|---|
| 1/Sinding | 1913 | Bernard | centennial/birth | France | phil. & biol.; |
| | 1965 | ″ | centennial/work | ″ | phil. & biol.; |
| | 1978 | ″ | centennial/death | ″ | hist. & biol. |
| 2/Sinding | 1922 | Pasteur | centennial/birth | France | immunology; |
| | 1995 | | centennial/death | ″ | epidemiology |
| 3/Haddad | 1922 | Tubercle bacillus/Koch | 40th anniv. | Germany | medicine |
| | 1932 | ″ | 50th anniv. | ″ | ″ |
| | 1932 | ″ | 50th anniv. | USA | ″ |
| 4/Barberis | 1925 | Charcot | centennial/birth | France | neurology |
| | 1993 | ″ | centennial/death | ″ | ″ |
| 5/Harvey | 1931 | Royer | centennial/birth | France | anthropology |
| 6/Beller | 1935 | Bohr | 50th anniv. | Denmark | quantum physics |
| | 1945 | ″ | 60th anniv. | ″ | ″ |
| | 1955 | ″ | 70th anniv. | ″ | ″ |
| | 1985 | ″ | centennial/birth | U.S.A. (D.C.) | ″ |
| 7/Elliott | 1936 | Harvard | tercentenary | U.S.A. (Mass.) | education |
| 8/Gingerich | 1943 | Copernicus | 400th d/w | Poland-occupied; | astronomy |
| | ″ | ″ | ″ | Germany | |
| | ″ | ″ | ″ | Poland-free | ″ |
| | 1973 | ″ | 500th birth | Poland; U.S.A. | ″ |
| 9/Hoffman | 1958 | Planck | centennial/birth | Germany | physics |
| 10/Smocovitis | 1959 | Darwin | centennial/work | U.S.A. (Illinois) | evol. biology |
| 11/Abir-Am | 1966 | phage course; Delbruck | 21st anniv.; 60th anniv. | U.S.A. (New York) ″ | mol. biology ″ |
| 12/Abir-Am | 1971 | Lwoff | 50th anniv. | France | mol. biology |
| 13/Pestre | 1979 | CERN | 25th anniv. | Europe | high-energy physics |
| 14/Seidel | 1981 | L. Berkeley Lab. | 50th anniv. | U.S.A. (Calif.) | nuclear physics |
| 15/Seidel | 1993 | Los Alamos Lab. | 50th anniv. | U.S.A. (N.M.) | nuclear physics |
| 16/Goldberg | 1995 | atomic bomb | 50th anniv. | U.S.A. (D.C.) | nuclear physics |

contexts (Polish, American, French, and German); disciplines (astronomy, biomedical sciences, and physics); and historical periods (from the 1743 bicentennial of Copernicus's death to the quinquecentennial of his birth in 1973).

In chapter 1, Owen Gingerich examines the long dispute over portraying Copernicus's illustrious heritage as either Polish or German. Following a survey of nineteenth-century confrontations on the issue (the nationalistic dimension of the Copernican commemorations was much greater in 1873 *vis-à-vis* in 1843), Gingerich notes the historiographical payoff of each commemoration. For example, the recovery of original Copernican manuscripts in a library in Prague in 1831 led to their translation into Polish for the 1843 tercentenary of the publication of *De revolutionibus,* and new interpretations of his book became available when it was translated into German for the 1873 tercentenary of his birth.[15]

Gingerich focuses in particular on recovering the political and nationalistic agendas embedded in the 1943 Copernicus quadricentennial, which occurred at a critical time since Poland was then occupied by Nazi Germany. That year, the Free Poland movement in New York City organized a major commemorative event that appealed to Copernicus's transnational status and the lasting value of his scientific achievements. These themes resonated well with American values on the central role of freedom in social life and in scientific inquiry, and on the need to combat the totalitarian regimes that were curtailing such freedoms at the time. Indeed, America's emergence as the leader and the scientific center of the free world resonated with its hosting of such a distinguished quadricentennial, which included the participation of Albert Einstein (a particularly fitting recipient of a Copernican honor as the most distinguished living heir of the Copernican legacy in cosmology), and President Franklin D. Roosevelt, who was present in spirit with his telegram being read to the audience.

The resonance of objective yet unique historical moments, including nationalistic agendas, with commemorative dates that remain arbitrary and grounded in subjective assessments of a scientist is also evident in chapter 2. Here, Christiane Sinding ponders the contrasts between several commemorations held for two eminent French nineteenth-century biomedical scientists, the physiologist and philosopher of medicine Claude Bernard and the microbiologist Louis Pasteur, founder of the Institut Pasteur and developer of the first vaccine against an infectious disease. These two scientists shared the same historical period, country, level of eminence, and political interest in helping their country deal with the defeat of 1871 to Germany and the loss of Alsace and Lorraine. Yet Bernard came to be commemorated mainly as a methodologist and philosopher of biomedical sciences, his reputation restricted to a narrow professional elite. Pasteur, on the other hand, was commemorated as a savior of humanity in a number of large-scale commemorative events held both in and outside France, culminating with the proclamation of 1995 as Pasteur's Year (*'l'Année Pasteur'*).[16]

[15] See Owen Gingerich, "Did Copernicus Owe a Debt to Aristarchos?" in *Proceedings of the Symposium Aristarchos of Samos* (Samos: University of Athens Press, 1980), pp. 26–32.

[16] For example, exhibitions and colloquia held at the National Institutes of Health, among twenty other centers of health sciences throughout the United States, were supported by the French consulates in the United States. On these as well as the numerous commemorative events held in France, see newspaper clippings with the public relations office of the Institut Pasteur in Paris. See also Christiane Sinding, "La grande année Pasteur: Echec du contre-culte?" in Abir-Am, *La mise en mém-*

In contrast to the intellectual preoccupation with issues of scientific rationality that dominated Bernard's commemorations in 1965 and 1978, Pasteur's commemorations, especially the centennial of his death in 1995, emphasized the wider social theme of public health or disease prevention. Hence, Pasteur's image as a savior of humanity from the scourge of infectious diseases was easily extended into the 1990s to the prevention of AIDS and even of genetic diseases. Bernard's commemorations were mostly held in France with Bernard eulogized as a French scientist; but Pasteur's accomplishments made him an international object of commemoration, as befits the notion that infectious diseases know no border.[17]

Sinding breaks new ground in her analysis of countercommemorative books produced in the context of a commemorative endeavor, such as François Dagognet's and Bruno Latour's efforts to deconstruct the Pasteurien legend by shifting the basis of his undisputed greatness from ideas and procedures (many of which have been shown to involve various misconceptions, sheer errors, and ethical problems) to social accomplishments, such as establishing a research institute that has had a lasting impact, for example, by nurturing the French school of molecular biology.[18] She concludes by exploring to what extent the differences between Bernard's and Pasteur's commemorations stemmed from the existence of an institutional player in the latter case (the Institut Pasteur), whose own agenda in maintaining its position as a leading research center is always reinforced by international attention to its founder.[19]

In chapter 3, Joy Harvey examines the intersection of feminist, socialist, liberal, and scientific agendas at the centennial of the first woman to be so commemorated in France (and possibly worldwide), the polymath Clemence Royer. Harvey contrasts Royer's centennial, held at the Sorbonne in 1931 in the presence of famous politicians, scientists, and feminist spokespersons, with Royer's more modest career. Royer's work was always overshadowed by the contradictions perceived in her time between her gender's social subordination and the visions of republican politics and positivist science, which claimed equality of rights and merit of the intellect, respectively, yet dabbled in subtle and not-so-subtle forms of gender oppression and discrimination.

While Royer's commemoration lacked the strong disciplinary dimension that characterized the centennials of Bernard and Pasteur, it resembled Copernicus's in the sense that a key political attribute (gender for the former and contested ethnicity or nationality for the latter) overshadowed the scientific component of the event.[20]

---

*oire* (cit. n. 1), pp. 289–310; *idem,* "La célébration des découvertes thérapeutiques: Des chasseurs de microbes aux manipulateurs de genes," *ibid.,* pp. 125–42.

[17] On the role of the May 1968 events for the leaders of the French school in molecular biology see chapter 15 in this volume of *Osiris.* See also Patrick Rambaud, *Les aventures de Mai* (Paris: Grasset, 1998); Alain Touraine, *The May Movement, Revolt and Reform* (New York: Random House, 1971).

[18] See chapter 15 in this volume of *Osiris;* see also the chapter on the French school in Michel Morange, *Histoire de la biologie moleculaire* (Paris: La Découverte, 1994) translated by Harvard University Press in 1998; and the essay review of this book, among others, by Pnina G. Abir-Am in *Historical Studies in the Physical and Biological Sciences,* 1995, *25:*167–96.

[19] See note 16, see also "Proceedings of the Centennial of the Pasteur Institute held in Paris in 1988," in Michael Morange, *L'Institut Pasteur: Contributions à son histoire* (Paris: La Découverte, 1991).

[20] See also Harvey's monograph *Almost a Man of Genius: Clemence Royer, Feminism and Nineteenth Century Science* (New Brunswick, N.J., and London: Rutgers University Press, 1997) and the preface by Pnina G. Abir-Am, pp. ix–xi.

If, as Gingerich argues, the Copernican commemorations were largely driven by the politics of nationalism, then Harvey's parallel argument is that Royer's commemoration was driven first and foremost by the gender politics of the Third Republic.

As do Sinding and Gingerich, Harvey documents a systematic link between the legacy of her commemorative object (accomplished in the last third of the nineteenth century) and the presentist agendas of the participants at her centennial in 1931. If the presence of feminist public figures (most notably Marguerite Durand) at Royer's centennial is understandable as they knew her personally, shared her ideas advocating equality for women, and helped her disseminate them, then the presence of famous scientists and politicians (Raymond Poincaré, Paul Painlevé, Paul Langevin) is less explicable, since they neither knew Royer nor represented any of the disciplines to which she had contributed.

That scientific commemorations can be used to belatedly affirm formerly unpopular political stances, as well as to construct diverse images of the honoree that resonate with presentist agendas, is also observed by Daniela S. Barberis. In chapter 4, Barberis compares the centennials of Jean-Martin Charcot's birth in 1925 and of his death in 1993. She argues that the first centennial created and circulated the stereotyped image of Charcot as a "father of neurology," while also commemorating France as a nation successfully regaining its international standing in the aftermath of a devastating great war. By contrast, the second centennial, as befits its occurance at the height of the postmodernist influence in cultural and historical studies, produced a fragmented, unheroic image of a physician whose once enormous fame shrank considerably under the scrutiny of a new, critical generation of historians of medicine. While Charcot remained a founder of neurology, his more obvious sources of fame (his writings and especially his public performances demonstrating hysteria, hypnosis, somnambulism, and related psychological phenomena) were criticized as charlatanistic.[21]

Barberis's essay also shows how the commemoration of Charcot as a founder of neurology purged his biography of connections with major political figures, such as Gambetta and Bourneville. Those connections had been crucial for building a neurological empire at the Salpêtrière Hospital complex in Paris, a once-modest geriatric facility that Charcot transformed into a world-renowned center of innovative treatment for hypnosis and hysteria, among other less spectacular neurological disorders.[22] As a result of this purge, Charcot's phenomenal public success was presented as the product solely of the logic and rigor of his work.

Unlike the 1925 dual centennial of Charcot's birth and the founding of the Societé de Neurologie, the 1993 centennial of his death featured neither commemorative sessions at the Académie de Médecine and the Sorbonne nor a division between symposia and public celebrations. Also, in contrast to the avoidance in 1925 of the controversial or failure-ridden issues of hypnosis and hysteria, Charcot's weak points were dealt with at the beginning of the 1993 program, as befit the era of deconstruc-

[21] For a recent, comprehensive, and authoritative treatment of Charcot including a brief discussion of his commemorations, see *Charcot: Constructing Neurology,* eds. Christopher Goetz, Michel Bonduelle, and Toby Gelfand (Oxford: Oxford University Press, 1995) especially the last chapter by Toby Gelfand.

[22] See Alain Lellouch, *Jean-Martin Charcot et les origines de la geriatrie* (Paris: Payot, 1992).

tion. Charcot's prominent political connections were also discussed, thus producing a less mythical portrait than that of 1925.

The centennials of Bernard, Pasteur, Royer, and Charcot illuminate the changing historical context of the rich and pioneering French commemorative tradition in the biomedical sciences. In that tradition, commemorative grandeur was invariably put to the service of the positivistic ideology that anchored the power and authority of science in society as a secular religion. It was also used to promote political agendas of the time, ranging from reaffirming France's international standing as a diplomatic center for new European countries to isolating Germany and German science, encouraging scientists to serve in the French colonies in the 1920s, and negotiating France's status as a biomedical superpower in the 1990s.[23]

While the case studies in chapters 2, 3, and 4 are useful for contrasting the political agendas interacting with French commemorative events in the 1920s and in the 1990s (the post-World War II and post-cold war eras), chapters 5 and 6 shed light on the political agendas of German case studies, especially in the 1930s and the 1950s (the era around World War II). Chapter 5 by George Haddad focuses on two successive commemorations of Robert Koch, the German "country doctor" who became a founder of bacteriology and, together with Pasteur, Joseph Lister, and others, revolutionized medicine by putting the diagnosis and treatment of infectious diseases on the causal, systematic basis of germ theory.

Haddad's topic ties particularly well with Sinding's, since Koch occasionally competed with and was often compared to Pasteur. Moreover, the fortieth anniversary of Koch's discovery in 1922 coincided with the centennial of Pasteur's birth. But Haddad's focus on two successive commemorations of Koch's discovery also parallels Barberis's comparison of two Charcot anniversaries, especially since they were all part of a medical tradition of commemoration.[24]

However, Haddad's essay also has a unique feature as far as commemorating personal grandeur in science is concerned, since it was not Robert Koch who was selected as a commemorative object, but rather his discovery of the tubercle bacillus as the cause of tuberculosis. The enormous importance of this discovery did not stem from any particular scientific challenge posed by this bacterium, though the methodological rigor with which Koch established it as a causative agent was much admired. Rather, the importance of the discovery, and the fame and political appeal it brought to Koch, stemmed from the social and cultural significance of tuberculosis.[25]

Furthermore, Koch's low social origins as a country doctor, the persistant gossip that he had never been accepted by the Berlin medical establishment, and his involvement in personal and professional scandals (such as a divorce and the premature announcement of tuberculin as a presumed cure for tuberculosis) made him a

[23] On the "scientific Locarno" lagging behind the political one, see Brigitte Schroeder-Gudehus, *Les scientifiques et la paix: La communauté scientifique internationale au cours des années 20* (Montreal: Presses de l'Université de Montréal, 1978).

[24] See George Weisz, "The Self-Made Mandarin: The Eloges of the French Academy of Medicine," *History of Science,* 1988, *26:*13–40.

[25] Barbara Gutman Rosenkrantz, "Koch's Bacillus: Was There a Technological Fix?" in *The Prism of Science,* ed. Edna Ullmann-Margalit (New York: The Humanities Press, 1986); see there also Pnina Abir-Am, "Koch's Bacillus: A Comment," pp. 162–64.

controversial figure unfit to be directly commemorated as a symbol of social esteem. The meaning of commemoration as an occasion for social and moral consensus explains why Koch's scientific achievement had to be detached from his controversial persona and made into a formal object of commemoration itself.[26]

Thus, unlike Sinding, Harvey, and Barberis, who took for granted the commemoration of their honorees' personas, Haddad has to demonstrate the centrality of Robert Koch to the medical commemorations that he examines. He highlights the special, even theatrical, efforts made by the organizers of the 1922 celebration to authentically recapture the historical moment in March 1882 when Koch presented his discovery before the Berliner Physiologische Gesellschaft while his assistant, Friedrich Löffler, provided technical demonstrations of bacterial cultures.[27]

As did Sinding and Barberis in their studies of centennials in the 1920s, Haddad notes the nostalgia of the audience (composed mostly of physicians) for the pre-World War I era, which not only symbolized the heroic phase of biomedicine but the presumed glory of a colonial era. In a similar vein, while the 1925 Charcot centennial was used to reaffirm France's position as a leader in international affairs, the 1922 anniversary of Koch's discovery remained an internal German affair, since German scientists and meetings were still boycotted by other countries as part of the sanctions included in the post-World War I settlement.

Haddad contrasts the 1922 anniversary and its focus on technical and professional authenticity with the golden anniversary in 1932, in which political overtones became more overt as the event intersected with the swan song of the German medical mandarins. The imminent demise of the Weimar Republic and the beginning of the Third Reich (which was to absorb the medical establishment into state service to the effect that Josef Mengele, rather than Koch, would become the best-known German doctor) greatly affected the golden anniversary. Many foreign invitees failed to come to Berlin, which was enduring daily street clashes between the Nazis and the Communists.[28] And the commemorative site was not the Berliner Physiologische Gesellschaft, as it had been in 1922, but the meeting place of the Prussian State Assembly. As Haddad notes, at the 1932 commemoration Koch's discovery shifted from being a landmark in the eradication of tuberculosis to a symbol of public health policy in and outside of Germany.

In Haddad's discussion of a fiftieth anniversary commemoration of Koch's discovery, celebrated in 1932 in the United States, he makes the interesting observation that a major theme was the detailed account of how the discovery reached America. This corroborates other observations in this volume and elsewhere on the American cul-

[26] As chapter 6 in this volume shows (in relation to a plan to commemorate the fiftieth anniversary of the theory of relativity in 1955 rather than its creator, Albert Einstein, in Germany), this procedure of detaching the discovery from the discoverer for commemorative purposes has been used in other cases in which a great scientist is professionally or socially controversial or otherwise marginal *vis-à-vis* the hegemonic value system of the organizers. For such an argument in relation to John D. Bernal (1901–1971), the Marxist founder of protein X-ray crystallography who was commemorated via a proto-discovery rather than in his own right, see Abir-Am, "Historical ethnography of scientific anniversaries" (cit. n. 14). See also chapter 15 in this volume of *Osiris.*

[27] For a similar argument of converting a scientific nonevent or a protodiscovery into the origins of a new discipline, see Abir-Am, "Historical ethnography of scientific anniversaries" (cit. n. 14).

[28] See Beno Muller-Hill, *Medical Genetics at Auschwitz* (Cologne: Verlag, 1988); Robert Proctor, *Racial Hygiene* (Cambridge, Mass.: Harvard University Press, 1993).

tural bias evident in the perception that the significance of a discovery begins, or at least becomes of concern for Americans, only from the moment it reaches America.[29]

Altogether, the essays on the commemorations of Bernard, Pasteur, Charcot, and Koch illuminate the biomedical tradition of commemoration in France, Germany, and the United States, especially in the first third of the twentieth century. These essays also clarify the key role of turn-of-the century revolutionary disciplines, most notably neurology and bacteriology, in producing and disseminating a collective memory of a heroic biomedical science. Yet, as these essays also demonstrate, revolutionary biomedical science always strove to be perceived as socially and politically responsive, whether in the context of regaining France's international standing, at the Charcot centennial in 1925; emphasizing the role of the German state in public health at Koch's half-centennial in 1932; or enunciating a new, post-cold war alliance between the biomedical scientific leadership, the government, and the biotech industry at Pasteur's centennial in 1995.

By contrast, in chapter 6, Dieter Hoffmann examines a case study in which the political use of scientific commemorations was the driving force in the construction of the events. Not unlike the Copernican commemorations held since the nineteenth century (which, as Gingerich demonstrates, were largely driven by the rivalry over appropriating Copernicus as an asset for either Polish or German nationalism), the Planck centennial in 1958 became a bone of contention between communist East Germany, which controlled the historically authentic sites in East Berlin, and West Germany, which represented a greater part of the German people, scientific community, and territory.

As Hoffmann's judiciously selected opening quotation suggests, "at the Planck celebration the independent role of the GDR [German Democratic Republic] should be emphasized and the initiatives in West Berlin undercut"), the political use of the Planck centennial was more evident to the East German government, whose leaders quickly grasped the unique opportunity it offered for rallying the scientific community to visit East Berlin and be seen as accepting, if not quite endorsing, its regime.[30] By then, Berlin had become a major site of cold war confrontation. Furthermore, the spring of 1957, when the preparations for the 1958 Planck centennial began, was close to landmarks such as the Soviet repression of the Hungarian rebellion in November 1956 and the XX Congress of the Soviet Communist Party, which denounced Stalinist repression. Within three years of the Planck centennial, the

[29] For a similar argument in connection with the discovery of the double helix (a discovery made in the United Kingdom that was portrayed as part of the American collective memory in molecular biology from the moment James D. Watson introduced it to the summer course at Cold Spring Harbor Laboratory), see Pnina G. Abir-Am, "Entre mémoire et histoire en biologie moleculaire: Les premiers rites commemoratifs pour les groupes fondateurs," in Abir-Am, *La mise en mémoire* (cit. n. 1), pp. 25–79.

The United States commemoration's emphasis on practical results, which resonates with American values, glossed over the fact that Koch's premature announcement of tuberculin was made under imperial pressure to show dramatic results by a state doctor. This situation resembles the United States' *Challenger* disaster in which an emphasis on technical factors obscured the role of presidential pressure not to delay the launching—pressure grounded in expectations for political gain. See Diane Vaughn, *The Challenger* (Chicago: University of Chicago Press, 1996), a book that marks the tenth anniversary of the disaster.

[30] See, for example, Charles S. Maier, *Dissolution: The Crisis of Communism and the End of East Germany* (Cambridge, Mass.: Harvard University Press, 1996).

erection of the Berlin Wall would finally transform Berlin into the most vivid and tangible symbol of the cold war.[31]

The idea that a scientific commemoration could contribute, even a little, to a "general easing of political tensions" had inspired the initiator of the Planck centennial, Max von Laue (1880–1965).[32] As Hoffmann details in this fascinating saga, in 1955 von Laue had seized upon another scientific commemoration—the half-centennial of the theory of relativity—as a vehicle for easing the cold-war tension centered on Berlin when he unsuccessfully tried to organize a joint celebration of the East and the West Berlin physical societies.

Unlike Koch and Einstein, who were indirectly commemorated in Germany in 1922/1932 and in 1955, respectively, through celebration of their most important scientific achievements, Planck was widely accepted as a unifying symbol of German science. Planck combined a great scientific achievement with proper social origins and state service through four regimes (Empire, Weimar Republic, Third Reich, and Federal Republic of West Germany). Though his precise stance during the Nazi regime remains disputed among scholars, his effort to distance himself from it after World War II was typical of scientists who had remained in Germany during the Nazi era. In this sense, Planck was very much a representative of what happened to German science in the first half of the twentieth century. Hence, as Hoffmann argues, his centennial was bound to become a major occasion on the political agendas of both Germanys.[33]

As noted by both Sinding (in relation to the Institut Pasteur) and by Barberis (relative to Charcot's affiliation with La Salpetrière), institutions may make a difference in the perception of a scientist's personal grandeur. Indeed, Sinding argues that Pasteur's major legacy may have well been the Institut Pasteur, especially its crucial role in launching the molecular revolution in biology, rather than his own discoveries.[34] The question thus persists as to the role of scientific institutions as objects of commemoration that complement, or sometimes substitute for, the more common commemorative choice of greatness of mind. This signifies the conflation, in the collective memory of science, of the material sites for the production of scientific facts with the social sites for the reproduction of scientific careers. This question is addressed by four essays in part II that examine commemorations of laboratories and universities in America and Europe.

## PART II: COMMEMORATING SCIENTIFIC INSTITUTIONS: THE RE/PRODUCTION SITES OF SCIENTIFIC PROGRESS

Commemorating the foundation of great scientific institutions, like honoring the personal legacy of great scientists, involves an interplay between historical and pres-

[31] See *The Cold War in Europe,* ed. Charles S. Maier (New York: Wiener, 1991); Daniel J. Kevles, "Cold War and Hot Physics: Science, Security, and the American State," *Hist. Stud. Phys. Biol. Sci.,* 1990, *20:*239–65.

[32] The quote is from chapter 6 of this volume of *Osiris.* See Alan Beyerchen, *Scientists under Hitler* (New York: Yale University Press, 1977); Ruth Sime, *Lise Meitner, A Life in Physics* (Berkeley: University of California Press, 1996).

[33] See John L. Heilbron, *Dilemmas of an Upright Man* (Berkeley: University of California Press, 1988). See also Krige and Pestre, *Science in the 20th Century* (cit. n. 9).

[34] See also Sinding, "La grande année Pasteur" (cit. n. 16); chapter 15 in this volume; Abir-Am, "Entre mémoire et histoire" (cit. n. 29).

entist—sometimes even futuristic—agendas. As detailed in part II through examples ranging from Harvard's tercentenary in 1936 to the half-centennial of the Los Alamos National Laboratory in 1993, the commemoration of an institution inevitably includes a self-celebratory dimension (institutions always orchestrate their own anniversaries) as well as its concern for its own future survival. Institutions also provide more formal, structured, and larger-scale opportunities to recruit the past in the service of political agendas in the present while appealing to a captive audience: academic and administrative, officers, alumni, students, and the science-education-conscious public.

The first paper, by Clark A. Elliott, focuses on the presence of science at the Harvard tercentenary in 1936. At this historical moment, the three-hundredth anniversary of the founding of America's oldest and one of its most prestigious universities intersected with a clear danger to freedom of thought in Central European universities, as a result of intervention by fascist regimes. The fact that Harvard's president since 1933, James Bryant Conant, was a scientist with a far-reaching reformist agenda for American education added to the magnitude of the occasion.

With the deft hand of a former Harvard archivist, Elliott unravels this impressive event which, under Conant's leadership, was devoted to promoting chiefly academic agendas. By contrast, the Harvard bicentennial in 1836 revolved around political figures, while its 250th anniversary in 1886 had a primarily literary cast.

The Harvard tercentenary was a major international event that received delegates from five hundred institutions—both universities and learned societies—from around the world. As Elliott argues, the tercentenary pursued multiple agendas, of which the evocation of the past was merely one. The most urgent agenda stemmed from threats to the freedom of learning, both in the domestic arena, dominated by the Great Depression, and in the international arena, dominated by the rapid entrenchment of fascism. Elliott also demonstrates that the Harvard tercentenary was not devoid of domestic politics, especially since an alumnus running for his second term as president, Franklin D. Roosevelt, was invited to speak on the last day.

While the tercentenary included many festive dimensions, the organizers considered its Conference of Arts and Sciences to be its most substantial feature. The conference's three integrated symposia were supposed to synthesize knowledge from specialized branches. Thus, the commemorative rite was used as an opportunity to rebalance the particularistic tendencies of various scientific disciplines with the medieval ideal of universalistic, unified, and international knowledge. More critical commentators, however, noted that the effort to integrate specialized branches of knowledge into a coherent system of thought "progressed no far than the titles of the symposia."

Elliott uncovers perhaps the most relevant side effect of the Harvard tercentenary for the readers of *Osiris:* its role in establishing a history of science concentration for undergraduate and doctoral students at Harvard. His discussion of the multiple agendas intersecting in the tercentenary is fascinating for anyone concerned with the dynamics of institutional commemorations, the history of science education in America, the history of Harvard, or with the eventful 1930s.[35]

Elliott's chapter is followed by two essays on the golden anniversaries of three

[35] On Lawrence Henderson see John T. Edsall, "Lawrence J. Henderson and George Sarton," *Isis,* March 1984, *75:*11–13; on Harlow Shapley see chapter 1 in this volume of *Osiris.*

interrelated American institutions: the Lawrence Berkeley National Laboratory, established in 1931; the Los Alamos National Laboratory, established in 1943—known as the "home of the atomic bomb"; and finally the atomic bomb itself, particularly the decision to drop it on Hiroshima and Nagasaki in August 1945. While the use of the first two atomic bombs inaugurated the nuclear age, particularly the United States's rise to superpower status on nuclear-technoscientific coattails, the large national laboratories were involved in both basic and applied big science while testing a vast nuclear arsenal that provided the strategic deterrence underlying half a century of cold war.

In chapter 8, Stanley Goldberg examines how the fiftieth anniversary of the decision to drop the atomic bomb from the airplane *Enola Gay* exploded into a public controversy over the shaping of America's collective memory of the conclusion of World War II. The commemoration was initially conceived as an exhibition of World War II artifacts with the fuselage that carried the bomb as its centerpiece, to be held at the National Air and Space Museum (hereafter NASM) in Washington, D.C., throughout 1995. But the controversy ultimately led to cancellation of the exhibition, resignation of the NASM director, and censoring of the American public memory by interest groups, most notably the American Legion and the Air Force Association but also some congressmen, journalists, and members of the public.[36]

The difficult problem facing Goldberg, like other historians who have tried to explain the incredible chain of events that led to the cancellation of the *Enola Gay* exhibition, is to explain why self-appointed custodians of the public American memory were able to censor a well-researched exhibition by the largest and best-endowed public museum complex in the United States. In the clash that developed between collective memory and critical history, why did the former win the day while the latter was slow and inept to even prepare a proper defense? Sadly, Stanley Goldberg, who resigned from the NASM advisory committee to protest the organization's having caved in to censorship, died prematurely in 1996 before being able to fully pursue this paradox which so consumed him, and which remains unresolved despite the recent publication of several impressive books and reviews. Nevertheless, Goldberg's paradox lies at the heart of our study of commemorative practices as the sociopolitical nexus of memory and history.

Ironically, light is indirectly shed on this paradox of American memory and history in chapter 9, written by Robert Seidel. Like Goldberg, Seidel was both a participant and observer in the golden jubilee of a spectacular outcome of World War II: the Los Alamos National Laboratory (hereafter LANL). He examines the scaled-down, partly secret, yet prolonged and future-oriented celebrations of the half-centennial of LANL in 1993. At the time, domestic antinuclear sentiment and foreign policy had converged to put an end to further nuclear testing, to the effect that the Department of Energy, the political patron of the laboratory, did not want a "birthday party for the bomb."

Seidel contrasts the LANL half-centennial in 1993 with the half-centennial of the Lawrence Berkeley National Laboratory (hereafter LBNL) in 1981, the year the institution opted to restrict itself to basic research. He also contrasts his own perspective on the laboratories' past (which he acquired as the coauthor of a history of

[36] See also note 1.

LBNL and as the LANL museum administrator and author of a historical column for its in-house newsletter throughout the 1980s) with the perspective of the commemorations' scientist participants.[37] The latter were not only much less aware of the bureaucratic and political history of their institutions, but were mainly interested in the scientific accomplishments achieved there, such as the discovery of new elements, subatomic particles, or types of particle accelerators.

As Seidel notes, the specificity of these two half-centennials was grounded in changes in the laboratories' original social and political missions. As a result of détente and hence of reduced budgets and increased regulation throughout the 1980s, which culminated with the end of the cold war in 1989, their very survival was at stake. Thus, their commemorative events became opportunities, or even necessities, for seeking new social missions. The laboratories had to play down those aspects of their past that hinged on war and nuclear weaponry, while highlighting their future in nuclear medicine, materials science, computer science, and bioengineering, among other multidisciplinary research efforts. As Seidel suggests, the need to justify the laboratories' traditional mission in a different political climate or even to find new missions was a major reason why the labs turned to professional historians to chronicle their past for commemorative purposes. Their own "na/t/ive" understanding of the past as a logical and inevitable stream of scientific discoveries was insufficient to elaborate social spin-offs for patrons' consumption in a post-cold war climate.

In contrast to personal commemorations of great people, Seidel notes that institutional anniversaries focus on the laboratory rather than its director, the team rather than its leader, the collective rather than the individual accomplishment. This structural feature of institutional anniversaries was also coupled with the historical reality that neither LANL nor LBNL would again have a charismatic director of the scientific and enterpreneurial stature of its first leader: Robert Oppenheimer and Ernest Lawrence, respectively. Only Edward Teller, commemorated in June 1998 on the occasion of his ninetieth birthday in a rodeo-style festival at Livermore (the township near the Lawrence-Livermore National Laboratory [hereafter LLNL], which he co-founded with Lawrence) approached the standing of Oppenheimer and Lawrence.[38] In fact, Seidel notes, Lawrence's widow, Molly, led a campaign to strip LLNL (the "prodigal son" of LBNL) of Lawrence's name because of its participation in President Ronald Reagan's "tragically ill-advised nuclear weapon build-up."

Seidel also shows that the political context of the early 1990s had the effect of drastically scaling down the planning for the LANL half-centennial. Among the casualties was the idea of opening all the laboratory's facilities to the public. Furthermore, the plan to mount a display of the test rack itself, along with a vast array of

[37] See John L. Heilbron, Robert W. Seidel, and Bruce Wheaton, "A Historian's View of the Lawrence Years," *LBL Newsmagazine,* Fall 1991, pp. 1–106; J. L. Heilbron and Robert W. Seidel, *A History of the Lawrence Berkeley Laboratory* (Berkeley: University of California Press, 1989). See also notes 6 and 9 of chapter 9 in this volume.

[38] See "Celebrating Edward Teller at 90," special issue of *Science and Technology Review,* Lawrence-Livermore National Laboratory, director's office publication, July–August 1998, pp. 1–29. I am endebted to Mary Singleton, former director of plutonium safety at Lawrence Berkeley National Laboratory and now my student at UC-Berkeley, for bringing this special issue to my attention. See also Herbert F. York, *The Advisors: Oppenheimer, Teller, and the Superbomb* (San Francisco: Freeman, 1989).

nuclear testing instrumentation, had to be cancelled because of personnel layoffs and programmatic budget cuts. Ironically, as Seidel notes, the celebration of a largely secret past became confined to largely secret (secure) spaces.

The ironies noted by Seidel at the golden anniversaries of LBNL and LANL, which sprang from the real-time political constraints of the post-cold war era on the commemorative times of 1981 and 1993, provide insight into the cancellation of the *Enola Gay* exhibition at NASM. Scheduled for August 1995, the exhibition was supposed to employ historical reflection and archival documentation educate the public about the complex decision to drop the bomb. Yet, while LANL's projected "birthday party for the bomb" was cancelled because of overt political pressure from its patron, as befits a largely secret facility, the *Enola Gay* exhibition was cancelled through indirect public pressure from interest groups, as befits NASM's ultrapublic mission.

The role of institutional commemorations in exposing the delicate balance between the moral necessities of memory and the pragmatic contingencies of history is also evident in chapter 10 by Dominique Pestre. This essay discusses the 1979 silver jubilee of CERN, the European Center for Nuclear Research, near Geneva, Switzerland, among its other commemorative practices. CERN, the oldest and largest supranational facility in Europe, combines cutting-edge research in high-energy physics with the political agenda of European cooperation, integration, and relative regional autonomy.

Much as Seidel contrasts the subcultures of LBNL and LANL, especially as manifested in their golden anniversaries, Pestre highlights the commemorative subculture of European big science and its unique institutional embodiment in CERN. Though the most persuasive justification for CERN pertains to economic pragmatism (i.e., sharing the high costs of accelerators), the institution was conceived in the late 1940s and emerged by the 1970s as a symbol of European cooperation. It is a technoscientific counterpart of the European common market. After the Maastricht agreement, CERN might become a symbol of European integration in the 1990s and beyond.

CERN's practising scientists were chiefly interested in using the center's silver jubilee to survey, for the benefit of their younger coworkers, the laboratory's progress in theoretical and empirical research as well as its association with illustrious scientist founders. However, the administrator organizers constructed the occasion as an opportunity to showcase CERN's achievements before politicians and the public so as to secure new budgets for the ever-increasing cost of high-energy physics research. They focused, therefore, on CERN's ability to sustain competitiveness with comparable facilities, most notably LBNL, Fermilab, and Brookhaven in the United States. As Pestre emphasizes, even at the modest CERN anniversaries held for the retirement or death of a member, the center's research-enabling conditions, rather than the research itself, were in the spotlight.

A dual agenda is reflected in the roll call of invitees to the CERN anniversaries, which Pestre documents as ranging from scientists and engineers to administrators and politicians and even to kings and presidents. On the one hand, for the participating scientists of various generations, the anniversaries were an opportunity to display their professional identity and share nostalgia for a time when discoveries could still be claimed by individuals. On the other hand, for both the government officials and the public at large, the main message of the anniversaries was the demonstration of scientific progress at a world-class laboratory which, by working at the cutting edge of physics with highly sophisticated technologies, was bringing unspecified benefits to society. As Pestre argues, CERN's institutional anniversaries functioned as fund-raising occasions targeting bureaucrats from relevant government ministries. At the same time, through extensive media coverage, the general public (whose taxes are used to finance the laboratory) was co-opted by CERN's institutional agenda of growth and continuity.

The institutional anniversaries examined in part II shed light on a new breed of event organizers and orchestrators. They are neither the disciples of a great scientist who commemorate his or her personal greatness, nor the spokespeople for disciplinary landmarks, but administrators who preside over commemorations as a means to secure a brighter future, or even the survival, of their large-scale scientific institutions. Commemorative events thus emerge as necessary occasions for scientific institutions to reinvent themselves while recasting the relationship between memory and history for ever-newer political uses.

Whether the declared agenda of the commemorating institution was the preservation of the unity and autonomy of knowledge in the 1930s, as in Harvard's tercentenary, or the quest for a new social and political mission in the 1990s, as in the big science laboratories' jubilees, the question persists as to how commemorative occasions negotiate the changing meaning of science across the span of time lapsing between the commemorated event and the present.[39] Since the changing meaning of science is embedded in its disciplines, or its legacies of objectivity, part III examines commemorative practices unique to science.

## PART III: THE COMMEMORATION OF SCIENTIFIC DISCIPLINES: MEMORIES OF OBJECTIVITY

Part III covers several major scientific disciplines—mathematics, quantum physics, evolutionary biology, biochemistry, and molecular biology—and includes both singular and serial commemorative events as well as serious and jocular ones. In chapter 11, Liliane Beaulieu discusses the systematic use of mathematical humor as a form of collective memory that evolved in the context of an informal group of mainly French mathematicians that was established in 1934 and is still active. This group published under the pseudonym Nicholas Bourbaki *Eléments de mathématique,* a series of revisionist overviews of many areas of mathematics that also emphasized their structural features and similarities, thus demonstrating the group's capacity to unify the disparate domains of this highly fragmented discipline.

As Beaulieu explains in fascinating detail based on more than a decade of meticulous research on the personal papers of key members, this anonymous group shunned commemorative events that would have required the unbearable sacrifice

[39] In this connection, it might be interesting to expand the range of scientific institutions that served as commemorative objects throughout the twentieth century to include not only universities and laboratories but also museums and academies, even libraries. Two examples of such commemorations that remain to be examined in future studies are the symposium held on the occasion of the bicentennial of the Museum National d'Histoire Naturelle in June 1993, (see *Le Museum au premier siècle de son histoire,* eds. Claude Blaenckart, Claudine Cohen, Pietro Corsi, and Jean-Louis Fischer [Paris:1997]); or that held on the occasion of the bicentennial of the Ecole Polytechnique in March 1994 (program available with the author).

of "going public."[40] Indeed, Beaulieu considers only two events in the group's sixty-plus years of existence to have been quasi-commemorative.[41]

In view of the Bourbaki group's paucity of major commemorative events, Beaulieu suggests that its collective memory was constructed in the various issues of its internal newsletter, "La Tribu" ("The Clan"). This newsletter recorded faithfully, in the manner of a satirical chronicle, the activities done and the plans discussed at the three to four meetings the group held per year, while encoding this collective memory in the form of mathematical humor. The humor consisted of a mathematical spin on the cultural, largely literary and philosophical, and jocular tradition that prevailed at the Ecole Normale Supérieure (hereafter ENS). As Beaulieu comments, the group's newsletter, which became its main "testimonial, overriding and unifying other memory-making devices" resembled the satirical revues staged annually at ENS graduations that featured word plays, puns, caricatures, risqué songs, and sketches with absurd plots. While the serious work that went into composing *Eléments de mathématique* (especially the numerous drafts that were criticized, revised, or discarded) was passed over in silence, "La Tribu" emphasized the fun activities, "systematically reversing the work-play ratio." In this manner, the collective memory most likely to be remembered was one of entertainment and leisure, rather than hard work and stern discipline.

In her analysis of the issues of "La Tribu" as devices for forging a collective memory and sense of group identity, Beaulieu highlights the role of specialized humorous lingo of both religious and political origins. Having come to regard humor as a superior habit of mind, the group used it as the main component of its memory-making devices, a decision that reflected a convergence of mathematical and cultural traditions (examples include works by Corneille and Racine, La Fontaine, Mallarmé, Valéry, Prévert, and Brassens) in creating Bourbaki's own "art of memory."

As Beaulieu explains, Bourbaki's humor was primarily verbal, mediating between mathematical concepts and concrete reality. She also notes that since the group was entirely male, a good deal of blue humor was in evidence. Mathematical terms, especially algebraic ones, were always redeployed in sexually suggestive ways. Though humor can occasionally set limits on memory, as when it is tied to circumstances difficult to recall by those who were absent, humor can also revise the events it captures, producing a mediated set of recollections. This was especially so when only the humorous parts of the newsletter were sent to former members; these parts became mementos that the members kept, read on occasion, and recalled as reality, rather than as the illusion of reality they actually were. The systematic use of humor thus created a collective memory that bracketed out problematic issues, such as hierarchical and competitive relationships within the group and the anger and aggression that they generated. Nevertheless, Beaulieu emphasizes that invective and banter were commonplace, since in this inversely ironic register, a strong insult was better than mild criticism. Even the group as a whole could be a target of sarcastic humor.

As Beaulieu convincingly argues, humor stood in for aggressive feelings that could not be expressed directly, and thus guarded the group venture from collapsing

[40] For the full story see Liliane Beaulieu, *Bourbaki: History and Legend* (Springer-Verlag, forthcoming).

[41] See Jean-François Sirinelli, *Ecole Normale Supérieure: Le livre du Bicentenaire* (Paris: Presses Universitaires de France, 1994).

under internal conflict. It worked so well because winning by wit was predicated on reciprocity (i.e., equality between group members, presumption of competence, and mutual esteem), but also because it was restricted to the collective endeavor only. In their individual work and private lives, members of the Bourbaki group did not play in the same universe of inverted humor. Nevertheless, the longevity or social reproduction of the group, as Beaulieu amply shows, depended to a great extent on using humor to create a collective memory that enabled members to identify with the enterprise, relish its solidarity, and ignore its frustrations and heavy demands. As Beaulieu concludes, "Mirth is always worthy of remembrance, especially when it saves the day: the wickedest jokes, set by the group's conventions and tirelessly repeated, are among Bourbaki's sweetest memories."

If Beaulieu demonstrates that humor lay at the core of the collective memory of perhaps the most prestigious group in pure mathematics, then Mara Beller accomplishes a similar feat in chapter 12 with regard to the most prestigious group in quantum physics, the disciples of Niels Bohr (1885–1963), who performed jocular commemorations of their venerated leader that also celebrated the famed "Copenhagen spirit."

Unlike the Bourbaki group, whose collective memory blended mathematical humor and French cultural resources while revolving around the fictitious persona of an eccentric mathematician, the disciples of the Copenhagen spirit created a collective memory that blended physics-derived humor and Central European cultural resources (for example, Goethe's *Faust*) while revolving around the life of a real physicist, Niels Bohr, the creator and hero of the Copenhagen spirit.

And while the Bourbaki group met several times a year and recorded each "congress" and its folklore (both humorous and technical) in an internal newsletter, the quantum physicist disciples of the Copenhagen spirit met every spring for weeks or months and recorded their own mix of humor and technical problems in a *Journal of Jocular Physics,* released in 1935, 1945, and 1955 to commemorate Bohr's fiftieth, sixtieth, and seventieth birthdays.

Like Beaulieu, Beller acknowledges both the subversive role of jocular representations and the role they play in actually enhancing the prevailing social and scientific hierarchies by providing a diversion from the group's frustrations and tensions. Unlike the Bourbaki group, which was essentially peripatetic (holding its meetings in various sites of the French countryside), the jocular events analyzed by Beller all took place at the Institute for Theoretical Physics in Copenhagen. And Beaulieu's group of jocular mathematicians was more homogeneous, most of them being not only French but graduates of the prestigious ENS. By contrast, the disciples of the Copenhagen spirit were a rather international group, including especially Central and Eastern Europeans and North Americans.[42]

As Beller recounts in her bold and fascinating analysis of the three jocular journals, as well as of the "serious" centennial celebrations of Bohr's birth in 1985, the generation of quantum physicists that came of age between 1930 and 1960 venerated Bohr to such an extent that they created an elaborate collective memory revolving around his ideas and personality. A focal point for these efforts was known as the "Copenhagen interpretation," a melange of scientific and philosophical concepts

[42] Among the better-known disciples one finds Germans, Russians, Americans, Swiss, Dutch, and, of course, Danes.

elaborating on the idea of complementarity in quantum physics (the supposed capacity of atomic entities to behave as both particles and waves under different experimental conditions). This interpretation, which dominated physics until recently, was also applied to other disciplines, most notably psychology and biology.

Beller examines how the extensive admiration, even worship, of Bohr, a great physicist with interests in world affairs as well as in the human condition, was manifest in the collective memory created by his disciples, and how it affected twentieth century physics by converting his scientific and institutional authority into wide acquiescence to the Copenhagen interpretation of physical reality. She suggests that the "inevitability of complementarity," long accepted as dogma among quantum physicsts, is increasingly questioned, whether in physics, philosophy, or other disciplines to which it was exported. In the aftermath of the death of both master and many disciples, the historical decontextualization of complementarity has exposed and corroded the protective belt of collective memory which had sustained the concept as part and parcel of the collective identity of quantum physicists.

As Beller explains, the identity of the most powerful scientific community of the twentieth century depended not only on scientific authority grounded in Bohr's iconic model of the atom, but also on the Copenhagen spirit of personal freedom, creativity, epistemological ambition, cultural aspiration, international camaraderie, social mobility, gentile paternalism, religious and political tolerance, and even an ethics of suffering as part of the incessant quest for knowledge and for truth. At least until the mid-1970s, when the quasi detente that culminated with the end of the cold war in 1989 can be said to have begun, there was no substitute for the Copenhagen spirit's offerings of scientific and moral guidance, packaged as uplifting collective memory.[43]

Beller accomplished two major feats in this essay, among many other insightful analyses of the sociopsychological dynamics of the group of Bohr's disciples. On the one hand, she meticulously details how the group's collective memory produced a quasi-mythical image that edited out all of Bohr's failings—scientific, philosophical, social, political.[44] Like all collective memories, it was constructed out of a few memorable actions by an inner core of self-appointed disciples who were capable of blending a legitimizing proximity to the leader with technical capability for cultural productions (ranging from oral scientific jokes to written stories in journals to staged performances of allegorical plays).

A second, major accomplishment is Beller's own jocular use of complementarity, the key concept in the Copenhagen spirit, to deconstruct its collective memory by applying it to such topics as "seriousness versus humor," "truth versus clarity," and "love versus justice." Beller suggests that jocularity could both challenge and reinforce the social order around Bohr's persona. On the one hand, she explains that Bohr was not amused by jocular critiques of complementarity published by George Gamow in 1955 in the *Journal of Jocular Physics.* Thus, jocular commemorations offered a nonthreatening way to express dissent, especially by those who could not contradict Bohr in the social context of his hegemonic position. At the same time,

[43] For an overview of twentieth century science see Krige and Pestre, *Science in the 20th Century* (cit. n. 9).

[44] Beller mentions only scientific and philosophical failings.

the jocular commemoration could also reinforce the authority structure in the group. Leon Rosenfeld's story, "A Voyage to Laplacia," also published on the occasion of Bohr's seventieth birthday in 1955, was a transparent tirade against determinism and an advertisement to abandon the classical interpretation of physics in favor of the nondeterministic one advocated by Bohr and his faithful disciple.

Beller's critique of the disciples' use of the complementarity principle to explain, or rather explain away, the ambiguity, opaqueness, and incomprehensibility of Bohr's verbal and written pronouncements is quite telling. Bohr's difficulties with communication were glossed over by his disciples with the rhetoric of the complementarity between truth and clarity: i.e., they claimed that his lesser clarity was the inevitable price to pay for his greater truth.

The ultimate irony that Beller observes pertains to the fact that Bohr's centennial symposium in 1985 (much like the centennial of Claude Bernard's famous book *Introduction à l'étude de la médecine expérimentale* in 1965, examined in chapter 2) became an opportunity for penetrating criticism of Bohr's philosophy by professional philosophers and historians of science. Complementarity came to be increasingly denied as an objective, long-term contribution to science, being viewed instead as an interpretation contingent upon Bohr's personality and his hegemony over a generation of quantum physicists at a critical time in the middle third of the twentieth century.

Commemorations of the Copenhagen spirit, or of a rebellious outlook in pure mathematics, signify the prevalence of a jocular tradition in creating collective memories for ultraselect groups of innovators in these highly prestigious but relatively esoteric disciplines. Both the jocular and the nonjocular commemorations in mathematics and quantum physics, as discussed in these papers, were meant for an audience of a relatively small number of scientists. Even when a commemoration was sponsored by a supradisciplinary body such as the Académie des Sciences in Paris (which awarded a prestigious prize to four founders of the Bourbaki group in 1967) or the American Academy of Arts and Sciences (which organized one of the 1985 centennials of Bohr's birth), it was addressed to a highly restricted elite, since membership in the sponsoring academies was limited to a small minority of distinguished veteran scientists.[45]

In contrast, chapter 13 by Vassiliki Betty Smocovitis examines a commemorative occasion that involved not only an unusually large number of distinguished scientists, but also the public. The 1959 centennial of the publication of Charles Darwin's epoch-making *On the Origin of Species* became a huge public affair, since by that time evolutionary biology (a discipline of consequence for the physical and social sciences as well as for religion and morality) had established itself as scientific fact and was represented by a well-organized professional community.

Attended by over 2,500 participants, including delegates from 184 institutions and fourteen countries, the Darwin centennial at the University of Chicago resembled in magnitude, intellectual ambition, and institutional pomp the Harvard tercentenary held in 1936. A comparison between them is interesting because it affords a clear contrast between a primarily institutional agenda (for Harvard's tercentenary) with a primarily scientific or disciplinary agenda (for the Darwin centennial). But it also

[45] See Harriet Zuckerman, *Scientific Elite* (New York: Free Press, 1977).

offers an example of commemorations as performance arenas that may have unintended consequences, since these two agendas were actually inverted during the complex process of their implementation.

The Harvard tercentenary was primarily meant to celebrate Harvard's position as the oldest United States university and as the country's foremost link to European academia, and its scientific symposium was only one aspect of a wide range of festivities. The primary rationale for holding the Darwin centennial, on the other hand, was its scientific symposium on the state of the art in evolutionary thinking across disciplines, while a secondary rationale was reaffirming the academic leadership of the hosting University of Chicago. Nevertheless, as the commemorative occasions unfolded, it became clear that just the opposite may have taken place. The academic symposia emerged as the most important and memorable dimension of the Harvard tercentenary, while the University of Chicago easily surpassed the scientific symposia at the Darwin centennial in gaining public attention and pursuing an impressive variety of political agendas.

As Smocovitis argues in her superbly documented and ethnographically perceptive study, the Darwin centennial's spectacular success depended not only on its good timing (following the completion of the great synthesis of evolutionary biology and population genetics), but also on creative organizers, particularly the visionary and managerially capable Sol Tax, a professor of anthropology at the University of Chicago. Yet the event's success in boosting the University of Chicago's public agendas should not gloss over the fact, perceptively observed by Smocovitis, that it still largely failed to accomplish Tax's disciplinary agenda of generating new insights on evolution through crossdisciplinary discussion.[46] His goal of bringing the topics of human and cultural evolution to the center of a reconfigured anthropological agenda, while enlisting evolutionary biology as a more scientific framework for recasting anthropology as a link between the natural and the social sciences, would not be accomplished until the 1980s. The missed opportunity for an intellectual accomplishment at the Darwin centennial was not only a byproduct of its "premature" timing, but also of a planning process that consistently prioritized institutional and political agendas set up by the chancellor of the University of Chicago.

Still, the disciplinary agenda of recasting anthropology as kin to evolutionary biology coupled with the strong institutional agenda of putting the University of Chicago on the world map of major commemorative events produced an extravaganza rarely surpassed in either scale or theatrical pomp. In analyzing all these complex facets, Smocovitis not only reaffirms her standing as a superb historian of evolutionary biology but also as a leader in the cultural studies of science.

She argues persuasively that the grandiose saga of the University of Chicago's Darwin centennial hinged on the organizers' co-optation of Julian Huxley, a leading spokesman and popularizer of evolutionary biology, who was also the grandson of Thomas Huxley, best known as "Darwin's bulldog" for his untiring propagandizing of evolution in Victorian society. Huxley could best incarnate historical authenticity as well as continuity—the bread and butter of commemorative occasions.

Huxley's involvement, secured through a visiting professorship at the University of Chicago, did not disappoint, since his keynote address contained a mid-twentieth

[46] To compare with other commemorative efforts presided over by scientific lightweights, see chapters 9, 12, and 15 in this volume of *Osiris*.

century version of his grandfather's famous repartee to Bishop Wilberforce in 1860 defending scientific truth against religious superstition. Pronouncing religion "an organ of evolution" and God an unnecessary accessory to evolutionary thinking, Julian Huxley's speech, delivered in the university chapel on Thanksgiving Day, challenged American religious sensibilities and overshadowed in emotion and public response the other aspects of the Darwin centennial.

International commemorations of the theory of evolution peaked in 1959, occasioned by the centennial of Darwin's seminal book, but commemorations in biology in the following decade focused on the new, macromolecular biology.[47] In chapter 14, Pnina G. Abir-Am compares the textual, social, and political dimensions of the first collective memories published in molecular biology in the United States between 1964 and 1966 and in France between 1968 and 1971. She also inquires into the role of the countercultural period known as "the Sixties" in constraining the historiographic and disciplinary messages embedded in each collective memory.[48]

As Abir-Am suggests, the contrasts between the two commemorative efforts that produced the two texts of collective memory began with their rationale, or pretext. Neither seemed to be an inevitable commemorative occasion, but reflected a deliberate decision by the organizers to use the culturally and socially acceptable commemorative veneer as a guise for implementing a nexus of presentist (essentially political) agendas. Though the two collective memories (commemorative volumes) *Phage and the Origins of Molecular Biology* (1966) and *Les microbes et la vie/ Of Microbes and Life* (1971) were similar in length and format, they provided very different social and scientific charters for the rise of molecular biology, a discipline that had increasingly become paradigmatic of the scientific imagination on the verge of the twenty-first century. In their turn, those charters revolved around the nexus of three elements: genealogical, disciplinary, and institutional.

Since both the American and the French collective memories projected a broad consensus (with almost three dozen scientists contributing to each) and considerable authority (they both featured a Nobelist contingent, among other leading scientists), Abir-Am asks whether differences in their assembly reflected alternative historical routes, or whether they merely reflected different cultural traditions of recall and commemoration. She also inquires whether the French collective memory, published five years after the American, inevitably came to function as a countermemory.[49] If so, how might a systematic comparison of the processes involved in assembling collective memories and countermemories contribute to a comparative history of molecular biology, especially one that retains the authenticity and diversity of all commemorative practices in science?

Abir-Am's study unfolds in three parts, with each devoted to a comparison within an analytical category: the symbolic-commemorative texts, the social pretexts, and

[47] On this topic see Pnina G. Abir-Am, "The Politics of Macromolecules: Biochemists, Molecular biologists, and rhetoric," *Osiris,* 1992, *7:*164–91. See also Arthur Kornberg's *For the Love of Enzymes: Adventures of a Biochemist* (Cambridge, Mass.: Harvard University Press, 1989); see also the dozen books reviewed in Pnina G. Abir-Am, "Nobelesse Oblige: Lives of Molecular Biologists," *Isis,* 1991, *86:*326–43; *idem.,* "New Trends in the History of Molecular Biology," *Hist. Stud. Phys. Biol. Sci.,* 1995, *26:*167–96.

[48] See Touraine, *The May Movement* (cit. n. 17).

[49] On countermemory see Michel Foucault, *Language, Counter-memory, Practice,* ed. Donald F. Bouchard (Ithaca, N.Y.: Cornell University Press, 1977); Zemon Davis and Starn, "Memory and Counter-Memory" (cit. n. 1).

the political contexts. First, she contrasts the highly structured, modernist, and monolingual American collective memory, whose key message is on the revolutionary origins of molecular biology in phage genetics, with the unstructured, postmodern, bilingual French collective memory, with its key message on the evolutionary origins of molecular biology in microbial physiology. These texts constitute different strategies of using commemorative occasions for producing collective memories or appeals to the past that are designed to play a role in the contested present.

In addition to contrasting the historiographical claims of the American and the French collective memories, Abir-Am also uncovers several types of memory that together precipitate the objectified effect of collective memory (i.e., a consensus on the past of the new discipline). These types include totemic memories, taboo memories, and visual memories that encode and condense the previous two in more memorable ways.

The analytical and empirical diversity of memories raises the question of who orchestrated these convoluted mosaics of private or subjective memories so as to create the desired objectifying effect. In response, Abir-Am uncovers the social pretexts deployed by the organizers in order to mobilize the contributors' memories for public use. She suggests that the major differences between the American and French collective memories reflected different relationships among the organizers. The coherence and clarity of the American collective memory was the product of a strategic, businesslike alliance between three organizers-turned-coeditors whose personal agendas complemented, even reinforced, each other's. By contrast, the French collective memory was less coherent and more opaque—even decadent in its disturbing lack of any temporal or spatial guiding order, because the social relationships between its organizers were grounded in a misalliance. Instead of cooperating, they barely communicated with each other, while splitting the editorial responsibilities along linguistic lines. But in both the American and French cases, the social pretexts used by the organizers constrained the symbolism of the commemorative texts.

Abir-Am then turns to the *political context* of each collective memory. In both cases, she demonstrates the converging role of scientific change, international science policy, and domestic sociocultural politics, though these key factors were different and interacted uniquely in each case.[50]

She concludes her Franco-American comparative study of collective memory in molecular biology with lessons for a history of the discipline. Such a venture, she argues, must take into account diverse American and French notions of historicity, as well as their contrasting forms of historiographic and ethnographic surplus (e.g., the meaningful role of nonpositivist commemorative objects, such as the spirit of the Phage Group or of the Pasteurien "attic, where the French founders of molecular biology were long located as a sign of their institutional marginality"). Abir-Am underscores the pertinence of French-American comparisons as useful probes into the ongoing rise of "biopower" well into the twenty-first century, while noting the royalty-sharing agreement on HIV drugs signed by the Institut Pasteur and the Na-

[50] For further details see Abir-Am, "Entre mémoire et histoire" (cit. n. 29); *idem., Research Schools of Molecular Biology in the United Kingdom, United States and France: Comparing Transnational Innovation in Twentieth-Century Science* (Berkeley: University of California Press, forthcoming).

tional Institutes of Health, in the mid-1990s when these key institutions were headed by second-generation molecular biologists.

Part III suggests that the tradition of selecting a discipline as a commemorative object is pervasive in science (the case studies include mathematics, quantum physics, evolutionary biology, and molecular biology). At the same time, the prevalence of nonpositivist commemorative objects, such as the spirit of self-styled revolutionary groups (the Bourbakists and the Pasteurien attic in France; the disciples of the Copenhagen spirit in Europe and America; the Phage Group in the United States) may relate to the postmodern quest to allow for a multiplicity of voices.

Part III includes events held in France, Denmark, and in the United States, thus evoking the transnational appeal of disciplinary commemorations. Out of five case studies, four involve international cultural politics, as the revolutionary groups in question included scientists from several countries.

In terms of historicity, the case studies in part III include a spectrum of temporal distances between the commemorated event and the commemorating occasion. While the centennial of Darwin's *On the Origin of Species* was the longest such span, the twenty-first anniversary of the first phage course at Cold Spring Harbor Laboratory was the shortest. The majority of the commemorative events involved fiftieth, sixtieth, or seventieth anniversaries for a major scientist. These figures suggest that commemorative spans shorter than half a centennial are rare and may occur only under unusual circumstances such as those described in chapter 14, when political events converged to create an urgent sense that the past, as well as the present and future, were "up for grabs."

In terms of timing, disciplinary commemorations as a genre unique to science appear to have become more widespread in the second half of the twentieth century. Except for the Bohr celebrations in 1935 and 1945, all the commemorative events discussed in part III took place after 1950: Bohr's seventieth birthday in 1955; the centennial of Darwin's *On the Origin of Species* in 1959; the twenty-first anniversary of the first phage course and Max Delbrück's sixtieth birthday in 1966; the Bourbaki's indirect commemorations in 1967 and 1994; and the fiftieth anniversary of Andre Lwoff's arrival at the Institut Pasteur in 1971. This pattern indicates the progress of science from a follower of commemorative traditions in society to a culturally hegemonic enterprise, creating its own commemorative style.

## CONCLUDING REMARKS AND ISSUES FOR FURTHER RESEARCH

Though commemorations are not the only cultural forms that store, as well as encode, collective memory in science (other forms include museums, archives, oral histories, and biographical dictionaries); they are particularly suitable for observing not only the outcomes, but the very process of assembling a collective memory. The case studies of censored commemorations, described in chapters 8 and 9, illustrate this well, since there the nonoutcomes exposed most clearly the vicissitudes of the process.

The performance of such rites retains traces of the concerted social action and negotiations involved in selecting suitable commemorative objects, recruiting participants, and planning a dual scientific and social event aimed at producing a final product (usually a collection of essays) that "goes public" with a particular range of memories. Therefore, these rites offer a unique window of opportunity to follow the complex relationship between the commemorating present—which appeals to the past for legitimation of diverse and even contrasting conceptual, social, political, and ethical agendas—and the commemorated past. Even if, as this collection attempts to show, presentist, invariably political agendas constrain the re-presentation of the past, still, the social dynamics of commemorative effort require that subsequent commemorations surpass their predecessors in producing usable artifacts of the "actual past."[51]

The custom of periodically taking stock of scientific progress (usually at fractions of centennials) provides an ample reservoir of occasions to promote problematic agendas in the present by forging ingenious links between those agendas and newly emphasized turning points from the past. Indeed, the shortening of the temporal interval considered appropriate for holding commemorative rites from the previously respectable centennial to the more recently celebrated fortieth, twenty-first, and tenth anniversaries of a discovery (and even the fifth anniversary of a major museum for science and industry) suggests that commemorative action is increasingly becoming a perpetual feature of consumerist mass society and science.[52] It provides facile means for drawing attention under a cloak of "exotic" temporality, captured by the phrase "the past is a foreign country." (Ironically, late capitalism as a modulator of commemorative behavior for the purpose of circulating collective memories lags behind early communism, which created an endless supply of commemorative occasions to be used by a political regime requiring perpetual forms of legitimation.)[53]

The convergence of major commemorative occasions with historical trends (e.g., the bicentennial of the French Revolution with a socialist government that saw itself as heir to the Revolution's legacy and sponsored numerous commemorative projects; or the half-centennial of the atomic bomb with a Republican majority in the United States Congress, which encouraged opposition to the *Enola Gay* exhibition at NASM) reflected a wider, societal upsurge in commemorative activity since the 1980s. In science, commemorative activity increased due to both external social trends and internal restructuring between disciplines. Most notably, the convergence of a decline in the social utility of physics in the post-cold war era, with a spectacular rise in the utility of molecular biology—especially via its applications in biotechnology and the human genome project—manifested in heightened commemorative activities, as captured in the essays by Goldberg, Seidel, Pestre, Beller, and Abir-Am.[54]

Another interesting question is under which historical or cultural circumstances

[51] For a most illuminating analysis of the relationship between collective memory and the actual past see Steven Knapp, "Collective Memory and the Actual Past," in *Representations,* Spring 1989, *26:*123–49.

[52] The double helix, published in 1953, was commemorated with special anniversary issues in 1963, 1974, and 1993; the discovery of the tubercle bacillus was also marked on its twenty-fifth, fortieth, and fiftieth anniversaries (in 1907, 1922, and 1932 respectively). Cité des Sciences et d'Industrie in Paris, a museum for science and industry, commemorated its fifth anniversary in 1994; the commemorative book, is a study of museology. See *The Industrial Society and its Museums,* ed. Brigitte Schroeder-Gudehus (Montreal: Presses de l'Université de Montréal, 1994).

[53] See David Lowenthal, *The Past is a Foreign Country* (Cambridge: Cambridge University Press, 1985); See also Irina Gouzevich, "La commémoration scientifique" (cit. n. 35).

[54] For an earlier example see Abir-Am, "Historical ethnography of scientific anniversaries" (cit. n. 14).

the scientific community finds it more productive to commemorate a disciplinary object, a great scientist, or an institution. Our case studies suggest that the commemorative focus tends to shift away from controversial scientists to their scientific discoveries, as in the case of Koch, while scientists are the foci when their accomplishments fit a wider social mission, as with the image of Pasteur as a savior of humanity from infectious diseases; of Royer as a paradigm for gender politics; or of Planck as a symbol of German science in the twentieth century.

This collection conveys a relative balance between the three commemorative objects that define its three parts: seven great scientists, five scientific institutions, and four disciplinary ethoses. Still, the question persists whether any of these three epistemological choices of commemorative objects tends to prevail in a given discipline, national culture, or historical period.

France, for example, especially since the Third Republic, has fostered a distinct tradition of commemorative pomp for great personages who became republican substitutes for royals and saints, the personalized icons of politics and religion.[55] By contrast, the American case studies indicate a preference for commemorations of scientific institutions. In the same vein, physics (notwithstanding the cult of Bohr) was responsible for more institutional commemorations than other disciplines, largely as a result of its evolution into a "big science" after World War II. At the same time, the biomedical sciences were responsible for more commemorations of great minds and discoveries as a result of their resonance with civic aspects such as public health and morality.

Commemorations of discoveries became more frequent in the second half of the twentieth century while those of great minds and institutions prevailed earlier in the century. Ironically, while personalized grandeur, however subjective, enables a wider identification with more social groups, more objective commemorative objects (such as disciplines and institutions) seem to appeal primarily to scientific elites concerned with controlling those objects' public image as contributors to scientific progress and social welfare.

The linkage between these three categories of commemorative foci not only means that many personalized rites have disciplinary overtones, but also that disciplinary commemorations often invoke the heroic status of founders. For example, the centennial of evolutionary biology invoked Darwin, whose book became the issue of a commemoration, and the commemorations of the Copenhagen spirit invoked its towering creator, Niels Bohr.

Furthermore, in some cases the linkage was triple as when the commemorative effort meshed a scientist hero with the discipline she or he helped found and a famous institution. This triple linkage was evident in the case studies examined by Sinding (Bernard, physiology, and the Collège de France); Barberis (Charcot, neurology, and the Salpetrière hospital); Haddad (Koch, preventive medicine, and the Imperial Health Ministry); Hoffmann (Planck, physics, and the Kaiser Wilhelm Institutes); Beller (Bohr, quantum physics, and the Institute for Theoretical Physics

[55] See Jean-Marie Goulemot and Eric Walter, "Les centenaires de Voltaire et de Rousseau," in Nora, *Les lieux de mémoire* (cit. n. 1), vol. 1, pp. 381–420; see also Ozouf, "Le Panthéon" (cit. n. 12); Ben Amos, "State Funerals in the Third Republic as Commemorative Rites," in Nora, *Les lieux de mémoire,* vol. 1.

in Copenhagen); and Abir-Am (Delbrück, molecular biology, and Cold Spring Harbor Laboratory; Lwoff, molecular biology, and the Institut Pasteur).

Another interesting pattern pertains to the national, cultural, and political context of scientific commemorations. French and American case studies emerged as preponderant. The French case studies include three individual and three disciplinary choices, but no institutional ones.[56] By contrast, the American case studies include four institutional and two disciplinary choices, but none of personal grandeur. This suggests that American commemorative rites tend to focus on institutions, reflecting the impact of a political culture that emphasizes institutions as manifestations of public will. By contrast, the French commemorative rites tend to focus on personal grandeur as a republican substitute for the royalist and religious iconography that long dominated the country's political culture.

The rest of the case studies include Polish, German, and international cultural context. All the crosscultural papers in this collection include the United States as one of the countries, thus enabling the reader to follow French-American, German-American, and Polish-German-American comparisons. This invariable American presence in the crosscultural papers indicates the emerging role of the United States as a major commemorative arena. While most American commemorations of American objects tend to focus on institutions or disciplines, on several occasions they became the stage for honoring foreign scientists with a transnational appeal, most notably in the case of Copernicus's quadri- and quinquecentennials in 1943 and 1973, the Darwin centennial in 1959, and the Pasteur centennial in 1995.

The national-cultural dimension is also responsible for the major political association of many scientific commemorations. For example, the Copernican quadricentennial included an address by President Roosevelt at its New York City site, while at its site in Nazi-occupied Poland, the head of the Third Reich's occupation forces made sure they were present. Pasteur's jubilee was conducted in 1892 in the presence of the president of the Republic, and in 1895 Pasteur, like Claude Bernard in 1878, received a state funeral. Royer's centennial in 1931 was attended by Raymond Poincaré and Paul Painlevé, former president and prime minister of France, respectively; Planck's centennial was attended by the general secretary of the East German Communist Party, Walter Ulbricht; the East German Prime Minister Otto Grotewohl; Willy Brandt, the mayor of West Berlin; and Theodor Heuss, the president of the Federal Republic of West Germany, in addition to the Soviet ambassador and the rest of the political apparatus.

This pattern of high-level political presence also obtained in the institutional commemorations, described in part II. The Harvard tercentenary featured both the governor of Massachusetts and the United States president and candidate for a second term, F. D. Roosevelt. The cancellation of the half-centennial of the decision to drop the atomic bomb was made in view of the Smithsonian's invasion by twenty-one angry congressmen. The half-centennial of Los Alamos National Laboratory featured representatives of both the George Bush and Bill Clinton administrations, as well as of many state and regional office holders and Washington, D.C. patrons. The CERN quarter-centennial featured government representatives from the leading member countries as well as from the European Parliament.

[56] For case studies of institutional commemorations in France, see the essays by Anousheh Karvar, Anne-Marie Moulin, Terry Shinn, and Christiane Sinding in Abir-Am, *La mise en mémoire* (cit. n. 1).

Even commemorations unique to science saw their share of political associations. The Darwin centennial in Chicago featured Adlai Stevenson, the governor of Illinois and a two-time presidential candidate; as well as Richard Daley, the mayor of Chicago and a leading figure in the Democratic Party.

Not all these political associations had a national or nationalistic dimension, though some did. The commemorations of Copernicus involved a Polish-German nationalistic rivalry, while those of Koch reflected German imperial nostalgia for the lost colonies in tropical Africa. Some of the French commemorations shortly after World War I featured the national agenda of boycotting German scientists. For example, they were not invited, despite having made major contributions to neurology, to the centennial of Charcot in 1925. Domestic political issues played a role in the centennial of Royer (the vote for women), of Planck (East-West ideological squabbles, with each side featuring Soviet or American sponsors); at the Harvard tercentenary (election politics); at the Los Alamos half-centennial (patronage in Washington, D.C., and in the state of New Mexico); at the CERN quarter-centennial (patronage in the European Council and other centers in Brussels); and at the Darwin centennial (Chicago civic and state democratic politics).

This pervasive association, which also appears in case studies explored elsewhere, suggests that commemorations serve as a bridge between science and politics.[57] Politicians seek associations with scientific commemorations to domestically convey a message of social progress and of rationality in decision-making, or to capitalize on science's international ethos in order to break through political isolation in foreign affairs. Scientists, meanwhile, cultivate political presence as a resource for budgets, prestige, and the appearance of power. The question is whether these macropolitical dimensions of commemorations were always there, or whether they emerged once the political culture became scientistic, as during the cold war (see Table 1).

Multiple commemorative occasions of the same object (see examples in Table 1) enable a comparison of the impact of two or more historical moments upon the contents and form of the commemorative practices, including their presentist concerns, which may emphasize different aspects of the "same past." These longitudinal comparisons are particularly evident in Sinding's study of three commemorations of Bernard and two of Pasteur; Barberis's study of two Charcot centennials; Haddad's study of two anniversaries of Koch's discovery of tubercle bacillus; and Beller's study of three jocular commemorations of Bohr's birthdays as well as the centennial of his birth.

Among the final questions that this collective study raises is whether commemorative activity not only tends to re/produce, display, and inculcate the collective memory of groups who are contending for leadership in a scientific discipline (and, by derivation, to spokesmanship for science in society), but also produces new insights into the past. Can new insights emerge, even though the scientists' appeals to the more or less remote past are invariably driven by social, political, or ethical agendas in the changing, contested present?

The case studies in this volume suggest that a surplus of genuine new knowledge of the past is generated by the very social dynamics of commemorative activity. On the one hand, in order to pursue diverse contested agendas in the present (be they

[57] See the essays by Rosalyn Rey, Irina Gouzevich, Anne-Marie Moulin, and Annie Petit in Abir-Am, *La mise en mémoire* (cit. n. 1).

boosting a group's morale, patching its divisiveness, or legitimizing new leadership) the orchestrators of such rites must ingeniously smooth out the immanent contradictions between the unpredictible direction of scientific progress in the present and the grounding of their group's identity, practices, and global position in events from the past. This smoothing often requires that they uncover formerly neglected scientific events, actors, concepts, archival documents, photographs, or oral testimonials, among an increasing range of artifacts of memory.

On the other hand, the performers and orchestrators of commemorative rites are invariably motivated to surpass the rites that have previously colonized their discipline's frontier of collective memory, so that their efforts will be accepted as a valid countermemory by both insiders and outsiders. This explains why scientific commemorations generate not only new meaning about the past, or historiographic surplus, but also new meaning about the custodians of collective memory in the present, or ethnographic surplus.[58]

Finally, an important aspect of this study that deserves greater attention in the future pertains to the reflexivity of authors *vis-à-vis* the commemorative events they examine and in which they also participate. For example, Gingerich reflected on his own participation in the Copernican quinquecentennial in Poland in 1973, reaching the conclusion that his being awarded a medal by the Polish government on that occasion might not only have been due to his status as a Copernican historian of science, but also to micro- and macropolitical considerations such as his nonparticipation at a competing commemoration in Australia and his past role in providing support to Poland (he went there in 1946 as part of a postwar relief effort, bringing much-needed supplies, especially horses).

The issue of reflexivity looms larger in cases of commemorations in high-prestige disciplines, where progress is fast, the stakes are high, and the honorees are still around. Here, the historian faces a more difficult situation than that presented by the textual analysis of eulogies to long-dead protagonists. This situation is alluded to, but not reflexively examined, by Goldberg, Seidel, Pestre, and Gingerich, historians who took part in both preparing and writing about major scientific commemorations. Each had to confront his or her own archivally informed and orally recorded sources of historical truth with the collective memory that the commemoration's organizers (invariably laboratory directors and other scientists in power) sought to impart to a younger generation of scientists or to political patrons.

Another type of reflexivity is implicit in the depth of historical insights available to authors who spend considerable time investigating the commemoration's honorees (e.g., Beaulieu and the Bourbaki founders, Smocovitis and the leading evolutionary biologists, or Abir-Am and the founders of molecular biology). A great deal remains to be said about such fateful encounters on the verge of posterity, perhaps in the form of a dialogic ethnography. As more historians come to terms with the ethnographic and literary turns in science studies, such a reflexivity might become more commonplace in the near future.

Meanwhile, let's hope that several other wonderful case studies that this volume sadly missed will be brought to publication in the near future. They include contradictory commemorations of Gregor Mendel by four Czech regimes; the commemo-

[58] For an earlier example see Abir-Am, "Historical ethnography of scientific anniversaries" (cit. n. 14).

rations of Alessandro Volta, Galileo Galilei, and Leonardo da Vinci during the fascist regime in Italy; the Ibero-American commemorations of Alexander von Humboldt; the Spanish commemoration of the Nobelist biochemist Severo Ochoa; and last, but not least, the commemorations of Alexandre Koyré by "Americans in Paris" and of Thomas Kuhn by "Americans in Cambridge, Mass.," among various other telling examples.

# *Part I. Commemorating Great Minds: Scientists as Cultural Heroes*

# The Copernican Quinquecentennial and its Predecessors
## Historical Insights and National Agendas

*By Owen Gingerich**

MY OPENING SCENE IS SET in Copernicus's birthplace in 1972, one year before the five hundredth anniversary of his birth. The international planning committee for the quinquecentennial symposium, of which I was a member, had been tracing the "Footsteps of Copernicus," a tour of northern Poland proposed for the meeting's delegates. After seeing conspicuous physical improvements in Toruń, including at the Uniwersytet Mikołaja Kopernika, Mikołaja Kopernica, the French representative remarked, "It's a pity every town doesn't have a Copernicus."

Her perceptive comment reminds us that every commemoration endorsed by historians or scientists invariably has background agendas other than simply the increase and diffusion of knowledge. This point was not lost on the planning committee's German representative, who reported to the Deutsche Forschungsgemeinschaft about the exemplary use of scientific anniversaries for funding scientific institutions.[1] But beyond the comparatively benign agenda of funding, there can also be strong nationalistic political agendas. The Copernican celebrations over the past two centuries, perhaps more than any other anniversaries, carried the heavy burden of political overtones that such occasions entail.

To set the stage, I shall first briefly sketch the worldwide observation of the 1973 Copernican celebration. Five hundredth anniversaries of major scientific figures have been rare, and the Copernican quinquecentennial attracted even more attention than the recent Einstein centenary. I shall then turn to its historical antecedents, especially the quadricentennial celebrations of Copernicus's death in 1943, when Poland was overrun by the Nazis. Finally, I will shift roles from that of historian to participant-actor as I describe some of the behind-the-scenes intrigue engendered by the 1973 anniversary.

The principal international quinquecentennial celebrations took place in Poland in September 1973 with the International Astronomical Union's (IAU) Extraordinary General Assembly, a related International Union for the History and Philosophy of Science (IUHPS)/IAU historical symposium, and three IUHPS symposia (see Appendix 1). The Poles inaugurated their *Nicholas Copernicus Complete Works* series

* Harvard-Smithsonian Center for Astrophysics, Cambridge, MA 02138.
Jerzy Dobrzycki assisted enormously in the research for the heart of this paper, and his collaboration is gratefully recognized.
[1] Personal memorandum from Bernhard Sticker to Jerzy Dobrzycki.

 0369-7827/99/1401-0003$02.00

with a multicolor facsimile edition of Copernicus's holograph manuscript, and augmented their recently established *Studia Copernicana* series.[2] In Poland Copernicus was everywhere, from the street corner kiosks filled with Copernican kitsch, to the posters, to the government ministries that wished to link Copernicus with their own specialties.

America, with its flourishing scientific establishment and several major ethnic Polish communities, was an especially fertile land for Copernicana. At the end of 1972 the American Association for the Advancement of Science sponsored a special symposium with an international cast of speakers, including Jerzy Dobrzycki from Poland, John North and Jerome Ravetz from England, and Alexander A. Mikhailov from the USSR.[3] The American Astronomical Society had a plenary lecture, the American Philosophical Society held a symposium,[4] and the Universities of Arizona, California, and Michigan each sponsored symposia that resulted in published volumes.[5] The Smithsonian Institution and the National Academy of Sciences joined forces for a three-day international symposium on the nature of science, which featured concerts with specially composed music, a Copernicus portrait commissioned from artist Leonard Baskin, and a triple-screen slide show by designer Charles Eames (see Appendix 2).[6]

Countless Copernican celebrations were held around the world, from Afghanistan to Zaire, many sponsored by UNESCO. The *Studia Copernicana* volume 17 of the Polish Academy of Sciences's Institute of the History of Science, Education, and Technology lists scores of lectures and seminars, the unveiling of monuments, and the renaming of streets. The tally of special exhibitions runs to thirty-two pages. And an outpouring of colorful commemorative stamps issued from a host of nations, although one might suspect that several countries were more interested in aug-

[2] The *Nicholas Copernicus Complete Works* series is bibliographically rather complex because it includes several parallel language editions, and particularly because the four English volumes are hardly complete without the Latin texts found in the *Nicolai Copernici opera omnia* edition (and the fourth Latin volume still remains to be published). The situation is further complicated by a Polish issue of the English volumes followed by a Macmillan (England) edition, which The Johns Hopkins University Press subsequently extended and reissued, together with a paperback reprint series containing new and controversial introductions. Suffice it to say that the first volume, the facsimile of Copernicus's manuscript of *De revolutionibus,* was ready in 1972, in time for the celebrations. It came also in a deluxe form with uneven edges in such close imitation of the original manuscript that occasionally observers have been fooled into thinking they were looking at the original holograph. The *Studia Copernicana* series of the Polish Academy of Sciences's Institute for the History of Science and Technology was inaugurated in 1970 with a large volume of articles in French by Alexander Birkenmajer (many of which were specially translated from Polish). This series eventually continued through vol. 24, Grażyna Rosińska's *Scientific Writings and Astronomical Tables in Cracow (XIVth–XVIth Centuries)* (Wrocław: Ossolineum, 1970–84). Included as a subset were the four *Colloquia Copernicana* volumes (*Studia Copernicana* vols. 5, 6, 13, and 14) and also the *Nicholas Copernicus Quincentenary Celebrations Final Report,* vol. 17 (1977).

[3] Recorded in *Copernicus: Yesterday and Today,* eds. Arthur Beer and K. Aa. Strand, vol. 17 of *Vistas in Astronomy* (Oxford: Pergamon Press, 1975).

[4] "Symposium on Copernicus," *Proceedings of the American Philosophical Society,* 1973, *117,* n. 6:413–550.

[5] *Man's Place in the Universe: Changing Concepts,* ed. David W. Corson (Tucson: College of Liberal Arts, University of Arizona, 1977); *The Copernican Achievement,* ed. Robert S. Westman (Berkeley and Los Angeles: University of California Press, 1975); *Science and Society: Past, Present, and Future,* ed. Nicholas H. Steneck (Ann Arbor: University of Michigan Press, 1975).

[6] *The Nature of Scientific Discovery,* ed. Owen Gingerich (Washington, D.C.: Smithsonian Institution Press, 1975).

menting their state treasuries at the expense of philatelists than in lauding the birth of the heliocentric cosmology.

Despite the superficial appearances of a splendid celebration for a founding father of modern science, there were, behind the scenes, multiple political agendas. To understand some of these, we must look back at the complex history of Copernican celebrations. There are two pivotal dates in Copernicus's lifetime: his birth in 1473 and his death in 1543, which is also the date of publication of his opus magnum, *De revolutionibus.* Because fifty-year anniversaries are sometimes celebrated as well as the centennials, we have a well-spaced series of dates for Copernican observances: 1823, 1843, 1873, 1893, and so on.

In the sixteenth century commemorative tablets were erected in Copernicus's honor, but apparently in those early centuries scholars were not especially enthusiastic about round-number anniversaries. However, we do find a lengthy memorial prepared by the German critic and pedant Johann Christoph Gottsched (1700–1766) for reading in the Frauenburg (today, Frombork) cathedral in 1743, the bicentennial of *De revolutionibus.*[7]

The observance years 1773, 1823, and 1843 passed comparatively unnoticed. It must be understood, however, that in the nineteenth century, Poland had been erased from the geopolitical map (see Figure 1). The Congress of Vienna in 1815 ratified the 1792–1795 partitioning of Poland among Germany, Austria, and Russia; the Poles became ethnic minorities in hostile lands, like the present-day Kurds or Armenians. This was a time of ethnic and cultural repression in Polish Russia. The university in Warsaw was closed after the uprising in 1831, although the observatory was preserved, and in 1843 the Warsaw astronomers initiated a new edition of *De revolutionibus* in Latin and in Polish, the first translation into any modern language.[8] In this edition, which was finally published in 1854, for the first time the original autograph of Copernicus's manuscript was used in the recension. It also included the first publication of Copernicus's "Letter against Werner," which had been found in Berlin by the Warsaw philologist Wacław Aleksander Maciejowski. One of the few substantial Polish works timed for the anniversary year was by Adrian Krzyżanowski, who had been a professor of mathematics when the university was closed. His monumental essay (over 900 pages including supplements) was entitled (in Polish), "Ancient Poland's Contribution to Mankind's Progress, Sketched in the Copernican Jubilee Year of 1843."[9]

## I. THE 1873 QUADRICENTENNIAL

The year 1873 was totally different in terms of conspicuous anniversary celebrations. By then, there was a flourishing Copernicus-Verein für Wissenschaft und Kunst in Thorn (today, Toruń), Copernicus's birthplace. Founded a few years earlier in anticipation of the anniversary, its leading intellectual lights were Maximilian

[7] "Gedächtnisrede auf Nicolaus Coppernicus," in *Gesammelte Schriften von Johann Christoph Gottsched,* vol. 6 (Berlin, c. 1910), pp. 123–48.

[8] *Nicolai Copernici Torunensis: De revolutionibus orbium coelestium libri sex: Accedit G. Joachimi Rhetici Narratio prima, cum Copernici nonnullis scriptis minoribus, nunc primum collectis, eiusque vita. Mikołaja Kopernika Toruńczyka O Obrotach Ciał Niebieskich ksiąg sześć* (Warsaw, 1854).

[9] Adrian Krzyżanowski, *Dawna Polska ze stanowiska jej udziału w dziejach postępującej ludzkości, skreślona w jubileuszowym Mikołaja Kopernika roku 1843* (Warsaw, 1844).

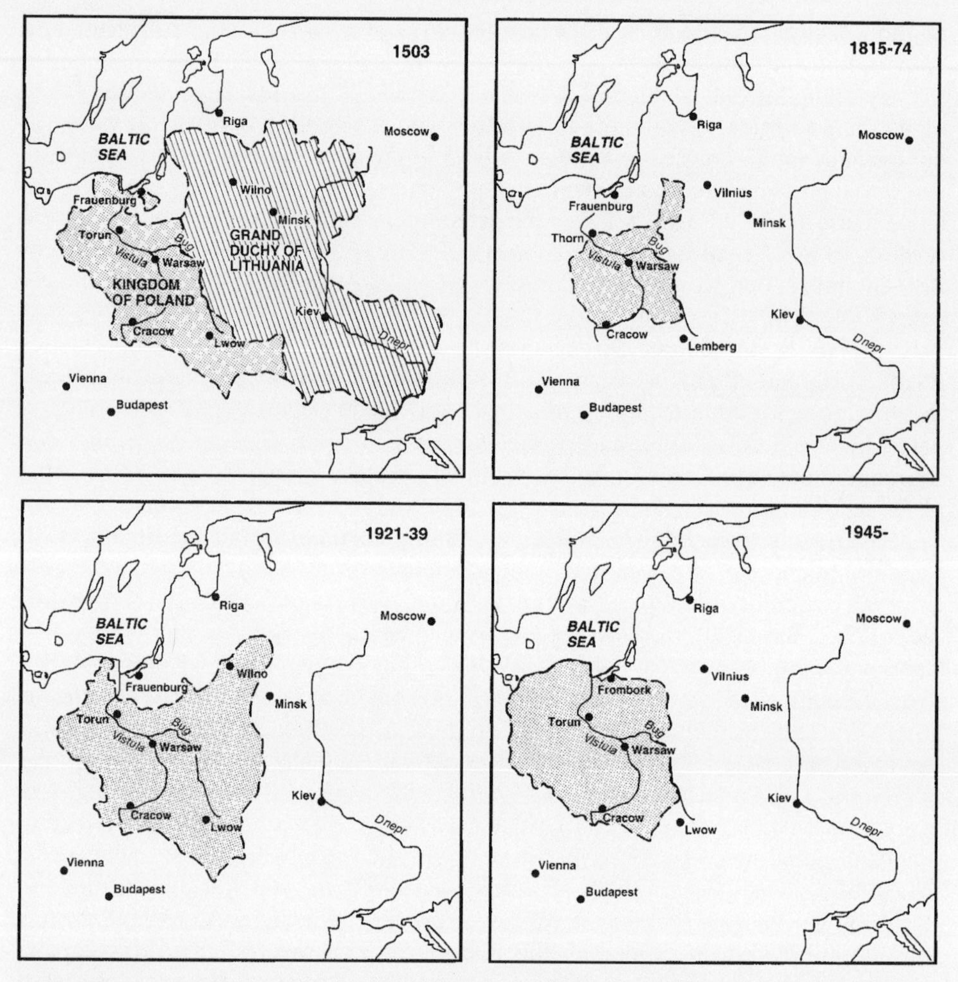


***Figure 1.*** *The maps show some of the changing boundaries of Poland from the time of Copernicus to the present. From 1875 through 1920 Poland was not on the map. Adapted from Norman Davies,* God's Playground *(Oxford, 1981) and Cz. Nanke, L. Piotrowicz, and Wł. Semkowicz,* Mały Atlas Historyczny *(Warsaw, 1972).*

Curtze (1837–1903), the mathematician and paleographer, and his older colleague at the Thorn Gymnasium, Leopold Prowe (1821–1887), whose multivolume Copernican biography would appear in 1883–1884. For the anniversary year the society undertook a lavish edition of *De revolutionibus* dedicated to the German Kaiser Wilhelm II, whose financial assistance had made it possible.[10] It also published a detailed account of the festivities held in Thorn on 18–19 February 1873, which included the extensive greetings received from around the world.[11]

The Copernicus-Verein was German-speaking and Prowe, speaking on behalf

[10] *De revolutionibus orbium coelestium libri VI. Ex auctoris autographo reduci curavit Societas Copernicana Thorunensis* (Thorn, 1873).

[11] *Die vierte Säcularfeier der Geburt von Nicolaus Copernicus Thorn 18. und 19. Februar 1873* (Thorn, 1874).

of the society, situated young Niklas Koppernigk in a Hanseatic merchant city. Although Toruń was in 1473 a leading town in a Polish state, Prowe maintained that by founding the town was German.[12] After all, the language of the Polish king's chancery fifty years before Copernicus's birth had been German, the language of northern commerce. By inclination and implication, Copernicus was one of theirs. The society chose to overlook the fact that in 1466, seven years before Copernicus's birth, Toruń had been added to Poland in the Peace of Toruń, and that Copernicus's political loyalty was first to the local region known as West or Royal Prussia and second to the Kingdom of Poland. Moreover, Copernicus's close personal links of long standing were to the scholars in Cracow, many of whom were of Polish birth. In 1504 Copernicus joined his uncle, Bishop Lucas Watzenrode, in pledging allegiance to King Alexander of Poland, and three years later he presumably accompanied his uncle to the coronation of King Sigismund I. To ignore these points was part of an aggressive cultural cleansing carried out by Prowe and his German colleagues. There was, nevertheless, a certain amount of ambiguity in the situation because the modern concept of nation-state was not yet fully developed. Copernicus was in reality a trans-national, but at the same time he was a German-speaking Pole.

In Warsaw, under Russian hegemony, there was little possibility of a public Polish celebration of the 1873 Copernican anniversary. There was a lecture in the town hall and an evening celebration in the opera theater, all under the strict vigilance of the Russian administration. In Toruń under the German rule, Polish-language celebrations were held by patriots determined to counter the Prussian cultural assertions and keep alive their own national heritage, which included a claim to Copernicus. In Gniezno, canon Ignacy Polkowski (1833–1888) published a 653-page book with many Copernican documents translated into Polish.[13] However, his report on the four-hundredth anniversary celebrations in Toruń was confiscated by police who searched the bookstores.[14] Polkowski was accused of sponsoring ideas that endangered the sovereignty of the Prussian state because he had included the text of a sermon read at a church ceremony that contained a veiled but unmistakable allusion to the Polish White Eagle. Polkowski as well as the preacher, Jażdżewski, were tried; Polkowski was sentenced to three weeks in prison or a fine of 30 talers, but Jażdżewski was released.

The quinquecentennial of the refounding of the Jagiellonian University in Cracow was in 1900, and in preparation the mathematics-teacher-turned-historian, Ludwik Antoni Birkenmajer (1855–1929), was apparently encouraged to prepare a major biography in Polish to counter the three-volume German work by Prowe. The result was volume I of the monumental, 700-page *Mikołaj Kopernik,* a somewhat episodic, rambling compendium of interesting, erudite, and often quite original Copernican scholarship.[15] Birkenmajer's major research efforts included an examination of as-

[12] *Ibid.,* p. 40.
[13] Ignacy Polkowski, *Kopernikijana* (Gniezno, 1873).
[14] See Henryk Baranowski, *Bibliografia Kopernikowska 1509–1955* (Warsaw: Państwowe Wydawnictwo Naukowe, 1958), and his comments on item #2878, Ignacy Polkowski, *Czterowiekowy jubileusz urodzin Mikołaja Kopernika w Toruniu dnia 19 lutego 1873* (*Quaternary Jubilee of the Birth of Nicholas Copernicus in Toruń on the 19th day of February 1873*) (Gniezno, 1873).
[15] L. A. Birkenmajer, *Mikołaj Kopernik* (Cracow, 1900); an English translation of the major part of this book is available from Bell and Howell Learning and Information, Ann Arbor, Michigan (formerly University Microfilms International).

tronomical manuscripts available to Copernicus in Cracow in the early 1490s, and the discovery in Sweden of many books from Copernicus's personal library. He also demonstrated that Copernicus's *Commentariolus* was a comparatively early work differing in many essential details from *De revolutionibus.* Though not the sought-for biography, *Mikołaj Kopernik* established a contemporary Polish presence in the ranks of serious Copernican researchers. A second volume was never produced in the same format, but for the 450th anniversary of Copernicus' birth in 1923, Birkenmajer published his more synthetic *Stromata Copernicana.*[16]

## II. THE RE-EMERGENCE OF POLAND AND THE GERMAN RESPONSE

By 1923, the political map was, of course, entirely different. Like a phoenix, Poland had re-emerged from the ashes of World War I. For the first time in 150 years an independent Polish state had come into existence. There was a veritable outpouring of Copernicana in Polish, and Baranowski's Copernican bibliography lists no fewer than twenty-five articles with the title "Mikołaj Kopernik"—that is, with the presumed original spelling of the name, which became a shibboleth for Copernicus's Polish origins.[17] Yet, apart from the Birkenmajer volume, these contributions contained far more enthusiasm and nationalism than memorable scholarship. Besides the many Polish publications, there were some in German, including a charmingly printed translation of parts of book I of *De revolutionibus* plus fragments of relevant correspondence.[18] Elsewhere, the anniversary passed almost unnoticed. *Nature* reported briefly that "Impressive ceremonies were held in Warsaw, Wilno, Poznan, Lodz, Wlockawek and Kielce," that a two-day meeting in Toruń included the inauguration of the first general meeting of the Polish Astronomical Society, and that a memorial tablet was unveiled on the house in Toruń where Copernicus was born.[19] In *Science* magazine the Smithsonian Institution's assistant secretary, C. G. Abbott, concluded a short encomium by announcing an informal meeting and brief address at the institution.[20]

The German response to this burst of Polish nationalism came slowly but steadily. Father Eugen Brachvogel, founder of the Copernicus Museum in Frauenburg (today, Frombork)—and I remind you that the cathedral where Copernicus wrote his *De revolutionibus* fell outside the Polish boundary—wrote an article nearly every February (the month of Copernicus's birth) entitled "Zum Geburtstag unseres Koppernikus" for the publication *Unsere ermländische Heimat.* Brachvogel avoided shrill claims about Copernicus's nationality, but with the approach of the Nazi period such declarations soon dotted the German landscape. In 1932 the *Jahrbuch* of the Thorn

[16] L. A. Birkenmajer, *Stromata Copernicana* (*A Copernican Counterpane*) (Cracow: Nakładem Polskiej Akademji Umiejęności, 1924).

[17] Baranowski, *Bibiografia* (cit. n. 14)—see most notably pp. 360–1, but also 152–3 and 362–71. Note also Henryk Baranowski, *Bibliografia Kopernikowska II 1956–1971* (Warsaw: Państwowe Wydawnictwo Naukowe, 1973) and the selected continuation for 1972–75 in *Studia Copernicana* (cit. n. 2), vol. 17, pp. 179–201.

[18] *Über die Umdrehungen der Himmelskörper: Aus seinen Schriften und Briefen* (Posen: Deutsche Bucherei, 1923).

[19] "Polish Celebrations of the 450th Anniversary of the Birth of Copernicus," *Nature,* 1923, *111:*515.

[20] C. G. Abbott, "The Four Hundred and Fiftieth Anniversary of the Birth of Copernicus," *Science,* 16 February 1923, *57:*196–7.

Heimatbund published "Nikolaus Coppernicus, der deutsche Forscher."[21] In 1937 a rash of articles on "Nikolaus Coppernicus—der Deutsche" appeared. What provoked this particular onslaught was apparently the Polish pavilion at the International Exposition in Paris, which displayed a statue of Copernicus as one of seven heroes of the Polish spirit. "This idea is a Polish falsification of history, not new and already long since disproved by Prowe using archival sources," complained Professors Kühn and Lockemann in the pages of the popular astronomy monthly *Die Sterne.*[22] They concluded by saying, "We categorically reject the renewed attempt from the Polish side to expunge Copernicus from German cultural life and to place him in the ranks of the Polish community."

Two years later, in 1939, when the German astronomers renamed the Astronomische Rechen-Institut at Berlin-Dahlem as the Coppernicus-Institut, Erich Krug from the party newspaper reported the event in *Die Sterne* with an article titled "Coppernicus, der grosse Deutsche."[23] To name the new institute after our great Forscher—our great scientific investigator—was particularly appropriate, he declared. "That this man was a German, stands firmly beyond doubt. Even in Poland one finds this more and more. It is a sign of the times that last year the Wiadamowski Literature Prize [*sic Wiadomosci Literackie,* a leading Polish cultural weekly] for the best book in Polish, went to the Polish historian Wasiutyński for his Copernicus biography in which the Germanness of the great researcher is stated without reservation."

It will perhaps be illuminating to briefly consider here this book, in which Jeremi Wasiutyński—actually an astrophysicist rather than a historian—presented Copernicus as the universal man who, like Einstein, rose above an easily pigeonholed nationality.[24] At the time the book was rather out of step with the Polish national line, although it comes close to our contemporary view of Copernicus. Stephen Mizwa, the energetic executive director of the Kosciuszko Foundation in New York City, had this to say about the book a few years later:

> A remarkable performance of a young man, who attempted to write a definitive biography of Copernicus with the aid of the new sociologico-psychological tools. The subject is perhaps overpsychoanalyzed. In trying to reach different conclusions from previous Copernican scholars, the author is often rasping. Reads like an historical novel, although its scientific value, for the most part, is much greater. Taken in conjunction with other works as correctives, it will have a permanent place in the Polish literature on the subject.[25]

While the Poles avidly defended the Polish origins of Kopernik, they did so in their national language, which was no longer a suppressed tongue. Only rarely did their case reach out to other languages. One example is when the Prussian State Archives in Berlin published in 1938 a collection of German translations of notices in the

[21] Cited in Baranowski, *Bibiografia* (cit. n. 14), item #3197.

[22] Kühn and Lockemann, "Coppernicus—ein deutscher Forscher!" *Die Sterne,* 1937, *17:*230. I wish to thank Prof. Mark Gingerich for his assistance with this translation.

[23] Erich Krug, "Coppernicus, der grosse Deutsche," *Die Sterne,* 1939, *19:*81–3.

[24] Jeremi Wasiutyński, *Kopernik: Twórca Nowego Nieba* (*Copernicus: Creator of the New Heavens*) (Warsaw: Wydawnictwo J. Przeworskiego, 1938).

[25] Stephen P. Mizwa, *Nicholas Copernicus:1543–1943* (New York: The Kosciuszko Foundation, 1943), p. 64.

Polish press under the title *Polnische Stimmenzur Volkszugehorigkeit des Coppernicus* (*Polish Voices on the Nationality of Copernicus*).[26]

### III. WAS COPERNICUS A GERMAN? CELEBRATING THE 1943 QUADRICENTENNIAL

It might be safe to say that during the 1930s, the Poles expected to take their viewpoint to an international audience at the 1943 quadricentennial of the publication of *De revolutionibus.* But the situation turned out very differently from what they had hoped. As of 1 September 1939 Poland became an occupied country, and it was the Nazi warlords who called the shots. Thus for the Poles, the quadricentennial was a tragic and highly politicized occasion. In central Europe the anniversary gave the controlling forces the opportunity to enforce the belief that Copernicus was German. (No doubt the Nazis firmly believed that no Pole could have been so clever, so how could the matter even be debated?) But outside the Axis powers, the free Polish community used the opportunity to exploit Copernicus as a rallying symbol for Polish independence. Few scientific anniversaries have ever been suffused with such contrasts in nationalistic fervor.

From the German side, two major contributions of lasting importance emerged. First was astronomer/historian Ernst Zinner's *Entstehung und Ausbreitung der Coppernicanischen Lehre.* Published as the seventy-fourth volume of the *Sitzungsberichte der physikalisch-medizinischen sozietät zu Erlangen,* it was—curiously enough—dedicated to the bicentennial jubilee of Friedrich-Alexander University in Erlangen, but Zinner (1886–1970) argued in the foreword that it was appropriate to celebrate both anniversaries together. Zinner's joint interests in astronomical bibliography and in the history of astronomical instruments made this a unique and valuable research contribution.[27] To his detailed history of Copernican astronomy and its antecedents (especially Germanic ones), he appended a discussion of the Copernican portraits, a list of the Copernican library, a checklist of locations of first editions of Copernicus's *De revolutionibus,* and a discussion of Copernicus's sundial, among other topics.

The second important advance in German Copernican scholarship was the beginning of a proposed nine-volume *Nikolaus Kopernikus Gesamtausgabe,* sponsored by the Deutsche Forschungsgemeinschaft and under the editorship of Fritz Kubach (1912–last seen in 1945, declared dead in 1957). The first volume was a facsimile of Copernicus's original autograph manuscript of *De revolutionibus,* which scholars had become aware of in the Nostitz Library in Prague in the previous century. In 1931 the astronomers at the Czechoslovakian National Observatory in Prague had launched a subscription for a facsimile edition following the recommendation of the IAU at its Rome meeting in 1922, but this had evidently come to naught during the Great Depression.[28] The 430 photographs required for the German publication were snatched from the flames in Stuttgart after a British bombing raid on 11 March 1943,

[26] Cited in Baranowski, *Bibliografia* (cit. n. 14), item #786.

[27] See the review of the reprint edition by Robert S. Westman, "Zinner, Copernicus, and the Nazis," *Journal for the History of Astronomy,* 1997, *28:*259–70.

[28] This proposed edition, to have been published under the direction of Q. Vetter and sold for $35.50, is documented in a flyer, "Invitation à la Souscription." This ephemera is found in the pamphlet files of the History of Science Library in Harvard's Widener Library.

and the volume appeared the following year.[29] The introductory text made it clear that the editors and sponsors saw this as a great national project, just as Antonio Favaro's turn-of-the-century *Opere di Galileo Galilei* is customarily referred to as "the National Edition." Kubach opened with the words, "Nikolaus Kopernikus, der grosse deutsche Astronom aus dem weichselländischen Preussen, . . . "—"Nicolaus Copernicus, the great German astronomer from Vistular Prussia, has, as the founder of our contemporary view of the cosmos, brought about a unique historical achievement. His name, which he has indelibly inscribed in the rolls of German and European scientific and intellectual history, has through his work been attached to a firm and imperishable conception."

Only a single further volume of this series was ever published, the critical text of *De revolutionibus,* in 1949. Most of the proposed work (for example, a census of copies of the first edition) never came to fruition. It is chilling to see the *Nachlass* from this effort, still preserved at the Copernicus-Forschungsstelle at the Deutsches Museum in Munich, an archive of letters invariably signed, "Heil Hitler." But out of the cinders of this project, the German scholars have now planned, with a gross dose of Teutonic thoroughness, an even more monumental Copernican edition: ten volumes in twenty-three parts, of which the fourth appeared in 1994.

Of course, many secondary scholarly works were published in 1943. *Kopernikus-Forschung,* a volume from Leipzig in the *Deutschland und der Osten* series of sources and research, printed eight articles with many illustrations of original documents. It followed what had by then become the official spelling of Copernicus adopted by the Nazi Party.[30] Party concern over how to spell the name had first surfaced in 1938. On 24 May 1941 Joseph Goebbels' Reich propagandaministerium, in agreement with Alfred Rosenberg of the party's educational office, advocated "Kopernikus" because the letter K looked more "natural" to a German. However, on 21 October 1941 the Ministerium für Wissenschaft, Erziehung und Volksbildung (Education Ministry) ordered "Coppernicus" because this spelling was impossible in Polish. Then Rosenberg, who had in the meantime become Reichsminister for the Occupied Eastern Regions, asked Ernst Zinner for an opinion, who offered still a third possibility, "Koppernik." On 9 September 1942 an anonymous bureaucrat in the Education Ministry noted that Hitler favored "Kopernikus," and this became the official ruling on 14 January 1943, reversing their previous opinion.

The lead article in *Kopernikus-Forschung* was Hans Schmauch's "Nicolaus Kopernikus—ein Deutscher," which had already appeared in 1937 but with the older spelling "Coppernicus," the form still retained by Zinner. (In any event, it would have been difficult to maintain the double *p* spelling in a volume that illustrated a number of original Copernican documents, all signed or written with a single *p.*) Schmauch's article was one of the most thorough of the volume's scholarly but highly polemical accounts of Copernicus's German heritage. He argued that Toruń

[29] The fire bombing is described in the "Nachbericht," p. xxiii of the facsimile, *Nikolaus Kopernikus Gesamtausgabe,* vol. 1, *Opus De revolutionibus caelestibus manu propria: Faksimile-Wiedergabe* (Munich and Berlin: Verlag R. Oldenbourg, 1944).

[30] See Hans Koeppen, "Die Schreibweise des Namens Copernicus. Betrachtungen zur Schreibung des Namens des großen Astronomen, ausgehend von der Kontroverse im Dritten Reich," in *Nicolaus Copernicus zum 500. Geburtstag,* eds. Friedrich Kaulbach, Udo Wilhelm Bargenda, and Jürgen Blühdorn (Cologne and Vienna: Bohlau, 1973), pp. 185–234. I wish to thank Richard Kremer for drawing my attention to this article.

*Figure 2. Pane of German occupation stamps with Copernicus and border of swastikas.*

was four-fifths German in the mid-fifteenth century and that Copernicus came from German lineage on both sides of his family. For a country increasingly obsessed with genealogies, a large fold-out diagram delineated Copernicus's ancestry. Schmauch stopped short of claiming that Copernicus was of pure Aryan blood, but the message was clear.[31]

Another secondary or even tertiary publication, rather elaborately printed in Krakau (Krakow or Cracow), Copernicus's university town, was by the short-lived Institut für Ostarbeit, which was organized by the occupying powers. It contained a reprint of a general Copernican article originally written by Potsdam astronomer Hans Kienle for *Naturwissenschaften,* and a new contribution from Erwin Hoff of Krakau.[32] Here there was no need to make a strident claim for Copernicus's German nationality—the whole appearance, from the decorative column of swastikas to the language and the sponsorship, made clear the message.

There were other ways to reinforce the German viewpoint in the quadricentennial year. For example, in Poland the Deutsche Post Osten issued a souvenir pane of semipostal stamps with a border of swastikas (see Figure 2). And at the dedication of a Copernicus monument in Toruń, Gauleiter Förster, the top-ranking Nazi commander of the region, was highly conspicuous among the sponsors (see Figure 3).

[31] "Würdig reiht sich dieser große Sohn des Preußenlandes in den Kreis der hervorragenden Astronomen deutschen Geblüts ein." Hans Schmauch, "Nikolaus Kopernikus—ein Deutscher" in *Kopernikus-Forschungen,* eds. Johannes Papritz and Hans Schmauch (Leipzig: S. Hirzel Verlag, 1943), p. 30.

[32] *Die Burg: Vierteljahresschrift des Instituts für Deutsche Ostarbeit, Krakau,* May 1943, *4,* n. 2.

***Figure 3.*** *Nazi Gauleiter Förster dedicating the Copernicus statue in Toruń during the 1943 celebrations. (Collection of Jerzy Dobrzycki.)*

## IV. THE POLISH OBSERVANCES OF 1943

In Poland there was, of course, no open way to celebrate Copernicus in Polish, although the astronomer remained a lightning rod for the country's aspirations. For example, the Nazis had placed a German plaque over the Polish inscription on the base of the Thorwaldsen statue of Copernicus that sits in front of the Staszic Palace (which is today the headquarters of the History of Science Institute in Warsaw); two young men of the Polish underground surreptitiously removed the plaque from under the nose of the guard.[33] The German police took their revenge by removing a historic statue of an eighteenth-century Polish soldier from the Old Town Quarter. Later, near the end of the occupation, the Copernicus statue itself was removed, but it was eventually recovered from a scrap-metal junkyard in Silesia in 1945.

The control center for the free Polish celebrations was at the Kosciuszko Foundation in New York City, whose executive director, Stephen Mizwa (1892–1971), was indefatigable in suggesting ways to commemorate Copernicus and in recording the celebrations that took place around the world. With the backing of the foundation, Mizwa put out a preliminary source book including outlines for commemorative

[33] Personal communication from Jerzy Dobrzycki.

***Figure 4.*** *Arthur Szyk's colorful, symbol-filled miniature of Copernicus. Arthur Szyk (1894–1951) was born in Lodz, Poland; he learned the historical art of illustration as a student in Paris; and in 1940 came to New York, where he became a prodigious book illustrator. His Copernicus portrait was commissioned in 1942 by the Kosciuszko Foundation and proclaims in Latin (lower right), "he was a Pole."*

programs.[34] Its cover was distinguished by a color reproduction of a symbol-filled miniature by Arthur Szyk, a refugee Polish artist (see Figure 4). In the background appears Cracow's Wawel castle as it looked in Copernicus's student days. In the foreground is the famed Jagiellonian globe, the earliest to show America, and in the lower right corner is a parchment fragment with the Latin inscription "He fixed the sun, he moved the earth, he was a Pole."

The big anniversary event was a gala affair at Carnegie Hall in New York on the evening of Monday, 24 May, the exact quadricentennial of Copernicus's death.[35] Harvard's Harlow Shapley, then perhaps the best-known astronomer in America, presided, and read greetings from Franklin Roosevelt. "Commemoration of the quadricentennial of the death of Copernicus naturally turns our thoughts to his native Poland, now in chains and prostrate under the evil power of the Nazi conquest," the president wrote;

[34] Mizwa, *Nicholas Copernicus* (cit. n. 25).

[35] Stephen P. Mizwa, *Nicholas Copernicus: A Tribute of the Nations* (New York: The Kosciuszko Foundation, 1945).

***Figure 5.*** *Albert Einstein responding to his Copernican citation at Carnegie Hall. At right, looking on, is astronomer Harlow Shapley, the chairman; Stephen P. Mizwa, the organizer, stands left with a box of additional citations.*

> Poland's plight today is indeed tragic . . . Although free institutions are suppressed temporarily in the land of Copernicus' birth and in other once happy lands, the dawn of a happier day is assured. It is therefore highly appropriate that in the midst of all-out war and the sacrifices which it demands, we pause for a moment to draw refreshment of mind and spirit by recalling the great contribution which Copernicus made. . . . Copernicus serves to remind us that small nations have given for common advantage of all peoples many of the great enduring concepts which have enriched the life of man. This lays on all of us so imperious a responsibility that we should pledge ourselves in the name of all venerated great men of ideas to strive to maintain that opportunity forever.

The program included numbers by the harpsichordist Wanda Landowska and the violinist Bronislaw Huberman. The audience had also been looking forward to a lecture on "Nicholas Copernicus, The Founder of Modern Astronomy" by a young instructor from the City College of New York, but as Shapley explained, "On the eve of this celebration, last Saturday afternoon in fact, Dr. Edward Rosen, the leading Copernicus scholar in America, was inducted into the Army of the United States. His training obviously is not yet completed and the commanding officer found it not expedient to grant him furlough for tonight." For most of the audience, the high point of the evening must have been the presentation of diplomas to ten illustrious "modern revolutionaries" including John Dewey, Walt Disney, Henry Ford, E. O. Lawrence, and Orville Wright, none of whom was present and to Albert Einstein, who was indeed there (see Figure 5).

Stephen Mizwa, in summarizing these events in his book *Nicholas Copernicus: A Tribute of Nations,* lists dozens of programs held at colleges and universities throughout America. To mention just three: George Sarton spoke at the University of Chicago, Alexandre Koyré at the University of Wisconsin, and astronomer Edwin Hubble at Midyear's Commencement at Johns Hopkins University.

A final act in these celebrations came on 3 February 1944, when Mizwa presented a magnificent portrait of Copernicus to Shapley for the Harvard College Observatory. By the Polish artist Maxim Kopf, it is perhaps the finest modern representation of Copernicus.[36]

## V. A PERSONAL VIEW OF THE 1973 QUINQUECENTENNIAL

Having laid the background of the earlier Copernican anniversaries, I now return to the 1973 quinquecentennial. World War II had passed, but Europe lay divided by an iron curtain. A cold war was in progress. Poland had fallen into the grip of a Soviet hegemony. The United States, while not wishing to challenge the Warsaw Pact directly, did what it could to encourage more Polish independence. The Poles, still reeling from the devastation of the war, were energetically trying to rebuild their cultural heritage and Copernicus had become a major icon in their national self-image.

Having lost the opportunity for an international celebration in 1943, the Polish embassies tried to arouse interest in 410th-year celebrations in 1953, but clearly 1973 was the real target date. Quinquecentennial celebrations were held the world over, but always the Polish-German political agenda lay close beneath the surface. It is not my intention to provide a comprehensive review of the 1973 anniversary. Rather, I shall go behind the scenes and discuss four specific cases to illustrate the nature of the underlying political agenda. Because I was heavily involved in some of the preparations for the quinquecentennial, I am now entering the schizophrenic, ambiguous world where I play the dangerous double role of both historian and actor.

For the Polish astronomers, the General Assembly of the International Astronomical Union offered the obvious international focus for a Copernican commemoration, and happily the IAU's triennial pace of meeting—1964, 1967, 1970—provided for a congress in 1973. Furthermore, the rough alternation of meeting sites between east and west—Moscow, Berkeley, Hamburg, Prague, Brighton—made a Warsaw General Assembly politically acceptable. Despite the harsh economic conditions still prevailing in Poland, Western astronomers realized that this could provide an excellent opportunity for meeting colleagues from behind the iron curtain. The Polish astronomers informally indicated their intention to host the congress, and here the machinations began.

Thinking they had an unchallengeable offer, the Poles were astonished to learn at Brighton in 1970 that another invitation, from the Australians, had taken priority over their own, which had been tendered by the USSR Academy of Sciences. Although they did not say so publicly, some Polish astronomers were convinced that the then-president of the IAU, the German Otto Heckmann, simply refused to concede that Copernicus was a Pole and had deliberately maneuvered the Australian

[36] This large painting currently hangs in Building A at the Harvard-Smithsonian Center for Astrophysics, 60 Garden Street, Cambridge, Mass.

invitation into consideration before Poland could officially issue its own. A row broke out in the executive committee (which, besides President Heckmann, included the six vice-presidents, who were from Argentina, Australia, India, Italy, the United States, and the USSR, plus the general secretary L. Perek from Czechoslovakia and the assistant general secretary C. de Jager from the Netherlands), and eventually a face-saving plan was adopted. The members of the executive committee noticed a little-used constitutional provision for an Extraordinary General Assembly, which the IAU scheduled for Warsaw in early September 1973 while the primary meeting was set for Sydney at the end of August.[37]

Meanwhile, in 1965 a Copernicus Committee was appointed under the auspices of the International Union for the History and Philosophy of Science to help the Poles plan an appropriately international historical celebration. The committee was at first headed by the ailing Aleksander Birkenmajer (like his father Ludwik, a distinguished historian and Copernican scholar) and its secretary was Jerzy Bukowski from Warsaw, who was effectively advised by Jerome Ravetz, a scholar and American expatriate living in England who had spent some time in Poland, and whose leftist sympathies were widely known. Later Bukowski became president of the committee and Ravetz one of the secretaries. The committee met almost annually in Poland. Its principal tasks were to organize a commemorative symposium on the reception of Copernicanism to be held in Toruń, and to publish the symposium's papers in the *Studia Copernicana* series of the History of Science and Technology Institute of the Polish Academy of Sciences. As Ravetz saw it, the most sensitive task was to find a distinguished German author who had not identified with the German claims on Copernicus's nationality and who could discuss the reception of the Copernican cosmology in Germany. The obvious choice was Willy Hartner, a historian of the exact sciences who had clearly distanced himself from the shrill forms of German nationalism. Hartner, who saw his own research agenda increasingly encroached upon by the demands of historical anniversaries, was a reluctant recruit.

In the interim, Robert Westman, an energetic younger American scholar, was completing a thesis on Kepler's adoption of the Copernican hypothesis, and when Edward Rosen and I commented enthusiastically about his work and also told him about the proposed *Studia Copernicana* volume, he informed Hartner of the obvious relevance of his own research. Hartner, who was keen for any excuse to get out of the agreement, quickly handed his assignment over to Westman, who was innocently oblivious to some of the issues at stake. Ravetz, who had become chairman of the working group for publications, was taken by surprise and wrote to Westman that, "We had been hoping to enrich our volume with a piece of Hartner's mastercraftsmanship. If he must now turn to something else, the volume will be slightly poorer, *and* it is anyone's guess whether he will ever publish the Kepler thing."[38] Westman promptly offered to withdraw, but the trajectory was irreversible, and Hartner never wrote an essay on Kepler and the Copernican reception. Westman's exten-

[37] A bland and unrevealing account is given in *Proceedings of the Fourteenth General Assembly Brighton, 1970,* eds. C. de Jager and A. Jappel, *Transactions of the International Astronomical Union,* 1971, *14B:*67–68. See also *Proceedings of the Fifteenth General Assembly Sydney, 1973 and Extraordinary General Assembly Poland, 1973,* eds. G. Contopoulos and A. Jappel, *Transactions of the International Astronomical Union,* 1974, *15B:*197–203.

[38] Jerome Ravetz to Robert Westman, 15 June 1970.

sively researched and important contribution quite appropriately filled the scholarly niche, but his authorship did not entirely satisfy the intended agenda.

As a third case, let me turn to the major American celebration. The Smithsonian Institution had been holding a series of international symposia, one every few years, and its Office of Seminars noticed that the forthcoming Copernican quinquecentennial would provide an excellent vehicle for a symposium on the nature of science. Considering the nature of the celebration, it would be appropriate to involve the National Academy of Sciences. In due course, a joint National Academy–Smithsonian program committee was set up to plan the event, under the chairmanship of physicist John Wheeler. It was natural to ask a distinguished Polish scholar to be honorary chairman of the symposium, which was done, but the committee felt that a truly international event required a distinguished scientist-philosopher from Germany as well. I suggested Carl Friedrich von Weizsäcker, a physicist who had major publications on the nature of science, but the committee hesitated. Would von Weizsäcker, whose family had been associated with the German government of World War II, be acceptable to the Poles or to Jewish Americans? John Wheeler, Gerald Holton, and Stephen Toulmin, all fellow committee members, carefully considered the sensitivities involved and decided that a call to Victor Weisskopf, then head of CERN in Geneva, was in order. Weisskopf reported back that it would be more appropriate to invite Werner Heisenberg, who accepted the invitation and became a star attraction at the symposium.

Finally, I turn back to the Polish National Organizing Committee of the IAU, which was preparing for the Extraordinary General Assembly in Warsaw. One of the three invited discourses was assigned to the historical aspects, and so the search for an appropriate commemorative speaker began. To set the proper international tone, it would be best not to choose a Pole. Too many sensitivities would be raised by having a German or Russian speaker in such a spot. The speaker should preferably be an astronomer and IAU member, hopefully a person with some sympathy for Poland, and if he or she actually knew something about Copernicus, so much the better. Eventually the committee chose me for this conspicuous role. I was highly honored to accept the invitation, but as I reflect on it now, I realize that their selection was, in reality, a political decision. The Poles knew that I had first visited their country in 1946 as a young seagoing cowboy—one of a crew of thirty—with a load of nearly 800 horses sent by the United Nations Relief and Rehabilitation Agency to a war-torn Poland.[39] They knew that I had tried hard to send a large surplus computer to Toruń, and they knew that I was not planning to attend the competing IAU General Assembly in Sydney. Their ambassador later conferred on me the Order of Merit, Commander Class, Poland's highest honor for foreigners who are not heads of state. The award was ostensibly given for scholarly reasons, but in a country where Copernicus is such a cultural icon, politics can scarcely be separated from scholarship.

## VI. LEARNING FROM THE COPERNICAN CELEBRATIONS

What lessons can we learn from the Copernican quinquecentennial and its predecessors? Every historical event or person worth celebrating has links with a discipline,

[39] Owen Gingerich, "The Return of the Seagoing Cowboy," *American Scholar,* 1999, *68,* n. 4:71–81.

an institution, or a nationality, or perhaps all of the above. The more significant the anniversary, the more laden the agenda. Historians may innocently propose a memorial and rightfully expect it to be an impetus for scholarship, but then be unwittingly captured by other and even competing interests. How can historians, as willing accomplices in the construction of anniversaries, achieve a balance between memory and history? We must be aware that we are treading on sensitive ground when the tensions between these two voices are heightened. The more explicitly cognizant we are of these agendas, the better we can adjust our expectations to the realities of commemoration and find the balance between memory and history.

As historians, we are continually called upon to assist in scientific anniversaries, and these can be important consciousness-raising events. For example, notice how Harlow Shapley summed up the 1943 Copernican quadricentennial:

> The imperishables among our human activities are hard to detect at close range. In 1543 Charles V, as Holy Roman Emperor and ruler of many states, was fighting a seemingly history-making war with Francis I of France. But little did that monarch realize that his own anniversaries would arouse no interest in days to come, and that he and his work would be a firefly's fitful flash compared with the enduring light that was set shining by a churchman in Frauenburg, whose hobby was planets and cosmic thought, and who was, at the time of the sack of Rome, arranging the sunrise of the scientific age on which civilization would build. . . .
>
> But presumably in [the 24th century], as now, the attention of mankind will be directed mainly to that forward segment of the time line and we will turn to the past then, as we should now, chiefly to discover the ways of avoiding error, and to derive from the spiritual and intellectual nobility of earlier days the brave Copernican inspirations that guide the present and design the future.[40]

From the vantage point of history of science, the impact of these anniversaries on Copernican scholarship has been highly significant. Frequently scholarship begun in connection with the round-number anniversaries has come to fruition in later years. The Latin edition and Polish translation of *De revolutionibus* set in motion in 1843 finally appeared in 1854. The valuable *Opera omnia* begun by the Poles for the quinquecentennial is only now reaching completion, and the *Studia Copernicana* series has extended over many years. Swerdlow and Neugebauer's *The Mathematical Astronomy of Copernicus'* De Revolutionibus had roots in the quinquecentennial discussions and confrontations, and my own *An Annotated Census of Copernicus'* De revolutionibus *(Nuremberg, 1543, and Basel, 1566)*, still to be published, began in connection with that anniversary year.

Nevertheless, other forces were also at work, ranging from pressures on the United States Post Office (resisted by the postmaster general) to issue a stamp bearing the Polish spelling "Kopernik," to the nature of a commemorative volume sponsored by the United States National Academy of Sciences and organized by a Polish-American astronomer who was determined that historians of science should have no hand in it.[41] The Polish and German embassies in Washington competed over which

[40] Harlow Shapley, "Foreword on Quadricentennials," in Mizwa, *Tribute of the Nations* (cit. n. 31), pp. vii–ix.

[41] *The Heritage of Copernicus: Theories "More Pleasing to the Mind,"* ed. Jerzy Neyman (Cambridge, Mass.: MIT Press, 1974).

could hold the most significant reception in overlapping time slots in connection with the Smithsonian–National Academy symposium. And, of course, there was a dark undercurrent of European ethnic rivalries about who could claim Copernicus as a cultural son. For better or worse, historians of science got caught up in these machinations and were often unwitting pawns in a cultural battle that was only dimly visible.

As historians of science, we must be prepared to use the anniversaries to advance our own goals, but we must enter into them with full knowledge that ours will not be the only agendas in play.

# Appendix 1

Colloquia Copernicana
sponsored by
The International Union for the History and Philosophy of Science
Academie Internationale de l'Histoire de la Science
Polish Academy of Sciences
UNESCO and the International Astronomical Union
Toruń, Poland, 7–12 September 1973

**Symposium I—The Astronomy of Copernicus and Its Background**

Chairman of the organizing committee: Owen Gingerich

*Session I, 7 September 1973: Astronomical Background*

Introductory Paper, *W. Hartner*
Current Research on Islamic Astronomical Tables, *D. King*
La tradition astronomique de l'Islam à Cracovie au XVe siècle, *G. Rosińska*
The Earth's Rotation in the Aryabhata, *W. Petri*
Indian Planetary Theories of Ancient and Medieval Times, *S. N. Sen*

*Session II, 7 September 1973: Astronomical Tradition of the Middle Ages*

Introductory Paper: The Main Stream of Mediaeval Latin Astronomy, *O. Pedersen*
Copernicus retroversus, *F. Krafft*
The Role of Neoplatonism, *Z. Horský*
Les equatoires, instruments de la théorie des planetes au moyen âge, *E. Poulle*
The Philosophical Background of the Cracow XVth-century Astronomy, *R. Palacz*
The Tradition of Antiquity in Copernicus' Astronomical Instruments, *T. Przypkowski*

*Session III, 8 September 1973: Copernicus' Astronomy*

Introductory Paper: The Achievement of Copernicus, *E. Rosen*
Copernicus' Solar Theory, *N. Swerdlow*
The Uppsala Notes, *J. Dobrzycki*
Le pythagorisme de Copernic, *B. Bilinski*
The Reluctant Revolutionaries: Astrology after Copernicus, *J. North*
Copernicus and Calendar Reform, *D. J. K. O'Connell*
The Influence of Copernicus on the Gregorian Calendar Reform, *H. J. Felber*

**Symposium II—Man and Cosmos**

Chairman: Włodzimierz Trzebiatowski

*10 September 1973*

Opening Addresses:
W. Łukaszewicz, rector of N. Copernicus University in Toruń
W. Trzebiatowski, president of Polish Academy of Sciences
J. Harrison, vice-president of UNESCO

Lectures:
L'homme dans le cosmos et l'homme sur la terre, *B. Suchodolski*
Man and Cosmos, *T. Araki*
Scientific Perspectives of Cosmic Flights, *J. Werle*
Les idées de Copernic et l'évolution de la mecanique, *L. I. Siedow*
La logica celestial de Copernico, *Eli de Gortari*
Utilité d'études spaciales, *J. F. Denisse*
Closing Address: *B. Leśnodorski*

**Symposium III—The Reception of the Heliocentric Theory**

Chairman: Henrik Sandblad

*Session I, 11 September 1973*

Review Paper: The Reception of Copernicus' Theory: Misunderstandings and Real Consequences from the Sociological Viewpoint, *P. Rybicki*
Review Paper: The Actual State of Research of the Reception of the Heliocentric Theory, *P. Costabel*
Discussion of papers published in volumes V and VI of *Studia Copernicana.*

*Session II, 12 September 1973*

Das Rad des Schicksals von Rhäticus, *J. O. Fleckenstein*
L'audience du copernicanisme au temps de la vie de Nicholas Copernic, *A. Kempfi*
Die Copernicus-Biographie von G. Ch. Lichtenberg, *D. Herrmann*
Beziehungen zwischen Copernicus und Leipzig, *H. Wussing*
Über die Stellung der deutschen Reformatoren zum Opus des Copernicus, *S. Hoyer*
Zur Rezeption und Verbreitung der Copernicanischen Lehre an der Universität Wittenberg im 16. Jahrhundert, *H. J. Bartmuss*
Copernicus in der Beurteilung der Reformatoren, *G. Hammann*
La reception de las ideas de Copernic en Cuba, *J. L. Sanchez*
Copernicus and Upper Austria, *A. Adam*
Théorie hèliocentrique dans l'interpretation des theologiens du XVIe siècle, *W. Wardęska*

**Symposium IV—Copernicus and the Development of Exact and Social Sciences**

Chairman: Bogdan Suchodolski

*Session I, 11 September 1973*

Copernicus and Galileo against Aristotle, *W. Krajewski*
Copernic et ses études italiennes, *C. Vasoli*
Copernic et Galilee, *F. Barone*
The Investigation of Proofs of the Earth's Motion, *A. A. Michailov*
Copernicus und die Newtonische Mechanik, *G. Jackisch*
Über den Einfluss des copernicanischen Systems auf die damaligen und späteren Anschauungen über die Physik der Atmosphäre, *H. G. Körber*
The Existence of a Reference Frame—The Basis of the Work of Copernicus and Kepler, *H. U. Sandig*

*Session II, 12 September 1973*

The Philosophical Meaning of the Discovery of Copernicus, *B. Kedrov*
Pensée humaniste de Copernic, *B. Leśnodorski*
Copernic, precurseur du positivisme, *P. Arnaud*
Système heliocentrique et les prémises de l'anthropocentrisme modern, *V. Voisé*
L'inscription de la cosmologie de Copernic dans la mécanique de Descartes, *S. Bachelard*
Copernicus und sein Weltbild im Spiegel von Humboldts Kosmos-Vorlesungen, *F. Herneck*
Fragmentation in the Community Structure of Science, *J. Smolicz*

# Appendix 2

The Nature of Scientific Discovery
sponsored by
The United States National Academy of Sciences and The Smithsonian Institution
in cooperation with UNESCO and The Copernicus Society of America
Washington, D.C., 22–25 April 1973

Janusz Groszkowski
Honorary Chairman of the Symposium
Polish Academy of Sciences

S. Dillon Ripley
Secretary
Smithsonian Institution

Philip Handler
President
National Academy of Sciences

John Archibald Wheeler
Program Chairman
Joseph Henry Professor of Physics, Princeton University

## I. The Festival

*Opening Ceremonies, Sunday, 22 April 1973*

Introduction, *Philip Hadler*
Responses, *S. Dillon Ripley, E. J. Piszek,* and *Janusz Groszkowski*
Musicale by the Gregg Smith Singers,
Including new music by Leo Smit with narration by Sir Fred Hoyle

*Monday, 23 April 1973*

Annual Awards Ceremony of the National Academy of Sciences

*The Symposium Dinner, Tuesday, 24 April 1973*

Baskin's Portrait of Copernicus, *S. Dillon Ripley*
A Copernican Appreciation and Toast, *Janusz Groszkowski*
Presentation of Special Awards to Edward Rosen and Jerzy Neyman by
*E. J. Piszek* (President, Copernicus Society of America)

*Wednesday, 25 April 1973*

Concert by Boston Symphony Orchestra members, conducted by Leon Kirchner,
presenting works of Webern, Messiaen, Schönberg, and Kirchner

## II. The Symposium

*The First Day: Monday, 23 April 1973*

Presentation by the Senior Assistant Postmaster General of the Stamp Commemorating the Copernican Quinquecentennial; Music by the U.S. Marine Band Brass Ensemble
Opening Remarks, *Janusz Groszkowski*
Introduction: The Nature of Scientific Discovery in the Sixteenth Century, *A. Rupert Hall*
Science and Society in the Age of Copernicus, *Owesi Temkin*
The Age of Copernicus (triple-screen slide presentation), *Charles Eames,* with musical score by Elmer Bernstein
Reformation and Revolution: Copernicus' Discovery in an Era of Change, *Heiko A. Oberman*
The Ascent of Man, Episode 6, *Jacob Bronowski.*

*The Second Day: Tuesday, 24 April 1973*

Introduction: The End of the Copernican Era? *Stephen Toulmin*
Mainsprings of Scientific Discovery, *Gerald Holton*
Tradition in Science, *Werner Heisenberg*

*The Third Day: Wednesday, 25 April 1973*

Introduction: Does Science Have a Future? *Owen Gingerich*
Quasars and the Universe, *Maarten Schmidt*
The Universe as Home for Man, *John Archibald Wheeler*

## III. The Collegia

*Collegium I: Science and Society in the Sixteenth Century*

*Session I, Monday, 23 April 1973*

The Spirit of Innovation in the Sixteenth Century, *Marie Boas Hall*
Discussion of M. B. Hall's Paper
Venice, Science, and the Index of Prohibited Books, *Paul F. Grendler*
Discussion of P. F. Grendler's Paper

*Session II, Tuesday, 24 April 1973*

The Quest for Certitude and the Books of Scripture, Nature, and Conscience, *Benjamin Nelson*
Discussion: Philosophical and Theological Backgrounds

*Session III, Wednesday, 25 April 1973*

The Wittenberg Interpretation of the Copernican Theory, *Robert S. Westman*
Discussion: The Reception of Heliocentrism in the Sixteenth Century

*Collegium II: The Interplay of Literature, Art, and Science*

Background Paper:
The Interplay of Literature, Art, and Science in the Time of Copernicus, *John U. Nef*

*Session I, Tuesday, 24 April 1973*

Discussion: Parallels and Contrasts of Science and Art

*Session II, Wednesday, 25 April 1973*

Discussion: Discovery in Art and Science
Discussion: Ramifications of the Word "Revolution"

*Collegium III: Science, Philosophy, and Religion in Historical Perspective*

*Session I, Tuesday, 24 April 1973*

Discussion: Claims to Truth of Faith and Science

*Session II, Wednesday, 25 April 1973*

The Structure of Academic Revolutions, *Langdon Gilkey*
Discussion: The Nature of Scientific Knowledge
Discussion with Professor Heisenberg

*Session III, Thursday, 26 April 1973*

Discussion on John Wheeler's "The Universe as Home for Man"
Discussion: Science and the Foundation of Ethics
Discussion: Are There Ethical Limits to Scientific Discovery?

# Claude Bernard and Louis Pasteur

## Contrasting Images through Public Commemorations

*By Christiane Sinding**

CLAUDE BERNARD (1813–1878) AND LOUIS PASTEUR (1822–1895) are both considered to be giants of science, equally productive and creative. Bernard is acclaimed for having founded a rational system of medicine based on experimental evidence, whereas Pasteur is credited, rightly or not,[1] with developing the foundation of microbiology as well as the first vaccines against infectious diseases. However, when one examines the number and scope of books, articles, movies, and public commemorations devoted to the two scientists, there is a marked inequality. Pasteur comes out ahead by far. One hundred years after his death he is still celebrated as the savior of humankind, the benefactor of humanity.[2]

In this paper I will first attempt to analyze the contrasting styles of commemoration for these two men. Second, I will comment on the tentative deconstruction of Pasteur's celebrations by modern historians. Third, I will suggest an interpretation of these differences in commemoration as they relate to the general issue of commemorations in the biomedical sciences. For reasons of space and coherence I will restrict my analysis to the French commemorations of both scientists.

Prior to analysing the case studies of Bernard and Pasteur, I would like to point out some general features of commemorative activity in France.[3] For more than 200

* Institut National de la Santé et de la Recherche Médicale, Centre National de la Recherche Scientifique, Unité de Recherche Associée 583. Send correspondence to: CERMES (Centre de Recherche, Médecine, Sciences, Santé et Société), 182 Boulevard de la Villette, 75019 Paris, France; e-mail sinding@ext.jussieu.fr.

I am grateful to Pnina G. Abir-Am for her stimulating comments on early drafts of the paper. I also wish to thank two anonymous referees for valuable suggestions and Diana Kenney for excellent professional assistance with the final version of the manuscript.

[1] This issue will not be fully discussed in this paper since I do not intend to deal with "what has really been done" by the two scientists. Nonetheless, I will make scattered points on the scientists' achievements.

[2] As demonstrated by the 1995 commemoration of his death coordinated by the Institut Pasteur in Paris and UNESCO.

[3] The impressive collective work on French memory, edited by Pierre Nora and published between 1984 and 1992, brings the richest—if not exhaustive—information and analysis available on this issue. According to Nora, this work has become an instrument of commemoration itself, although it was originally conceived as a critical and countercommemorative work. See Pierre Nora, "L'Ere de la commémoration," in *Les lieux de mémoire,* ed. Pierre Nora, vol. 3, *Les France,* part 3, "De l'archive à l'emblème" (Paris: Gallimard, 1986), pp. 977–1012, quotation on p. 977. Incidentally, it is remarkable that science has almost no place within the seven volumes of *Les Lieux de mémoire.* This absence indicates that history of science still does not have full status in France, especially at the universities.

 0369-7827/99/1401-0004$02.00

years this activity has probably been more intense than in any other country, although there was some decline after World War II followed by a strong reemergence more than twenty years later. According to Pierre Nora, this interest in commemoration can be explained by three factors: the richness of French history, the social rupture introduced by the French Revolution, and a "memorial rumination" ("rumination mémorielle") caused by the fact that the country had become a second-rank nation by the mid-twentieth century. Generally speaking, the commemorative activity in France has been politically oriented, led by leftist governments, and linked to the traditions of the Enlightment, the French Revolution and secularity.[4]

The French Revolution replaced the worship of the king and the church with the worship of great men. In 1791 the Catholic church Sainte-Genièeve was transformed into the Panthéon, a temple devoted to the worship of these men, who were not kings, nobles, or gentlemen, but ordinary men of the masses. The Panthéon was rendered to Catholic worship in 1806, made a secular Temple of Glory in 1830, rendered to Catholic worship again by Napoleon III, and finally again made a temple for the memory of great men in 1885 on the occasion of the state funeral for Victor Hugo. On its pediment one can still read the inscription that replaced the previous frieze in 1891: *Aux Grands Hommes la Patrie Reconnaissante* (To Great Men, the Grateful Motherland).[5]

Centennial celebrations of great men were first introduced during the early years of the Third Republic.[6] At the same time, the organization of state funerals became an art and an opportunity for a ritual inculcation of republican ideals, which culminated in Victor Hugo's funeral. It must be recalled that between 1789 (the French Revolution) and 1871 (instauration of the Third Republic) a succession of political events, both internal (revolutions in 1830 and 1848 and revolts like the terrible Commune in 1871) and external (Napoleonic wars, the defeat in the war with Germany) made France a wounded and divided country. National commemorations served to reconcile and unite the nation around republican values. In 1878, the same year Bernard died, the centennials of Voltaire and Rousseau allowed the nation to appropriate these two philosophers for a celebration of the Republic. More than Rousseau, Voltaire could be associated with contemporary issues that were heavily debated between 1877–1882, like the status of the Catholic Church and the problem of the laicization of schools.[7]

For Voltaire's centennial, Victor Hugo, who was celebrated as the father of the Republic, a great national poet, and the pope of the civil religion, was asked to deliver a discourse.[8] Beyond Voltaire, Hugo honored the eighteenth century and its philosophers, particularly their battle for human rights with its final outcome, the Revolution. He celebrated the direct filiation between the Enlightment, the Revolution and the Third Republic. As is always the case with French commemorations of

[4] *Ibid.*

[5] On the Panthéon and its political use for commemorations, see Mona Ozouf, "Le Panthéon: L'Ecole Normale des morts," in Nora, *Les lieux de mémoire* (cit. n. 3), vol. 1, *La République,* pp. 139–66.

[6] *Ibid.*, p. 982.

[7] The revival of the battle between the French state and Catholic schools during the Mitterrand years demonstrates that many disputes dating from the French Revolution are still unresolved in France.

[8] See Jean-Marie Goulemot and Eric Walter, "Les centenaires de Voltaire et de Rousseau: Les deux lampions des lumières," in Nora, *Les lieux de mémoire* (cit. n. 3), vol. 1, pp. 381–420.

the Revolution, the war's dreadful aspects, such as the Terror, were carefully omitted from this "masterpiece of republican messianism."[9] Victor Hugo's own funeral in 1885 was a "big event"[10] that gave rise to a battle between the laic Republic and the Catholic and royalist church. For the Republic it became the best opportunity to reopen the Pantheon for the worship of great men. The extreme-left newspaper *La Bataille* (*The Battle*) wrote, "Priests, make room. Clear your gods from this sanctuary so that we can put our men in it."[11] The Catholic newspaper *l'Univers* (*The Universe*) called for a demonstration against this "blasphemy," and tried to mobilize the "Christian consciousness" to oppose the "madness of paganism."[12] There were some violent debates at the Chambre des Députés and the Sénat, but the republican majority won the votes. The government also feared extreme-left demonstrations and the reappearance of red flags, so the ceremony was carefully planned and controlled. The funeral finally took place without major trouble, although conservative and Catholic newspapers described it as a "funereal orgy," "disgraceful Bacchanalia" and a "feast of the mad." The ostentatious ceremony deeply impressed the public, according to the testimony of contemporary authors: an estimated one million spectators watched the procession, which featured 1,168 various societies and delegations and regular discharges of guns. Apartment balconies along the route were rented for high prices, especially in the rue Soufflot, which leads to the Panthéon. The army took such an important part in the ceremony, allowing dreams of revenge against Germany, that it almost diverted the spectators' attention from the hearse. The pilgrimage continued for a few days: people visited the Pantheon and Hugo's statues, bringing thousands of pictures and medals that had been manufactured by the newborn commemoration industry. Political tensions were briefly supended and the funeral was considered to be a massive, egalitarian manifestation that paved the way for similar ceremonies in the future.

In these secular commemorations and ceremonies, which retained and adapted an aspect of religious ritual, scientists occupied an important place. Their role would increase between 1878, the year of Bernard's national funeral, and 1895, the year of Pasteur's. The laic "religion" of the Republic included the positivist worship of science and progress, and republican governments were quick to appropriate scientists to celebrate the nation. Christophe Charle has shown that after the death of Victor Hugo, the figure of the scientist progressively replaced that of the écrivain.[13] He points out that scientists acquired political legitimacy at the end of the nineteenth century because they were considered to be rational and disinterested seekers of truth. Indeed, some of them became politicians, including Paul Bert

[9] *Ibid.*, p. 406.

[10] "*Evénement monstre*" ("monster event") is the term coined by Nora to characterize this kind of event. Pierre Nora, "Le retour de l'évènement," in *Faire de l'histoire,* eds. Jacques Le Goff and Pierre Nora, vol. 1 (Paris: Gallimard, 1973), pp. 210–28.

[11] "Curés, faites place. Débarrassez cet asile de vox dieux, que nous puissions y mettre nos hommes." Quoted by A. Ben Amos, "Les funérailles de Victor Hugo," in Nora, *Les lieux de mémoire* (cit. n. 3), vol. 1, pp. 473–522.

[12] Quoted by Goulemot and Walter in "Les centenaires de Voltaire" (cit. n. 8).

[13] Christophe Charle, *Naissance des intellectuels* (Paris: Minuit, 1990), pp. 28–34. It should be noted, however, that Goulemot and Walter pointed to the persistent sanctification of literature in France and the "strange complicity" that unites the politician, the author, and the statesman in this country. See Goulemot and Walter "Les centenaires de Voltaire" (cit. n. 8), p. 417.

(1833–1890), Bernard's most well known pupil, and Marcellin Berthelot (1827–1907).[14] Bernard himself, although he kept his distance from politics until his death, had been made a senator by Napoleon III. From 1878 to 1907 four scientists (Claude Bernard, Paul Bert, Louis Pasteur, and Marcellin Berthelot) had national funerals as compared with only two writers (Victor Hugo [1802–1885] and Ernest Renan [1823–1892]); before 1878 only politicians and members of the military were honored in this manner. During the debate about Bernard's funeral at the Chambre des Députés, Bernard was characterized by the minister of national education as a disinterested seeker of truth. Only one deputy voted against the state funeral, demonstrating that a consensus could be obtained much more easily about a scientist than about a philosopher or poet. Writers made scientists important characters in their novels, and Emile Zola went as far as to write a *roman experimental* (experimental novel) inspired by Bernard's writings.[15] Among the scientists, Pasteur certainly best personified the power of the human intellect when devoted to progress. On the other hand, Bernard personified scientific reason and, furthermore, had the admirable social status of being an author. Incidently, Bernard was not buried at the Panthéon, which was still a Catholic church when he died. And Pasteur's family opposed his burial at the Panthéon, so he was finally buried underneath the Institut Pasteur in 1886. Thus, ironically, neither great scientist lies in the Panthéon.

## I. THE METHODOLOGIST

Claude Bernard was celebrated in France, as well as in other countries, much more as a philosopher and methodologist than as a scientist.[16] This becomes especially evident when one compares the twentieth-century Bernard commemorations with the celebrations devoted to Pasteur, the majority of which focused on his scientific work despite the fact that his work has not been analyzed as thoroughly as Bernard's.[17]

[14] Marcellin Berthelot, now almost forgotten, was a chemist and a prominent figure in positivism and science; he became a minister of public education and later of foreign affairs. In the early 1890s it was widely believed that scientists had moral and political authority.

[15] Reino Virtanen, "Claude Bernard et le roman expérimental," in *Philosophie et méthodologie scientifiques chez Claude Bernard,* Actes du Colloque International organisé par la Fondation Singer-Polignac en 1965 (Paris: Masson, 1967), pp. 49–55. The idea of an "experimental novel" was invented by Emile Zola, who admired Bernard. Zola stated that a novel had to be based on observation and experiment; the writer had to submit his characters to different kinds of experiences. Moreover, according to Zola, the art of the novel had to become a science.

[16] It is interesting to note that many children in France still have to comment on some part of *Introduction à l'étude de la médecine expérimentale* (Paris: Baillère, 1865) for the philosophical examination given at the end of their school years. Mirko Grmek wrote, "Almost paradoxically, the 'philosophical' aspects of Bernard's work resulted in a bibliography much larger than that of his strictly scientific work." See "Bernard, Claude", Charles C. Gillispie, *Dictionary of Scientific Biography,* vol. 2, (New York: Macmillan, 1970), pp. 24–33, quotation on p. 31.

[17] This point is seldom acknowledged, probably because the proliferation of books devoted to Pasteur leave the impression that everything has already been said or written about his work. But the two impressive books produced in the mid-1970s on Bernard's scientific work (Mirko D. Grmek, *Raisonnement expérimental et recherches toxicologiques chez Claude Bernard* [Genève: Droz, 1973] and F. L. Holmes, *Claude Bernard and Animal Chemistry* [Cambridge, Mass.: Harvard University Press, 1974]) have no equivalent in the Pasteur "industry", which is overcrowded with biographical studies that more or less analyze his scientific work. The word "industry," which I borrow from Olga Amsterdamska (See Olga Amsterdamska, "The Historiography of the Claude Bernard Industry," *Hist. Sci.,* 1978, *16:*214–21), now applies much better to studies devoted to Pasteur, whereas Bernard has been neglected in recent history of science. The most complete analysis of Pasteur's work still remains the entry by Gerald Geison, "Pasteur, Louis," in Gillispie, *Dictionary of Scientific Biography* (cit. n. 16), vol. 10, pp. 350–416.

While Pasteur was mainly celebrated for the practical outcomes of his work, Bernard was honored for having personified experimental rationality. The philosophical side of Claude Bernard explains why the scientists who organized the commemorative ceremonies always invited philosophers.[18] Although the latter offered a rather critical analysis of Bernard's science, they often were the scientists' allies in that they shared their rhetoric of truth and rationality.

Even politicians were fascinated by Bernard's method. At his birth centennial in 1913, just before the beginning of World War I, the minister of the *instruction publique* compared France to a laboratory in which the experimental method made it possible to establish better legislation. Politicians, he added, were animated by passions, whereas science brought them "indestructible truth." Scientists themselves had often claimed that politics should benefit from science and the scientific method. In 1871 after the French defeat by the Prussians, Bernard wrote, "We were defeated scientifically." Introducing rationality to politics became an important aim, and Bernard was appropriated by "*moderate progressists*" who used his idea of biological equilibrium as a political model.[19] Bernard himself was politically moderate and tried to stay away from the political scene, although he was made a senator by the emperor. As mentioned, he was the first scientist to have had a national funeral, although it did not take place at Notre-Dame like Pasteur's, but "only" at Saint-Sulpice.

### *Philosophers' Images of Claude Bernard as Evident in Commemorative Discourses*

One of the strongest appropriations of Claude Bernard by a philosopher was made during the 1913 commemoration by Henri Bergson (1859–1941), a professor at the Collège de France. It is important to understand the political and scientific context of this celebration: France had lost its scientific supremacy over the previous fifty years. Pasteur and Bernard themselves had resented this loss, as had most French scientists. They had both, in their own ways, emphasized the country's need to develop science.[20] In this context, the centennial afforded the opportunity to simultaneously celebrate a man (Claude Bernard), a discipline (philosophy), the birth of

[18] A notable exception was the last colloquium on Claude Bernard, which was organized by a political scientist from Lyon and held in December 1989 at Saint-Julien en Beaujolais in Bernard's former house, now the Musée Claude Bernard. See *La nécessité de Claude Bernard,* ed. Jacques Michel (Paris: Méridiens Klinsieck, 1991).

[19] "*Moderate progressists*" is a term used by Paul Bacot, a political scientist. This term refers to leftist politicians who rejected the idea of a political revolution and pleaded for a slow "experimental" transformation of society. Strangely enough, those politicians have for a long time been called "*radical-socialistes,*" although there is nothing radical about their position. See Paul Bacot, "L'affaire Claude Bernard: De quelques hommages publics à leur scientificisation et de leur politisation," *ibid.*, pp. 199–228.

[20] Bernard, for instance, had used a report requested in 1867 by the minister of public education (*Rappport sur les progrès et la marche de la physiologie générale en France*) as an opportunity to state his opinions on the matter. Pasteur also made numerous statements on the link between the French defeat and the lack of will to develop science in the country, and he emphasized the fact that Germany, unlike France, had encouraged the creation of universities and laboratories and favored emulation between those institutions. See René Vallery-Radot, *La vie de Pasteur* (Paris: Flammarion, 1900), pp. 255–8. For a general discussion, see Harry W. Paul, *The Sorcerer's Apprentice: The French Scientist's Image of German Science, 1840–1919* (Gainesville: University of Florida Press, 1972).

another discipline (experimental medicine), an institution (the Collège de France), and above all, a nation (France). Bergson's lecture took place a few months before World War I, which for France was the second war against Germany, and his discourse had a strong nationalistic tone.

Bernard had taught "experimental medicine" at the Collège de France for more than thirty years, having officially succeeded his mentor François Magendie (1783–1855) in 1855.[21] The Collège de France is a unique institution that was founded in 1529 to provide stimulating teaching and allow for innovation. Freedom and creativity are the mottos of the institution, which is open to lay people.[22] Various disciplines are taught, ranging from the natural and social sciences to literature and philosophy. No examinations are given. Christophe Charle has used religious terminology to construct a typology of teachers and courses at the Collège de France, classifying the teachers into three groups: the hermits, the prophets, and the sect leaders. Among the hermits, who had very few students, he cited Ernest Renan and Marcellin Berthelot. (There was nothing pejorative in being a hermit since the Collège de France offered a number of specialized courses that attracted few people.) Henri Bergson belonged to the prophets and his courses were overflowing with people who often came to counterbalance the neighboring Sorbonne, which was ruled by the positivists. Among the sect leaders Charle listed Claude Bernard and his pupil Louis Antoine Ranvier (1835–1922), pointing out that they both conducted collaborative research and were regarded by their students as being mentors. In sharp contrast to the dogmatic education provided in the medical schools, the sect leaders offered nonconformist teaching and research.[23]

When Henri Bergson gave his ceremonial discourse on Bernard in 1913, he had just spent ten years studying biology, particularly evolutionary theory. He was an admirer of Herbert Spencer and thought that, given the development of the biological sciences, philosophy needed to make profound changes. His philosophy was vitalist and spiritualist: he called himself a philosopher of the "vital impulse" (*élan vital.*) (The vital impulse was a force that explained and shaped individual life and the development of new species.) Bergson also produced a detailed critical analysis of intellect, intelligence and understanding. To him, human knowledge was mainly dependent on intuition rather than intellect, and the idea of *durée* (duration) replaced that of time as measured by mathematicians. Generally speaking, Bergson's philosophy belonged to the antipositivist movement of the turn of the nineteenth century,

[21] Bernard had also given some courses at the Sorbonne as well as the Museum d'Histoire Naturelle, but these places were not as prestigious as the Collège de France. The recent renovation of the museum for its 1994 bicentennial brought it out of the shadows. F. L. Holmes gave a paper on Bernard's teaching at the museum at the bicentennial colloquium. See F. L. Holmes, "Claude Bernard et le Museum d'Histoire Naturelle," in *Le Museum au premier siècle de son histoire,* eds. Claude Blanckaert, Claudine Cohen, Pietro Corsi and Jean-Louis Fischer (Paris: Editions du Museum, 1997), pp. 403–24.

[22] Christophe Charle, "Le Collège de France," in Nora, *Les lieux de mémoire* (cit. n. 3), vol. 2 (*La Nation*), part 3 "La gloire. Les mots," pp. 389–424.

[23] Among the scientists who held a chair at the Collège de France were André-Marie Ampère (1775–1836), René Laënnec (1781–1826), Paul Langevin (1872–1946), Marcellin Berthelot (1827–1907), Claude Bernard (1813–1878), and Frederic Joliot-Curie (1900–1958). Among the professors of literature, social sciences, and philosophy, were Jean-François Champollion (1790–1832), Jules Michelet (1798–1874), Marcel Mauss (1872–1950), Ernest Renan (1823–1892), and Lucien Febvre (1878–1955). Among the famous "heretics" who found their way at the Collège, Michel Foucault (1926–1984) and Roland Barthes (1915–1980) were the latest.

which included the impressionists in painting and Marcel Proust (a cousin of Bergson's wife) in literature. This movement, sometimes related to a so-called *névrose fin de siècle,* was not specific to France but extended more or less to all of western Europe.[24]

In his opening discourse, Bergson compared Bernard's *Introduction à l'étude de la médecine expérimentale* with *Le discours de la méthode* by René Descartes.[25] What Descartes did for the science of matter, Bernard did for the science of life, Bergson claimed. He added that only twice in the history of modern science—both times in France—had a genius made brilliant discoveries possible. The appeal to the well-known figure of Descartes might have been surprising, coming from a philosopher who praised intuition much more than intellect and rationality. In my opinion, the comparison was purely rhetorical, and can be explained by the strong desire to celebrate the nation through these famous figures.

With regard to the more specific issue of Claude Bernard's commemoration, what Bergson celebrated above all was Bernard's philosophy of life and the fact that the great physiologist had avoided the pitfalls of mechanism as well as of vitalism. As with many commemorators, Bergson dwelled on the aspects of Bernard's writings that helped to reinforce his own philosophy, like the idea of spontaneity *idée directrice* (leading or driving idea) which is close to Bergson's *élan vital.*[26] Bernard had developed this concept in order to emphasize that life could not be completely understood by reducing it to chemical or physical terms. Bergson appropriated Bernard's claim of the irreducible originality of living organisms to legitimize, at least in part, his own philosophy.[27]

The second issue Bergson analyzed concerned Bernard's epistemology. Here the philosopher used Bernard's positions as a weapon against positivism. This great scientist, claimed Bergson, had been able to see that there was no difference between a good observation and a "well-founded theory."[28] *Invention* presides over the gathering of data, and the idea that science proceeds from discovering hidden laws just waiting to be described is meaningless. Although Bernard did not go that far in his theory of experimental knowledge, he did claim that there was no complete separa-

[24] See Philippe Joutard, "L'ouverture des connaissances et les mutations culturelles," in *Histoire de la France de 1852 à nos jours: Paris, 1871–1914,* ed. Georges Duby (Paris: Larousse, 1987), pp. 213–19. See also Pascal Ory and Jean-François Sirinelli, *Les intellectuels en France de l'affaire Dreyfus à nos jours* (Paris: Armand Colin, 1986).

[25] Henri Bergson, "Discours prononcé le 30 décembre 1913 à l'occasion du centenaire de Claude Bernard," reprinted in E. Dhubout, *Claude Bernard: Sa vie, son oeuvre* (Paris: Presses Universitaires de France, 1947), pp. 22–3.

[26] Bernard also used terms other than *idée directrice* to express the same idea: *élan primitif, plan vital, dessein vital* (primitive impulse, vital plan, vital design.) See Georges Canguilhem, "Le concept et la vie," in *Etudes d'histoire et de philosophie des sciences* (Paris: Vrin, 1979), pp. 335–64. From these terms Bergson chose *idée organisatrice et créatrice* (organizing and creative idea) which is the closest approximation of his own *élan vital,* except that for Claude Bernard this idea applied only to individuals, whereas for Bergson it also applied to the evolution of species.

[27] The vitalism/mechanism controversy is a recurrent issue in Claude Bernard celebrations, and each camp still uses Bernard's writing to legitimize their claims. My opinion is that Bernard's ideas on this matter were not only carefully nuanced but flexible, and they adapted to the various polemics in which he engaged. See Christiane Sinding, "Auseinandersetzung um die Forschungsleitenden Paradimata: Vitalism vs. Reduktionismus, organic life vs. individual life," in *Physiologie und industrielle Gesellschaft. Studien zur Verwissenschaftligung des Körpers im 19. Und 20. Jahrundert,* eds. Phillip Sarasin and Jakob Tanner (Basel: Reidel, 1998), pp. 76–98.

[28] "Il n'y pas de différence entre une observation bien prise et une généralisation bien fondée." In Dhubout, *Claude Bernard: Sa vie, son oeuvre* (cit. n. 25), quotation at p. 22.

tion between theory and facts. But Bergson maximized Bernard's claims in order to defend his own antipositivist philosophy (although the word "positivism" did not even appear in his discourse), without considering Bernard's own position towards positivism.[29]

### *Celebrations of Claude Bernard by Physicians and Scientists*

Generally speaking, biologists have appropriated Claude Bernard as the creator of physiology as a discipline autonomous from medicine (which still remains for them an art or an applied science), whereas physicians refer to Bernard as the creator of "scientific medicine." But both have celebrated Bernard for having introduced "The Method"[30] to medicine, allowing a scientific orientation to replace the old empirical medicine that Bernard despised and so often criticized.

This was the major claim made in discourses given by two members of the *Institut de France*[31] on 10 February 1978 for the centennial of Bernard's death.[32] Robert Debré (1882–1979), a well-known professor of pediatrics, gave a classical hagiographic discourse that portrayed Bernard as a martyr who suffered the hate of his wife and their two daughters, who had allied with the antivivisectionist movement.[33] The second discourse was given by Etienne Wolff (1904–1996), an embryologist and historian of embryology, professor at the Collège de France and a member of the Académie de Médecine as well as the Académie de France.

Wolff began his talk with a strong hagiographic, nationalistic and presentist statement: two great men had changed the world in the second half of the nineteenth century, and they were Pasteur and Bernard. Without them, the present biological revolution would not have been possible.[34] Wolff also commented on the classical

[29] Bernard is often considered a positivist, although he frequently criticized Auguste Comte. Bernard's positivism is a controversial issue. See Annie Petit, "Heurs et malheurs du positivisme comtien: Philosophie des sciences et politique scientifique chez Auguste Comte, et ses premiers disciples (1820–1900)" (Thèse de Doctorat d'Etat de philosophie, Paris-I Sorbonne, 1993), and Annie Petit, "Claude Bernard and the History of Science," *Isis,* 1987, *78:*201–19.

[30] As if there were some general method, valid for any matter, whether scientific or nonscientific.

[31] The Institut de France (often called l'Institut) has royal as well as republican origins. It was founded as a republican institute in 1795. Today it is comprised of five academies including the Académie Française and the Académie des Sciences. Although all of the academies aren't located here, the symbolic center of the Institut is the Palais Conti with its famous Coupole located at the quai Conti, facing the Louvre.

[32] *Hommage à Claude Bernard à l'occasion du centenaire de sa mort: Discours prononcé sous la Coupole le vendredi 10 février 1978 par Robert Debré et Etienne Wolff* (Paris: Typographie de Firmin-Didot et Cie, 1978).

[33] Debré had an important role in the development of pediatrics in France, and encouraged the introduction of biological methods into the field.

Robert Debré, "Un simple récit: La vie de Claude Bernard," in *Hommage à Claude Bernard* (cit. n. 32), pp. 3–17. Bernard's wife was antivivisectionist and raised their two daughters in the same spirit. There would probably be more sympathy towards Mrs. Bernard nowadays. However, to my knowledge, there has been no recent study devoted to the couple in relation to the problem of vivisection. Similarly, although scattered points have been made about Bernard and vivisection, it seems that the issue has never been extensively studied, although Bernard explicitly founded his experimental medicine using this technique. For recent work on vivisection and the laboratory, see Stewart Richards, "Anaesthetics, Ethics and Aesthetics: Vivisection in the Late Nineteenth-Century British Laboratory," in *The Laboratory Revolution,* eds. Andrew Cunningham and Perry Williams (Cambridge: Cambridge University Press, 1992), pp. 142–69.

[34] The talk was given in 1978, but "the present biological revolution" about which biologists love to speak apparently is a permanent one, since every day we hear about a new revolutionary development in the biomedical sciences. This, of course, raises problems about what the biological revolution is, if it exists at all.

comparison between Descartes and Bernard. Of course, he said, we French are used to being characterized as cartesian, which is sometimes more critical than complimentary. However, he went on, Descartes' system did not apply well to the sciences of nature, and what's more, Descartes made such extravagant statements about biology that only Bernard can be called the creator of method in biology.

Then Wolff took advantage of the ceremony to make a strong criticism of the French system of education. Since mathematics is emphasized in the schools and universities, he said, including those dedicated to biology and medicine, the country is producing people who are good mathematicians but who know nothing about experimentation and have no taste for observation. Here Wolff was displaying his vitalist belief that the life sciences cannot be reduced to mathematical and physical laws. It should be pointed out that other biologists have, on the contrary, insisted that the experimental method as introduced by Bernard to medicine and the life sciences did not, in fact, differ from the mathematical method applied in the physical sciences. Roger Guillemin, for example, claimed that this was Bernard's major contribution to modern medicine.[35] These contrasting opinions reflect the many and subtle nuances that Bernard brought to his writing, but also the views of the speakers and the social context that induced such statements: Wolff was concerned about the inappropriate education of biologists and physicians, whereas Guillemin wanted to make the hierarchical status of biology equal to that of the physical sciences.

The centennial of the publication of *Introduction à l'étude de la médecine expérimentale* in 1965 allowed Raoul Kourilsky (1900–1977), a professor of medicine, to challenge the commonly held idea that Claude Bernard was in no way interested in the art or technique of healing or in empirical medicine, but only in physiology and the mechanisms of disease.[36] Kourilsky's thesis was based on an analysis of Bernard's practical training as a physician, as well as on numerous statements found in *Introduction à l'étude de la médecine expérimentale.* In a detailed study, Kourilsky recalled some important facts about Bernard's medical education, his love for dissection and surgery, and his early interest in diabetes and neurology. Then he analysed the various biases introduced by many commentators on Bernard's book, and stated that the physiologist had always remained deeply interested in clinical

---

Etienne Wolff, "L'oeuvre de Claude Bernard replacée dans son temps," in *Hommage à Claude Bernard* (cit. n. 32), pp. 19–32.

[35] Roger Guillemin, "Ouverture," in Michel, *La nécessité de Claude Bernard* (cit. n. 18), pp. 3–6. Born in France in 1924, Roger Guillemin, a member of the Salk Institute and an American citizen, shared the Nobel Prize with his rival Schally for elucidation of the structure of TRH (thyrotropin releasing hormone). The story of this rivalry was told by Nicholas Wade in *Nobel Duel* (New York: Doubleday, 1981). Bruno Latour spent a year in Guillemin's laboratory, and the now famous *Laboratory Life,* first published in 1979 (Steve Woolgar and Bruno Latour, *Laboratory Life: The Social Construction of Scientific Facts,* 2nd ed. [Princeton: Princeton University Press, 1986]), was based on the work Latour conducted in this laboratory.

[36] Raoul Kourilsky, "La médecine clinique vue par Claude Bernard," in *Philosophie et méthodologie scientifiques chez Claude Bernard* (cit. n. 15), pp. 65–84. This paper, although outdated, is especially interesting because it challenges (without knowing it, apparently) the hard critical analysis made by Georges Canguilhem in *Le normal et le pathologique* (Paris: Presses Universitaires de France, 1966). Incidentally, the strong attacks Bernard made against "empirical medicine" were probably intended, at least in part, to support his own position as a founder of scientific medicine. John Lesch has emphasized Bernard's indebtedness to Paris medicine and the medical milieu of the first half of the century. See John E. Lesch, *Science and Medicine in France: The Emergence of Experimental Physiology, 1790–1855* (Cambridge, Mass.: Harvard University Press, 1984), pp. 197–224.

medicine. His aim was to make practical medicine more scientific, which was not possible if physiology and medicine remained separated. Finally, Kourilsky made presentist claims about the value of using the experimental method in clinical medicine, and, just as Wolff would do almost ten years later, he recommended that medical studies be reformed with this in mind.

### *The Centennial of a Bestseller*

Bernard's most famous published work, *Introduction à l'étude de médecine expérimentale,* was not immediately recognized by the scientific community as a great book. In fact, one of the book's first admirers was Pasteur. In 1866 Pasteur published an article devoted to the importance of Bernard's work in the *Moniteur Universel,* the official newspaper of the Empire. In this article he praised Bernard's book as a masterpiece that would serve as a guide for future generations. However, his enthusiasm dimmed after Bernard's death, when Marcellin Berthelot published some of Bernard's laboratory notes that described experiments he had performed in order to refute Pasteur's theory on fermentation.[37]

The centennial celebration of the publication of *Introduction à l'étude de médecine expérimentale* was directed by a professor of medicine, Bernard Halpern, who held the chair of experimental medicine at the Collège de France. The colloquium, which was held in Paris in 1965, is especially interesting because commemoration of a book's publication is not a common event.[38] Georges Canguilhem,[39] who participated in the meeting, gave a paper the same year at the Palais de la Découverte in which he asserted that commemorations of men are fruitful because they stimulate studies and research on the honorees.[40] However, he said, these ceremonies make no sense in terms of history of science, because they are subject to the contingency of dates of birth or death as well as the contingency of the choices made about who will be commemorated and when (ten years after the birth or death, fifty, one hundred . . . ?) Commemoration of the date of publication of a *book* makes much more sense, insisted Canguilhem. To him, 1865 was a very important date because Bernard knew at that time that he had paved a new road, and because, Canguilhem

[37] Although Pasteur and Bernard offered each other many gestures of friendship and respect, these two strikingly different personalities were never very close. The fact that Bernard was older than Pasteur and was at the peak of his career much earlier in the century than Pasteur, partly explains their distant relationship. See Mirko Grmek, "Louis Pasteur, Claude Bernard et la méthode expérimentale," in *L'Institut Pasteur: Contributions à son histoire,* ed. Michel Morange (Paris: La Découverte, 1991).

[38] The proceedings of the colloquium were published in 1967. See *Philosophie et méthodologie scientifiques chez Claude Bernard* (cit. n. 15). Another book that has been the focus of commemorative activity is Darwin's *On the Origin of Species.*

[39] The philosopher Georges Canguilhem (1904–1995) was mainly interested in the philosophy of life sciences. Probably the best essay on Canguilhem's work was written by his pupil Michel Foucault in his introduction to the English translation of Canguilhem's *Le normal et le pathologique* (Dordrecht: Reidel, 1978). Gary Gutting has made some interesting comments on Canguilhem's work in *Michel Foucault's Archaeology of Scientific Reason* (Cambridge: Cambridge University Press, 1989), pp. 32–52.

[40] Indeed, in 1966, one year after the commemoration of Bernard's book, Canguilhem wrote the preface to the new edition of Bernard's *Leçons sur les phenomènes de la vie communs aux animaux et aux végétaux* (Paris: Baillère, 1878) in which he pointed out that this book was in fact a text of biological philosophy, whereas François Dagognet, also a philosopher of biology, wrote the preface for a new edition (1966) of *Introduction à l'etude de la médecine expérimentale.*

argued, this road was still being explored.[41] With this presentist statement Canguilhem helped to legitimize not only the experimental life sciences, but also the role of philosophers as thinkers who are looking for meaning and who, in effect, *give* meaning to historical scientific events.

At the colloquium devoted to Bernard's book, most of the papers were clearly hagiographic with the exception of Mirko Grmek's detailed historical and technical study of the development of the concept of the milieu intérieur (internal environment).[42] These reverent contributions were made not only by the French participants but also by invited guests from Eastern Europe, (most Eastern European countries were Francophile). Generally speaking, the now-declining "Bernard industry" has always been clearly hagiographic with the notable exception of analyses by Grmek and Frederic Lawrence Holmes, who have discussed Bernard's achievements and legacy in more critical terms. By 1965, when the colloquium was held, the professional history of medicine had started.[43] Canguilhem took both sides: he admired Bernard, but also published a critical assessment of Bernard's theoretical views, notably in his book *Le normal et le pathologique.*[44] Canguilhem again gave a critical assessment of Bernard's theoretical work at a symposium held in 1978 for the commemoration of Bernard's death.

### *A Critical Commemoration*

In 1978 Georges Canguilhem and Yvette Conry, one of his former philosophy students, presented critical analyses of Bernard's work and writing for the centennial of his death. This conference took place in the framework of a regular seminar series *Séminaire sur les Fondements des Sciences,* held by physicians at the Faculty of Medicine in Strasbourg.[45] There were only six speakers: four physicians and the two philosophers. The latter brought a critical tone to the commemoration, which was probably facilitated by the intimate setting. The philosophers were careful to announce at the beginning of their talks that they would not conform to the "genre" of celebration, and that they would consider Bernard's shortcomings as well as his positive contributions. This may suggest that it is almost a moral and professional duty for a philosopher to make a critical evaluation of a scientist's work, even at commemorations. It implies that the act of commemorating introduces some bias in history, and that somehow philosophers are the guardians of "truth."

[41] Georges Canguilhem, "L'idée de médecine expérimentale selon Claude Bernard," conference held 6 February 1965, reprinted in *Etudes d'histoire et de philosophie des sciences* (Paris: Vrin, 1979), pp. 127–41. François Dagognet made similar presentist statements when he wrote that Pasteur, despite his errors, initiated a scientific movement and founded a school that are still of primary importance.

[42] Mirko D. Grmek, "Evolution des conceptions de Claude Bernard sur le milieu intérieur," in *Philosophie et méthodologie scientifiques chez Claude Bernard* (cit. n. 15), pp. 117–50.

[43] A colloquium for the celebration of Bernard's "bestseller" was held in the United States the same year, where Holmes gave his own interpretation of the *milieu intérieur* and challenged the commonly held idea that this famous Bernardian concept was immediately accepted and utilized by his French and foreign colleagues. See Frederic Lawrence Holmes, "Origins of the Concept of the Milieu Intérieur," in *Claude Bernard and Experimental Medicine,* eds. Francisco Grande and Maurice B. Visscher, (Cambridge: Schenfman, 1967), pp. 179–91.

[44] See "Claude Bernard et la pathologie expérimentale," in Canguilhem, *Le normal et le pathologique* (cit. n. 39), pp. 32–51.

[45] The papers from the conference weren't published until 1984. See *Médecine, science et technique: Recueil d'études rédigées à l'occasion du centenaire de la mort de Claude Bernard (1813–1878),* ed. Charles Marx (Paris: Editions du CNRS, 1984).

Conry characterized Bernard's vision of medicine as utopian, the primary function of which (like all utopias) was polemical: Bernard always fought empirical and descriptive medicine as well as the ontological view of disease. Then Bernard had to build a unitary theory, also a necessary component of utopias. The concept of the *milieu intérieur* served this purpose. Finally, disease would be cured only by acting on this milieu, since disease resulted from a loss of internal regulation rather than from an external cause such as a microbe.

Bernard's biological philosophy, according to Conry, prevented him from making therapeutic discoveries. For her, the most convincing refutation of Bernard's theoretical position occurred at the beginning of the twentieth century, when antimicrobial chemotherapy was discovered through a technical "ruse" (*"par la ruse d'une technique"*) instead of through theoretical deduction. What was important to Conry was not Bernard's failure to make pharmaceutical discoveries, but his adoption of positivism which gave primacy to theory and offered the illusion of an easy and direct passage from knowledge to action.[46]

Canguilhem, for his part, presented a comparative assessment of Western rationality in medicine, present and past. He criticized Bernard's physiological conception of disease which, he believed, had prevented the physiologist from understanding the significance of germ theory.[47] Generally speaking, Canguilhem applied Kant's critical method to the biomedical sciences and performed what I would call a "critique of medical reason." Finally, both philosophers pinned down Bernard's failure at the most basic level: his philosophical views. Their critical assessment of Bernard's methodological principles is devastating, because it aims straight at rationality and method, the priceless tools of both scientists and philosophers. This kind of critique, which can be managed properly only by philosophers, legitimizes their role in history of science. It should be pointed out that in France, until recently, history of science has been taught only in philosophy departments. (Auguste Comte asked for the creation of the first chair in history of science, which was established in 1892 after his death and assigned to the positivist Pierre Lafitte.) Thus, in France, the role of philosophers in history of science is rooted in historical and institutional events. Recently the situation has begun to change. On the one hand, the growing importance of science in society has led more scientists to write their own histories. On the other hand, the slow introduction of social history and social studies of science into France should help to develop modern history of science in the country.

To summarize, no French scientist of the nineteenth century has received as much attention from philosophers as Claude Bernard. It has been customary to study Bernard's scientific and philosophical work separately, a division that is highly problematic.[48] Nevertheless, this separation is significant, and it might be related to the peculiar French attitude towards rationality and method, the role of French philosophers of science, and the unusual status of Bernard as an author. It also explains in part the rather intimate and intellectual tone of Bernard's celebrations: until very recently,

[46] Yvette Conry, "Le 'point de vue' de la médecine expérimentale selon Claude Bernard: Une utopie positive?" *ibid.,* pp. 17–38.

[47] Georges Canguilhem, "Puissance et limites de la rationalité en médecine," *ibid.,* pp. 109–30.

[48] Amsterdamska, "The Historiography of the Claude Bernard Industry" (cit. n. 17), and Christiane Sinding, "Literary Genres and the Construction of Knowledge: Semantic Shifts and Scientific Change," *Social Studies of Science,* 1996, *26:*43–70.

French "thinkers" hated to appear in public. Indeed, when we turn to Pasteur, we find that the general tone of his commemorations was much more grandiose.

## II. THE SAVIOR

For reasons of space, it is not possible to discuss how the myth of Pasteur was constructed during his life and after his death.[49] It might be useful however, to quote the following passage by Albert Delaunay from the French *Encyclopaedia universalis,* as it is can be considered as both an historical statement and a mythical construction, demonstrating how difficult it is to unravel one from the other:

> "There are three main reasons why Louis Pasteur is a great figure of humanity. The first has to do with his scientific work. Pasteur has renewed entire chapters of physics and chemistry; maybe more than any one else, he has revealed the importance of the microbial world, either as an equilibrium factor at the surface of the globe or in terms of its responsibility for animal and human disease. Secondly, one owes to his genius techniques that have transformed entire industries and the development of important vaccines. Finally, in contemplating Pasteur's life, one can only be struck by its moral quality."[50]

As far as the "moral quality" of Pasteur's life is concerned (a moral life is a common feature in the construction of heroes in science), it has been challenged by some of Pasteur's contemporaries as well as modern historians. As far as his scientific work is concerned, there is no doubt that Pasteur *participated* in great lines of research. He performed important work in basic science, mainly crystallography, stereochemistry, and fermentation, as well as on the physiology of microbes. He also contributed practical discoveries, among which the invention of vaccination is the most well known. But one of the questions raised by modern historians concerns the extent of Pasteur's contribution, which according to some has been greatly overestimated. This question is related to one of the major theses of the new history of science: that scientific work is a collective enterprise and a social construction. This central feature of the scientific enterprise is concealed by scientists themselves, who tend to erase all traces of their predecessors and minimize the importance of their colleagues. Thus, although most of the new historians of science admit that Pasteur did some great work,[51] their main criticism is aimed at the legend, constructed by

[49] The major works on this subject are: Bruno Latour, *Microbes, guerre et paix* (Paris: A. Métailié, 1984), which was translated by Alan Sheridan and John Law under the title *The Pasteurization of France* (Cambridge, Mass.: Harvard University Press, 1988); and Gerald Geison, *The Private Science of Louis Pasteur* (Princeton: Princeton University Press, 1995).

[50] "Trois raisons font de Louis Pasteur une grande figure de l'humanité. La première tient à son œuvre scientifique. Pasteur a renouvelé des chapitres entiers de la physique et de la chimie; plus qu'aucun autre peut-être, il a révélé l'importance du monde microbien, soit comme facteur d'équilibre à la surface du globe, soit comme responsable de maladies animales et humaines. En second lieu, on doit à son génie des techniques qui ont transformé des industries entières et la mise au point de vaccinations importantes. Enfin, vient-on à méditer sur sa vie et sur son œuvre, on ne peut qu'être frappé par la qualité morale qui s'en dégage." Albert Delaunay, "Pasteur, L.," in *Encyclopaedia universalis,* vol. 7 (1996), p. 627. My translation.

[51] An example is this quotation from Geison: "The myth of Pasteur, like all the myths, embodies important elements of the truth. After all, Pasteur's scientific work was enormously important and fertile, and some of his principles continue to guide us today. As Bruno Latour and others have recently reminded us, it would be folly to deny the fruits of the Pastorian enterprise, and there was obviously something like a Pastorian "revolution," with consequences like the "pasteurization of France." Geison, *The Private Science of Louis Pasteur* (cit. n. 49), p. 277.

Pasteur himself as well as his early biographers, that he did it all alone, without contributions from his predecessors or contemporaries.

Before analysing the construction and deconstruction of this myth, it might be of interest to look briefly at the most obvious appropriations of Pasteur by scientists and nonscientists. The way in which general practitioners appropriated Pasteur's ideas for their own purposes after his death in 1895 has been analyzed by Bruno Latour. Before 1895 physicians were somewhat skeptical and critical of Pasteurism because its emphasis on the prevention of disease, it appeared to lead to the dissolution of the medical profession—no more diseases, no more doctors. But after the discovery in 1895 of the antidiphtheria serum which, unlike vaccines, had to be given to sick people, the doctors adopted Pasteurism without hesitation because they were the only ones allowed to use this therapeutic tool. Moreover, "they *immediately altered the chronology* so as to include Pasteur among other elements of the old, at last triumphant medicine."[52] This provocative analysis by Latour should be softened, as some physicians acclaimed Pasteur as a savior of humanity as early as 1882.[53] Nevertheless the argument still holds: physicians did not adopt or appropriate Pasteur because they were less or more clever than others, or because they suddenly understood that Pasteurism would change medicine, but due to practical and financial reasons.

Today physicians readily appropriate Pasteur's ideas for all matters related to infectious or even genetic diseases. Pasteurian biologists and physicians have appropriated Pasteur in many ways and on a number of occasions, including the centennial in 1995. According to Claire Salomon-Bayet, Pasteurians constitute a "family" that has "its own habits and rites, its own patrimony, the consciousness of its importance which is sometimes looked upon as arrogance."[54] Among the rites, the biannual ritual in which all members of the Institut Pasteur participate has been described with an ethnographical eye by François Jacob:

> I followed the crowd of people emerging from their laboratories and going to the garden toward the institute's oldest building, where Pasteur was buried. A sudden hush signaled the arrival of the dignitaries . . . The director's brief address reminded the personnel of the virtues on which "our house," its continuity and traditions were founded.
>
> Then, in silence, the descent into the crypt began, in Indian file, in hierarchical order: the director and the board; the council; the department heads, eldest first; the heads of laboratories and their collaborators; the technicians and assistants; and finally, the cleaning woman and lab boys. Each went slowly down some steps before passing in front of the tomb . . . At the entrance, over the whole of the vault, mosaics depicted, in the manner of scenes from the life of Christ, those from the life of Pasteur . . . And at the summit was the supreme image, the struggle of a child with a furious dog, to glorify the most decisive battle, that against rabies.[55]

[52] Latour, *The Pasteurization of France* (cit. n. 49), p. 133. Italics are the author's.

[53] Jacques Leonard, "Comment peut-on être pastorien?" in *Pasteur et la révolution pastorienne,* ed. Claire Salomon-Bayet (Paris: Payot, 1986), pp. 143–79.

[54] Claire Salomon-Bayet, "Penser la révolution pastorienne," in Salomon-Bayet, *Pasteur et la révolution pastorienne* (cit. n. 53), pp. 17–62.

[55] François Jacob, *La statue intérieure* (Paris: Odile Jacob, 1987), pp. 328–37. Translated by Philip Franklin under the title *The Statue Within: An Autobiography* (New York: Basic Books, 1988), pp. 244–7.

Pharmacists of the end of the nineteenth century appropriated Pasteur in order to improve their status and redefine their role in society. They fought for the privilege of making medical analyses and to be regarded as scientists.[56] Politicians appropriated Pasteur, as they did Bernard, to unite the nation around Republican values and mobilize it against Germany. Like Bernard, Pasteur had voiced his opposition to Germany many times in speech and writing; furthermore, he had returned a diploma that he had received from the University of Bonn. Whereas the Bernard centennial of 1913 was used as a means of mobilizing the French against Germany before the war, Pasteur's 1923 centennial was used to comfort the nation after the war, which had been won at a terrible price. The "cure" from rabies of the young Joseph Meister, an Alsacian boy, permitted the concurrent celebration of Pasteur and the recovery of Alsace-Lorraine from Germany. But the strongest appropriation of Pasteur by politicians was made with the help of hygienists, physicians and legislators, who redefined health policy in the name of microbial eradication. This aspect of Pasteur's appropriation has received the full attention it deserves, and I won't go further into it.[57] However, it should also be remembered that the "Pasteurization" of France has taken many different forms, and that it has partly failed.[58]

### *The Myth Still Holds*

As mentioned, Pasteur was credited with inventing vaccination, the first preventive measure for infectious diseases (which were often identified at the end of the nineteenth century with disease in general).[59] Pasteur was celebrated as a saint, a figure both compassionate and disinterested. Not only was he celebrated in France but in other countries as well. The 1995 centennial, dubbed "*L'Année Pasteur*," showed that more than twenty years of critical history had barely harmed the great myth, which in fact seems to have been reinforced, as it obviously fulfills a social and political need. Therefore, I will give much attention to this recent celebration, particularly to a special issue of *Le Figaro* (a conservative newspaper) that was devoted to the commemoration.[60] This issue had two sections: the man and his time and the institute today. Four pages were devoted to "the first section" and six to the institute today; evidently the newspaper did not dissociate the celebration of the man from that of the institute. The presentist dimension of the commemoration was apparent: one could read subtitles such as "Rage ou SIDA: La même démarche" ("Rabies or AIDS: The same Approach") along with more moderate commentaries by Maxime Schwartz, the present director of the Institut Pasteur. Pierre-Gilles de Gennes, a recent Nobel

[56] Viviane Thévenin, "Les pharmaciens et l'officine: Les pharmaciens de 1971 à 1919," in Salomon-Bayet, *Pasteur et la révolution pastorienne* (cit. n. 53), pp. 183–214.

[57] See Ann-Louise Shapiro, "Private Rights, Public Interests and Professional Jurisdiction: The French Public Health Law of 1902," *Bulletin of the History of Medicine,* 1980, 54:4–22; William Coleman, *Death is a Social Disease: Public Health and Political Economy in Early Industrial France* (Madison, Wisc.: University Press, 1982); Bernard-Pierre Lécuyer, "L'hygiène en France avant Pasteur," in Salomon-Bayet, *Pasteur et la révolution pastorienne* (cit. n. 53), pp. 65–139; Robert Carvais, "La maladie, la loi et les moeurs," *ibid.,* pp. 279–330; Latour, *The Pasteurization of France* (cit. n. 49).

[58] Lion Murard and Patrick Zylberman, *L'hygiène dans la République* (Paris: Fayard, 1996).

[59] Pasteur's role in the invention of vaccination has been reassessed by some of his biographers. For a recent evaluation see Anne-Marie Moulin, "La métaphore vaccine," in *L'aventure de la vaccination,*" ed. A. M. Moulin (Paris: Fayard, 1996), pp. 125–42.

[60] "Pasteur: La légende du siècle," *Le Figaro,* special issue devoted to Pasteur, 17 January 1995.

Prize winner in physics, wrote that Pasteur had discovered the "asymmetry of life"; de Gennes made conjectures on the origin of life, adding that the basic foundations of this problem had been laid by Pasteur.[61] In a section entitled "Consequences of Pasteurian discoveries: No hygiene without clean hands," Georges Ducel, a professor of medicine, explained how Pasteur had laid the principles of hygiene and asepsis that are still used in hospitals and mentioned Lister's admiration for Pasteur.

The first page of the newspaper displayed a reproduction of a portrait of Pasteur made at the end of his life. In an editorial, Franz-Olivier Giesbert, the publisher of *Le Figaro,* gave an account of his recent trip to Teheran where he observed that all the names of the streets had been changed. To his astonishment, he found one street that had kept its name: the rue Pasteur! Pasteur was such a great scientist, the journalist added, that the mullahs themselves had not dared change the name of the street. Giesbert recognized that Pasteur had not really saved mankind nor cured disease, but claimed that he had laid the foundation for microbiology and immunology. Everything had passed through Pasteur's hands: milk, yeast, cows, hens, cheese, beer, rabid dogs, foxes, silkworms, infectious diseases, microbes—is it not right to celebrate such a man? concluded Giesbert.[62] Besides its nationalistic tone, this statement was obviously intended to recall the universal value of science.

Page two of the newspaper was based on extracts from a biography of Pasteur by Maurice Valléry-Radot that was first published in 1985 and reissued in 1995 with a preface by Luc Montagnier, the modern hero fighting HIV.[63] One can easily see that Valléry-Radot's book is hagiographic just by looking at the chapter titles, which allude to Pasteur's numerous laudable qualities (his love of truth, his tenacity, his sense of duty . . . ), with the exception of his love of honors. An article by a journalist, Anne Muratori-Philip, gave an account of the controversy between Pouchet and Pasteur, but did not allude to modern studies on the topic.[64] Latour's work was not quoted except for his recent book, which was coedited and funded by the Institut Pasteur for the centennial,[65] and designated the "official album of the centennial" by *Le Figaro.*[66] Thus any controversial aspect of Pasteur's life or work was omitted. This raises the question of whether a conservative newspaper can be anything other than hagiographic and nationalistic, as well as the difficult issue of the political uses of history in science.

Other French newspapers ran articles on Pasteur's commemoration but none of them had a special issue. For instance, on March 24, almost three months after the

[61] The term "asymmetry of life" comes from the subtitle of the article, which was written probably by the editor. If Pasteur discovered anything of the sort, it would obviously be the asymmetry of *organic molecules* (not life!) as opposed to the symmetry of nonorganic molecules.

[62] "Pasteur: La légende du siècle" (cit. n. 60), p. 4.

[63] Maurice Vallery-Radot, *Pasteur: Un génie au service de l'homme* (Paris: Perrin, 1994). Maurice Vallery-Radot is the great-grandson of Louis Pasteur; René Vallery-Radot, famous for his hagiographic biography of Pasteur, was his son-in-law.

[64] This cannot only be related to a language barrier, since several studies on this issue were published in French. See, for instance, John Farley and Gerald Geison, "Le débat entre Pasteur et Pouchet: Science, politique et génération spontanée au XIXe siècle en France," in *La science telle qu'elle se fait,* eds. Michel Callon and Bruno Latour (Paris: La Découverte, 1991); Bruno Latour, "Pasteur et Pouchet: Hétérogenèse de l'histoire des sciences," in *Eléments d'histoire des sciences,* ed. Michel Serres (Paris: Bordas, 1989); G. Geison, "Les à-côtés de l'expérience," in *Pasteur: La tumultueuse naissance de la biologie moderne, Les Cahiers de Science et Vie,* n. 4, 1991.

[65] Bruno Latour, *Pasteur: Une science, un style, un siècle* (Paris: Perrin-Institut Pasteur, 1995).

[66] "Pasteur: La légende du siècle" (cit. n. 60).

opening of *L'Année Pasteur,* the daily newspaper *Le Monde* printed reviews of some of the books on Pasteur published or republished in 1995, together with an article on Pasteur by the philosopher Dominique Lecourt. Although not as clearly hagiographic as the special issue of *Le Figaro,* none of the other newspapers alluded to the critical works published on Pasteur, except for *Le Quotidien du Médecin.* This weekly newspaper for general practitioners covered Pasteur's commemoration under the title *"Le mythe Pasteur tient bon"* ("The myth of Pasteur still holds").[67] The author of the article was obviously aware of the critical work published on Pasteur and, to my knowledge, was the only one who commented on it.

Movies on Pasteur's life were shown on television. In one movie produced by French television for the centennial, interviews with scientists from the Institut Pasteur were heavily loaded with commentary on the institute's present accomplishments in molecular biology and in the battle against AIDS. Interviewees including Luc Montagnier, the codiscover of the AIDS virus and Gallo's rival, emphasized the difficulties of their work, and their statements were very different from the triumphant claims made fifteen years earlier (not only in France) when HIV was identified. On the other hand, strong presentist claims were made: showing us coliform bacillus, the commentator credited Pasteur for having opened the door to the discovery of chromosomes and genes and finally to molecular biology, by allowing the bacillus to be cultivated![68] The problem of AIDS was present throughout the movie, and the commentary tended to make it clear that the disease has brought us back to the epoch of the great epidemics, thus reinforcing the appeal to Pasteur's image as a savior and a founder. The centennial exhibition at the Institut Pasteur also reflected this need for a founder: Pasteur was not the only scientist represented (one could see others, such as Lister and Koch), but his image as a benefactor of humanity had not changed. In general, the Pasteur Institute strongly appropriated Pasteur for its own celebration.[69]

The 1995 celebration of Pasteur displayed all the classical features of scientific commemorations: the myth of the founder, the nationalistic tone and the stress on the continuity of science. The tension between change and continuity in science is resolved by the idea that the hero paved new roads or invented new lines of work, so that the actual scientific content of the work can be easily dismissed by commemorators or reduced to short, vague statements. What is more specific to biomedical celebrations is the emphasis on lifesaving and disease eradication. The hero is viewed as the savior of humanity, and the religious aspect of the commemoration appears more clearly than in other scientific celebrations.[70] This explains why de-

[67] Marie-Françoise de Pange, "Le mythe Pasteur tient bon," *Le Quotidien du Médecin,* n. 5518 (21 November 1994), pp. 34–6.

[68] Igor Barrère and Jean-Pierre Fleury, "Sur les traces de Pasteur: Des microbes à la thérapie génique" ("On Pasteur's Trail: From Microbes to Genetic Therapy,") 1994. The title emphasizes the continuity between Pasteur's enterprise and the present and future work of the institute.

No allusion was made to Koch's work on microbiological cultures in the video.

[69] This 1995 exhibition, entitled "Une vie pour la vie" ("A Life for Life") took place at the Institut Pasteur from 17 January to 15 March. Afterwards, it was displayed in several French cities.

[70] It is not possible to comment on all the books, articles, exhibitions, television movies, etc., that were produced for the 1995 celebration. I can only list some of them, without pretending to be exhaustive. Among the books were *Cahiers d'un savant,* eds. Françoise Balibar et Marie-Laure Prévot (Paris: Zulma-CNRS, 1995); François Dagognet, *Pasteur sans la légende* (Paris: Les Empêcheurs de penser en rond, 1995), originally published under the title *Méthodes et doctrine dans l'oeuvre de Pasteur* (Paris: Presses Universitaires de France, 1967); Pierre Darmon, *Louis Pasteur* (Paris: Fayard,

constructing the myth of Pasteur is especially difficult and has encountered strong resistance.

### *Deconstruction by Philosophers*

There are several ways to deconstruct myths and legends, and philosophers have their own. François Dagognet, a philosopher of biology, recently revised his pioneering book on Pasteur giving it the new title *Pasteur sans la légende.*[71] One of the main theses of the book is that, contrary to a commonly held view, Pasteur's work was not contingent upon industrial demand. In Dagognet's opinion, Pasteur had a method and a biological philosophy that allowed him to treat practical problems as tools for gaining access to the knowledge of nature. For instance, even before he "solved" the problem of fermentation, Pasteur knew that living systems had to be at work in fermentation. His vitalistic philosophy, together with his choice of solvable problems (his "preconceived ideas", as he liked to say), explain his success. Thus, just as Canguilhem, also a philosopher, did with Claude Bernard, Dagognet explained Pasteur by the constancy of his method and his philosophical options, ruling out contingency as an important factor of discovery.

The religious aspects of scientific commemorations have often been analyzed. But it seems to me that no scientist has been as greatly sanctified as Pasteur. In the issue of *Le Figaro* commemorating Pasteur, François Dagognet noted that Pasteur was already canonized during his lifetime. In 1882, a jubilee was held for his seventieth birthday that was attended by legislative representatives, ambassadors, and delegates from the main European cities. The Republican Guard played a triumphal march as Pasteur entered the great amphitheater of the Sorbonne alongside the president of the Republic, Sadi Carnot, and Lister opened his arms to him. Why this triumph Dagognet asked. His answer was that Pasteur had discovered the cause of infectious diseases, developed aseptic technique, and invented vaccination. Indeed, the hero had made errors, added Dagognet, and most of his ideas have been modified or proven wrong. But errors contribute to progress when they are refuted. What is important, Dagognet concluded, is the movement Pasteur initiated as well as his founding of a research school in microbiology. Thus, although Dagognet attempted to deconstruct the classical legend, he nevertheless participated in the classical ritual of commemorations and made Pasteur a great founder. He also legitimized the role of philosophers in history of science.

It would seem, then, that Bruno Latour remains the sole French scholar who has succeeded in deconstructing Pasteur's legend. Since the last book he authored

1995); Patrice Debré, *Louis Pasteur* (Paris: Flammarion, 1995); René Dubos, *Pasteur: Franc-tireur de la science* (with a preface by Bruno Latour) (Paris: La Découverte, 1995), originally published in English under the title *Pasteur: Free-lance of Science* (Boston: Little, Brown, & Co., 1950); Geison, *The Private Science of Louis Pasteur* (cit. n. 49); Claude Jaugey, *Sur les chemins de Pasteur* (Paris: Barthélémy, 1995); Latour, *Pasteur: Une science, un style, un siècle* (cit. n. 65); Richard Moreau, *Louis Pasteur: La recherche d'une voie* (Paris: L'Harmattan, 1995); André Pichot, *Pasteur: Ecrits scientifiques et médicaux* (Paris: G. F. Flammarion, 1994); Daniel Raichwarg, *Pasteur: L'empire du microbe* (Paris: Gallimard, 1995); Maurice Vallery-Radot, *Pasteur: Une génie au service de l'homme* (Paris: Perrin, 1994). Among the movies and videos was a video produced for Pasteur's exhibition: "Une vie pour la vie" (1995). One should aso mention the opening session of L'Année Pasteur at the UNESCO (17 January 1995) and a "Séance solenelle" at the Académie des Sciences (Paris, 23 May 1995).

[71] Dagognet, *Pasteur sans la légende* (cit. n. 70).

(*Pasteur: Une science, un style un siècle*) was sponsored by the Institut Pasteur, this may be surprising. However, in the book's introduction, Latour makes it clear that the time has passed for "hyperbolic celebrations" and that nobody believes anymore in great men or in native countries full of gratitude. It appeared that Latour planned to renew this genre, and that we were being prepared to read a commemorative book that was countercommemorative. How is this possible?

### *History Against Memory*

In the introduction to the first volume of *Les lieux de mémoire,* Pierre Nora contrasted memory and history. Memory is alive, he wrote, carried by living groups of people so that it is always changing always open to the dialectic between remembrance and amnesia, and sustained by emotions and symbols.[72] Conversely, history analyzes and offers a critical discourse against memory, which seeks to sanctify events and men. History wants to eliminate particular memories that belong to groups and replace them with a universal discourse.

In France, Nora wrote, history for a long time had a political function, which was to create and rectify a sense of the nation. After the 1930 crisis, historical interest was displaced from nation to society, and historians became interested in particular memories of different social groups. History then lost its pedagogical function (transmission of national values) and became iconoclast, seeking to instill doubt about old myths. Now that critical history has won the battle against memory-history, every social group, including scientists, has begun to reconstitute its own particular memory and look for founders, heroes and continuity.[73]

If Nora is right, Latour can be seen not only as a researcher in science studies, but as the inheritor of this new critical and iconoclastic history. Indeed, when Latour published his first works on Pasteur in 1984 and 1986, he was perceived as an iconoclast by scientists as well as historians of science.[74] As analyzed by Pierre Thuillier, a philosopher of science interested in sociohistorical methods, the earliest readers of Latour were mainly struck by statements such as "there are no theories," "the sciences do not exist," "there never was such a thing as deduction," "there are no ideas," "no scientific thought," and so on. Pasteur appeared suddenly in Latour's oeuvre as a man who loved power, failed to quote his predecessors, loved controversies because he was a talented polemicist, and did not discover anything but instead transformed existing tools and ideas and imposed them by force, personal power and the use of rhetoric. Nobody in France was prepared to accept such an image. After this "liquidation" or "great cleaning," what remains of Pasteur? asked Pierre Thuillier.[75]

But then how could Latour write *Pasteur: Une science, un style, un siècle* or, to put it differently, how could the iconoclast become a commemorator of Pasteur? Did he change his mind? Had the scientists from the most respected scientific institute in France suddenly been converted to the spirit of social studies of science? Indeed, scientists have become more sensitive to the importance of social and economic

[72] Nora, *Les lieux de mémoire* (cit. n. 3).
[73] See Nora, "Entre mémoire et histoire: La problématique des lieux," *ibid.,* vol. 1, pp. xvii–xlii.
[74] Latour, *Microbes, guerre et paix* (cit. n. 49); see also Bruno Latour, "Le théâtre de la preuve," in Salomon-Bayet, *Pasteur et la révolution pastorienne* (cit. n. 53), pp. 337–84.
[75] Pierre Thuillier, "La science existe-t-elle?" *La Recherche,* 1985, *18:*500–11.

factors in disease principally because of AIDS. Their therapeutic failure against this "new" disease, together with the re-emergence of infectious diseases, the problem of antibiotic resistance, and the lack of medication against the major parasitical diseases of the Third World prevent them from making too many claims of victory over disease. But this is not enough to turn scientists into sociologists of science.

It should also be pointed out that, although basically Latour did not change his theses in the commemorative book, he slightly modified his style and suppressed such statements as "there are no theories." He stated in his introduction that new styles of celebration were needed and that, in order to understand Pasteur, one has to forget about statues, heroes, and hagiographies, and look at what Pasteur actually did. Despite the fact that Latour, like other researchers in science studies, would never use words such as "truth" or statements such as "what really happened," his plan really looks like a tentative restatement of what Pasteur "really" was and "really" did. For instance, he reports that Pasteur recreated parts of the outside world in his laboratory and conversely transformed parts of his laboratory into farms, breweries or vineyards. He knew how to interest more and more people in his work, how to get funds from the emperor, how to create a school, how to get collaborators, and how to create an institute that brought him as much fame as his work.

All this was already present in Latour's first works on Pasteur as components of a new account of the scientist, one which maintained that Pasteur was a great man; if not for the reasons usually alleged by classical hagiographies.[76] Furthermore, Latour's emphasis on Pasteur's ability to make allies in wider society made Latour a good ally of the Institut Pasteur. In our time, scientific institutions are calling for more private funds and support from the general public, especially when outsiders (and sometimes scientists themselves) express doubts or criticisms against the scientific enterprise. And there is no better figure than the new Pasteur, as reconstructed by Latour, to buttress the present and future of science. The image of Pasteur working closely with nonscientists and other professionals who asked for his help convinces lay people that scientists work on behalf of their best interests. And this is especially needed at the end of the present century, when so many people have become skeptical about science.

Finally, it should be pointed out that Latour disregarded the meaning or even the necessity of commemoration (or memory-history, as Nora calls it). Pnina Abir-Am has emphasized the importance and function of scientific anniversaries as occasions to "reconcile the contradiction between the relativism implied by discarded convictions with the pretense of present scientific progress to an absolutist truth."[77] The myth of the founder serves to unite the scientific community around accounts of heroic discoveries and support the idea of continuous scientific progress at times of no obvious progress. Past and present are brought together so that scientists can project themselves into the future. Geison has pointed out the useful functions of the legend of Pasteur:

> Like all myths, the standard legend of Pasteur has served several useful functions. Especially in the form purveyed in René Vallery-Radot's *La vie* and the children's books

[76] Latour, *Microbes guerre et paix* (cit. n. 49).
[77] Pnina Abir-Am, "A Historical Ethnography of a Scientific Anniversary in Molecular Biology: The First Protein X-ray Photograph (1984, 1934)," *Social Epistemology,* 1992, *6:*323–353, special issue, "The Historical Ethnography of Scientific Rituals."

derived from it, the legend served as a valuable resevoir of homilies for schoolteachers and French patriots, and as a source of inspiration for young would-be scientists. It has also provided a sense of human drama and excitement as opposed to the impersonal, collective sense of science about which so many complain today. Rarely has science been made so wonderfully simple, or so wonderfully grand and useful at once.[78]

## III. STYLES OF COMMEMORATION, STYLES OF RESEARCH

### *Styles of Commemoration*

The different styles of commemorating Bernard and Pasteur will now be juxtaposed for a more systematic comparison. Public commemorations devoted to Bernard displayed a rather moderate and often intimate style. This relative lack of pomp, especially when compared with Pasteur's celebrations, became more obvious with time. During his lifetime Bernard was much honored and received many honorary titles, and he was the first scientist to have a national funeral. In 1913 the centennial of his birth was celebrated with some distinction, although it was not as publicized as Pasteur's in 1922.[79] By the same token, the centennial of Bernard's death in 1978, just twenty years ago, was barely noticed. Articles appeared here and there, mainly in medical journals. Some exhibitions were organized in Paris and Lyon and small ceremonies took place at the Musée Claude Bernard,[80] the Académie Nationale de Médecine, the Collège de France, and the Musée d'Histoire Naturelle in Paris, but nothing to compare with the recent one-year celebration of Pasteur's death.

Pasteur's commemorations were always full of pomp and involved the whole community, especially politicians, scientists and lay people, whereas Bernard's usually only involved scientists, politicians, and intellectuals. Commemorations of Pasteur usually took place at the Institut Pasteur which was often celebrated itself as a leading French scientific institution. The institute's significant resources allow it to produce multiple ceremonies, exhibitions, and colloquia. Conversely, the Collège de France did not play the same role for Bernard. The Collège is a public institution

[78] Geison, *The Private Science of Louis Pasteur* (cit. n. 49). Interestingly, whereas some might have found Geison's book to be too critical, Bruno Latour found it disappointing. His major criticism of the book is that, "contrary to all habits of the new social history of science, Geison has chosen to concentrate only on the work of Pasteur alone, isolated from its context, the international research network, the hygiene of the time, the revolution of the medical profession, the reception of its work, the historical evolution of France" (my translation). Latour pointed out that Geison, who made aggressive inquiries into Pasteur's faults, quickly proceeded to minimize them. The only real scoop of the book, claimed Latour, was in Geison's study of two therapeutical trials that preceded the young Meister's vaccination. Nowadays, he added, those experiments would have landed Pasteur in jail. But all Geison did was admire Pasteur's luck, nerve and intuition. In the end, Latour reproached Geison for remaining fascinated by the idol he wanted to knock down. See Bruno Latour, "A propos de 'La science privée de Louis Pasteur': Les six 'révélations' du livre de Geison," in *La Recherche,* 1995, *281*:35–6. Since I found Latour himself fascinated by Pasteur, one can only conclude that there is still some need to deconstruct Pasteur's image as acknowledged by Geison himself at the end of his book.

[79] For the centennial of Pasteur's birth, schoolchildren and young women sold vignettes of Pasteur created by well-known artists; the city of Paris sent 160,000 vignettes to schools, as well as 50,000 copies of two lectures by Pasteur. Various ceremonies took place during a full week of celebration; the president of the Republic traveled in a special train throughout the country. The *Petit Journal,* the first mass daily newspaper in France, reported: "From 1800 to 1900, France committed a lot of faults and was punished by a lot of disasters. But she had Pasteur. Pasteur is dead but his work is still alive." Quoted by Raichwarg, *Pasteur: L'empire du microbe* (cit. n. 70). There was nothing comparable for Bernard's birth centennial.

[80] Located near Lyon; see citation 18.

with many professors in various disciplines, whereas the Institut Pasteur is a private institution—founded by Pasteur himself—that is devoted exclusively to biology and medicine, and is well known abroad. The creation of sattelite institutes abroad has enhanced the fame of the Paris institute and of Pasteur himself. If we consider the magnitude of Pasteur's commemorations, it might be said that he was the only celebrant until 1923. After that date, the institute appropriated the name of the hero for its own celebration.

It is common for the biographers of both scientists to compare their institutional achievement. Pasteur is always credited with the creation of the institute that bears his name, whereas Bernard is designated as the man who was not able to create an institute of his own. This "lack" could very well be an artifact of retrospective history, at least in part. If Pasteur hadn't founded an institute (which he did only at the very end of his life), nobody would blame Bernard for not having founded one. It has also been written that there was no Bernardian school. It seems to me that this point should be reconsidered, not only because the physiologist had some pupils who continued to work at the Collège de France, but also because the question of whether Pasteur created a real school has arisen. Paul Weindling recently pointed out that "certain Pasteurians such as Roux and Yersin recognized the need to deploy observational methods of German bacteriology," and that this "raised the question of whether (Pasteur) merely set certain aims and provided an institutional setting."[81] It should be recalled that France compensated scientists very poorly in the second half of the nineteenth century except for a short period after 1871, when the minister of public education, Victor Duruy, attempted to improve the situation.[82] In any case, Pasteur worked for more than thirty years not at the Institut Pasteur, but at the Ecole Normale Supérieure located on the rue d'Ulm (the "rue d'Ulm" is a common designation for this prestigious school), so that the best place to commemorate him would have been the "rue d'Ulm," had he not created the institute at the end of his life.

## *Research Styles*

These contrasting commemorative styles are somewhat related to the scientists' research styles, which in turn are related to their characters. Bernard and Pasteur differed strikingly. Bernard was rather shy, and although he was eager to have pupils, he did not have many collaborators. He was devoted to a "pure" quest for knowledge, and despite his remarkable technical skills as an experimenter, he did not display much interest in the practical and technical problems that incited Pasteur's curiosity. He showed only disdain for empirical or descriptive medicine, as well as for empirical remedies, although this disdain was partly rhetorical and served his claim that he was the founder of a new type of medicine. He was a thinker as well as a scientist, and his experimental discoveries are difficult to explain to lay people. Moreover,

[81] See Paul Weindling, "Pasteur's and Koch's Institutes Compared," in Cunningham and Williams, *The Laboratory Revolution* (cit. n. 33), pp. 170–88.

[82] See *Organization of Science and Technology in France,* eds. Robert Fox and George Weisz (London: Cambridge University Press and Paris: Maison des Sciences de l'Homme, 1980). Pasteur and Bernard kept complaining about the situation. See Ashley Miles, "Reports by Louis Pasteur and Claude Bernard on the Organization of Scientific Teaching and Research," *Notes and Records of the Royal Society of London,* 1982, *37:*101–18.

they do not fit the common representation of disease, which was mainly associated with epidemics until the discovery of sulfa drugs and penicillin.[83] This well-known fact is of primary importance when one analyzes Pasteur's fame and image. What is more, despite (or because of) his training as a physician, Bernard made no therapeutic discoveries. Like many physicians of his time, he was skeptical about the usefulness of empirical medicine, and his obsession was to introduce rationality, knowledge, and truth into medicine. Again, my contention is that these deficiencies in Bernard's research and institutional affiliation are partly artifactual and the products of retrospective evaluation. At the same time, commemorations depend upon these socially constructed images.

In contrast to Bernard, Pasteur loved public controversies, debates, and honors; he had many collaborators; he knew how to deal with the press;[84] and he created an institute that made him even more famous. Above all, he was viewed as the man who saved humanity from infectious diseases.[85] But the main point is that Pasteur, a chemist by training and not a physician, did not display as much interest as Bernard in rationality in medicine.[86] Of course, his aim was to introduce some rational basis for understanding infectious diseases, but his interest in medicine came in the latter half of his life, and he never knew a lot about it. On the contrary, from the very beginning of his medical education, Bernard was obsessed with the idea of making medicine rational and scientific. Pasteur's admiration for Bernard was certainly sincere, especially in his early appraisal of the *Introduction à l'étude de la médecine expérimentale.* He attended some of Bernard's courses at the Collège de France and at the Sorbonne and took meticulous notes.[87] But it was also in his best interest to admire such a well-known man. After all, Bernard was his senior, and he could help introduce him to the medical and biological milieu of Paris. In fact, in 1862, Bernard helped Pasteur win two prizes and be elected to the Académie des Sciences. In addition, the public tribute Pasteur paid to Bernard during his life allowed him to appropriate the idea of introducing the scientific method into medicine. Finally, I would agree with Geison's statement that "In truth, Pasteur did not think very deeply about

[83] Many books and papers have been devoted to this issue, including: William H. McNeil, *Plagues and People* (Oxford: Oxford University Press, 1977); François Delaporte, *Disease and Civilization: The Cholera in Paris, 1832* (Cambridge, Mass.: MIT Press, 1986); *Peurs et terreurs face à la contagion,* eds. J. P. Bardet, P. Boudelais, P. Guillaume, F. Lebrun, and C. Quetel (Paris: Fayard, 1988).

[84] See Bernadette Bensaude-Vincent, "Pasteur face à la presse scientifique," in Morange, "*L'Institut Pasteur*" (cit. n. 37), pp. 75–88.

[85] Pasteur's contribution to vaccination has been overestimated. He borrowed the practical idea from Jenner and others, and then went further and helped to elucidate the biological mechanisms of vaccination. This, however, is a common pattern in therapeutical discoveries: the so-called discoverers of new medications have only contributed to the discovery. See Christiane Sinding, "La célébration des découvertes thérapeutiques: Des chasseurs de microbes aux manipulateurs de gènes," in *La mise en mémoire de la science: Pour une ethnographie des rites commémoratifs,* ed. Pnina Abir-Am (Paris: Archives Contemporaines, 1997), pp. 133–52. In addition, as far as infectious diseases are concerned, a curative remedy would be found only in the early 1930s with sulfa drugs, and later with penicillin and other antibiotics. Pasteur of course, contributed to the foundation for these future discoveries.

[86] This fact was used against Pasteur by some of his contemporaries, but was also used by others to make him seem greater: physicians could only be despised for their lack of therapeutic discoveries when compared to Pasteur, who was not a physician but a "real" scientist! A hagiographic biography published in 1899 was titled "A physician without diploma: Pasteur." Xavier de Préville, *Un médecin sans diplôme: Pasteur* (Paris: Tolra, 1899).

[87] See Grmek, "Louis Pasteur, Claude Bernard" (cit. n. 37).

questions of Scientific Method," and with his comment on Pasteur's "scattered and inconsistent remarks about Scientific Method."[88]

On the other hand, contrary to Bernard, Pasteur was interested in empirical discoveries and methods, traditional skills, and practical problems. This, to me, is a major point and it is what made him great. Claire Salomon-Bayet recently pointed out that part of Pasteur's work was ethnographic because he made inquiries about techniques, vocabulary, and recipes to guilds representing dyers, coopers, silkworm breeders and others.[89] Pasteur conducted a sort of practical science that Latour described as being the ability to transfer external problems into the laboratory, and conversely to create a "mini" laboratory in settings outside of scientific institutions. This attitude was strikingly different from Bernard's, who shared the positivist assumption that knowledge comes first and application second. Along the same lines, Bernard believed that understanding physiology would directly lead to the understanding of disease. He never acknowledged his debt towards empirical discoveries, whether pharmaceutical or medical, despite his frequent use of empirical discoveries (such as curare), or practical problems of pathology (such as diabetes or carbon monoxide poisoning) as starting points for his experiments. As mentioned, he remained in touch with medical practice and physicians to a much greater extent than is usually thought.[90] If one wants to understand how Bernard worked, one should abandon stereotypical and retrospective views of the man. Latour, who is such a brilliant analyst of Pasteur, is mistaken when he compares Bernard's and Pasteur's research strategies using Bernard and his collaborators as a control group for his analysis of Pasteur:

> The control group was provided, even at the time, by the displacement of Claude Bernard. Experimental medicine was already an application of the scientific laboratory to the hospital, but the success of the Pasteurians, it will be readily admitted, bore no relation to that of the physiologists, who wanted a strict *separation* between a physiology, proud of its status as an exact science, and a medicine that was expected to change slowly . . . The laboratory of Claude Bernard at the Collège de France was in serene and polite juxtaposition with the art of medicine; that of the rue d'Ulm claimed to dictate its solutions directly to pathology.[91]

What Latour surprisingly fails to do here is apply his own methods and principles to Bernard's discourse, and to compare the way Bernard *acted* with what he *said.* Experimental medicine *was not* an application of the scientific laboratory to the hospital, nor was it in serene juxtaposition with the art of medicine. This is what Bernard pretended, but what he actually did was transfer practical knowledge and skills from medicine and surgery to the laboratory. In a note, Latour refers to an article by W. Coleman that he characterizes as "remarkable" despite its "bizarre dichotomy between 'cognitive' and 'social factors.' He ends the note with the remark, "Had Coleman studied Pasteur, this clean distinction would have been developed in an entirely different way." Taking lessons from Latour himself, I would fight against stereotypical and retrospective history, and write: had Latour studied Ber-

[88] See Geison, *The Private Science of Louis Pasteur* (cit. n. 49), p. 20.

[89] Claire Salomon-Bayet, "Introduction," in Balibar and Prévot, *Cahiers d'un savant* (cit. n. 70), pp. 5–31.

[90] Lesch, *Science and Medicine in France* (cit. n. 36).

[91] Latour, *The Pasteurization of France* (cit. n. 49), p. 61.

nard, this clean portrayal of Bernard and experimental medicine would have been developed in an entirely different way. Bernard's *Introduction à l'étude de la médecine expérimentale* was, above all, *a polemical discourse* that aimed to establish a new discipline;[92] this polemical discourse does not say anything about Bernard's "action."[93] Even philosophers agree that the physiologist's account of his own work in the book was retrospective and reconstructed in such a way as to solidify his position among scientists.

### *Applied versus Basic Science*

The praise of "pure science" is not unique to Claude Bernard. On the contrary, it was and is very common among biologists and physicians of his time and ours. It is related to the enduring controversy over basic versus applied science. Pasteur himself stated that there was no applied science, but only science and its applications, and regretted at the end of his life that he had not pursued pure science. However, what he practiced would be characterized as applied science by most scientists. The constant exchange between empirical and scientific knowledge, and scientists' reworking of practical and technical problems and successes, serve to erase the distinction between applied and pure science. But when Pasteur's commemorators—whether scientists, philosophers or historians—allude to the practical and empirical aspects of his work, they just point out that he was brilliantly able to handle this constant exchange and avoid the issue of the origin of empirical knowledge, because it would bring them to the troublesome question of the boundaries between science and nonscience, and between scientists and nonscientists.

Thus, Bernard's celebrations are more difficult to handle than Pasteur's, because scientists do not want to challenge the idea that basic science leads to true knowledge, which in turn gives birth to applied science, which leads to the solution of all human problems. But this was not the case with Bernard. His basic science did not lead to the cure of any disease, contrary to what he stated in the programmatic parts of his writing. Moreover, although Pasteur is mainly celebrated for the practical outcomes of his work, one has to emphasize the fact that discovering and acquiring detailed knowledge about microorganisms did not lead directly to the discovery of antibiotics.[94] One could go further and say that, in our time, the discovery of HIV has not, so far, led to the discovery of a remedy or even of a vaccine against the disease.

Thus, critical and even iconoclastic history, as well as the deconstruction of myths, are needed in order to bring to light such occulted issues. At the same time, commemorations gain some social justification by uniting various communities and preserving hope for the future. Critical history should not destroy memory, but at the same time it has the task of preventing commemorations from distorting history.

[92] Sinding, "Literary Genres" (cit. n. 48).
[93] Bruno Latour, *Science in Action: How to Follow Scientists and Engineers through Society* (Cambridge: Harvard University Press, 1987).
[94] This knowledge can be labeled basic or applied science, depending on the point of view of the historian but also of the scientist. If the scientist's intent is to study a pathogenic microorganism in relation to the causation of disease, it is called applied science. If the scientist wants to study DNA function in the same microorganism, it is called pure or basic science. Basic and applied science are just convenient and flexible designations without any ontological status.

# A Focal Point for Feminism, Politics, and Science in France

## The Clémence Royer Centennial Celebration of 1930

*By Joy Harvey**

VERY RARELY HAS A WOMAN SCIENTIST been celebrated with the kind of extensive ceremonies commonly held for great male scientists. Therefore, it may seem all the more surprising that a woman with no formal training in science was celebrated by noted French scientists one hundred years after her birth.

Clémence Royer, the subject of this commemoration, had been recognized in her time, as she is today, as Charles Darwin's first authorized French translator of *On the Origin of Species.*[1] Yet Royer's considerable writings on economics, anthropology, physics, and cosmology had been read primarily by members of scientific, economic, and philosophical societies. Her wider audience had come only through a regular column for a daily feminist newspaper that she wrote at the end of her life. She had no students and left neither monetary endowments nor an institution bearing her name. Why, then, was Royer commemorated at the Université de Paris (Sorbonne) in 1930?

This paper will make the argument that the Royer centennial served as an important focal point for feminists, politicians, and scientists of the Third Republic who wished to *signal* recent political events or correct what they saw as social errors or pressing wrongs. Since the topics discussed by the centennial's speakers are known but their full speeches are not, this paper relies on an examination of each participant's history and his or her connection to Royer's life and ideas.

Royer had been celebrated twice during her lifetime, once at a banquet instigated by her colleagues at the feminist newspaper *La Fronde* in 1897, and again upon her Légion d'Honneur award in 1900. At those banquets, she was honored with poems and speeches by scientific and feminist colleagues. She was praised not only by officers of the scientific organizations to which she belonged, but by representatives of her masonic group and even by Madeleine Brès, the dean of French women physicians. Her funeral had followed a similar pattern, with orations by distinguished associates from different aspects of her life, following the French style of formal obsequies.

* 29 Kidder Avenue, Somerville, MA 02144.

[1] See my biography of Clémence Royer: Joy Harvey, *Almost a Man of Genius: Clémence Royer, Feminism and Nineteenth Century Science* (New Brunswick, N.J.: Rutgers University Press, 1997).

Superficially, the centennial celebration of 1930 appeared to reiterate themes from both the funeral orations and the earlier Royer celebrations, although by this time, there were far fewer individuals present who had known her personally. Among these, however, were several important figures, many of whom were reaching the end of their own lives: Marguerite Durand, Raymond Poincaré, Adrien de Mortillet, Blanche Edwards-Pilliet, and Léopold Lacour. Others knew of her only by reputation, and the reason for their participation in the celebration is less immediately obvious. For example, why did the outstanding French physicist Paul Langevin, a close friend and former associate of both Pierre and Marie Curie, speak at this celebration?

The history of the centennial's sponsor, the feminist daily *La Fronde,* provides some clues about the celebration's background. Marguerite Durand, the newspaper's founder, was an actress and later a journalist who became a spokesperson for middle-class feminism in the late nineteenth century.[2] She founded *La Fronde* in 1897, turning to women writers including Royer, Séverine, Ghénia (Adrienne) Avril de Sainte-Croix, and Mary Léopold Lacour to strongly express anticlerical, republican sentiments typical of the liberalized Third Republic. Notable as a daily of large format that was directed and printed as well as written by women, *La Fronde* claimed a large circulation but, for financial reasons, had to be reduced to a weekly by 1903 and after 1906 it practically disappeared.

When *La Fronde* was revived in 1914 at the beginning of World War I, Durand responded to the plea of the president of the Republic (Raymond Poincaré) to slacken the paper's feminist appeals in order to reduce antagonism and unify the country during the war, promising to renew her feminism in the postwar period. After the war, the paper—now featuring male as well as female authors—continued to make appeals for women's suffrage and served as an organ for the small party of Republican Socialists. In 1926 there was a brave attempt to revive it as a weekly. During that year, it celebrated advances of women in various fields and pointed out the dangers of rising fascism in Italy. Otherwise, the articles primarily reminded readers of the large feminist archive and library that Durand had compiled and placed in the mayor's office of the fifth arrondissement of Paris. It also regularly advertised the pet cemetery Durand had created.

By the end of 1926 the newspaper was defunct, and was revived only for celebrations of individuals from its more glorious past, like a meeting held to commemorate the death of the journalist Séverine. And on 19 June 1930, *La Fronde* dedicated an entire issue to Clémence Royer with a motto taken from her writing.[3] The issue included a front-page reproduction of a painting of Royer at eighteen, and a summary of her writings and theories. A full account of the forthcoming centennial celebration was detailed on the back page.

## I. WHO WAS CLÉMENCE ROYER?

Clémence Royer was born on 21 April 1830 in Nantes, Brittany to a Breton mother, Josephine-Gabrielle Audouard. Her father, Augustin-René Royer, was an army cap-

[2] For a biography of Durand, see Jean Rabaut, *Marguerite Durand (1864–1936): "La Fronde" feministe ou "Le Temps" en jupons* (Paris: L'Harmattan, 1996).

[3] A copy of this issue is in the Marguerite Durand Library (Bibliothèque Marguerite Durand), in Paris.

tain from Le Mans. Throughout her life Clémence believed children "belonged to their mothers," and therefore she considered herself to be a true daughter of Brittany, though she lived there only for the first months of her life. Royer learned needlework and lace-making from her mother, very much in the tradition of Brittany, and her short stature, dark hair, and blue eyes gave her a strongly Breton appearance.

Royer's grandfather had been a distinguished naval captain who had fought in the Napoleonic wars and continued to serve in the merchant marine in Brittany throughout the peacetime. Her father participated in the Bourbon uprising of 1832 that was heavily supported by the Bretons. Following the Royalist defeat, Royer's family fled to Switzerland for four years. Upon their return, Clémence spent a short but disastrous period of time in a convent school, after which she was educated at home, learning mathematics from her father.

Royer became interested in republicanism during the revolution of 1848 that led to the short-lived Second Republic in France. Following her father's death in 1849 she trained as a secondary school teacher, obtaining certificates in French, music, and mathematics, and went for a year (1853–1854) to Haverfordwest in Wales to teach French and mathematics and to learn English. Upon her return, she found herself in a state of turmoil over her shaken religious beliefs, and after a short period of teaching she left suddenly for Lausanne, Switzerland, determined to make a new life for herself.

Royer spent two years educating herself in science and philosophy in the Lausanne public library. With the encouragement of the women of Lausanne, she began to lecture on science to women in 1859. In these remarkable lectures, she emphasized the need for women to embrace science as a source of knowledge and to transform it by creating a "female science." The following year she began to write for the political science journal *Le Nouvel Economiste,* edited by Pascal Duprat, a former French republican deputy in exile. Duprat would soon become her lover and the father of her child.

During this period in Switzerland, Royer won a prize offered by the canton of Vaud for her book on the income tax and began to write reviews and articles for the widely read *Journal des Economistes.*[4] She spent her thirties lecturing on science, economics, and Darwinism throughout much of Switzerland, and reporting on the Social Science Conferences held each summer in a different Swiss city. She also wrote the first draft of a melodramatic novel that contained some interesting claims about the future role of women in society, and included one section of scientific and psychological insights.

In relation to her lectures on Lamarck, Royer read Darwin's *On the Origin of Species* and immediately recognized its importance as an evolutionary theory that incorporated modern economic theories, notably those of Malthus. She then obtained the consent of her publisher, Guillaumin, to publish a translation of Darwin (although Guillaumin primarily published social science and economics books and journals). Obtaining help on the scientific aspects of the *Origin of Species* from the naturalist Edouard Claparède, who had enthusiastically reviewed Darwin's book for a Swiss journal, Royer produced a translation accompanied by copious explanatory notes and a lengthy preface. In her preface, she challenged religious authority and

[4] Her prize-winning book is Clémence Royer, *Théorie de l'impôt ou le dîme social,* 2 vols. (Paris: Guillaumin, 1862).

suggested that human beings were unwittingly and negatively influencing the normal course of human evolution by preserving the weak and the sick, and by marriage choices that favored passive, less intelligent women.

Although Darwin had authorized the translation, he had some objections to her notes and was both amused and startled by her preface. For the second edition in 1866, he made significant changes to her language, in particular replacing the word "election" which she had borrowed from Claparède with "selection." But he could not suppress her preface, which she reprinted with some modification to her severe eugenic claims.

In 1864 Royer finally finished her novel, *Les jumeaux d'Hellas.* Set in Italy and Switzerland, the book made strong claims about women's rights. It was published to no great success by Victor Hugo's publishers, although Royer would claim that her anticlerical and revolutionary sentiments had landed it on the prohibited book Index of the Catholic Church, which prevented its sale under Napoleon III's empire.[5]

In 1865 Royer went to live openly with Duprat in Italy. There she continued her lecturing on Darwinism, wrote a book on the evolution of human society, contributed to various journals and reviews, published a series of articles on Lamarck, and gave birth to her only son, René.[6] She also prepared a third edition of Darwin's *Origin,* but made the mistake of adding a new preface that criticized Darwin's hereditarian theory, pangenesis. Darwin took the opportunity to reject her edition and obtain a new French translation that, for various reasons, did not appear until three years later.

When the political climate improved in France, Royer and Duprat returned to Paris where Duprat re-entered politics and established a short-lived newspaper, to which Royer contributed. The two were never able to marry, since Duprat was already married and the divorce laws did not come into effect until just before his death.

In 1870 Royer was elected to the previously all-male Société d'Anthropologie, headed by Paul Broca. This influential scientific society engaged in a debate on Darwinism in 1870, to which Royer contributed her views on evolution. Years later, at a banquet in her honor in 1897, her colleague Charles Letourneau, the secretary-general of the Société d'Anthropologie, recalled the uproar that had accompanied the election of this woman who had "shattered the windows" with her Darwin translation. Royer contributed a large number of papers and discussions to the anthropological society over the following years, most of which were published in the society's bulletins and in Broca's own *Revue d'Anthropologie.* One remarkable discussion of the place of women in French society, "Sur la natalité," was suppressed by the society and has only recently come to light. This contribution challenged many elements in the civil code, questioned the presumed passivity of women, and explained the reduced growth of the French population as due to the decision of women to limit the number of their children through birth control or abortion.[7]

[5] Clémence Royer, *Les jumeaux d'Hellas* (Paris and Brussels: A. Lacroix and Verboeckhoven, 1864).

[6] Clémence Royer, "Lamarck: Sa vie, ses travaux et sa système," *La Philosophie Positive,* 1868, *3:*173–209; 242–72; *4*(1869):5–30.

[7] See Joy Harvey, "'Strangers to Each Other': Male and Female Relationships in the Life and Work of Clémence Royer", in *Uneasy Careers and Intimate Lives,* ed. Pnina Abir-Am and Dorinda Outram (New Brunswick, N.J.: Rutgers University Press, 1987), pp. 147–71; 322–30. For the text of this paper, see Albert Ducros and Claude Blanckaert, "'L'animal de la création que l'homme connaît le moins': Le mémoire refusé de Clémence Royer sur la femme et la natalité," *Bulletins et Mémoires*

Royer wrote several other books, including one on ethics and society in response to Herbert Spencer's study of ethics and evolution. She began to lose her audience, partly because the Paris police and the minister of culture deliberately prevented her from commenting on morality before public audiences, although she was still free to speak behind the closed doors of scientific societies.

In the mid-1880s, Royer lost her aging companion Duprat when he traveled as a diplomat to Chile, fell ill there, and died during his return voyage to France. She raised their son by herself, sending him to the prestigious Ecole Polytechnique to train as an engineer. Threatened by increasing poverty and aided only by a small grant from the Ministère d'Instruction Publique, Royer submitted a number of lengthy monographs on scientific and political questions to various branches of the Institut de France (the Académie des Sciences and the Académie des Sciences Morales et Politiques). She won only a couple of minor awards, for her submissions on public assistance to the poor and on the history of atomism.

Reduced to an almost pitiful level of poverty, Royer was rescued from increasing social isolation in the 1890s by a husband-and-wife journalist team, Léopold and Mary Lacour. The pair drew attention to the manner in which her repeated submissions for prizes made her one of the few women in France "knocking at the doors of the Institut de France."[8] With the help of her new friends, Royer obtained a position in a retirement home in a genteel suburb of Paris. She found a new community of women supporters who enlisted her aid in forming a new feminist newspaper, *La Fronde,* for which she began to write a daily column on scientific and political themes. Royer continued to write articles on science and the structure of matter for various scientific journals, and corresponded with mathematicians and philosophers on these topics. In 1900, she pulled together her compositions on the structure of matter and on cosmology in a massive volume, *La constitution du monde.*[9] In 1900 she was awarded the Légion d'Honneur and two years later she died, surrounded by friends and colleagues like Mary Lacour, Ghenia Avril de Sainte-Croix, and Marguerite Durand, founder and editor of *La Fronde.*

Royer's central ideas, as repeated in her articles for scientific, economic, and feminist journals, emphasized a number of themes common to lay republicans of the late-nineteenth century. Among these were her pacifist and anticlerical beliefs and her strong feminism that insisted on social and professional rights for women while hesitating at universal women's suffrage. Above all, she emphasized the importance of science as a basis for philosophy and as a guide to personal morality.

## II. FRENCH FEMINISM AND ROYER

American and English feminists have often conflated feminism with women's suffrage. But the two were not interchangeable in France, where most women were educated and influenced by the conservative Catholic Church, and the fear of losing the Republic paralyzed feminists from advocating for full voting rights. The term feminism was first used by the French social philosopher Charles Fourier, but came

---

*de la Société d'Anthropologie,* n. s., 1991, *3:*131–44; and Harvey, "Appendix," *Almost a Man of Genius* (cit. n. 1).

[8] Léopold Lacour, "Les femmes à l'institut," *L'Evenement,* 18 December 1892. See also Harvey, *Almost a Man of Genius* (cit. n. 1), pp. 167–8.

[9] Clémence Royer, *La constitution du monde: natura rerum* (Paris: Schleicher, frères, 1900).

into general use only after Hubertine Auclert called herself a feminist in 1882 and the first feminist congress (so-called) was held in 1892. A call for women's legal rights and for equality—and even for recognition of the superiority of women—became the clarion cry of early French feminists.[10] (Modern historians of French feminism speak of "feminisms," acknowledging the multiple aspects of the movement.)

Claire Goldberg Moses, among others, has pointed out the discontinuity in French feminism, which fluctuated with the politics of the time.[11] Feminist politics became stabilized only after 1879, when liberal republicanism permitted legislative changes in divorce laws and reorganized secondary education for young women. During the five years following Royer's death, women's rights were liberalized and extended through changes in the civil code. Royer had argued for these changes—the right of mothers to have equal say as fathers over their minor children; the right of illegitimate children to be recognized by and inherit from their fathers; and the right of married women to be economically independent—since the early 1860s in her articles for the *Journal des Economistes,* in her prize-winning study of taxation, in her discussions before the Société d'Anthropologie, and even in her novel *Les jumeaux d'Hellas.* Like the Saint-Simonists of an earlier period, she endorsed motherhood as a duty for women, but unlike both her scientific associates and many of her contemporary feminists, she argued that women held a genius within themselves that could evolve to advance the entire human race, extending to a possible social reorganization in which women would take the leading role, rewrite the civil code, and break with the masculine errors of the past. To do this, they needed to hold important economic positions. Royer opposed the view of woman as passive, unintellectual, and inactive, arguing throughout her life that "woman is not made like this."[12]

During the period that she wrote for *La Fronde* (1897–1902), Royer used both the noun and the adjective "feminist" in her letters. She wrote to Léopold Lacour, praising his book *L'humanisme intégrale* (which argued for the equal role of male and female) as "a real trap in which to catch the antifeminist wolves. . . . Only a man could have told them such blunt truths."[13] She regretted, however, that he had omitted the work of Jenny d'Héricourt. "Only the rarity [of d'Héricourt's book] excuses the ignorance of our contemporary feminists about this fundamental classic of feminism," she wrote.[14]

### III. THE CLÉMENCE ROYER COMMEMORATION

Two weeks before the centennial, held 19 June 1930, the critic Maurice Wolff hailed the forthcoming "centennial of a savant, Clémence Royer" in a major critical re-

[10] Christine Bard, *Les filles de Marianne: Histoire des feminismes 1914–1940* (Paris: Fayard, 1995).

[11] Claire Goldberg Moses, *French Feminism in the 19th Century* (Albany, N.Y.: State University of New York Press, 1984).

[12] See Harvey, *Almost a Man of Genuis* (cit. n. 1), p. 125.

[13] Clémence Royer to Mary Léopold Lacour, undated [1897] letter beginning: "Carina, René m'a donné une machine à écrire." Her reference is to Léopold Lacour, *'humanisme intégrale: Le duel des sexes* (Paris: Stock, 1897).

[14] A subsequent letter from Clémence Royer to Léopold Lacour in 1897 added detailed information about the men and women, as well as booksellers, who might own a copy of d'Hericourt's book, *La femme affranchie: Réponse à MM Michelet, Proudhon, E. de Girardin, A. Comte et aux autres novateurs modernes* (1860). For a discussion of this see Harvey, *Almost a Man of Genius* (cit. n. 1), p. 174.

view.[15] Wolff began his article by celebrating Royer's intelligence with the strong claim that "no life was more qualified to give a victorious answer to the principal objections about the capacity of the female brain or to serve as witness that some chosen women equal the best masculine minds, capable of embracing the immensity of science and aiding the scientific patrimony of humanity through their research!"[16] Wolff mentioned Royer's early interest in "solitary thought" and mathematics as well as her resolution to educate herself in the Lausanne library, and "to organize all science within her brain in an admirable manner." He discussed her lectures to women, her publication on taxation and, above all, her translation of Darwin's *On the Origin of Species.* Although Wolff apparently based his article on Albert Milice's 1926 biography (even to the extent of characterizing her novel, *Les jumeaux d'Hellas,* as a paean to science), he provided none of his sources. He also followed Milice's interpretation of Royer's social ideas by emphasizing a hierarchy of human beings based on intelligence.

Wolff reiterated Royer's appeal for a future matriarchy that would give mothers more control over their children's development and discovered in this "an echo of that rancor of the scientific woman [to] whom the masculine scientists refused too long to render justice."[17] Wolff ended his article with a discussion of Royer's idea of individual, intelligent atoms and quoted her "lyrical peroration" to science and reason from her novel: "O science, O truth, O reason, laws of thought and being, you are my gods. I adore you" (a quotation that would also grace the centennial issue of *La Fronde*). He also endorsed the commemoration by adding that "Clémence Royer certainly merited the voice of those authoritative savants who will come to exalt her immense scientific labor at the Sorbonne, which never welcomed her during her lifetime . . .'"[18]

*L'Intransigeant,* a newspaper that supported radical republicanism and had taken up many causes that *La Fronde* had earlier supported, carried reports on many of the feminist activities of Royer's old friends in the months before her celebration. It announced the creation of a new feminist organization by Ghénia Avril de Sainte-Croix, Les Etats Générales Féminines, and noted her attendance at the International Feminist Congress in Vienna at the end of May. The newspaper also quoted from a speech Léopold Lacour gave for the fiftieth anniversary of a Parisian school that trained young women in technical and vocational arts. The day before the Royer centennial, the newspaper briefly cited Wolff's article and mentioned the speakers who would be presenting their memorials at the Sorbonne the next day. Unfortunately, like most of the other Parisian papers, *L'Intransigeant* failed to cover the actual event.

A few days before the centennial celebration, *La Fronde* published its special issue on Royer. This included short biographical articles, excerpts from Royer's writings (particularly emphasizing the articles she had written for *La Fronde*), and her basic credo. Rather than using the better-known photographs of Royer showing the forthright and rather strong-jawed woman of her more mature years, the newspaper chose

[15] Maurice Wolff, "Clémence Royer, centennaire d'une savante," *Revue Politique et Litteraire, Revue Bleu,* 7 June 1930, *68:*324–9. It is interesting to note that this appeared in the political and literary journal, *Revue Bleu* and not in the related scientific journal *Revue Scientifique, Revue Rose.*

[16] Ibid., p. 324.

[17] Ibid., p. 329.

[18] Ibid.

instead a flattering portrait of her painted at the age of eighteen in the style of her heroine of that period, the actress Rachel. The program for the commemoration was emblazoned on the back of the journal in large type.

The commemoration was held in the Grand Amphitheatre of the Sorbonne at 8:45 p.m., with Paul Painlevé (a member of the Institut de France, the Chambre des Déutés and a former premier of France) presiding. The driving force behind the commemoration had been Marguerite Durand working in cooperation with "the great feminist associations," as the program put it: Avril de Sainte-Croix, president of the Conseil National des Femmes Françiases (CNFF); Cécile Brunschvicg, president of the Union Françaises pour le Suffrage des Femmes (UFSF); E. Fonseque, president of the Société pour l'Amélioration du Sort de la Femme; and Maria Vérone, one of the first female lawyers and president of the Ligue Française pour le Droit des Femmes (LFDF). The organization committee also included Harlor, an associate of Marguerite Durand and later the first archivist of her feminist library; Royer's biographer, Albert Milice; a representative of Le Droit Humaine; Royer's former masonic lodge; and others. The program listed the members of the celebration's honorary council: the former president of the Republic, Raymond Poincaré; the president of the Sénate, Paul Doumer; and the head of the Légion d'Honneur, General Dubail.

The celebration began at the Sorbonne with a welcome from the mathematician and politician Paul Painlevé. His greeting was followed by ancient Greek songs adapted from those found in Delphi and sung by the Université de Paris chorale under the direction of Mme. Samuel. The songs were followed by a description of Royer's Breton background delivered by a merchant marine representative from Nantes who was familiar with her grandfather's role in the Napoleonic wars and her father's military history.[19] Marguerite Durand followed with a description of Royer as a woman, a journalist, and a feminist, as she had known her in her final seven years.

Durand was followed by Charles Filliard who, as dean of the faculty of letters at the Académie de Lausanne, had known Royer's early lectures and writing in Switzerland. It was unforgettable that Royer had placed second to the great socialist and antifeminist Pierre Joseph Proudhon when she won the economics prize on taxation in Lausanne.

Etienne Rabaud, professor of experimental biology at the Sorbonne, then spoke of Royer as a biologist and of her contributions to evolution through her translation of Darwin. Like many French scientists, he welcomed the preface to her translation of *On the Origin of Species,* especially since it reintroduced the neglected French scientist Lamarck into Darwinian evolution.

A second musical interlude followed, with the university chorale performing excerpts from Schumann's opera *Faust.* When the commemoration resumed, it was to hear Adrien de Mortillet, honorary president of the Prehistoric Society. Mortillet, who was active in preserving prehistoric monuments throughout France, knew of Royer's interest in prehistoric caves and the tools of early humans.

The atomic physicist Paul Langevin, a friend and former colleague of Marie Curie, spoke next. He focused not on Royer's writings on cosmology and physics, which she had published in her last few years, but on her deep dedication to rational thought and her opposition to mystical explanations of the universe. Like Royer,

[19] No other information on this gentleman (Monsieur Farineau) is available.

Langevin believed that an adherence to rational explanation and scientific thought provided an important model for women's education, which had been under the control of the Catholic Church during the nineteenth century.[20]

An important woman physician and feminist, Blanche Edwards-Pilliet, then spoke on Royer's wide-ranging knowledge, calling her an *encyclopédiste.* (This word had also been used to describe Royer by Georges Clemenceau, who in 1897 praised her "precise clear thought" as being similar to the finest female minds of the eighteenth century.[21]) Edwards-Pilliet's presentation was followed by a series of Serbian dances from the opera *La Fontaine de Pritchina* performed by the university chorale.

Albert Milice then spoke of the reception of Royer's scientific writings within "official science." Four years before, in 1926, Milice had published a long biography of Royer that celebrated not only her life and work but what he termed "Royerian philosophy." The next speaker was Léopold Lacour, who with his wife Mary had alerted the public to Royer's work in the early 1890s.[22] For the celebration Lacour discussed Royer's philosophy as well as her attempts to enter the Académie des Sciences Morales et Politiques, a part of the Institut de France.

During the next break in the program, a female and a male soloist performed three folk songs (including "Le Canard Blanc," usually sung in Canada but also a popular folk song of Brittany). The poem *La Recherche* (Research), written by the Belgian Emile Verhaeren (1855–1916), was recited by Colonna Romana, a member of the Comédie Française. The celebration ended with the chorale performing a song from the opera *Hippollyte and Aricie* by the eighteenth century composer Jean-Philippe Rameau.

## IV. MOTIVATIONS FOR COMMEMORATION

Since a commemoration often reflects concerns that extend beyond that of the individual or event immediately celebrated (as many of the papers in this volume have emphasized), it is worth analyzing what additional motives each speaker had to appear in public to praise Royer and celebrate her life. Although Royer's ideas and writings were forgotten or ignored, for the most part, by the general public (with the exception of her Darwin translation that continued to be reprinted), some of the participants recalled being influenced by the clear and direct ideas of this unusual woman. Others had less obvious motives that are worth analyzing in some detail.

The presence of politicians like Raymond Poincaré, who sat on the committee of honor, was guaranteed by his long friendship with Marguerite Durand who, along with many other feminists, had urged him to support women's suffrage just before the First World War. Although he and other members of the Chambre des Députés had voted for women's suffrage, the conservative Sénat blocked any possibility of it becoming law. For a variety of reasons, Poincaré chose not to use his power during his presidency to push the decision through, although he depicited himself publicly as a feminist. During the war, all the feminist organizations agreed to downplay their demands and support the government. In the postwar period, feminist groups revived

[20] Bernadette Bensaude-Vincent, *Langevin: Science et vigilance* (Paris: Belin, 1987).

[21] Georges Clemenceau, "Madame Clémence Royer," *L'Illustration,* 13 March 1897, 194–95.

[22] One of the few biographical sources on Lacour is a pamphlet written by a close friend of Marguerite Durand, Harlor (Thérèse Hammer), later the first archivist of the Marguerite Durand library. Harlor, *Léopold Lacour: Biographie critique, suivie d'opinions* (Paris: E. Sansot, [1914]).

the debate and sent delegations to Poincaré requesting his support. Only one year before the Royer commemoration, the League for the Rights of Women (Ligue pour les Droits des Femmes), of which Blanche Edwards-Pilliet served as an officer, led a march to once again urge Poincaré, as president of the Conseil de Cabinet, to support votes for women. His public association with the Royer celebration sponsored by well-known feminists like Durand, Avril de Sainte-Croix, and Edwards-Pilliet appeared to reassure his critics.[23]

In her speech, Marguerite Durand reminded the audience of the importance of late nineteenth-century feminists and their contributions. (A major speaker at Royer's funeral, Durand had extolled her as a woman as well as a thinker and laid bouquets of forget-me-nots on her coffin. Just the previous year, in 1929, Durand had sponsored a commemoration, on the occasion of the death of another longtime contributor and friend, the radical writer Séverine.) Durand had an organization to promote as well, the feminist library that she had just established in Paris. This library had an archive representing many important French women, especially those associated with *La Fronde,* and it included correspondence and both published and unpublished writing by Royer.

Durand saw no discontinuity between Royer's feminism and her own. The feminist position Royer had adopted in the late nineteenth century was echoed by Durand and most of her contemporaries. They generally agreed that "premature" women's suffrage could risk the Republic by placing the government into the hands of conservatives and Catholic militants. But by the second decade of the twentieth century, both Durand and Avril de Sainte-Croix strongly supported the vote for women as girls' education was secularized and the Republic appeared secure.

When Blanche Edwards-Pilliet rose to speak on Royer's "encyclopedic" knowledge, she also expressed her long-held admiration for her. As a young physician, she had attended the earlier Royer celebrations: the banquet sponsored by *La Fronde* in 1897 and a second banquet three years later hosted by the Bleus de Bretagne (an honorary society of republican women from Brittany) on the occasion of the presentation to Royer of the Légion d'Honneur cross. Edwards-Pilliet had been a colleague of Royer's in the Société d'Anthropologie, having been elected to the society soon after she presented her thesis on hemiplegia, written under Jean Martin Charcot in 1888. She listened to Royer's regular discussions before the society and attended her public lectures on mental evolution in 1889 given as part of the society's annual "Conferences Transformistes" (transformism being the French term for evolution that the society had adopted, preferring it to the term Darwinist). Edwards-Pilliet had personal knowledge of Royer's wide-ranging interests in anthropology, biology, prehistoric archaeology, economics, and cosmology and admired her struggle to have her ideas heard. As a young medical student and externe of the hospitals of Paris, she had been involved in her own struggle to change attitudes towards women physicians. She had agitated for four years to obtain permission for herself and another young woman to take the examination for the prestigious *internat* (equivalent to a hospital residency) in 1888. She succeeded in obtaining a position in one of the Paris hospitals, although not without public outcry, and taught a successful course

[23] Steven C. Hause with Anne R. Kenney, *Women's Suffrage and Social Politics in the French Third Republic* (Princeton, N.J.: Princeton University Press, 1984). Poincaré had also provided legal advice to both Marie Curie and Paul Langevin at the time of the Langevin-Marie Curie affair.

for nurses, replacing her husband in that position following his death. She also developed a flourishing private practice and helped to found an organization that provided free breast milk for infants.

Edwards-Pilliet, like Royer, spoke at the 1889 Congress for the Rights of Women (Congrès des Droits des Femmes) in Paris. She later chaired the medical section of the Congress on Women's Work (Congreś des Oeuvres et des Institutions Feminines) in one of the two feminist conferences held in Paris in 1900, urging the improvement of women's education at all levels and unsuccessfully advocating for the creation of a hospital organized by women based on the English model of the New Hospital for Women, established by Elizabeth Garrett Anderson.[24] She also led a delegation of women from that congress to the bedside of the ailing Royer, honorary president of the congress, in order to pay respects to her as "the greatest French woman scientist."[25]

The biologist Etienne Rabaud chose to speak at the centennial about Royer as an evolutionary thinker. Trained in Toulouse and then in Paris, Rabaud had obtained doctoral degrees in both medicine and science. He was also a student of the prominent neo-Lamarckian Alfred Giard, the first occupant of the chair of evolution at the Université de Paris. Although Giard had died in 1908, Rabaud followed in his footsteps: he acted as an adjunct director for the marine biological station founded by Giard at Wimereux, took over the editorship of his journal *Bulletin Scientifique de France et Belgique,* and in 1923, obtained a chair in experimental biology at the Faculté des Sciences at the Université de Paris. His teacher's enthusiasm for Royer may help to explain Rabaud's willingness to celebrate her.

Many years earlier, Giard, who had been unable to attend the Royer banquet organized by Ghénia Avril de Sainte-Croix and Marguerite Durand in 1897, sent a letter to Avril de Sainte-Croix, in which he praised Royer's translation of *On the Origin of Species.* The book, he wrote, had "an enormous effect upon my life and my science. Without ever having met her personally, I contracted an enormous debt of recognition towards Madame Clémence Royer. But I do not forget the services of a more general order that this anthropological writer has rendered to science, to the socialist ideal, to free thought, and I regret not being able to join on the tenth of May to witness my gratitude and lively admiration."[26]

Aside from his wish to remind his audience of his teacher's evolutionary ideas, in 1930 Rabaud was in the process of writing a new book on evolution, *Le transformisme.* In his later writings, he went even further than Royer or Giard in questioning the usefulness of Darwinian evolution in comparison with Lamarckian concepts, and he severely attacked adaptation as providing too teleological an explanation for change.[27] He even hesitated at depicting plants and animals as originating from the same primordial form.[28] Yet Rabaud had another personal reason for joining in this

[24] For a detailed account of her life by two of her grandchildren, see François Leguay and Claude Barbizet, *Blanche Edwards-Pilliet: Femme et médecin, 1858–1941* (Le Mans: Editions Cénomane, 1988).

[25] Harvey, *Almost a Man of Genius* (cit. n. 1), p. 181.

[26] Alfred Giard to G. Avril de Sainte-Croix, undated [February 1897]. For more information on Giard's placement in the first chair for teaching evolution in France, see Marc Vire, "La création de la chaire de l'évolution des êtres organisées à la Sorbonne en 1888," *Revue de Synthèse,* 1979, *95–96:*377–91. For a biography of Giard see Georges Bohn, *Alfred Giard et son oeuvre* (Paris: Mercure de France, 1910).

[27] See *La Grande Encyclopédie,* Larousse, 1978, s.v. "adaptation".

[28] Etienne Rabaud, *Introduction aux sciences biologiques* (Paris: Armand Colin, 1941).

celebration that stemmed from his association with Langevin and Painlevé in a new organization devoted to rationalist thinking, the Union Rationaliste (about which more will be said later).

A very different motivation to speak may have driven Adrien de Mortillet who, like his famous father, the prehistoric archaeologist Gabriel de Mortillet, had been a colleague of Royer's in the Société d'Anthropologie. His father, who had appreciated the cogency of some of her ideas, often criticized her tendency to jump to speculation before full evidence was available. The son, while not wishing to revive some of his father's criticisms, took this opportunity to praise a woman anthropologist who had stood with his father's scientific materialist friends against the growing importance of Catholic anthropology in France.[29] Royer had been an early writer on human evolution in her book *L'origin de l'homme et des sociétés,* publishing on the topic even before Darwin. Adrien de Mortilett was also interested in gaining wider support for the protection of prehistoric sites in France, a topic that had also interested Royer. She had often visited caves and grottos and spoke on the subject of human prehistory before the Société d'Anthropologie.

For Albert Milice, the commemoration provided a crucial opportunity to establish "Royerian philosophy" as he had developed it in his biography. Milice's father had known Royer, as he had not, and Milice had published two previous studies of Royer's thought. The first was an adoption of her view of the fluid atom in his book *La constitution de l'univers,* published under his pseudonym Aristede Pratelle in 1912.[30] Following the First World War, he had praised her view of Darwinism as pacifist in tone and her term *concurrence vitale* (vital competition) as far preferable to the Darwinian "struggle for survival" that the German scientists had adopted as a justification for war. In his 1926 biography, he had interpreted Royer's idea of *élection naturelle* (her translation of "natural selection") as an appeal to the development of a human intellectual elite that would rise to the top of society and rule as a meritocracy. He also saw her cosmologies and theories of a fluid atom not as speculative theories providing occasionally stimulating ideas, but as major scientific truths. Milice had invested a great deal of intellectual and personal energy in establishing a school of Royerian philosophy. The commemoration of Royer offered an opportunity to make her name (and, incidentally, his own) more widely known. (In any event, he died in poverty and alone during the German occupation of France.)

Léopold Lacour took the opportunity of the centennial to express his interest in feminism, but also to pay homage to an old friend. Many years before, when Royer received her belated honors, she had expressed her gratitude to Lacour for having helped her to escape from poverty in 1892 when (as she put it) he offered her the "Herculean golden bough" after she had "descended alive into Hades."[31] The La-

[29] For a vivid evocation of Gabriel de Mortillet and his scientific materialism, see Michael Hammond, "Anthropology as a Weapon of Social Combat in Late Nineteenth Century France," *Journal of the History of Behavioral Sciences,* 1980, *16:*118–32.

[30] Aristede Pratelle, *La constitution de l'univers: L'atome fluide, moteur du monde, eléments de philosophie dynamiste* (Paris: P. De la Salle, 1912). According to Jean Maitron, the last name Milice was also a pseudonym. He was born Albert Préau in Paris in 1877, and published on the workers' movement and on scientific topics under the name Pratelle, *Dictionnaire biographique du movement ouvrier français,* ed. Jean Maitron, vol. 14, 1976. (Préau's exact date of birth and death were added in a later volume [1991]).

[31] Clémence Royer to Léopold Lacour, 18 August [1900], Marguerite Durand Library, Paris. Quoted in Harvey, *Almost a Man of Genius* (cit. n. 1).

cours brought her name before the French public when she thought herself forgotten. Royer, in turn, advised Léopold on further feminist readings upon the publication of his book, *L'humanisme intégrale.*[32] Léopold and Mary exchanged visits and letters with Royer, and Mary rushed to her side in her final hours. Like Avril de Sainte-Croix and Durand, the two Lacours were present at her funeral.

Paul Langevin's motivation was more complex. In celebrating Royer not as a physicist but as a rationalist, he emphasized the goals of a group he had just helped to found. (This organization, the Union Rationaliste, was dedicated to the promotion of science and rationality among the general public. Supported by Henri Roger, dean of the Paris Faculté de Médecine, the organization included among its first adherents a number of professors at the Faculté de Médecine, the Sorbonne, and the Collège de France. Langevin had just delivered three lectures on modern developments in physics before the Union Rationaliste in May and June of 1930.)[33]

In praising Royer, Langevin was able to celebrate a woman who had made rationality and science her philosophy of life. The year after his speech at the Royer centennial Langevin suggested—Royer had declared in her first Darwin preface and later in her book *Le bien et la loi morale*—that science could serve as a basis for a new, more humane morality to replace Christian belief. Einstein said of Langevin at his death that "Reason was his religion; it would bring not only light but redemption as well," a comment that could be aptly applied to Royer.[34]

There are strong similarities between Royer's view of Darwinism and the description of evolution Langevin promoted in 1934. Both emphasized nonviolent competition.[35] But he may have been less eager to embrace other aspects of Royer's thinking. She had, for example, been the first writer in France (and perhaps in the world) to take evolutionary theory, for good or ill, into the realm of human eugenics. In the preface to the first edition of her Darwin translation, Royer suggested not only that marriage practices needed to be rethought in the light of natural selection, but that the care of the weak "weighed on the arms of the strong."[36] (The potential for the misuse of social Darwinism would be demonstrated only too graphically in the decade to come.)

Given the presence of Langevin at the celebration and his well-known link to Marie Curie, it is worth questioning to what degree the celebration of Royer as a woman scientist and profound thinker might have served to remind the audience

[32] Lacour, *L'humanisme intégrale* (cit. n. 13). Her letter is quoted in Harvey, *Almost a Man of Genius* (cit. n. 1), p. 174.

[33] For a discussion of Langevin's lectures and the Union Rationaliste, see *Mercure de France* 11 July 1930, *68:*436–9, and Bensaude-Vincent, *Langevin* (cit. n. 20), p. 121–7, which discusses and lists Langevin's lectures on this topic.

[34] Albert Einstein, "Hommage à Langevin," quoted in Bensaude-Vincent, *Langevin* (cit. n. 20), p. 121. See also Paul Langevin, "Science et laïcité," *Groupe fraternel de l'enseignement* (Paris: Deshayes, 1931), pp. 7–28.

[35] Royer's view of *concurrence vitale* was emphasized by Aristide Pratelle in 1919 as a more humane view of evolution. See Aristide Pratelle, "Clémence Royer, notice biographique," *Revue Anthropologique,* 1918, 263–76. Albert Milice (the same writer, according to Geneviève Fraisse) made the same point in his extensive biography of Royer: Albert Milice, *Clémence Royer et sa doctrine de la vie* (Paris: J. Peyronnet, 1926). For the identity of Milice and Pratelle see Geneviève Fraisse, *Clémence Royer* (Paris: La Découverte, 1985). Langevin's comments on evolution are cited in Bensaude-Vincent, *Langevin* (cit. n. 20), pp. 118–19.

[36] Clémence Royer, "Preface" to Charles Darwin, *De l' origine des espèces ou des lois du progrès chez les êtres organisés,* 1st French ed., trans. Clémence Royer (Paris: Guillaumin et Cie & Victor Masson, fils, 1862).

(if only indirectly and by analogy) of Curie, who won two Nobel Prizes but whose modesty prevented a similar celebration. Marie Sklovdowska Curie was, of course, Polish by birth, and she had been unpleasantly attacked at times by the French press as a "foreign woman." Just as Royer had unsuccessfully beat at doors of the Institut de France, Marie Curie had been proposed for membership in the Académie des Sciences in the fall of 1910 (which, had it occurred, would have made her the first woman in the Institut de France), but her candidacy failed narrowly partly due to a scandal aggravated by the Paris yellow press linking the widow Curie to her friend and colleague, Paul Langevin, a married man. Curie regularly published in the journal of the Académie des Sciences, but the academy required that her articles be formally presented to the academy by a member, which was done by her close colleague and friend Jean Perrin. When Royer similarly found that no submissions to or prizes from the Académie des Sciences would result in her acceptance into this very conservative organization, she commented that she thought she would not have gained entrance even if she had been a man, because she was not content to nibble at "a single leaf of the tree of science. I have chewed on several, plentifully."[37]

If, indeed, Royer's celebration evoked Marie Curie, one should remember that Curie, like her husband Pierre, had avowed a dislike for personal commemorations. The only exception she had made was a ceremony held during her trip to the United States in 1921 to receive a gift of radium from American women. A second trip to the United States had taken place only a year before the Royer centennial, that time to obtain radium not for the French radium institute but for the Warsaw radium institute recently named after her. When less than seven years before, on 26 December 1923, a celebration of the twenty-fifth anniversary of the discovery of radium took place, the focus of commemoration was on the element, not on Marie and Pierre Curie, who had made the discovery. Marie Curie had used that opportunity to make her only public commemoration of her husband, urging support for the newly created French radium institute.[38] (Held in the same room [the grand amphitheater of the Sorbonne] as the Royer celebration, the radium anniversary was attended by the president of the Republic, who presented a substantial governmental pension to Marie Curie. Speeches were delivered by the noted atomic scientist Jean Perrin and the Dutch Nobel Prize winner in physics, Hendrik A. Lorentz, among others.)[39]

Aside from Langevin's motives to promote science to the general public and the inclusion of science in women's education, he was working with Jean Perrin to develop an important series of memos to the government that requested support for science as a crucial part of social action. This advocacy resulted in the creation of the Centres Nationales Recherches Scientifiques in the 1930s, a government organization that still funds most of French science.

Paul Painlevé, who chaired the Royer celebration, was another scientist-politician.

[37] Quoted in Harvey, *Almost a Man of Genius* (cit. n. 1), p. 168.

[38] For a discussion of Curie's attitude towards commemoration, see Susan Quinn, *Marie Curie: A Life* (Reading, Mass.: Addison-Wesley, 1995), p. 422. For a description of this celebration see Eve Curie, *Madame Curie* (New York: Doubleday, 1937), pp. 345–6; and Richard Reid, *Marie Curie* (New York: Saturday Review Press, 1974). This anniversary offers us an example of another celebration of a physical object (radium), like the celebration of the discovery of the tubercule bacillus by Koch, discussed in this volume by George Haddad.

[39] It is worth noting that three women were brought into the French cabinet in 1936, ten years before women's suffrage was declared. One of them was Irene Joliot-Curie, daughter of Marie Curie.

Like Royer, Painlevé had supported Alfred Dreyfus (the target of the notorious "Dreyfus affair") in the late 1890s when defining oneself as pro-Dreyfus in France was an important statement about one's liberal orientation. During World War I, Painlevé was involved in the government both as premier and as minister of war. He transferred his old friend, Paul Langevin, into the Ministère des Inventions to assist on scientific problems. Painlevé also had a strong personal link to Langevin. At the time of the Langevin-Curie affair, he had gone out of his way in one of his speeches to praise the intelligence of women and of Marie Curie in particular, comparing her to the heroines of classical antiquity.[40] He had served as a second to Langevin during his duel with the journalist Gustave Téry, who had accused Langevin of "hiding behind Marie Curie's skirts." For the 1930 Royer centennial, Painlevé again joined with Langevin to honor a woman.

Many commemorations in France serve to remind former colleagues and students of the importance of their affiliation to the sponsoring institution, and to bring together a number of generations of scientists. This was true, for example, of the centennial commemoration of the birth of the physiologist Etienne Jules Marey, which was celebrated by the Académie de Médecine in May 1930, just before the Royer celebration. The academy used the Marey celebration (as it later did with the Paul Broca celebration) to remind the public of its important role in the life of French science. (Although there would be no mention of it during his celebration, Marey had a link to Royer. During the last ten years of her life he regularly wrote on her behalf for receipt of an annual grant from the Ministère d'Instruction Publique.)[41]

Unlike her former colleague Paul Broca, who founded both institutions and journals, Royer left behind no durable institution.[42] Broca not only founded the Société d'Anthropologie in 1859, but he was an important professor in the Faculté de Médecine, serving as both a neurosurgeon and neuroanatomist and as a member of the Académie de Médecine in Paris. He is known even today for having pinpointed the area of the brain associated with speech through both clinical and pathological observations. Shortly before his death in 1882, he was appointed to the Sénat of the Third Republic, where he actively supported the revitalization of secular secondary schools for young women. A dramatic centennial commemoration of his death in 1982 included, in one location, an international meeting of anthropologists and prehistoric archaeologists sponsored by the Société d'Anthropologie de Paris, and a concurrent celebration at the Académie de Médecine.[43] A bronze medal was even struck in his honor by the government mint.

In the case of Royer, neither the feminists who organized the celebration nor the scientists who participated in it had formal institutions to celebrate (other than the

[40] See Quinn, *Marie Curie: A Life* (cit. n. 33), p. 323.

[41] See Harvey, *Almost a Man of Genius* (cit. n. 1), p. 164.

[42] Royer attempted to start a society to study philosophy and morality in the 1880s. Although it lasted a few years and published a journal, it eventually moved away from Royer's directorial control and she abandoned it. Broca established the Société d'Anthropologie and its accompanying bulletins as well his own anthropological review, *Revue d'Anthropologie.*

[43] *Paul Broca, 1824–1880: Exposition* (Paris: Académie Nationale de Médecine, 1980). The Société d'Anthropologie also published a commemorative publication the same year. The important sponsoring role of these institutions to which Broca belonged is evident in the centennial commemorations for other famous French scientists such as those for Louis Pasteur, discussed in this volume by Christiane Sinding, and for Jean Martin Charcot, analyzed here by Daniela Barberis. Here, memorialization plays a dual role of reification of the organization and reinforcement of the memory of the individual.

nearly defunct *La Fronde* and the new feminist library). However, they did have political, intellectual, and personal reasons to bring her name once again before the public. Significant scientific and political figures participated in Royer's commemoration not to endorse an institution, but to recognize a woman who had been rooted in the politics and intellectual life of the Third Republic, and who had consistently upheld a rational view of the universe.

# Changing Practices of Commemoration in Neurology
## Comparing Charcot's 1925 and 1993 Centennials

*By Daniela S. Barberis**

HISTORIANS HAVE TURNED THEIR ATTENTION to the topic of memory's relationship to history fairly recently, generating a revival of interest in Maurice Halbwachs' work on collective memory.[1] Collective memory—social, living, spontaneous, adapting itself constantly and imperceptibly to the changing needs of the present—is not the same as history. The individual's relationship with the past is such that she still has a portion of her life experience invested in that past. History, by contrast, can only begin when that personal tie ceases to exist. "At the heart of history is a critical discourse that is antithetical to spontaneous memory."[2]

The gap between memory and history has been rapidly growing since the nineteenth century. The turning point came when societies began to sense that they were no longer immersed in living traditions. A familiar world was being replaced by the new realities of industrialization, urbanization, and the emerging nation-states. Feeling the past irremediably slipping away, social groups sought to retain it through commemorative practices. But while the nineteenth-century pace of change evoked nostalgia for a traditional world that was being lost, places of memory still anchored the present in the past. As Pierre Nora argues, although historians of this period established a critical distance between themselves and the events and personalities they examined, these scholars still used collective memory as reference points to orient their investigations. While gathering materials from archives and other sources carefully and critically, these historians still believed that it was their duty to reconstruct history for the entire community. And, as Nathan Wachtel has offered, "it was still a univocal history to the glory of the nation that was imposed by positivist

* Committee on the Conceptual Foundations of Science, University of Chicago, 1126 E. 59th Street, SS. 205, Chicago, Illinois 60637

I would like to thank Robert Richards for his comments on a previous version of this paper. I am especially grateful to Elisa Barrios for guiding me through the literature on collective memory, lending me her own work on the subject, and discussing some aspects of the paper with me. I would also like to thank Andrea Murschel for her comments on the final version of the paper.

[1] Patrick H. Hutton, *History as an Art of Memory* (Hanover, N.H., and London: Univ. Press of New England, 1993). See also the special issue, "Memory and Counter-Memory," ed. Natalie Zemon Davis and Randolf Starn, *Representations,* 1989, *26.* Maurice Halbwachs, *Les cadres sociaux de la mémoire* (Paris: Alcan, 1925) and *La mémoire collective* (Paris: P.U.F., 1950).

[2] Pierre Nora, "Between Memory and History: *Les lieux de mémoire,*" *Representations,* 1989, *26:*7–25, on p. 9.

 0369-7827/99/1401-0006$02.00

history, whose method was typified at the end of the nineteenth century by the works of Langlois and Seignobos: the ideology of the Third Republic (at once scientific, unitary and patriotic) triumphed in the textbooks of E. Lavisse."[3]

Nora connects the interest of the contemporary world in memory to changes so precipitous that recourse to collective memory has diminishing importance. Patrick H. Hutton has concurred: "The wisdom of the past in which communities once trusted is immaterial to a culture in which today's improvisations pass quickly into tomorrow's obsolescence. In a world of future shock, places of memory disappear, and history surrenders its mediating role between past and future." But, as Wachtel points out, if historians are no longer the recognized custodians of national memory, this crisis of legitimacy has brought with it a raising of consciousness. They are now aware of the multiplicity of memories and have turned their attention to the politics of memory and especially the politics of commemoration. Turning to Hutton once again: "The commemorative practices of ages past . . . may be disassembled today with a detachment that would have been impossible a century ago."[4]

The two anniversaries of Jean-Martin Charcot I will analyze in this paper offer an opportunity to examine changing attitudes toward the past as reflected in commemorative practices. The earlier of the two anniversary celebrations, the centenary of Charcot's birth, took place in 1925; the latter on the centenary of his death, 1993. I will argue that the first anniversary successfully recreated and circulated the stereotypical images of Charcot that give form to collective memories and generated *communitas* among participants. Significantly, the 1925 commemoration was also a celebration of France as a nation. The second anniversary, on the other hand, displayed the syncretism that distinguishes our postmodern age—there was no consensus over the value or interest of Charcot's work, no unified biographical account of his life.

Scientific anniversaries have been described as moments of ritual celebration—as occasions for the reinforcement of group identity through formalized social interaction and the invocation of a common disciplinary past. Whether an individual or an institution serves as the object of the celebration, participation in the event is generally considered a moral obligation by both the performers (delivering lectures or speeches) and the audience. When an individual's achievements provide the focal point, as in the anniversaries of Charcot, they are generally reconstructed by the celebrants to transform the individual into a "scientific hero," who sets an example to be followed by all in the group (especially its younger members).[5]

Scientific anniversaries also present a specific characteristic that distinguishes them from other commemorations. As Pnina Abir-Am has pointed out, participants "seek to reconcile the contradiction between the relativism implied by discarded convictions with the pretense of present scientific progress to absolutist truth" and combine "the official noble goal of paying tribute to either influential living

[3] Nora, *ibid.*, pp. 10–11; and Nathan Wachtel, "Memory and History: Introduction" in the special volume, "Between Memory and History," *History and Anthropology,* 1986, *2*:207–224, on p. 217. See also Philippe Ariès, *The Hour of Our Death* (New York: Knopf, 1981); and Eric Hobsbawm and Terence Ranger, eds., *The Invention of Tradition* (Cambridge: Cambridge Univ. Press, 1983), esp. Hobsbawm's introduction.

[4] Hutton, *History as an Art of Memory,* p. 9; and Wachtel, "Memory and History," p. 218.

[5] Jean La Fontaine, "Invisible Custom: Public Lectures as Ceremonials," *Anthropology Today,* 1986, *2*:3–9, on p. 6. She points out that attendance at ceremonial lectures should be interpreted as a gesture of social solidarity.

colleagues or the notable dead with the less explicit one of circulating a certified conception of science."[6]

While the 1925 anniversary of Charcot accurately fits the mold of a scientific commemoration described above, the 1993 anniversary fails to do so. Charcot has been the object of intense scrutiny by historians since at least the 1970s.[7] The work of such scholars could not be ignored at Charcot's latest (1993) anniversary celebration, and that event had among its speakers a mélange of historians and critics whose irreverence would not have been tolerated at the earlier celebration. The presence of those outside the discipline of neurology, which claims Charcot as its founder jeopardized the identity-reinforcing function of the ceremony. There is a reason why scientists prefer to write their own disciplinary history: as with other groups, they seek to find moorings in the past.[8] Anniversary volumes have been the locus par excellence of scientists writing their own history. A comparison of these two Charcot anniversary celebrations will provide a powerful lens to see what remains of the basic characteristics of this kind of ritual when its presuppositions are stretched to their limits.

## 1925: CHARCOT'S BIRTH CENTENNIAL AND THE TWENTY-FIFTH ANNIVERSARY OF THE *SOCIÉTÉ NEUROLOGIE DE PARIS*

Charcot's death was greeted by an outpouring of eulogies and obituaries in the French and international medical and lay presses.[9] Most of his former students wrote lengthy accounts of the man and his work. Pierre Janet and Sigmund Freud were among those who paid their last respects in writing.[10] A subscription was soon started for a statue to perpetuate Charcot's memory. His sculpted likeness was placed in front of the Salpêtrière hospital and inaugurated 4 December 1898, during a ceremony in which the dean of the medical faculty, Paul Brouardel, Fulgence Raymond (successor to Charcot's chair), and Victor Cornil (representing the Academy of Medicine and the Anatomical Society), made speeches glorifying "the high moral character of the man and the imperishable work of the doctor." The Minister of Education

[6] Pnina G. Abir-Am, "A Historical Ethnography of a Scientific Anniversary in Molecular Biology: The First Protein X-Ray Photograph (1984–1934)," *Social Epistemology,* 1992, *4:*322–354, on p. 322; and Abir-Am, "How Scientists View Their Heroes: Some Remarks on the Mechanism of Myth Construction," *Journal of the History of Biology,* 1982, *15*(2):281–315, on p. 281.

[7] Henry Ellenberger's *The Discovery of the Unconscious* (New York: Basic Books, 1970) marked a turning point in the rather hagiographic literature on Charcot. Ellenberger spurred a new, broader literature on hysteria. Intellectual, social, and cultural historians; feminist historians; historians of science and medicine; art historians; and literary critics have developed a varied and specialized literature on hysteria, which is still growing. This renewed interest in Charcot has more recently moved from hysteria to other aspects of his work. For a complete review of this literature see Mark Micale, *Approaching Hysteria: Disease and Its Interpretations* (Princeton: Princeton Univ. Press, 1995).

[8] Nora, "Between Memory and History" (cit. n. 2), p. 15.

[9] There were over 70 eulogies published in France (Bourneville, Gilles de la Tourette, Féré, Joffroy, Raymond), England, Scotland, Austria (Freud), Germany (Leyden, Mendel, Erb, Strümpel and Mobius), Hungary (Jendrassik), Spain, the Netherlands, Poland, and the United States (Osler) between 1893 and 1894. From Christopher Goetz, Michel Bonduelle, and Toby Gelfand. *Charcot: Constructing Neurology* (Oxford: Oxford Univ. Press, 1995), p. 31.

[10] Pierre Janet, "J.-M. Charcot: Son ouevre psychologique," *Revue Philosophique,* 1895, *39:*569–604; and Sigmund Freud, "Charcot," *Wiener Medizinische Wochenschrift,* 1893, *43:*1530–1540.

and a representative of the Mayor of Paris were present to receive the statue and render homage to the "great savant and great citizen."[11]

The 1925 anniversary celebration of Charcot's birth was an imposing event, divided into two main parts, the *"travaux"* and the *"cérémonies."* The *travaux* consisted of the Sixth Annual International Neurology Meeting and a special session of the Société de Neurologie on its twenty-fifth anniversary, where foreign guests presented papers. All sessions of the *travaux* were centered on neurological syndromes that Charcot had either first described or had done important work on, such as amyotrophic lateral sclerosis ("Charcot's disease") and migraine. The speakers described the latest developments in these areas and, thus, created a feeling of continuity between past and present, and they emphasized scientific progress.

The ceremonies devoted to the memory of Charcot were various and impressive: a special meeting of the Academy of Medicine; a solemn session at the Sorbonne, in the presence of the President of the Republic; a visit to the old "service" of Charcot at the Salpêtrière; a banquet offered for the foreign delegates at the Palais d'Orsay; and a reception at the Hôtel de Ville hosted by both the Municipal Council of Paris and the General Council of the Seine. The organizing committee was largely composed of Charcot's former students and associates.[12] This committee obtained the patronage of a long list of important political and administrative figures including the President of the Republic, the Presidents of the Senate and of the Chamber of Deputies, and several ministers.

Allen Starr, an American delegate present at the celebration, reported that the reception at the Hôtel de Ville was very elaborate. All of the salons were thrown open and entertainment was provided by singers from the Paris opera. He also described a visit to Neuilly, Charcot's country residence, which remained in the possession of the family (unlike the Paris residence, the Hôtel de Varangeville, which was sold on Mme. Charcot's death). Charcot's son and daughter showed the foreign visitors the salons and library of their father and presented them with Charcot's photograph and copies of some of his pen and ink sketches. The Neurological Society also gave the participants souvenirs of the celebration: a bronze medal of Charcot and photographs of the Salpêtrière and the men who made it famous.[13]

The choice of simultaneously celebrating Charcot's centenary and the twenty-fifth anniversary of the Société de Neurologie was strategic.[14] Charcot was being honored as a neurologist, as a disciplinary hero of the reunited international "clan" of neurologists. Charcot's disciplinary affiliation was, however, not altogether obvious. There was no such thing as neurology during Charcot's student days, and he dedicated himself mainly to anatomical pathology and to clinical work. Charcot came to neurology relatively late in his career. His two theses (his doctoral thesis and his

[11] "Inauguration du monument élévé à la mémoire du Professeur J.-M. Charcot," *Nouvelle Iconographie de la Salpêtrière,* 1898, *6:*401–418 on p. 402.

[12] The honorary presidents were Pierre Marie, Pitres, and Paul Richer. Joseph Babinski was the president, Achille Souques was the secretary and the committee itself was formed by Blin, Colin, Dutil, Guinon, Hallion, Paul Londe (former photographer of Charcot's service), Henry Meige, Parmentier, and Jean Charcot (Charcot's son and former student).

[13] Allen Starr, "Boston Medical History Club: Charcot Centenary Meeting," *Boston Medical and Surgical Journal,* 1926, *194*(1):10–20, on p. 13–14.

[14] The Société de Neurologie was founded in 1899, and therefore the two anniversaries do not coincide precisely.

*agrégation* thesis) and his early publications were on non-neurological—rather eclectic—topics. He published on gout and rheumatism, on the vascular and respiratory systems, on diseases of the kidney, and on infectious diseases. Until he received nomination as *médecin de la Salpêtrière* in 1862, his interests followed no single path. From that point on, he focused primarily on the pathological anatomy of the nervous system. However, only later would this become the sole topic of his researches. His nomination to the faculty chair of pathological anatomy (1872) required him to lecture on a broad range of topics, including rheumatic diseases and other ailments found in the elderly, and diseases of the liver, lungs, and kidneys.[15] It was not until 1882 that Charcot's neurological work was given official recognition by the creation of a professorial chair for him in diseases of the nervous system. Only this new chair allowed him to devote his research energy exclusively to this topic. By 1882 he was already deep into the study of hysteria and hypnosis, which occupied much of his time until his death.

Recent literature on Charcot often refers to him as a psychiatrist—because most of these scholars have arrived at Charcot through an interest in hysteria.[16] Weighty arguments exist for placing Charcot in several disciplinary camps.[17] The organizers of the 1925 centenary celebration, however, specifically aimed to transform Charcot into the "founding father" of neurology. To that end, Charcot's biography was streamlined and some of his interests, particularly those in psychology and psychiatry, were obliterated. Most significantly, Pierre Janet, who, during Charcot's last years and shortly after Charcot's death, had been considered one of his most important disciples (and is still considered so by those who claim Charcot as a precursor of Freud or a significant figure in the "discovery of the unconscious"), was absent from Charcot's 1925 anniversary. Janet had written an obituary, "J.-M. Charcot. Son oeuvre psychologique," which outlined his teacher's contributions to psychology. Charcot's psychological interests were something to be forgotten in the eyes of most neurologists at Charcot's anniversary. The same can be said of the large amount of work done by Charcot on what would be termed today geriatric illnesses. Alain Lellouch has claimed for Charcot the position of founder of this medical specialty in his recent book, *Jean Martin Charcot et les origines de la gériatrie*.[18] In short, Charcot's interests and work do not self-evidently and unequivocally define him as the founding father of neurology.

Other nineteenth-century physicians, contemporaries of Charcot, could justly claim the title of "founding-father" of neurology if the claim were based solely on neurological work. So much so that Charcot's latest biographers, Christopher Goetz, Michel Bonduelle, and Toby Gelfand, felt the need to justify attributing this honor

[15] The Salpêtrière was the "Hospice de la Viellesse—Femmes" (Hospice of the Elderly, Women's Division) in Charcot's time and therefore contained an exceptionally large elderly patient population. Charcot's work on geriatrics can be attributed to the presence of this population. It should also be noted that the Salpêtrière was reserved exclusively for women. Charcot was responsible for the creation of the first outpatient ward that accepted men.

[16] Two influential books that consider Charcot as a psychiatrist are Jan E. Goldstein, *Console and Classify: The French Psychiatric Profession in the Nineteenth Century* (Cambridge: Cambridge Univ. Press, 1987); and Elaine Showalter, *The Female Malady: Women, Madness, and English Culture, 1830–1980* (New York: Pantheon, 1985). Most authors influenced by either psychoanalysis or Michel Foucault's *Histoire de la sexualité. 1-La volonté de savoir* (Paris: Gallimard, 1976) tend to see Charcot as a psychologist/psychiatrist.

[17] Goetz, Bonduelle, and Gelfand, *Charcot* (cit. n. 9), pp. 208–213.

[18] Alain Lellouch, *Jean Martin Charcot et les origines de la gériatrie* (Paris: Payot, 1992).

to him rather than to Paul Broca, Karl Westphal, Wilhem-Heinrich Erb, John Hughlings Jackson, William Hammond, or Silas Weir Mitchell. Goetz, Bonduelle, and Gelfand argue that Charcot's founding-father status should be based not only on his significant early work on the discipline but on other grounds as well. He held the first university chair in the study of the diseases of the nervous system; he created a school (and thus had followers who had a stake in supporting Charcot as a disciplinary founder); he obtained the physical facilities necessary for the development of this discipline (teaching and treatment facilities, outpatient service, research laboratory, photographic studio, museum of pathological anatomy, etc.); and he founded several journals in the field. Goetz, Bonduelle, and Gelfand's *Charcot: Constructing Neurology,* as its title indicates, is part of the effort, begun shortly after Charcot's death, to define Charcot as primarily a neurologist.[19]

Every facility installed in the Salpêtrière hospital was acquired through years of patient negotiations with the relevant administrative and political authorities. Part of this progress was smoothed by Charcot's political connections. Désiré-Magloire Bourneville, his student and associate, active in the Paris city council, had been an important asset in the gradual construction of Charcot's neurological empire at the Salpêtrière, as was Charcot's relationship with another important political figure, Léon Gambetta.[20] However, in a way typical of anniversary commemorations, speakers at the 1925 celebration presented the creation of Charcot's chair and service as the logical crowning of many years of patient scientific work—that is, as the result of the simple recognition by the relevant authorities of Charcot's scientific excellence. Charcot's achievements were purified of any political or social connections and presented as solely due to his personal talents, thus enhancing the myth of the scientist-hero.

In all the 1925 commemorative speeches, Charcot's life was reduced to the pursuit of his scientific interests. This reinforced the myth of the scientist-hero, who is "stripped down to his scientific garb while his many other nonscientific interests (artistic, social, political) are carefully relegated to the background."[21] The biographic information presented on Charcot took the form Michael Mulkay might call the "golden biography": "parents are kind and considerate, teachers are enthusiastic and stimulating, wives and children are loving and supportive, and there is no trace at all of the prejudice, bitterness or competition which appear persistently in scientist's . . . descriptions of their career in other contexts."[22] Mme. Charcot and his

[19] There is a clear rejection of a view of Charcot as psychiatrist by Gelfand in the chapter dedicated to hysteria in *Charcot* (cit. n. 9). Gelfand dedicates a chapter subsection to the intersection of psychiatry and Charcot's work on hysteria, insisting that Charcot's disciplinary affiliation was strictly neurological (pp. 209–210).

[20] The efforts of the French medical profession to gain political power during the latter part of the nineteenth century are placed in context by Theodore Zeldin, *France, 1848–1945,* Vol. 1: *Ambition, Love, and Politics* (Oxford: Clarendon Press, 1973), ch. 2, "Doctors." At this time, a considerable number of French doctors were either personally involved in politics or had very influential political connections. The large majority, however, remained in modest positions and had little opportunity to achieve financial success as physicians.

[21] Abir-Am, "How Scientists View Their Heroes" (cit. n. 6), p. 281.

[22] Michael Mulkay, "The Ultimate Compliment: A Sociological Analysis of Ceremonial Discourse," *Sociology,* 1984, *18:*531–549, on p. 544. While Mulkay is analyzing Nobel Laureates' discourse during Nobel Prize ceremonies, I believe some features of anniversary celebrations are quite similar. Mulkay's "golden autobiography" is transformed into the "golden biography" during anniversary celebrations.

children "adored" the head of their family, his students "admired" and "respected" their master, to whom they "owed much." The less idyllic aspects of Charcot's professional and family life never appear.

The keynote speakers generally tried to mix a sense of historical authenticity with a discourse on scientific progress. They either provided accounts of their association with Charcot or else this association was so well known that some pious remarks about their love and admiration for Charcot sufficed to instill a feeling of continuity with the past. Having studied with Charcot, the speakers were living witnesses of this past, which made them ideal intermediaries between past and present. This duality of speakers, who represent both the "heroic past" and recent scientific progress, is one feature of anniversary meetings that distinguishes such events from ordinary meetings.

The 1925 commemoration organizers demonstrated a concern for choosing celebration sites closely connected with Charcot by holding the scientific part of the conference at the Salpêtrière hospital and locating other official ceremonies at institutions where Charcot had been a member (the Academy of Medicine and the Sorbonne). Further, the official guided tour of Charcot's old "service" at the Salpêtrière was obviously meant as a visit to the site of his great discoveries. The organizing committee went to considerable trouble to recreate Charcot's consulting room as it had been in 1893. Visitors were taken to the library containing Charcot's personal books, to the museum of pathological anatomy assembled by Charcot, and, finally, to the Chapel of the Salpêtrière, where his funeral had taken place.

The keynote speakers were also successors to various institutional posts Charcot had held. During the ceremonies, those links were emphasized, while the rapid fragmentation of the "school of the Salpêtrière" after the death of its leader was veiled in discrete silence. The message implicitly delivered to the audience—especially its younger members—was that faithful disciples (as the keynote speakers had once been themselves) eventually would become disciplinary leaders by following a steady path of institutional advancement. What remained unsaid, however, shrouded untimely facts: at Charcot's death there was but a single professorship in neurology and no specialty neurological units in the hospital system; moreover, Charcot's youngest students (who were not established at the time of his death) never became neurology professors. The loss of their mentor seriously affected the career prospects of most of Charcot's former students.

Finally, the presence of Charcot's son and daughter added a further element of authenticity to the whole ceremony. Their willingness to open their home, Charcot's country residence at Neuilly, to the foreign visitors and to show the guests his drawings and personal papers provided concrete connections to a past, which the speakers were only able to discuss in the abstract. The presence of Charcot's immediate descendants also tended to push the whole celebration in the direction of a memorial service. In the company of Charcot's family, all speakers were evidently constrained to speaking only in laudatory terms about the paterfamilias.

Charcot's 1925 anniversary had broader political implications as well. The ceremony that took place in the amphitheater of the Sorbonne, in the presence of the President of the Republic and the Minister of Education, brought foreign delegates together to honor Charcot, the great French neurologist. This commemoration simultaneously served as an occasion to celebrate the greatness of France. Besides foreign neurologists, several countries sent consuls or ambassadors. Many of these delegates took the opportunity to affirm the allegiance of their countries to France and, furthermore, to recognize France as a scientific and cultural leader. The Minister of the Colonies presided over the banquet offered for the foreign delegates, further wrapping Charcot's science in the French Tricolor.[23]

The foreign representatives displayed little scientific acumen but considerable political awareness: Charcot was consistently compared to Louis Pasteur and (slightly less often) to Claude Bernard, the other great figures of French nineteenth-century medicine.[24] Charcot was a pioneer, an exceptional scientist who explored virgin domains, "the creator of the Neurology of the nineteenth century."[25] He was not only an outstanding scientist but also an extraordinary man of noble character and high integrity. The foreign delegates acknowledged a debt to this great French scientist and, by extension, to France. To quote one of the foreign speakers who was, in turn, quoting Pasteur: "Though science has no country, the scientist has one" (Si la science n'a pas de patrie, le savant en a une). Charcot's birth centennial thus became an occasion for exalting the French nation, "which gave not only science but also freedom to small nations."[26] Frequent references were made to the Declaration of the Rights of Man and to the Great War, which had recently finished and during which France had "fought for the Rights of Nations."[27] During the banquet, toasts were often made to the memory of Charcot and to the glory of France.

According to unspoken rules of diplomacy, delegates from the least powerful nations offered the strongest statements of admiration and debt toward France. By contrast, the speech made by Allen Starr, the delegate from the United States, while bowing to the greatness of Charcot, avoided any expression of debt toward French culture or toward France as a nation. Neither did Starr refer to the greatness of French science. As he said later, once back in the United States, "it may be justly said that France honored herself by conferring honors upon him [Charcot], for there are few Frenchmen who did as much as he to support the waning French authority in matters of science or to preserve the fading reputation of the French school of medicine."[28]

Thus, this scientific conference after the Great War was an occasion for diplomatic rhetoric to manifest itself. While the participants of this centennial adhered in principle to the myth of science as being independent from a social and political context (and affirmed it even during those same ceremonies and banquets where they drank to the glory of France), in practice the conference had political and symbolic significance and served to outline national allegiances. The French government appropriated this ceremony to project an image of political and scientific power, both of which it had in rather short supply. Though France was no longer at the forefront of medicine (or neurology) in 1925, the gathering of foreign neurologists who recognized Charcot as the founding father of a medical specialty, gave France a nostalgic glow of scientific eminence.

[23] Spain, Belgium, Brazil, Great Britain, Poland, USSR, Turkey, the Netherlands, Colombia, Czechoslovakia, Haiti, Portugal, Argentina, and Romania all sent official representatives to the event.

[24] See, in this volume, Christiane Sinding, "Claude Bernard and Louis Pasteur: Contrasting Images through Public Commemorations."

[25] "Discours de M. le Professeur Dr C. Winkler," *Revue Neurologique,* 1925, *1*(6):1,123.

[26] "Discours de M. le Professeur Poussep au nom de l'Esthonie," *ibid.,* p. 1,174.

[27] "Discours déposée par la Tchéco-Slovaquie," *ibid.,* p. 1,139.

[28] Starr, "Charcot Centenary Meeting" (cit. n. 13), pp. 12–13.

Scientific heroes, unlike political ones, have the potential of being acceptable to an entire nation and, even further, worldwide. Hobsbawm has pointed out that "unlike the USA and the Latin American states, the French Republic . . . shied away from the cult of the Founding Fathers."[29] The government preferred general symbols, such as the figure of Marianne, which were representative of the Republic itself, because each brand of Republicanism has had its own heroes and villains in the Revolutionary pantheon. Thus, French political personages could not sustain the same unifying power as a "politically neutral" scientific figure, such as Pasteur or Charcot. As the Minister of the Colonies put it, "it seemed desirable to the Government of the Republic to take part in this double manifestation [Charcot's centenary and the twenty-fifth anniversary of the Société de Neurologie]."[30] The political advantage in presenting scientific figures as national icons, when political ones are controversial, is clear.

Most commemorative speeches connected to Charcot's centennial took the standard form of scientific biography. The two longest speeches, which open the special number of the *Revue Neurologique,* were presented by Charcot's most successful followers, Pierre Marie and Joseph Babinski. These speeches followed a pattern set in Charcot's obituaries and repeated certain stereotypical images—even in the same order. Charcot was portrayed as a self-made man who rose from humble origins, found his true calling when he turned his attention to neurology, and achieved the highest honors of his chosen profession. Enumerations of Charcot's scientific works formed the largest parts of these biographies: his doctoral thesis on gout and rheumatoid arthritis; his lectures on pathological anatomy; the work on cerebral localizations and spinal lesions, on muscular atrophies (amyotrophic lateral sclerosis, progressive amyotrophies), on tabetic arthropathies; and, finally, his explorations of the vexed question of hysteria and hypnosis. Then, there was fond attention given to Charcot's method, his particular clinical intuition, and "*coup d'oeil.*" His pedagogical qualities, the quality and quantity of his students (his capacity to attract foreign students was a sub-topic), and the fidelity of students to the master merged to form another theme. The time when the Salpêtrière was the "Mecca of neurologists" was evoked with a touch of nostalgia. Finally, Marie and Babinski depicted Charcot's international fame, especially his relations with England and Germany. The relationship with the former served to emphasize Charcot's internationalism and the ideal of science *sans frontières.* Charcot's stance toward Germany, by contrast, was used to emphasize his patriotism by reference to Charcot's refusal to set foot in Germany after 1870. His very cordial relations with German colleagues, who visited him and sent him students, when mentioned, were offered as manifestations of the universality of science.[31]

These versions of Charcot's life and work established a biographical tradition that influenced later accounts of Charcot. As Dorinda Outram pointed out in the case of Georges Cuvier, an early biographical tradition raises the questions of how far these

[29] Hobsbawm, "Mass Producing Traditions: Europe, 1870–1914," in *Invention of Tradition* (cit. n. 3), p. 272.

[30] *Revue neurologique,* 1925, *1*(6):1,183–1,186, on p. 1,183. The Minister also took the opportunity to exhort the young doctors present to offer their services to the colonial health services, and, to inspire them in that arduous enterprise, he recommended that they take Charcot as their model.

[31] Pierre Marie, "Éloge de J.-M. Charcot," *Revue Neurologique,* 1925, *1*(6):731–745; and Joseph Babinski, "Éloge de J.-M. Charcot," *ibid.,* pp. 746–756.

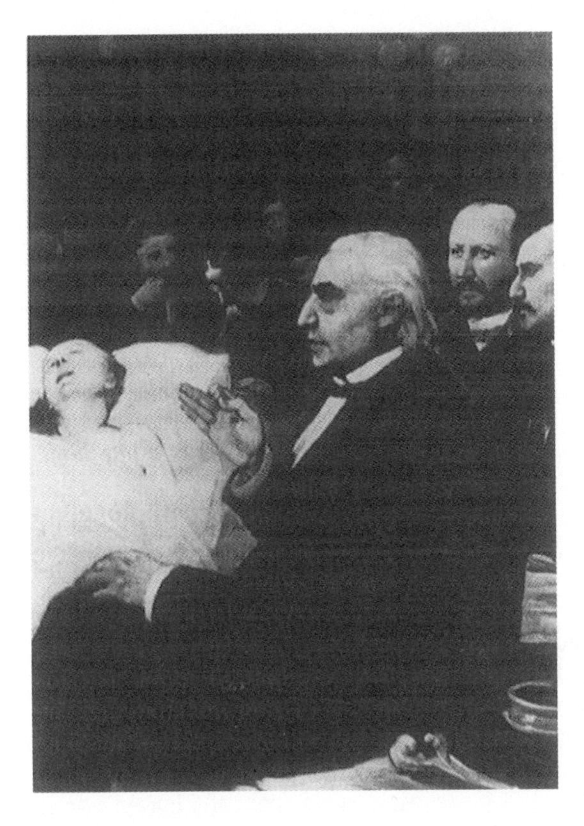

***Figure 1.*** *Professor Charcot lecturing at the Salpêtrière. Next to him are his students Joseph Babinski and Paul Richer. Oil painting by Jean Geoffroy, 1892.*

images of Charcot were influenced by his own self-presentation, as well as by his student-biographers' preoccupations. Charcot, like Cuvier, was enormously concerned with publicity and with projecting a certain image and his biographers also displayed a "fascination with his working habits, appearance, mannerisms, and domestic arrangements."[32]

Marie and Babinski each tried to achieve smooth narratives of scientific progress. The main stone in the path of a victorious life story was Charcot's work on hysteria and hypnosis. Babinski had aggressively rejected his former teacher's work on the subject, especially Charcot's rigid description of the four-phase hysterical attack. Babinski had disassembled Charcot's construct and even proposed renaming the syndrome "pithiatism," considering the term "hysteria" tainted by inappropriate usage. By so doing, Babinski had initiated the decline of hysteria as a diagnostic category and legitimate field of study.[33] Yet one would never know of this harsh criticism from Babinski's commemorative speech, which began with a declaration of "filial affection" for Charcot. During the entire address, Babinski made no reference to his own work on pithiatism. Babinski stated that even though the ideas of most neurologists had changed in relation to "certain important points relative to hysteria" (which he did not specify), Charcot's investigations in that domain had been productive. He emphasized Charcot's clinical descriptions of hysterical symptoms and the valida-

[32] Dorinda Outram, "Scientific Biography and the Case of Georges Cuvier, with a Critical Bibliography," *History of Science,* 1976, *14:*101–137, on p. 112.
[33] See Mark Micale, "On the 'Disappearance of Hysteria,'" *Isis,* 1993, *84:*496–526.

tion of the existence of male hysteria, which were, in fact, relatively minor elements of Charcot's total work on the subject. By arguing that the value of scientific work should not be measured by the immediate advantages derived from it but rather from the beneficial changes in mindset that it occasions, Babinski was able to preserve the importance of Charcot's contribution while denying certain specific constructs (such as hysteria) all but limited value.

Marie dealt with hysteria in a more direct way—probably because he had not explicitly discredited Charcot's thinking on hysteria (as Babinski had done).[34] For Marie, Charcot's work on hysteria was a "slight lapse" (légère défaillance) due to a "series of regrettable circumstances," in an otherwise immaculate scientific career. Charcot had, as it were, been forced to direct his attention to hysteria due to events that were beyond his control. Marie recounted a story that many in his audience knew well. The Salpêtrière administrators evacuated a building because of its state of disrepair. This building had been a psychiatric ward and contained, mixed up with the "aliénées," hysteric and epileptic patients. Since patients who fell into the latter two categories both had convulsive crises, the administration considered it logical to unite them in a special separate service. And, since Charcot was the senior physician at the Salpêtrière, this new ward was "naturally" assigned to him. Charcot was thus involuntarily "plongé en pleine hysterie."[35]

Once driven by destiny to the study of hysteria, Charcot had made the mistake of letting himself be overtaken by his tendency to classify syndromes and illnesses (a tendency that had served him so well in regard to organic neurological diseases). He had tried, as Marie delicately pointed out, to contain hysteria inside a rigid nosological scheme. His failure had been the unfortunate consequence of an otherwise excellent methodology. Marie concluded, however, that this part of Charcot's work was a small portion of his outstanding oeuvre and could be quickly dismissed. Scientific impartiality obliged Marie to acquaint his audience with this episode. The admission of this "slight failing" of Charcot enhanced Marie's reliability as a historical source. He had stated at the beginning of his speech that it was as "a historian that one must make the effort of speaking here." Having set aside the questionable aspects of Charcot's work on hysteria, Marie was then free to extol what were then considered the scientifically valid aspects and thus to end his list of Charcot's multiple achievements and discoveries on a triumphant note. The image of science as the discovery of absolute truths was preserved for the audience, which was presented a highly idealized version of scientific careers. Charcot's life, as another speaker put it, was offered "as an example for the young scholars."[36]

The disagreements between speakers and the speakers' critical views about Charcot were cast aside during this ceremony in order to achieve a moment of *communitas.* The desire for group cohesion was great enough that individuals buried their intellectual hatchets, if only for the duration of the ceremonies, and perhaps only during the official performances (unofficial gossip remains unrecorded). Babinski's

[34] As Goetz, Bonduelle, and Gelfand wrote in *Charcot,* Marie was an aging man when he finally returned to the Salpêtrière as the successor of Dejerine to hold Charcot's chair and "spent his career wheedling away many of his teacher's major tenets," (cit. n. 9), p. 321.

[35] Marie, "Éloge," (cit. n. 31), p. 741.

[36] "Discours prononcée au nom de l'Académie des Sciences à la cérémonie du centenaire de Charcot, par M. Ch. Lallemand, Vice-Président de l'Académie," *Revue Neurologique* 1925, *1*(6):1,142–1,146, on p. 1,146.

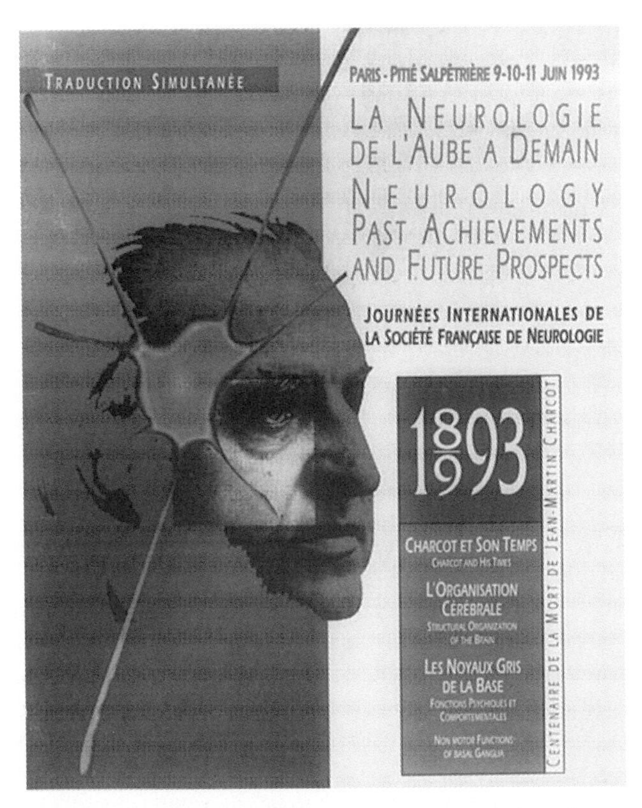


***Figure 2.*** *Cover of the program for the 1993 Charcot centennial conference. The superimposition of a neuron on Charcot's portrait suggests his neurological affiliation. The emphasis on his eye may have been intended to imply superior observational abilities.*

belabored avoidance of direct criticism of Charcot's notion of hysteria, for example, must have been obvious to all neurologists present and suggests a certain complicity between speakers and audience. There was, as it were, a pact to accept the mythical account of the scientist's life, despite widespread insider knowledge of the "dirty details."

A sign of the vitality of this commemoration is the successful circulation of a highly idealized image of Charcot that went unquestioned. The past was transformed for present purposes by the performers. Charcot was claimed as their own by the clan of neurologists that emerged fortified through the affirmation of their particular world view. It was a view that included the myths of progress and of the separation of science and political influences. All of these characteristics were notably absent from the 1993 centenary celebration of Charcot.

## 1993: "NEUROLOGY: PAST ACHIEVEMENTS AND FUTURE PROSPECTS"

The celebration of the centenary of Charcot's death was again an international neurology conference. The conference was organized by the French Society of Neurology in conjunction with the French Society for the History of Neurology. It had the patronage of the president of the Republic and also that of the ministers of Health and Culture, the mayor of Paris, the president of the Université Paris VI Pierre et Marie Curie, and the Director of the Assistance Publique. It is interesting to note that the ministers sponsoring the 1993 event were directly linked to the topics of the conference (health and culture). There was no attempt by the French government to use this event as a diplomatic venue for fostering relations with other countries nor were there diplomatic representatives from other countries present. An entire day was devoted to "Charcot and his time" and the following day, dedicated to brain organization, opened with a historical session focusing on localization theory before and after Charcot. The third day was dedicated to the behavioral and cognitive functions of basal ganglia. As evening entertainment, the participants were offered an organ concert in the Salpêtrière Chapel, followed by cocktails; a reception in the "Mairie de Paris"; and a closing dinner at the "Maison de l'Amérique Latine" (Hôtel de Varangeville), Charcot's former residence on the boulevard Saint-Germain.

The 1993 centennial differed markedly from the 1925 celebration in that it included no occasions designed exclusively to laud Charcot's greatness. There were no special sessions of the Academy of Medicine or solemn sessions at the Sorbonne. There was a day dedicated to the presentation of papers centered around Charcot's person, but the day did not have the ritual celebratory overtones of the 1925 events. Rather, the attention devoted to Charcot was just one part of the neurology conference. There was no division into "travaux" and "cérémonies." The same was true of the published papers. They were published with no special introduction; they did not form a special number of the *Revue Neurologique* but were simply published together.[37] Apparently, there was a distrust of ceremony among the organizers of the 1993 conference.

The straightforwardness of the speakers in dealing with the controversial topic of hysteria contrasts with the strategies adopted in 1925, when the subject was generally relegated to the end of speeches centered on Charcot's neurological discoveries and successes. In 1993, the first three papers were "Charcot: scientifique bifrons," "Charcot et l'hystérie," "Charcot and *les névroses traumatiques*." All of them directly addressed Charcot's scientific weak point.

The author of the first paper, Christopher Goetz, developed a dichotomy between anatomy and physiology. Anatomy corresponded with observation; physiology with experimentation. Charcot—an exemplary scientist as long as he remained in the terrain of anatomy (work on organic neurological diseases)—became non-scientific when he ventured into physiology. That was the simple reason for his failure in the field of hysteria—his limited capacity and inexperience in physiology. "Seduced by the most illusory and controversial neurosis," Charcot fell into a trap that drew him into scientific marginality. Seduced also by the cutting-edge status of physiology during the 1880s and 90s, he attempted to use hypnosis as an experimental instrument, treated his patients as mere laboratory objects, and committed many scientific errors. Thus, what remains of Charcot's work on hysteria a hundred years after his death? For Goetz, only the clinical descriptions of hysteria.[38]

The second article, by Daniel Widlöcher and N. Dantchev, "Charcot et l'hystérie," attempted to refute what the authors considered the main criticisms of Charcot's work on hysteria. They considered especially the problem of simulation and the

[37] *Revue Neurologique,* 1994, *150* (8–9):485–683.
[38] Goetz, "Charcot: scientifique bifrons," *ibid.,* pp. 485–489.

excessive rigidity of the four-phase description of the hysterical attack. While they admitted that this model of hysteria was inadequate, they believed that it represented an early stage of Charcot's work on hysteria. Since Charcot later progressed to a psychological theory of hysteria, they felt that criticism of this early work was invalid. The authors depict Charcot as eventually discovering the mechanism for the formation of hysterical symptoms. Charcot arrived at this result by comparing hysterical symptoms with similar symptoms provoked through hypnosis. Janet, Breuer, and Freud departed from this insight to develop different conceptual systems. Thus, according to the authors, a hundred years later, Charcot's mechanism of symptom formation remains the basis of the current psychopathology of hysteria.[39]

Mark Micale argued in the third paper that dismissal of Charcot's work on hysteria as a whole has been a serious mistake and that the case histories on "traumatic hysteria" (which would be called today the "neuropsychology of trauma") showed "the same superb qualities that distinguished Charcot's work on other topics." Micale then pointed out the revival of interest in psycho-traumatic phenomena in the last fifteen years and the relevant aspects of Charcot's work. While he was careful to point out Charcot's failures as well, Micale concludes that "it is not hagiographical but simply accurate historically to acknowledge Charcot's considerable place within the medical history of the trauma concept." Thus, for Micale, it is Charcot on traumatic hysteria that remains valuable and relevant.[40]

The three articles evaluate Charcot according to different criteria. Goetz, an M.D. and a professor of neurological science at Rush University, interestingly, comes closest to the traditional account of Charcot's investigation of hysteria. Goetz values only the rigorous clinical neurology. The main lines of Goetz's paper very much bring to mind Marie's argument in the 1925 anniversary: Charcot's work on hysteria was an unfortunate mistake caused by a generally sound methodology badly applied, in an otherwise impeccable career. For Widlöcher and Dantchev, psychiatrists at the Salpêtrière, Charcot's psychological work (in the tradition of dynamic psychology) is of interest. Micale, a historian of hysteria, considers Charcot's work interesting because Charcot paid attention to both psychogenic symptom-formation and the complex sequelae of bodily injury associated with traumatic hysteria. According to Micale, the long period during which psychoanalysis (or other related forms of dynamic psychology) dominated was responsible for the negative appraisal of Charcot's work on hysteria. Now that a revival of neuroscience is in progress, Charcot's non-neurological oeuvre is being reappraised. Micale's paper is part of this reaction against Freudianism. Thus, it is a historian who made the argument that neurologists should reclaim hysteria (or certain aspects of it) for their discipline.

A second contrast with Charcot's first centennial becomes clear: in the second centennial there is no unity of view. Unlike the first, there is no single biographical account of Charcot presented over and over again in different words but containing basically the same message. Rather, the 1993 congress consisted of a series of presentations dealing with different, well-defined, and restricted topics. Besides the three articles on hysteria, there were articles on Charcot's work on geriatric diseases,

[39] Widlöcher and Dantchev, "Charcot et l'hystérie." *Ibid.*, pp. 490–497. Both authors are in the Psychiatric service of the Salpêtrière Hospital.

[40] Micale. "Charcot et *les névroses traumatiques:* Historical and Scientific Reflections," *ibid.*, pp. 498–505, final quotation on p. 503.

on his religious views, on the role played by his international reputation in his career, on his inner circle of relationships, on his work on aphasia, and on his contribution to the study of the motor cortical area.

Further contrasts with the 1925 centennial abound. Charcot was not portrayed as a "mere scientist" insofar as other of his interests (those in religion, for instance) were treated at length. The political implications of his anti-clerical views were made explicit by Jacqueline Lalouette, who examined the reactions of the Catholic press to Charcot. Charcot's political (republican) views were also treated by Michel Bonduelle in his article on Charcot's relationships and personality. The micro-politics of academic institutions were recognized and scrutinized. Gelfand's paper showed that "Charcot's international success was not merely a passive honorific consequence of his work but a strategy actively pursued by the chief of the Salpêtrière in order to advance his Salpêtrière school at home and defeat opposition to his programme by rivals." Thus, Charcot's success was not perceived as due uniquely to his outstanding personal qualities but as shaped by the academic institutions of which he was a member.[41]

Also, speakers in 1993 had no personal links to Charcot—too much time had passed for them to be his students or even students of his students. Moreover, many speakers were not neurologists. Was the 1993 anniversary a commemoration in the same sense as the 1925 event? In 1993, Charcot was cut down to a more human size: with failings, including a rather unpleasant, authoritarian character; and with assets, such as political contacts and abilities. While he was not portrayed as a scientific hero, or an incarnation of scientific progress, Charcot's great stature as a scientist was affirmed by all present. His political allegiances and institutional maneuvers were not seen as weaknesses or failures—as they would have been, had they been admitted, by the participants of the 1925 ceremonies. Rather, they were often seen by the 1993 participants as part of an organizational "genius"—which allowed Charcot to build a unique neurological service that became the "Mecca of neurology" in his time.

Charcot's scientific weaknesses posed less of a problem in the 1993 anniversary. Perhaps simply because the conception of science of the participants was different in 1993. In 1925, a triumphant vision of historical progress still predominated. In 1993, even a common-sense conception of science no longer included attainment of absolute truth but rather a partial truth that would eventually be superseded in the march towards an unattainable (Popperian) truth. In 1993, Charcot's work formed a whole that had to be understood on its own, historical, terms. In 1925, the participants were still too close to Charcot to attempt to provide a balanced picture. They were the direct descendants of Charcot and had a stake in his status as a "scientist-hero." To the 1993 participants, Charcot's life was far enough in the past that they could allow themselves the luxury of an outsider's view, a historically critical assessment. Nevertheless, Charcot remained, despite his mistakes and authoritarianism, a founding father. The fact that the French Neurological Society chose to celebrate the centenary of his death demonstrates this point.

As compared to the previous anniversary, the 1993 one lost the capacity to gener-

[41] Lalouette, "Charcot au coeur des problèmes religieux de son temps. A propos de 'La Foi qui guérit,'" *ibid.*, pp. 511–516; Bonduelle, "Charcot intime," *ibid.*, pp. 524–528, esp. pp. 526–527. Gelfand, "Charcot médecin international," *ibid.*, pp. 517–523 on p. 517.

ate an acceptable “collective memory.” The vision of Charcot it offered its participants was fragmented, not canonical. Multiple and sometimes conflicting perspectives were presented. Regarding hysteria alone, there were three different interpretations of the worth of Charcot’s work. There was no longer an unquestioning identification with Charcot’s heritage. The different disciplinary affiliations of the participants were partly responsible for the diversity of points of view. But if unity was lost, the participants in the conference may have derived comfort from their greater sophistication. They clearly did not subscribe to the myth of “progress” or succumb to any naïve, “hagiographical” temptations. Yet, after many years as a topic of predilection among historians and psychoanalysts, Charcot was again reclaimed by neurology at an international meeting.

# Medicine and the Culture of Commemoration

## Representing Robert Koch's Discovery of the Tubercle Bacillus

*By George E. Haddad**

"THERE ARE INDEED GOOD REASONS why I've invited you to this narrow room rather than [to] our large lecture hall," began Bruno Heymann (1871–1943), director of the Hygienesches Institut der Universität Berlin. "It alone is the proper spot for the celebration that we intend [to conduct]. It alone [served] as the stage of the memorable events that took place between these four walls—forty years ago today." He then solemnly pronounced, "Here, on 24 March 1882, Robert Koch held his lecture, 'On Tuberculosis,' before the Berliner Physiologische Gesellschaft—the first public announcement of the discovery of the tubercle bacillus."[1]

Heymann and members of the Berliner Mikrobiologische Gesellschaft had gathered in the historically restored reading room of their institute for their commemoration of Robert Koch (1843–1910). For them, remembering and celebrating Koch's historic announcement took on a particular form—it meant recapitulation and representation. The celebrants sat around the same table as had the original audience, and in what amounted to an evocation of collective memory and a collective re-enactment, Heymann led his audience back to the evening of 24 March 1882 (see Figures 1 and 2).

These celebrants in 1922 commemorated much more than that single lecture by Koch, however. They lionized Koch and his achievements because, in the professional narrative of medicine, he had come to embody the transition to a new scientific epoch in medicine—what historically minded German physicians called the "aetiological period." And, of course, their very commemoration contributed to the work of extolling and constructing a posthumous Koch. Although it took decades of debate and discussion to form a consensus on what bacteriology and germ theory meant for medicine, early twentieth-century physicians, almost without exception,

*Department of History of Medicine and Science and Department of Internal Medicine, L-130 Sterling Hall of Medicine, Yale University, P.O. Box 208015, New Haven, CT 06520.

For their helpful comments on earlier drafts, I wish to thank Eileen Cheng, Christoph Gradmann, Frederic L. Holmes, Michael Holquist, Stanley Jackson, Heinrich von Staden, and especially John Harley Warner, whose scholarly example and insight greatly influence the way I write cultural history. Also, I appreciate the assistance of Toby Appel and the staff of the Cushing-Whitney Medical Library of Yale University.

[1] Bruno Heymann, "Zum 40-jährigen Gedenktage der Entdeckung des Tuberkelbacillus," *Centralblatt für Bakteriologie,* 1922, *88,* n. 5:337.

***Figure 1.*** *Robert Koch around 1882 when he announced before the Berlin Physiological Society his discovery of the tubercle bacillus. From Bruno Heymann's commemorative biography,* Robert Koch, 1843–1910, *vol. 1 (Leipzig: Akademische Verlags-gesellschaft, 1932).*

***Figure 2.*** *Site of Robert Koch's announcement of the discovery of the tubercle bacillus—and the site of Bruno Heymann's fortieth anniversary commemoration and re-enactment of the event. Found in* Klinische Wochenschrift, *1932,* 11*(12):489.*

pointed to Koch's announcement before the Berlin Physiological Society as the watershed moment heralding a new era in medicine.

Doctors have always had their heroes, and a nineteenth-century culture—sometimes even cult—of commemoration revolved around many of the canonical figures of medical history. Glorifying each other and their heroes of the past was one way in which physicians promoted ideological reform and fashioned a collective professional identity. While there were many medical figures who served as objects of commemoration, few specific and circumscribed nineteenth-century historical events come to mind that were the objects of collective remembrance. Contenders, among others, include 16 October 1846, the day William Thomas Morton (1819–1868) first publicly demonstrated the use of ether as an anesthetic, and 8 November 1895, the day Wilhelm Conrad Roentgen (1845–1923) discovered x-rays. But the evening of 24 March 1882 stands alone in the historical consciousness of Western medicine. More than any other event, Koch's announcement came to represent the scientific modernization of biomedical culture in the late nineteenth century.

In 1932 there were important fiftieth-anniversary commemorations of Koch's announcement in Germany, the United States, Japan, France, and a few other countries.[2] Each of these commemorative efforts provides a valuable window into the cultural contexts of medicine. More than in other countries, however, memories of Koch were especially important for the professional identity of German and American physicians. During the tumultuous interwar period, German doctors sentimentally looked back to a more stable and heroic time. They celebrated Koch as a matter of national pride, for the announcement represented all the positive aspects of bacteriology and the new laboratory-based medicine. Americans too had compelling reasons to commemorate Koch. Many of the American doctors who remembered Koch in 1932 had spent time at German-speaking universities or studied under those who had gone abroad, and these experiences were central to the transformation of American medicine in the late nineteenth and early twentieth centuries.

For both Germans and Americans, Koch's announcement fit conveniently and elegantly into a grand professional narrative of medical origins. Pointing to the epistemological, esthetic, practical, and historical significance of the discovery of the tubercle bacillus, commemorators of Koch in Germany and the United States retold the story of mid- to late nineteenth century medicine. They situated the Berlin announcement at the fulcrum of a sea change in medical culture. Employing Koch's announcement as an historiographical prism, celebrants told two related professional narratives—what happened before and what happened after the evening of 24 March 1882. In their commemorative representations, the historical events and figures that predated Koch's announcement teleologically led to that momentous evening in Berlin. And their narratives of what happened after that evening granted Koch's announcement the unique and powerful privilege of originator.

On the one hand, commemorations were opportunities to reaffirm this grand story of the scientific transformation of medicine. Yet, on the other hand, celebrants tended to promote their individual ideologies and agendas. As the many celebrations of Koch in Germany and the United States reveal, the cultural puissance of commemorative representations may be the opportunity they provide both to reaffirm a collective professional past and to champion present concerns.

[2] See citations n. 21 and 22 for references to Koch commemorations in Japan and France.

## I. GENESIS OF AN ORIGINS STORY

I return again to the events that took place in the reading room of the Berlin Hygiene Institute on 24 March 1922.

> "Ladies and gentlemen," requested Bruno Heymann, in this historical room, which I have restored as much as possible to the condition it was in at the time, please allow me the following attempt: let me construct an image of that meeting as faithfully as the evidence allows—namely, the accounts of still-living eyewitnesses, both published and unpublished sources, and demonstration materials that are still at hand.[3]

Speaking in the present tense and using the first-person plural, Heymann repeatedly reminded his audience that they, themselves, were active participants in this commemoration of Koch. He set the stage for the public spectacle by describing the weather on that fateful day. He then took his audience on a figurative promenade through Bismarck's Berlin, filling them with nostalgia for a "time that had the pomp of marble and gold."[4] After entering the three-story building that housed the meeting site of the Berlin Physiological Society, the celebrants took an imaginary walk beside paintings depicting Albrecht von Haller (1708–1777) and Johannes Müller (1801–1858). In effect, in Heymann's theatrical narrative, the celebrants were passing the icons of Germany's medical past on their way to the coronation of a new figure in medical history.

Heymann attempted to build the tension and expectation in his audience that the original audience must have experienced. "Attracted by the sound of our footsteps," he narrated, "the young building attendant of the institute comes up from the basement and busily hastens us, hopefully directing us to the reading room where the lecture is to take place."[5] Koch had modestly—and strategically—called his talk "On Tuberculosis," foregoing, according to Heymann, a title that better reflected the revolutionary content of the presentation. Koch's audience did not know what to expect from a talk with such an unspecific title, yet they expected a great deal from the former country doctor who had earlier impressed the medical community with his bacteriological isolation techniques for anthrax (1876) and other bacilli.

Heymann then set the stage for a collective re-enactment, bringing his imaginary historical actors into the reading room one by one. Robert Koch and his assistant Friedrich Löffler (1852–1915) entered, and Heymann instructed the celebrants to imagine they heard Koch saying, "No, no, my colleague Löffler; it will require at least a year-long battle before clinicians recognize my discovery."[6] With this quote, Heymann was making a direct—but not verbatim—reference to a commemorative essay that Löffler had written on the occasion of the twenty-fifth anniversary of Koch's discovery.[7] In the culture of medical commemoration, each successive com-

[3] Heymann, "Zum 40-jährigen Gedenktage" (cit. n. 1), p. 337.

[4] *Ibid.*, p. 339.

[5] Heymann, "Zum 40-jährigen Gedenktage" (cit. n. 1). This "young building attendant" was not just a theatrical device employed by Heymann. In the first footnote to his commemorative address (p. 343), Heymann pointed out that he had actually spoken with the man who had been the building attendant at the time of Koch's announcement. The quote is from p. 339.

[6] *Ibid.*, p. 341.

[7] Löffler used the following words to describe the situation: "On the way to the meeting, he [Koch] was still opposed to me, claiming that it would require a year-long battle before his discovery of the

memorative event itself becomes an historical artifact and the source of future stories and memories.

The article to which Heymann referred, Löffler's 1907 commemorative essay in the *Deutsche Medizinische Wochenschrift,* can rightfully be called the original or *ur*-commemoration. That is, the anecdotes and historical explanations that Koch's closest collaborator provided there became the origin and well-spring of stories and memories that later doctors repeated in their own commemorations. Koch was still alive when the essay was published, and it was Löffler's chance to pay him tribute by reminding a new generation of physicians of how important the Berlin announcement was. Löffler based his commemorative representation on fewer than two pages of minutes from the original meeting and his own recollections.[8]

For the fortieth-anniversary commemoration, Heymann did make an effort to find firsthand oral and written accounts of the announcement other than Löffler's, but, by and large, commemorations were not a time to present original historical research. On the contrary, celebrants tended to construct their own stories by capitalizing on earlier anecdotes. The more an anecdote was repeated, the more established it seemed in the collective memory of the profession. This layer-upon-layer process of storytelling is, of course, an important element of most types of historical discourse. However, it is the detached rhythm of the calendar that regulates the sequential aspect of commemorative events. This cold logic allows us to see how, anniversary after anniversary and stratum upon stratum, medical commemorations repeat and perpetuate certain central anecdotes, while at the same time allow for new refractions of the same old stories.

Before Löffler's 1907 commemorative representation, much of the lore surrounding the original meeting was still part of medicine's culture of oral storytelling. But by the time of the twenty-fifth anniversary, Löffler felt it was time to publish some of the details concerning the original meeting. "Without doubt," claimed Löffler, "the discovery of the tubercle bacillus belongs among those earth-moving events, and therefore the circumstances surrounding the meeting and the accompanying details claim a very special interest."[9]

Löffler first explained why the great discovery was announced at the Berlin Physiological Society rather than at any of the other large medical societies in Berlin. Rudolf Virchow (1821–1902), the founder of cellular pathology, dominated the medical community in Berlin at the time, and in light of the cold reception he had given Koch's earlier work, it made little sense to present this sort of material before a prejudiced and hostile audience. Löffler then described how Koch conducted his presentation: "The words faltered and came out of his mouth slowly, but what he said was clear, simple, built on powerful logic—pure, unadulterated gold."[10]

With one brief anecdote after another, Löffler provided future celebrants with the kernels and foundation for a multitude of stories and myths. For example, after

---

tubercle bacillus would achieve recognition by clinicians." In Friedrich Löffler, "Zum 25 jährigen Gedenktage der Entdeckung des Tuberkelbacillus," *Deutsche Medizinische Wochenschrift,* 1907, *33,* n. 12:449.

[8] See Heymann's description of the original "Sitzungsbericht" (meeting report) of the Berlin Physiological Society in footnote n. 1 on p. 343 of Heymann, "Zum 40-jährigen Gedenktage" (cit. n. 1).

[9] Löffler, "Zum 40-jährigen Gedenktage" (cit. n. 7), p. 449.

[10] *Ibid.,* p. 450.

Koch's presentation, there was no thunderous applause. "No, everything remained still," wrote Löffler. "But there was not a single person in the room, who—even against his will—was not convinced by the validity of what he had heard and seen, and did not have the feeling that he had been present at an epoch-creating announcement of colossal consequences that would rattle the entire medical world."[11] Almost without exception, later celebrants—especially at the fiftieth-anniversary commemorations—would describe that evening as an epoch-making event. Even though some made blunders when citing dates and names, these later celebrants all recapitulated Löffler's commemorative representation of the awe and stillness experienced at the end of Koch's announcement.

Fifteen years after Löffler's article, at the fortieth-anniversary commemoration led by Heymann, the memories of the Berlin announcement were no longer in the hands of a direct witness of the meeting. It was a chance for the next generation of German bacteriologists and hygienists to claim their lineage from that original announcement. Heymann described in great detail the slides, cultures, and pathological specimens that Koch and Löffler unpacked before their audience. As part of his own theatrical commemoration, Heymann also presented these very items, many of which were "in their original glasses and some with Koch's own signature."[12] These treasured relics of an heroic past only added to the authenticity of the collective re-enactment. After the imaginary Koch and Löffler arranged the microscopes and other materials, Heymann recounted how Emil Du Bois-Reymond (1818–1896), head of the Berlin Physiological Society, stood and offered a few words of introduction. Echoing Löffler's words from 1907, Heymann then described Koch's first utterances: "His words come out a bit choppy and faltering, but they are simple and crystal-clear (*goldklar*)."[13] Heymann provided the celebrants with many details about Koch's presentation, but he also made a great effort to lead his audience through the emotions of the original historical actors. "Koch surprises us with the explanation that he has declared it as his goal—as an especially 'urgent duty' of the Imperial Health Office (Reichsgesundheitsamt)—the solution of this problem,"[14] narrated Heymann. And, near the close of the re-enactment, Heymann uttered, "Koch is at the end; he summarizes briefly and turns around to take his seat. But in the room everything remains still. It is as if no one can cease pondering this amazing accomplishment."[15]

By the time of this commemoration in 1922, Koch's ideas were, for the most part, unquestioned in biomedical culture, and the central doctrines of germ theory and bacteriology were taken for granted. By provoking in his audience the emotions of witnessing Koch's lecture, Heymann hoped to revive the wonder and amazement of that original experience. The forty years that had passed since the original audience gathered at the Berlin Physiological Society had diminished much of the brutal novelty and singularity of Koch's announcement. The 1922 commemoration was, in the end, an effort to revive collectively that epoch-creating moment. I suggest that Heymann's narrative spectacle was the sincerest form of commemorative representa-

[11] *Ibid.,* p. 450.
[12] Heymann, "Zum 40-jährigen Gedenktage" (cit. n. 1), footnote n. 4 on p. 341.
[13] *Ibid.,* p. 342.
[14] *Ibid.,* p. 343.
[15] *Ibid.,* p. 347.

tion, for he attempted to make the original meeting present again—he re-*present*ed Koch's announcement.[16]

Ten years later, the detached logic of the calendar again called on physicians to remember and celebrate that fateful Friday evening in Berlin, and 1932 was indeed *the* year of Koch commemorations. The golden anniversary of the announcement alone may have been enough to encourage greater commemorative activity. Yet there were other factors that made the fiftieth anniversary an especially important occasion. These were the twilight years of the generation of physicians who could still remember the time before Koch's announcement. Many of Koch's own students in Germany and America were still alive, and this was one of their last chances to harken back to an event that had defined an entire generation of physicians.

However, this very ability of an historical event to define the collective identity of a generation is precisely what deserves questioning and exploration. Common expressions such as "generation-defining event" reflect an historical consciousness that grants a specific, transcendent power and cultural authority to the past. But this way of thinking about the past elides the active, constructive cultural practices that endow memories with the ability to influence thinking and doing. Once the memory of Koch's announcement had passed into the hands of those who did not witness it themselves, these later celebrants *chose* to preserve this memory and its cultural authority. It cannot be denied that Koch's memory was still perpetuated in 1932 as a central part of professional medical culture, but this cultural process of mnemonic inheritance should not be taken for granted. What, at first glance, appears to be the passive inheritance and repetition of a commemorative tradition was, nonetheless, still a matter of cultural choice.

## II. LATE-WEIMAR MEMORIES OF AN HEROIC PAST

In the 1932 German commemorations of Koch and his discovery, a blatant sort of professional nostalgia is unmistakable. German doctors recounted stories of the time when Berlin, Vienna, and German-controlled Strasburg were the principal pillars of the new scientific ethos in medicine. Koch's memory symbolized that time, when German-speaking universities were at the center of Western medicine. No doubt, many German physicians viewed Koch as a valuable icon of German pride and nationalism, but many also saw him as a symbol of German contributions to the international community. Boycotted by much of the scientific world in the early 1920s, German physicians hoped that commemorations of Koch would encourage a restoration of their former strong ties to the international medical community.

The grandest of the German fiftieth-anniversary commemorations took place on 5 March 1932 in Berlin at the former Prussian House of Lords (Herrenhaus), the

[16] Despite the theatricality of Heymann's commemorative representation, it is probably the most valuable account that we have of the Berlin announcement. Heymann used an unaltered version of his fortieth-anniversary address as the last chapter in vol. 1 of his biography, *Robert Koch 1843–1882* (Leipzig: Akademische Verlagsgesellschaft, 1932), pp. 310–30. (A second volume was never published.) Biographies written since Heymann have relied, almost exclusively, on Löffler's and Heymann's representations. See, for example, the best English language biography, Thomas D. Brock, *Robert Koch: A Life in Medicine and Bacteriology* (Madison: Science Tech Pub., 1988), pp. 126–9; footnote n. 13, p. 329. See also Wolfgang Genschorek, *Robert Koch: Leben, Werk, Zeit* (Leipzig: S. Hirzel Verlag, 1975), pp. 103–8; notes on pp. 234–6.

meeting place of the Prussian State Assembly (preußischer Staatrat).[17] Two days before the commemoration, the assembly announced its approval to change the name of one of Berlin's city squares, Luisenplatz, to Robert Koch-Platz. Members of Koch's family and his former students were among the group of physicians, public health workers, and government officials who gathered in his honor. The Berlin Physician-Orchestra began the ceremony with a performance of the overture to Händel's *Rodelinde.*

The leader of the Prussian medical system, Heinrich Schopohl (1877–1963), offered an apologetic introduction to the day's event. It had been the intention of the organizing committee to invite a good number of foreign guests to the commemoration, but, as Schopohl explained, "the general situation in Germany does not permit such an arrangement."[18] Indeed, street fighting between Nazis and Communists and general economic hardships had made a more international gathering impossible. Even though few foreign guests attended the commemoration, Fred Neufeld (1869–1945) proudly published in a commemorative booklet the numerous letters and telegrams he had received from foreign scholars and officials. One after another, individuals, institutes, and organizations throughout the world had paid verbal tribute to Koch. William Bulloch and John Charles Grant Ledingham of the Lister Institute of Preventive Medicine in London wrote, "We recall with pleasure that the name our Institute bears was the friend and admirer of Robert Koch. The high traditions set by Lister, we, his humble successors, strive to foster and with him we share the frankest admiration for the pioneer achievements of your great compatriot."[19] Albert Calmette (1863–1933), one of the developers of the BCG vaccine and France's greatest supporter of Koch, apologized for not being able to attend the Berlin commemoration, but announced that the Académie de Médecine would dedicate one of its meetings to Koch.[20] In similar fashion, Sahachiro Hata (1873–1938) of the Kitasato Institute for Infectious Diseases apologized for not sending a representative of his institute to Berlin but commented, "We also will arrange in Tokyo a 50th Year Commemoration of Koch's great discovery."[21] Americans also sent a number of letters and telegrams to Berlin on the occasion of the Robert Koch-Ehrung. William Henry Welch (1850–1934), on behalf of President Herbert Hoover, the Johns Hopkins University, the Rockefeller Institute, and the whole American nation sent a message to recognize "the great Benefactor Scientist and Founder of the Modern Science of Bacteriology on [the] fiftieth anniversary of [the] discovery of [the] Tubercle Bacillus."[22]

After Schopohl's speech, the Prussian minister for public welfare (Volkswohl-

[17] *Die Robert Koch-Ehrung in Berlin, am 5. März 1932* (Berlin: Verlagsbuchhandlung von Richard Schoetz, 1932), p. 5.

[18] Heinrich Schopol in *Die Robert Koch-Ehrung* (cit. n. 17), p. 7.

[19] John Charles Grant Ledingham to Fred Neufeld, 11 March 1932, reprinted in *Die Robert Koch-Ehrung* (cit. n. 17), p. 37.

[20] Albert Calmette to Fred Neufeld, 15 February 1932, *ibid.,* p. 38. For a published version of Calmette's address see M. A. Calmette, "Robert Koch: Le cinquantième anniversaire de la découverte du bacille tuberculeux," *Bulletin de l'Académie de Médecine,* 8 March 1932, *107,* n. 10:346–56.

[21] Sahachiro Hata to Fred Neufeld, 16 February 1932, in *Die Robert Koch-Ehrung* (cit. n. 17), pp. 39–40. A Japanese publisher printed a commemorative pamphlet in German for this occasion. See *Die Gedenkfeier anlässich der Robert Kochs Entdeckung des Tuberkelbazillus am 24, März, 1932 in Tokio, Japan* (Tokyo: Kokusai Shuppan Insatsusha, 1932).

[22] William H. Welch to Fred Neufeld, 23 March 1932, in *Die Robert Koch-Ehrung* (cit. n. 17), p. 44.

fahrt) and the president of the Imperial Health Office took the stage. These government officials took the opportunity to emphasize Koch's importance for the establishment of national laws to combat epidemic diseases. They pointed out that when Koch had originally come to Berlin, he was appointed as an official in the Imperial Health Office. Remembering Koch's career as a government health official was a way to underscore the connection between medical science, public health, and the government. And they gave Koch much of the credit for encouraging more government involvement in public health reform.

The last speaker at the Berlin commemoration was Richard Pfeiffer (1858–1945), who at 74 years of age was one of Koch's oldest living students. Pfeiffer took this chance to provide the audience with a "living" memory of Koch. He related the memory he had of first encountering, as a student, the "epoch-making" publication of Koch's announcement: "The weight of the facts filled me with enthusiasm, and the puzzles of tuberculosis infection were solved with a simple and convincing form of presentation—so much so that it was impossible for even the greatest skeptic to raise a sincere objection."[23] The next year, in 1883, Pfeiffer visited the first Berliner Hygienesche Ausstellung (Berlin Hygiene Exhibition) and experienced something tantamount to a religious conversion. There, in the pavilion of the Imperial Health Office, he examined a display of Robert Koch's microphotographs of bacteria, and for the first time he witnessed an actual culture of the tubercle bacillus.[24] Pfeiffer solemnly described this experience as a "revelation" (*Offenbarung*). "It then became clear to me," explained Pfeiffer, "that the greatest goal of my life would be to make personal acquaintance with the brilliant man, that prince of science whom we thank for these wonderful achievements, and I modestly hoped to participate in the development of the new aetiological science."[25] Pfeiffer, of course, soon had a chance to fulfill his dream as Koch's student and colleague.

With these final, emotional reminiscences of Pfeiffer, the celebrants ended their commemoration of Koch. According to the commemorative pamphlet, "This dignified celebration made a deep impression on all who were present, [with] a brilliant rendition of the Larghetto from Beethoven's Second Symphony providing the conclusion."[26]

Other former students and coworkers of Koch participated in additional commemorations of the discovery of the tubercle bacillus in 1932. Paul Uhlenhuth (1870–1957) was one of Koch's colleagues who had worked on immunological techniques of protein identification and antimony treatments for certain tropical diseases. Placing the fiftieth-anniversary commemoration of Koch's announcement within the context of other commemorations of "great men," he insisted that "in life we need *rays of hope* (*Lichtblicke*)—especially in this gloomy time in which political strife and economic necessities rattle the foundations of our culture and threaten to unsettle

[23] Richard Pfeiffer in *Die Robert Koch-Ehrung* (cit. n. 17), p. 27.

[24] Koch was one of the first bacteriologists to use microphotography as a means to provide "objective" representation of bacteria. See Thomas Schlich, "Repräsentationen von Krankheitserregern: Wie Robert Koch Bakterien als Krankheitsursache dargestellt hat," in *Räume des Wissens: Spur, Codierung, Repräsentation,* eds. Hans-Jörg Rheinberger, Michael Hagner, and Bettina Wahrig-Schmidt (Berlin: Akademie-Verlag, 1996), pp. 165–90.

[25] Pfeiffer in *Die Robert Koch-Ehrung* (cit. n. 17), p. 27.

[26] *Die Robert Koch-Ehrung* (cit. n. 17), p. 35.

us."[27] Uhlenhuth claimed that "great men"—through the power of their personalities and their brilliant achievements—embodied the unity and connectedness of humanity, qualities that were so necessary in times of political and economic upheaval. And commemorations of great men were particularly important, since they were these necessary "rays of hope," he insisted.

Uhlenhuth was quick to point out, however, that his personal mentor, Koch, was unlike most of the great men of history books. Koch had not been a military or political leader; rather, he had "forged weapons in the struggle against the common enemy of humanity—the causes of epidemics." Uhlenhuth maintained that this sort of struggle was "more useful for humanity than the battles of peoples among one another"[28] and that Koch's greatness resulted from his ability to transcend national boundaries. "All peoples of the world are one with us Germans in offering thanks to the man who fifty years ago announced the pioneering discovery of the *etiology of tuberculosis*,"[29] claimed Uhlenhuth.

Even when remembering Koch's greatest research failure—the premature and erroneous announcement at the 1890 Berlin International Medical Congress that "tuberculin" was a curative agent—Uhlenhuth defended his teacher to the end. "We today know," said Uhlenhuth, "that Robert Koch was correct on all the essential points and through careful and skillful application and correct dosage [of tuberculin], in many cases a brilliant curative effect can be obtained."[30] Most German commemorators in 1932 glossed over the tuberculin failure or tended to emphasize tuberculin's diagnostic value. Even Koch's most ardent supporters rarely claimed that tuberculin had a curative value. But Uhlenhuth proudly recalled the "unforgettable happy time" when Koch personally led him through the application of tuberculin and convinced him of its success. By defending tuberculin in this way, Uhlenhuth defined his professional identity against a younger generation of German doctors who had given up on tuberculin's therapeutic uses. The fiftieth-anniversary commemoration was, in part, a chance for Uhlenhuth's generation to once again take center stage in professional medical culture and celebrate its youthful accomplishments, even if it meant celebrating an apparent failure.

More than most other German physicians, Wilhelm Kolle (1868–1935) emphasized memories of Koch as a valuable symbol of German national pride. "Although Koch's efforts have been honored by all nations," insisted Kolle in a commemorative essay, "and all peoples have in general benefited from the results of his research, *Koch* was, and still remains a German scientist, a pioneer of the German spirit and the German method."[31] In the 1890s Kolle had investigated the immunology of typhus and cholera at the Robert Koch-Institut in Berlin, and as another one of Koch's remaining students in 1932, Kolle "cherished" the opportunity to once again expound on the meaning of his master's discovery. He believed that "every German

[27] Paul Uhlenhuth, "Die Auswirkung der Kochschen Entdeckung auf die Heilkunde," in the "Zum 50. Jahrestag (24. März 1882) der Bekanntgabe der Entdeckung des Tuberkelbazillus durch Robert Koch" issue of *Medizinische Klinik,* 18 March 1932, *28,* n. 12:387.

[28] *Ibid.,* p. 387.

[29] *Ibid.,* p. 387.

[30] *Ibid.,* p. 392.

[31] Wilhelm Kolle, "Robert Koch—Gedenkworte zur 50. Wiederkehr des Tages, an dem Robert koch die Entdeckung des Tuberkelbazillus bekannt gab," *Die Medizinische Welt,* 5 March 1932, *6,* n. 10:328.

doctor and researcher can be proud that *Robert Koch* was a German and will therefore honor the memory of his dominating spirit."[32] Kolle recapitulated earlier descriptions of the announcement as a "turning point" in the history of medicine and insisted that Koch be placed in the company of Vesalius, Morgagni, Harvey, and Malpighi—other men who had inaugurated new eras in medicine. Furthermore, in his opinion, even the discoveries of Jenner, Lister, Pasteur, Virchow, Helmholtz, and Roentgen paled in comparison to Koch's accomplishments.

In his essay, Kolle nostalgically recalled Koch's research expeditions to tropical climates, and he associated Koch's memory with the brief period of German colonialism. Kolle himself had enthusiastically participated in various expeditions to the tropics, studying leprosy and cattle-plague in South Africa, Egypt, and the Sudan.[33] For Kolle, the fiftieth anniversary of the Berlin announcement was a chance to revive pre-Great War German nationalism and colonial aspirations:

> In the last decades of his life, *Robert Koch* turned to the study of tropical diseases . . . In him was the conviction that Germany needed colonies, namely the German colonies in Africa. This wholehearted interest always was with him, and he conducted his research in order to free the fertile German regions of East and West Africa of the dreaded diseases and prepare them for German settlement. He did this with no little interest in his fatherland.[34]

By celebrating Koch's efforts in tropical medicine and connecting them with past colonial aspirations, Kolle provided his not-too-subtle support for the emerging *realpolitik* of late-Weimar Germany. "I must in this essay make reference to this great tragedy," explained Kolle, "that a people with such a surplus population should lose, because of the outcome of the war, these hopes for a colonial empire—even the very ideas and developing prospects which Robert Koch so enthusiastically inflamed with his efforts."[35]

Kolle wrote much of his essay in this personal, sentimental style. He claimed that the "fateful tragedy" of the last twenty years tended to make Germans forget the truly great events of the past; therefore, it was especially important for the current generation of German physicians not to forget Koch's great accomplishments. He saw Koch's announcement as the singular event of his generation, and he insisted that Koch's memory would continue to exert an influence in medicine. Even centuries later when new questions concerning tuberculosis would arise, Kolle predicted, Koch's discovery would still "connect everything." "Perhaps," speculated Kolle, "historians will at some later time, after studying the medical literature and publications, establish the enormity of the impact that this discovery made on the entire inhabited earth."[36]

German-speaking physicians in 1932 participated in a wide variety of commemorative activities. In addition to the grand commemoration in Berlin, special meetings of a professional society in Vienna and commemorative lectures at the Universität

[32] *Ibid.*, p. 328.

[33] Manfred Stürzbrecher, "Wilhelm Kolle," *Neue Deutsche Biographie,* vol. 12 (Berlin: Duncker & Humboldt, 1980), pp. 464–5.

[34] Kolle, "Robert Koch—Gedenkworte zur 50" (cit. n. 31), p. 331.

[35] *Ibid.*

[36] *Ibid.*

München were held. Also, a number of medical journals dedicated a special issue to the memory of Koch's discovery.[37] While doctors chose to commemorate Koch in greatly diverse ways, there was, conversely, an astonishing degree of repetition and similarity in the structure of the stories that celebrants told.

In a lecture given before the Society of Physicians (Aerzlicher Verein) in Munich, Karl Kißkalt (1875–1962) structured his commemoration of Koch around the history of tuberculosis, and in many ways his address was typical of the 1932 commemorative representations. Kißkalt suggested that it was difficult to grasp the meaning of Koch's discovery without first examining "what was wrong or lacking" prior to 24 March 1882.[38] This sort of evocation was a common way to begin a speech or essay commemorating Koch. Another celebrant wrote in the *Deutsche Medizinische Wochesnschrift,* "If one wants to appreciate the meaning and effect of a great but long-past discovery on medicine and its specialties, one must first go back to the 'spirit of the time.'"[39] Yet another physician wrote, "On a memorial day it is indeed customary to glance back, and it is therefore necessary to place oneself back in the spirit and thoughts of the past time."[40]

While it is not altogether implausible that celebrants borrowed certain expressions from each other, I suggest that these evocations were more than a mere rhetorical convention, for they reveal an important facet of the culture of medical commemoration. In Heymann's theatrical re-enactment, there was an obvious attempt to make the past present again. And this sort of re-*present*ation (even if not as obvious) was crucial for other commemorative efforts, too, such as these evocations of a time before Koch's announcement. Commemorations were occasions to reflect on the meaning of Koch's discovery, but, as Kißkalt and other commemorators repeatedly insisted, this was not altogether possible in 1932. The full meaning of Koch's memory depended on a well-articulated historical narrative, and in order to recover the novelty and singularity of Koch's discovery, celebrants carefully revived the forgotten story of events and figures that predated him.

After his evocation, Kißkalt turned his attention to tuberculosis research before Koch. Relying on published histories of the disease, he recounted numerous mid-nineteenth century theories concerning the causes of consumption, including "getting a cold, unfavorable climate, poor nutrition and habitation, dangerous occupation, emotions, over-stressing the lungs."[41] (Preparing the historical terrain for Koch's discovery by recounting prior tuberculosis research was a very typical feature of commemorative representations in 1932.) In one historical account after another, physicians described the confused atmosphere with different and contradictory theo-

[37] See, for example, *Zeitschrift für Tuberkulose. Robert Koch-Heft,* 1932, *64:*1–126; *Deutsche Medizinische Wochenschrift,* 1932, *58:*474–510; *Münchner Medizinische Wochenschrift,* 1932, *79:*497–505; *Medizinische Klinik,* 1932, *28:*387–407; *Klinische Wochenschrift* 1932, *11:*489–92; *Die Medizinische Welt* 1932, *6:*325–64; *Wiener Klinische Wochenschrift* 1932, *45:*417–22; *Angewandte Chemie,* 1932, *45:*273–6.

[38] Karl Kißkalt, "Die Entdeckung des Tuberkelbazillus (Rede, gehalten im Aertzlichen Verein München, zur Feier des 50 jährigen Jubiläums)," *Münchner Medizinische Wochenschrift,* 25 March 1932, *79,* n. 13:497.

[39] E. Payr, "Über den Einfluß der Tuberkuloselehre R. Kochs auf die Chirurgie," *Deutsche Medizinische Wochenschrift,* 25 March 1932, *58,* n. 13:481.

[40] G. Seiffert, "Die Tuberkulose als übertragbare Krankheit und ihre Bekämpfung vor Robert Koch," *Münchner Medizinische Wochenschrift,* 25 March 1932, *79,* n. 13:501.

[41] Kißkalt, "Die Entdeckung des Tuberkelbazillus" (cit. n. 38), p. 497.

ries of tuberculosis that existed when Koch began his research. They reminded their audiences and readers that it was difficult for the current generation of physicians to imagine how baffling the situation was in the mid-nineteenth century.

When Koch finally achieved his goal, argued Kißkalt, "The first perception must have been that with this, a centuries-long yearning had been fulfilled—that finally a secure foundation was at hand with which the cure and eradication of this illness could be pursued."[42] While much of Kißkalt's oration focused on Koch's discovery as an intellectual and scientific accomplishment, he also underscored the "high esthetic pleasure of the compelling logic of the report," and he suggested that by formulating a new conception of disease, Koch had, in fact, discovered an entirely new disease. Kißkalt ended the first part of his commemorative narrative with Koch's originating discovery, and then he turned to the story of everything that resulted from it.

Many commemorative essays focused on the practical value of Koch's discovery and its impact on public health. By enthroning Koch as the originator of the scientific understanding of tuberculosis, commemorators could give him ultimate credit for the remarkable decline in morbidity and mortality from tuberculosis in the fifty years since the Berlin announcement. In what amounted to a roll call of medical specialties, the heads of medical departments at many German universities paid tribute to Koch's discovery. Seeing his influence in almost all aspects of contemporary medicine, they wrote articles such as, "Effects of Koch's Discovery of the Tubercle Bacillus on Pediatrics" and "The Decline of Tuberculosis since the Discovery of the Tubercle Bacillus and its Causes."[43] Ultimately, the celebrants told two commemorative narratives with two ends in mind. The first story began with the murky state of tuberculosis research prior to Koch and teleologically led to him as originator, and the second story began with the Berlin announcement and ended with its ongoing effects in public health and Koch's widespread influence in 1932. Although celebrants always advanced the pretense that commemorations were dedicated to the past, by deriving their professional lineage from Koch's announcement and telling a careful tale of their *own* origins, they had a more selfish end in mind.

The fiftieth anniversary of the 1882 announcement transpired at the cusp of great tumult in German medicine and in the country at large. This was the twilight of the Weimar Republic and the eve of National Socialism in Germany. Soon, the term "hygiene" would take on its particularly heinous Nazi refraction. These were also the last days of a German professional medical culture that could, in all sincerity, claim its lineage from the time of Virchow and Koch. Historically minded Nazi physicians, too, would claim their cultural lineage from these times, but the 1930s intellectual exodus from Germany and physicians' active participation in both "racial hygiene" and the Nazi genocide machine altered and perverted German medical culture to an almost unrecognizable form.[44] The rhythm of commemoration follows a logic of its own; however, the fact that the fiftieth anniversary of Koch's announcement fell on 1932 has the undeniable poignancy of a swan song.

[42] *Ibid.,* p. 499.

[43] H. Kleinschmidt, "Auswirkungen der Kochschen Entdeckung des Tuberkelbazillus in der Pädiatrie," *Medzinische Klinik,* 18 March 1932, *28,* n. 12:398–400; B. Möllers, "Der Rückgang der Tuberkulose seit der Entdeckung des Tuberkelbazillus und seine Ursachen," *Die Medizinische Welt,* 5 March 1932, *6,* n. 10:340–1.

[44] Heinrich Zeiss, the head of the Berlin Hygiene Institute during the period of National Socialism, was especially concerned with the role of commemoration and medical historiography in Nazi

### III. MEMORIES OF ROBERT KOCH AND AMERICAN PROFESSIONAL IDENTITY

American physicians who commemorated Koch in 1932, by and large, saw Germany as the source of much of their professional identity. In his classic account, Thomas Neville Bonner illustrates how many of the eventual leaders of late nineteenth- and early twentieth-century American medicine spent time studying in German hospitals and laboratories.[45] In his words, "Between 1870 and 1914 a high proportion of the most talented and ambitious American medical men studied abroad in German universities. Upon their return they exerted a powerful influence on the direction of the science and practice of medicine."[46] In a more nuanced and comprehensive study of professional culture and foreign education, John Harley Warner shows how, in the early to mid-nineteenth century, American experiences in France—or, more specifically, the stories that derived from these experiences—were central to the formation of American professional identity.[47]

Taken together, these efforts by Warner and Bonner sketch the grand narrative of the transformations of American medical culture. Bonner ended his story with the First World War and the assumption that American medicine was on its way towards cultural independence and ascendancy, while German medicine was entering a period of decline. Certainly, such generalizations are necessary for drawing the grand narrative of medicine, but the culture of commemoration does not necessarily obey these simple demarcations. As Warner suggests, even in the period of German ascendency in the late nineteenth century—long after study in Paris had lost its allure for most Americans—the shared memories and stories of experiences in France still provided a generation of American doctors with their cultural coherence.

So, too, with the leaders of American medicine in the early twentieth century. At a time when Weimar Germany had lost much of its popularity and Americans were turning to their own much-improved universities for training, memories of experiences in Germany were still alive in many of America's most eminent physicians. And with these memories also came the epistemological, esthetic, ethical, and professional convictions that we tend to associate with modern American medicine. Remembering Koch in 1932 meant much more to Americans than a chance to laud a foreign medical researcher. By placing Koch at the center of the transformation of late-nineteenth-century medicine, Americans were not just recapitulating a foreign grand narrative. They were celebrating and reaffirming their own origins: they were, in the end, commemorating *themselves.*

In the United States as in Germany, March and April of 1932 were filled with

---

medical culture. Nazi uses of the medical past deserve further study. For an account of Zeiss see Paul Weindling, "Heinrich Zeiss, Hygiene and the Holocaust," in *Doctors, Politics and Society: Historical Essays,* eds. Dorothy Porter and Roy Porter (Amsterdam: Ridopi B. V., 1993), pp. 174–87. See also George E. Haddad, "Heinrich Zeiss as *Vertrauensmann* (Confidant) of the Nazi Party and Medical Historiography," 1994, unpublished manuscript in author's possession. For a study of the 1930s intellectual exodus from Germany, see Peter Schneck, "Sozialhygiene und Rassenhygiene in Berlin: Die Schüler Alfred Grotjahns und ihr Schicksal unter dem NS-Regime," in *Exodus von Wissenschaften aus Berlin: Fragestellungen-Ergebnisse-Desiderate Entwicklung vor und nach 1933* (Berlin/New York: Walter de Gruyter, 1994), pp. 494–509.

[45] Thomas Neville Bonner, *American Doctors and German Universities: A Chapter in International Intellectual Relations, 1870–1914* (Lincoln: University of Nebraska Press, 1963).

[46] *Ibid.,* p. vii.

[47] John Harley Warner, *Against the Spirit of System: The French Impulse in Nineteenth-Century American Medicine* (Princeton: Princeton University Press, 1998).

different sorts of remembrances of Koch's discovery of the tubercle bacillus. The Philadelphia College of Physicians and the New York Academy of Medicine each dedicated an evening to the anniversary of Koch's announcement. The Laënnec Society of the Johns Hopkins Hospital held a commemoration in honor of Koch, and for a week "Books, Specimens, Cultures and other material" pertaining to the discovery were displayed on the floor of the Welch Medical Library.[48] At Yale University, John F. Fulton led a discussion of the significance of Koch's discovery, and he displayed a number of German and American articles pertaining to Koch.[49]

Henry R. M. Landis (1872–1937), director of the Henry Phipps Institute, opened the commemoration in Philadelphia with a paper entitled, "The Reception of Koch's Discovery in the United States." Landis first offered a step-by-step account of how news of Koch's announcement reached the United States:

> A copy of his address was sent by Koch to the English scientist, John Tyndall, who at once recognized the "serious public import" of the communication and forthwith embodied its salient features in a letter to the *London Times* under the date of Saturday, April 22, 1882. The following day, Sunday, April 23, 1882, the *New York World* contained a brief cable dispatch calling attention to Koch's discovery.[50]

Offering a detailed account of how the news reached America was important, especially since, in Landis' portrayal, this announcement had led to the beginnings of American bacteriology. Before March of 1882, claimed Landis, American physicians were ill prepared for these new ideas flooding in from Europe, since the serious study of pathology and bacteriology was still in a nascent stage. Better-informed Americans had heard of the efforts of Louis Pasteur (1822–1895), Joseph Lister (1827–1912), and the early German bacteriologists; however, it was not, according to Landis, "until Koch made known his discovery of the tubercle bacillus that the medical profession in this country began to take interest in the 'infinitely little.'"[51]

Landis advanced a negative representation of early American bacteriology because he wished to enthrone Koch's announcement as the decisive factor that had established bacteriology in the United States. For Landis and other physicians, a commemoration of Koch was an occasion to reaffirm a professional narrative that wholeheartedly embraced the image of the United States as a scientific backwater of mid-nineteenth-century medical culture. In a recent convincing article, Nancy J. Tomes contests this interpretation of mid- to late-nineteenth-century American medicine—what she calls the "myth of American backwardness."[52] Why, then, did

[48] Harvey Cushing scrapbook, "William H. Welch Memorabilia, 1930–1934," Historical Library, Cushing-Whitney Medical Library, Yale University, p. 76.

[49] John F. Fulton Diaries, vol. V, 20 August 1931–27 July 1932, Historical Library, Cushing-Whitney Medical Library, Yale University, p. 162.

[50] Henry R. M. Landis, "The Reception of Koch's Discovery in the United States," *Annals of Medical History,* 1932, *4:*531.

[51] *Ibid.,* p. 532.

[52] Nancy J. Tomes contends that this myth arose out of a narrative that structured itself around certain "ahistorical premises." She charges an earlier generation of historians of medicine with invoking the "phrase 'germ theory of disease' as if it had an existence independent of a specific historical context" and endowing "experimentalism with a similar transcendent authority." This sort of revisionist enterprise is, indeed, long overdue, but Tomes does much more than provide a new interpretive framework for early American bacteriology. Her revisionist understanding also puts the older professional narrative of medicine in an altogether new and more useful light. See Nancy J. Tomes, "American Attitudes toward the Germ Theory of Disease: Phyllis Allen Richmond Revisited," *Journal of the History of Medicine and Allied Sciences,* 1997, *52,* n. 1:17–50. See also Nancy J. Tomes and John

Landis champion such an interpretation, if Americans were not so backward after all? It is not that Landis and other American physicians were poor historians. Rather, they believed that certain watershed advances in mid- to late-nineteenth-century Germany had decisively placed medicine on a new scientific footing, and the stories they told about pre-Koch American backwardness went hand-in-hand with this belief.[53]

In order to attribute the origin of laboratory-based scientific medicine to Koch, Landis, like his German colleagues, prepared the pre-24 March 1882 historiographical terrain. He retold the story of pre-Koch bacteriology in such a way that these events inexorably led to the Berlin announcement:

> Villemin was the first to partially break through these barriers by showing that tuberculosis could be conveyed to animals, which although received many verifications, also met with well-founded opposition. And the question of whether tuberculosis was an infectious disease still remained debatable. More headway was made by Cohnheim and Salomonson with their inoculation experiments into the anterior chamber of the eye of rabbits, and later by the inhalation experiments of Tappenheimer and others. Thus the stage was set for the discovery of the final link in the chain of evidence.[54]

In effect, the efforts of Villemin, Cohnheim, Salomonson, and Tappenheimer, among others, are realized in this narrative only insofar as they paved the way for the critical event in establishing the bacteriologic revolution—Koch's announcement.[55] There is, of course, a contingency to this sort of narrative. For example, one could retell this story with Villemin's inoculation experiments at the focal point of an altogether different professional narrative, and, in turn, the events before and after Villemin's discovery would be refracted through a different historiographical prism. In this hypothetical alternative narrative, Koch would merely be a subsequent supporter of the efforts that Villemin "originated." Certainly, a countless number of stories could be told. This does not mean that all stories are equal or carry the same cultural cachet, however. Some stories win out over others, and it is safe to say that in 1932, the professional narrative that placed Koch's announcement at center was the one that triumphed in Germany and the United States. And even though, today, we can tell the story of late-nineteenth-century bacteriology and germ theory in a more "historical" way, we should not turn our backs entirely on this older form of professional narrative in medicine. Rather, we should embrace these older stories and examine the cultural contexts and practices that eventually led to an hegemonic professional narrative. We need to ask, What made one story better than another?

The year 1932 was the centennial of another significant event in German history—the death of Johann Wolfgang von Goethe (1749–1832). At first glance, the anniversaries of Goethe's death and the discovery of the tubercle bacillus may appear hope-

---

Harley Warner, "Introduction to Special Issue on Rethinking the Reception of the Germ Theory of Disease" in this same issue for an outline for revisionist interpretations of the early development of bacteriology.

[53] John Harley Warner, in a series of essays, has explored the various uses and refractions of "science" in nineteenth-century medical culture. For a start, see John Harley Warner, "The History of Science and the Sciences of Medicine," *Osiris*, 1995, *10*:164–93.

[54] Landis, "The Reception of Koch's Discovery" (cit. n. 50), p. 535.

[55] Jean-Antoine Villemin (1827–1892) in the years 1865–69 proved that tuberculosis was contagious by infecting laboratory animals with tuberculosis. Julius Cohnheim (1839–1884), one of Rudolf Virchow's students, successfully inoculated the anterior chamber of a rabbit with tuberculosis in 1877.

lessly unrelated. In the culture of commemoration, however, this apparent happenstance was endowed with specific meaning, and the simultaneity of the anniversaries became a springboard from which both men were commemorated. At the annual banquet of the National Tuberculosis Association in Colorado Springs, Colorado, Gerald B. Webb (1871–1948) attempted such a parallel commemoration. (A veteran of medical commemorative events, Webb had been a United States delegate to the 1926 Laënnec-Centenary in Paris.[56]) Webb called Goethe a "poet-naturalist" and reminded his audience that Goethe had announced his discovery of the intermaxillary bone on 27 March 1784, which was about one hundred years before Koch's announcement.

Webb then offered an anecdote-filled account of Koch's life. Even as a young boy, he explained, Koch displayed an impassioned interest in nature, spending much of his time roaming the Harz Mountains collecting minerals, plants, insects, and animals.[57] Goethe, too, had roamed these mountains, but he had gone in search of the stuff of poetry. Webb then quoted from Goethe's "The Harz Mountains":

> Sure a God hath
> Unto each his path
> Fore-appointed,
> Which the fortunate
> Swift to happiest
> Goal pursues[58]

Moving his audience back to Koch, Webb insisted, "That Robert Koch pursued the goal of his fore-appointed path, swiftly, is evidenced by his graduation *eximina cum laude* in 1866."[59] In Webb's narrative, this path gradually and inescapably led to the momentous announcement before the Berlin Physiological Society.

As with the many other commemorations in the United States, Webb's description of Koch's lecture relied almost exclusively on Löffler's account, and his creative representation entailed repeating and reliving anecdotes that were established by previous commemorations. "Löffler described the paper as simple, clear, logical and pure gold," reported Webb. "The demonstrations were convincing. Yet when Koch was finished there was no storm of approval and while the audience was astounded, there was no applause."[60] Webb's address was extremely anecdotal—almost gossipy. He described his personal good fortune to have met Koch in Philadelphia, and spoke of the public scandal surrounding Koch's divorce and second marriage.

Quoting from Goethe's poetry, making rather ostentatious use of German words and expressions, and spreading old gossip may simply have been some of Webb's idiosyncrasies, but the character of his oration served other purposes as well. This was Webb's way of reveling in the German origins of his professional identity. In the 1930s, studying in Germany—or simply being able to read a German scholarly article—no longer carried the professional cachet that it did for Webb's generation.

[56] "Gerald Bertram Webb," *Who Was Who in America,* vol. 2, 1943–1950 (Chicago: The A. N. Marquis Company, 1950), p. 563.
[57] Gerald B. Webb, "Robert Koch [1843–1910]," *Annals of Medical History,* n. s., 1932, *4:*510. Webb read this commemorative address in Colorado Springs on 6 June 1932.
[58] *Ibid.,* p. 510.
[59] *Ibid.,* p. 510.
[60] *Ibid.,* p. 517.

***Figure 3.*** *Group photograph of participants in the 1932 Johns Hopkins commemoration of Robert Koch's discovery of the tubercle bacillus. Lawrason Brown stands at left; William H. Welch sits at right. (Courtesy of the Historical Library, Cushing-Whitney Medical Library, Yale University.)*

But the memories and stories that made up American commemorations of Koch in 1932 did not belong to the young. For the most part, an older generation was guardian of the commemorative version of medicine's cultural memory, and the golden anniversary of Koch's announcement was an opportunity for Webb and his cohort to engage in a bit of chatty professional nostalgia.

March, April, and May of 1932 were busy months for Lawrason Brown (1871–1937), founder of a journal for tuberculosis patients, the *Journal of Outdoor Life,* and head of the Saranac Lake Sanitorium. On 14 March he spoke at the College of Physicians in Philadelphia; at another Koch commemoration on 11 April he spoke at the Johns Hopkins University; and at yet another Koch commemoration on 5 May, Brown presented a lecture at the New York Academy of Medicine (see Figure 3). Taken together, these orations comprised the most sizable effort by a single American to remember and celebrate Koch. Brown's orations were typical of tributes paid to great doctors: he combined an anecdotal synopsis of Koch's life with assessment and commentary on his overall importance to medicine.

In German accounts of Koch's life, commemorators never missed the opportunity to point out that Koch had been a *Kreisarzt* (country doctor) when he made his original discoveries concerning the anthrax bacillus. But in Brown's American tribute, Koch's rapid rise from rustic origins to urbane success took on an unmistakable flavor of Horatio Alger. "It is rumored that the Kochs were never received socially in Berlin," suggested Brown. "In the first place he had made a grievous mistake. It was quite unforgivable for a country doctor or indeed for any physician to make the discoveries, to rise to the heights that Koch reached, when he had not gone through the usual channels and advanced through the ordinary university positions."[61] In Brown's interpretation, Koch was an extremely practical man who cared little for the social conventions of a community of medical elites. Brown carefully described how Koch had attempted to call on the older, esteemed Emil Du Bois-Reymond, but because Koch had left his personal card at Du Bois-Reymond's residence rather than his office, the card was ignominiously returned. "They order these things differently in Germany," Brown explained, "and it is difficult for us in America to grasp their importance, but it remains that Koch was *persona non grata* in Berlin, which naturally embittered him."[62] At first glance, this sort of anecdote seems a bit out of place in a commemoration of Koch. But while this brief yarn did little to celebrate the medical discovery itself, it made Koch's persona much more appealing for an American audience. Without denying Koch's foreignness, Brown provided a characteristically American refraction of Koch's medical life.[63]

Underscoring another aspect of American identity, Brown discussed Koch's interest in producing usable results from the study of bacteriology, claiming, "Koch's great eagerness to make practical use of his discoveries played a prominent part in all of his work."[64] And it was this eagerness, claimed Brown apologetically, that had led Koch prematurely to pronounce tuberculin a cure for tuberculosis. In a commemoration of Koch, the dramatic failure of the so-called "tuberculin cure" was difficult to ignore. But rather than ignoring this negative memory of Koch completely, Brown explained it away by emphasizing Koch's overwhelming interest in practicality. (He also resorted to one of the more usual explanations: "It was rumored that high authorities—even Emperor William, some whispered—had forced Koch to make a premature announcement.")[65] A public lecture was not an opportune setting for an exhaustive account of Koch's life. Rather, it was a time to focus on certain meaningful anecdotes in a public affirmation of the ideal medical life, and in Brown's case, this meant appropriating memories of Koch for a characteristically American portrayal of that ideal.

Doctors are indeed storytellers. The things we typically associate with the practice of medicine—diagnosis, therapeutics, surgery—are but part of the story of professional medical culture. It is also a culture of *memory*—or, more precisely, a culture that selectively perpetuates specific interpretations of the past that harmonize with and promote certain epistemological, esthetic, ethical, and professional convictions. Physicians do not just diagnose, prescribe medicines, and perform surgeries; they also tell stories about themselves and their collective sense of professional identity. A medical commemoration provides just that opportunity: it is a forum for professional storytelling and for a ritualized reaffirmation and active construction of a

[61] Lawrason Brown, "Robert Koch," *Bulletin of the New York Academy of Medicine,* September 1932, *8,* n. 8:575.

[62] *Ibid.,* 575–6.

[63] For other examples of American physicians negotiating foreignness—especially "Germanness"—with their own American identity, see George E. Haddad, "Germ Theories, Scientific Medicine, and the Buffalo Medical Community," in *Medical History in Buffalo 1846–1996,* ed. Lilli Sentz (Buffalo, N.Y.: School of Medicine and Biomedical Sciences, State University of New York at Buffalo, 1996), pp. 129–34.

[64] Brown, "Robert Koch" (cit. n. 61), p. 573.

[65] *Ibid.,* p. 573.

collective professional past. In the end, German and American commemorations of Koch's announcement were multilayered opportunities to "make" history. The celebrants contributed to the construction of the profession's historical consciousness and collective memory, while they themselves—knowingly or unknowingly—were participating in an historical event.

# The Divided Centennial
## The 1958 Max Planck Celebration(s) in Berlin

*By Dieter Hoffmann**

> The government simply cannot accept a keynote address at the State Opera on the anniversary of the Deutsche Akademie der Wissenschaften zu Berlin (German Academy of Sciences) delivered by anyone other than a well-known representative of the German Democratic Republic. A leading GDR scientist must give the main speech at the celebrations being organized by our academy. Maybe we can persuade Professor Hertz to turn down his speaking engagement in West Berlin, in favor of [giving an address at] the State Opera . . . The Office of the Central Committee has forwarded a recommendation to the Politburo, based on this principle: the independent role of the GDR should be emphasized at Planck's jubilee, and the initiatives in West Berlin should be undercut . . . We must talk to the comrades at the Physikalische Gesellschaft der DDR (Physical Society in the GDR) and the German Academy of Sciences about proper party policy in such matters, and alert them to their wrong-headed notion of an undivided German physics. This mistaken concept has caused them to look forward to joint celebrations, in which GDR scientists were to go to West Germany, and the West German scientists would come to the GDR.[1]

Kurt Hager, dean of cultural and scientific affairs for the Central Committee, the most powerful state organ in the GDR, wrote this to GDR deputy prime minister Fritz Selbmann in January 1958. Hager's words served not only to rebuke Selbmann; they also asserted party control over the country's physicists. Even in the sciences, where there was a general consensus that East-West cooperation should be maintained, cooperation without conditions across the boundaries of the cold war was impossible.

What was this all about? German physicists had begun to draw plans for Max Planck's one hundredth birthday celebration in the spring of 1957. The initiator was primarily Max von Laue, undoubtedly Max Planck's favorite and most eminent student. Laue's enthusiasm did not stem solely from an obligation to his former teacher; he was also an outspoken proponent of an all-German body of thought. Despite his reservations, Laue served as a committed member of the German Academy of Sciences, working persistently to encourage contact between the scientists—and especially the physicists—in East and West Germany. Von Laue had initiated a joint celebration of the fiftieth anniversary of the theory of relativity in 1955 in Berlin,

*Max Planck Institute for the History of Science, Wilhelmstraße 44, D-10117 Berlin.
This paper was based on lectures given in 1993 in Mainz and Paris. The German version was translated with the help of Sophie Hoffmann. I thank her and Brian Plane, who revised the English translation, very much for their assistance.

[1] Kurt Hager to Fritz Selbmann, 28 January 1958. Stiftung Archiv der Parteien und Massenorganisationen der DDR (hereafter cited as SAPMO) IV 2/2.024/25, pp. 57–9.

and Berlin's two physical societies had extended a common invitation to Albert Einstein to attend. In a private letter to Einstein, Laue laid bare his designs. An important initial step, he explained,

> could be achieved by securing signatures from the two physical societies with seats in Berlin. That would be the old Physikalische Gesellschaft zu Berlin (Physical Society of Berlin) within the Verband der Physikalischen Gesellschaften (Association of German Physical Societies)—the society I sign for—and the Physical Society in the German Democratic Republic (the society Gustav Hertz probably will sign for). Until now the two societies have been separated by the iron curtain. And long and quite difficult negotiations have led to the present cooperation. Now there seems to be good will on both sides and hopes that this is just the beginning of a future cooperation. Thus we can make a contribution, even if it is a small one, to a general easing of political tensions.[2]

Although regretting that "age" and "sickness" prohibited a long journey to Berlin, Einstein extended his thanks for the invitation, noting his pleasure at being "cause for fraternal cooperation, and not for controversy, in this extraordinary case."[3] Yet it must be remarked that the "fraternal cooperation" that inspired Einstein proved to have its limits. Political conditions in the divided "frontier city" hampered the organization of a common celebration; two separate ceremonies honoring Einstein were eventually held in 1955 in Berlin: one on 18 March in the big hall of the Technische Universität Berlin, and one the following day in the chamber of the Academy of Sciences in East Berlin. Still, mention was made in both invitations of the concurrent celebration being held by the other society.

Thus, the Einstein commemoration can be seen as an important attempt by von Laue to build closer cooperation between the physical societies in East and West Berlin, and it probably encouraged Laue to continue his efforts on the occasion of the Max Planck centennial. More than simply being Laue's mentor and friend, Max Planck had stood as a guiding and unifying force in German physics for over half a century. Planck's stature was based not only on scientific reputation—internationally established with his epoch-making hypothesis of the quantum of energy (1900)—but also on his political commitment and personal integrity. Long respected by German physicists, Planck's status as spokesman was solidified by his courageous stand against National Socialism.[4] Planck capped his career by assuming the provisional presidency of the Kaiser-Wilhelm Gesellschaft in 1945, and agreed to renaming it the Max-Planck-Gesellschaft. Thus Planck contributed to the preservation of the great tradition of German science, and became an ideal figure to unify the physicists' community in the postwar period—especially for physicists who were separated by National Socialist expulsion and exile or by the barricades of the cold war.

The notion of using the Planck centennial for an "all-German national celebration" (Laue's words) first surfaced on 4 March 1957 at a board meeting of the Physical Society of Berlin. On this occasion, von Laue called for "a commemoration for Max Planck's one hundredth birthday organized for 23 April 1958, if possible together with the Physical Society in the GDR."[5] Six weeks later, Laue took the idea

[2] Max von Laue to Albert Einstein, 16 January 1955. Einstein Papers, Boston, 16207–1.

[3] Albert Einstein to Max von Laue, 2 March 1955. Einstein Papers, Boston, 16207–1.

[4] See John Heilbron, *Max Planck: An Upright Man* (Berkeley: University of California Press, 1986).

[5] Minutes of meeting of the board of the Physical Society of Berlin (West), 4 March 1957. Archiv der Deutschen Physikalischen Gesellschaft, Berlin (hereafter cited as ADPG).

to East Berlin and discussed it in detail on the fringes of the German Academy of Science's celebration of the 250th birthday of the great mathematician Leonard Euler. "Some of the members of the Physical Society of the GDR made a suggestion" to organize a joint function of both physical societies, to be held in West Berlin on the day after the planned centennial celebration for Planck.[6]

Max von Laue and GDR physicist Robert Rompe agreed upon a temporary program on 25 April 1957, probably at a German Academy of Sciences meeting, of which both were members. Laue, from West Berlin, would speak at the commemoration organized by the academy in the east on 24 April 1958; on the following day, Gustav Hertz representing the GDR would speak at the West Berlin celebration; another speaker would be Lise Meitner from Stockholm. Still, von Laue noted in a memo, although Rompe "expected consent" from political officials he refused to be firmly pinned down, since "the academy hadn't given its opinion on this plan yet."[7] Rompe had good reasons for reservation: approval of such an exchange from party and government authorities—who held the final word—was still unofficial. With this in mind, Rompe and Physical Society Secretary Alfred Büchner probably consulted the Central Committee Office for Academic Affairs soon after conversing with Laue.[8]

That Rompe undertook negotiations and became the key East German figure on this matter is indicative of the "leading role of the party" in the GDR. Gustav Hertz was in many ways more prominent, being a Nobel Prize winner and official board spokesman for the GDR Physical Society. But as a longstanding party executive and Sozialistische Einheitspartei Deutschlands (SED) Central Committee (ZK) member, Rompe enjoyed good connections and confidence within the government and party. For his part, Hertz would be made the scientific *Gallionsfigur* (figurehead) of the venture and be asked to chair a committee in preparation for the Planck celebration. Meanwhile, Rompe and Búchner worked behind the scenes on the details of the all-German celebration, delivering a detailed report to the ZK apparatus (but not the Ministry) in May 1957 that requested passage of the following resolutions:

1. Max Planck's one hundredth birthday will be celebrated. The main celebration will take place in Berlin. The German Academy of Sciences and the Physical Society in the German Democratic Republic will organize this event.

2. The German Academy of Sciences, the Physical Society in the GDR, and the Association of German Physical Societies in West Germany will jointly sponsor a central celebration.

3. The celebration will have an international character . . . "[9]

While the party machinery took its time reaching a decision on the resolutions, events in East and West took their course and the Planck centennial began to take

[6] Ibid.

[7] Memo written by Max von Laue, 25 April 1957. Archiv der Max-Planck-Gesellschaft, Nachlaß Max von Laue (hereafter cited as MPGA-L), III/50/1535, p. 7.

[8] Memo about the planned Planck ceremony, 25 January 1958, SAPMO IV 2/2.024/25, p. 47.

[9] Report from the Physical Society of the GDR to the ZK, 22 May 1957, SAPMO IV 2/2.024/25, p. 1.

shape. Over the summer and fall of 1957, Laue and Büchner met several times to iron out the details of the celebration. Laue informed Hermann Ebert, secretary of the Association of (West) German Physical Societies, about the course of the talks and negotiations, reporting that even though Büchner "is not authorized to make decisive promises, I (still) have the impression that the colleagues from the Soviet-occupied zone are *willing* to follow our suggestions."[10]

But by November 1957 Laue's confidence had begun to wane. In a letter to Otto Hahn, president of the Max-Planck Gesellschaft, Laue complained that there had still been no action on his request "for the conversion of agreements on the April 1958 Planck celebration, which have been unofficially made several times, into definite resolutions."[11] Büchner could only give evasive answers, since the ZK was still withholding a green light for the venture. A study of party records indicates that, up to this point, GDR physicists—and particularly Rompe and Büchner—were still acting without political backing. In December 1957, with the planned celebration just four months away, GDR Deputy Prime Minister Fritz Selbmann noted that Secretary Kurt Hager, top authority for scientific policy in the GDR, was just "getting familiarized" by Büchner with the plans for the Planck celebration. Not wanting to be caught unprepared, Selbmann encouraged "the office of the ZK to attend to the suggestions of the Physical Society in the GDR, which is already busy realizing a big part of the venture, and tell me how we are going to react."[12]

Hager finally—and emphatically—attended to Selbmann's request in his fiery letter of January 1958. In the passage cited at the beginning of this article, Hager took particular aim at those physicists who hold "mistaken notions of a unified German physics." In an official report to the Central Committee, Selbmann warned that "preliminary talks with the Physical Society [in the GDR] had progressed much further than laying down drafts . . . The main mistake is that the comrades of the Physical Society have started from the wrong position: [they are discussing] a real joint function, where the GDR scientists go to West Germany and the West German scientists to the GDR. This has not been discussed with the comrades of our office and has never been accepted."[13]

By this point there was probably no going back on the planned celebration; at least not without political loss of face. To improve the "political conditions in connection with the organization of the celebration," the ZK began to emphasize that the "comrades of the Physical Society organize the celebration so that the main function takes place in the GDR. A second celebration in West Berlin will be organized as a subsidiary event."[14] In concrete terms, the ZK demanded that a GDR scientist—possibly the vice president of the German Academy of Sciences, Hans Frühauf—kick off the celebration in the State Opera; that the government be invited to the function; and that during the event a government representative present the Physical Society with the keys to Berlin's Magnus House, where the society would take up residence.

Finally, after months of procrastinating and with the Planck celebrations just around the corner, the ZK had laid down a clear policy. Immediately, the loyal party

[10] Max von Laue to Hermann Ebert, October 1957, MPGA-L III/50/1536, p. 9.
[11] Max von Laue to Otto Hahn, MPGA-L III/50/1536, p. 10.
[12] Fritz Selbmann to Kurt Hager, 18 December 1957, SAPMO IV 2/2.024/25, p. 4.
[13] Ibid., p. 47.
[14] Ibid.

guard jumped to attention. Within the German Academy of Sciences, party stalwarts called SED members of the centennial organizing committee together for a preliminary meeting in February 1958, explaining the new party line concerning the event. Yet, as party officials reported back to the Central Committee, all sides did not greet the new dictates with enthusiasm. For instance, academy President Max Volmer insisted—in the presence of Rompe—that "he—after talking things over with Herr von Laue—had decided to speak himself at the Max Planck celebrations." Comrade Rompe, to his credit, "pointed out that Mr. Volmer's actions in this matter were a bit disconcerting, since decisions in this matter were to be made by the academy's Max Planck committee and not by Volmer himself. Comrade Rompe also made reference to comrade Selbmann's reasonable suggestion that comrade Frühauf should speak at the event in the State Opera."[15]

The choice of Frühauf was not made by chance. On the one hand, he was an established physicist and engineer with some international reputation (but compared with Hertz or Laue, of second rank, of course); on the other hand he had been engaged in developing socialism in the GDR from the very beginning and had taken over some official positions at the Technische Universität Dresden, where he had a professorship, and at the German Academy of Sciences, of which he had been a member since 1953. As an early member of the SED, he accepted the general policy of the party and could also have been controlled by it. Last but not least, he was *Geheimer Informant* and as such he was trusted highly by the authorities.[16]

After further discussions at the board meeting of the GDR Physical Society, a compromise was passed: Volmer, as academy president, would open the function in the State Opera; Frühauf would still give a talk; and programmed between them would be the speech by Laue. A fierce discussion ensued during which many sides reproached Rompe for "politicization of the Max Planck celebration."[17] The central issue of these debates is clearly documented, since Laue and Hertz had already set the board meeting's agenda in prior consultations: they were to discuss a plan to give the jubilee a strictly scientific character. Politics and politicians would be left out; neither the mayor of Berlin nor the president of West Germany had been invited. Any meetings between scientists and politicians would have to take place apart from the scientific events. Even official inquiries of diplomatic corps were declined to avoid risking the East-West cooperation and aggravating the peculiar situation in the divided city of Berlin.[18]

The ZK apparatus, naturally, was unimpressed by the plans for a "scientific" centennial as envisioned by Laue, Hertz, and others in the GDR Physical Society. A meeting at the headquarters of the Central Committee—with representatives of the GDR Physical Society and the German Academy of Sciences (Rompe, Günter Rienäcker, and Robert Havemann) on one side and comrades of the ZK's Department of Science (Hannes Hörnig, Hager, and others) on the other—concluded that "the Planck celebration is for us—as much as for the enemy—an event that takes on very considerable political character, over and above scientific questions." Rompe clearly observed the political dictates of the time, remarking that "it can be shown

[15] Report of Robert Dewey, party secretary, to the ZK, 14 February, SAPMO IV 2/2.024/25, p. 8.
[16] Archiv Bundesbeuaftragter für die Unterlagen des Ministeriums für Staatssicherheit der ehemaligen DDR, 1107/59, vol. 1, p. 2.
[17] SAPMO IV 2/2.024/25, p. 9.
[18] Max von Laue to Hermann Ebert, 13 March 1958, MPGA-L III/50/1539, p. 12.

that Planck is ours and not the fascists' in West Germany."[19] A team of comrades "supervising the preparations for the party," Rompe argued, should be established and charged with the following agenda:

> On 23 April Professor Hertz will talk on the GDR's broadcasting station Deutschlandsender. Take a look at the speech! . . . We have to take care of the participant list for the State Opera . . . The function must not become the business of the German Academy of Sciences . . . The seating plan must be fixed according to protocol. Representatives of the government will be sitting in the honor's box of the Opera. On the stage will be seated members of the official committee of the [German] Academy [of Sciences] and the board of the Physical Society of the GDR.[20]

In future meetings, GDR officials left little to chance as they planned to harness Planck, his centennial, and his legacy for the GDR camp. Organizers took special pains to ensure that speeches, toasts, and other ceremonies would not become opportunities for "demonstrations from the West"—truly an SED leader's trauma.[21] Wreaths placed on Planck's grave in Göttingen would have ribbons on which "the word 'GDR' has to come out clearly." Toasts for the planned reception were scripted in advance,

> along the lines of: Science and defense of peace, creating a nuclear-free weapon zone, science and the masses, progress. There should be no talk about alliances between East and West. We should avoid making this a scientific affair. Take advantage of the chance to connect Planck's personality to our conception of science. The toasts should be written together. Even here it must not come to demonstrations for the West.[22]

The student torchlight procession planned for the night of 24 April would be cancelled since "great personalities from the West are going to be present in the State Opera. There must not be any unpleasant demonstrations on the street Unter den Linden that evening. A few reliable comrades should be standing in front of the State Opera. It must not come to demonstrations for Hahn or Heisenberg. No torchlight procession. The university should organize something big. But not at night. Maybe the next day at eleven o'clock."[23] And that's how it happened. In a letter, Laue grumbled ironically that the torchlight procession had been cancelled "allegedly because of the rector's protest. Maybe the university will at least be lit by floodlights, instead."[24]

Not only in the East, but also in the West, politicians and scientists voiced reservations about the Planck celebrations and their all-German impetus. Colleagues complained in writing about Max von Laue's naïve conduct during talks and negotiations. Politicians like Ernst Lemmer, federal minister for all-German questions, did emphasize the Planck conference as an example of successful German-German cooperation in a February 1958 interview. Nonetheless, Lemmer also confessed "that at times there have been many doubts, even on the West side, about these contacts.

[19] SAPMO IV 2/2.024/25, p. 11.
[20] Ibid.
[21] Ibid.
[22] Ibid.
[23] Ibid.
[24] Max von Laue to Otto Hahn, 24 February 1958, MPGA-L III/50/1538, p. 18.

There's been worry about danger of infection."[25] That March, the Federal Ministry for all German Questions reported in a letter to the Association of German Physical Societies that it had been informed

> from an absolutely reliable source that the first secretary of the SED, [Walter] Ulbricht, has raised strong objections to the agreement [in which von Laue speaks at the academy function] and demands that a loyal party speaker sponsored by the Central Committee appear before the Academy of Sciences in East Berlin. He reportedly said it is out of question that a scientist from the Federal Republic speak in the academy on this occasion. I write to inform you about this situation, and to request your inquire at the Academy of Sciences to see if the program agreed to with Professor Rompe stays or not.[26]

Rumors of communist sabatoge of the Planck centennial, however, were easily quelled. In spite of general cold war anxieties and the recent appeal of the "Göttingen Eighteen," nervousness and mistrust probably had little tangible effect on the attitudes of Western scientists and politicians towards the conference. The gentlemen of the All-German Ministry could be reassured by the physicists with the comment that "Professor Frühauf from Dresden, one of the vice presidents of the academy, will deliver an additional speech, in which Planck's administrative work at the Prussian Academy of Science will be acknowledged." Otherwise everything "remains unchanged."[27] Later, Bonn officials inquired about a second matter. They feared that the gift of a bust of Planck to the academy would be equivalent to a kind of trading and could therefore turn into a political act.[28] All of this helped to nourish the "many doubts" Lemmer had mentioned.

Eastern "cold warriors," on the other hand, grasped Lemmer's February 1958 interview as an excuse to rally opposition to the all-German character of the Planck celebration. Even the GDR Politburo put the Planck centennial on the agenda of its regular meetings. On 1 April the GDR press was instructed to respond to Lemmer's interview "in the sense that he shouldn't interfere in questions of science."[29] Indeed, a couple such comments were published in the GDR press during the following weeks. But even some Western scientists complained that Lemmer's involvement was bothersome and not conducive to the organization of the Planck event.[30]

Thus, despite reservations, conference preparation moved ahead at full steam. GDR officials, for their party, worked to tie the Planck centennial into ongoing efforts to solidify the prestige of GDR science. The Physical Society in the GDR, for instance, put the finishing touches on additional plans for a Leipzig colloquium on scientific theory, which hopefully would draw in participants from the Berlin Planck celebrations. Scheduled for 27 April, immediately following the Berlin events, the Leipzig colloquium ostensibly dovetailed with the annual general meeting of the Physical Society. In reality, the conference was skillfully timed so as to prevent an International Theoretical Colloquium that had been in the works for West Berlin.[31]

[25] *Süddeutsche Zeitung,* 22 February 1958, p. 3.
[26] MPGA-L III/50/1539, p. 35.
[27] Ibid., p. 45.
[28] MPGA-L III/50/1540, p. 11.
[29] SAPMO J IV/2/587.
[30] SAPMO IV 2/2.024/25, p. 12.
[31] Hermann Ebert to Ferdinand Trendelenburg, 25 November 1957, MPGA-L II/50/1535, p. 19; see also the minutes of the board of the Physical Society, ADPG.

The conference in Leipzig would center on the basic questions of quantum mechanics, and given that Werner Heisenberg, Paul A. M. Dirac, von Laue, Victor Weisskopf, Paul Vigier, and Lajos Janossy all planned to participate, this promised to be an interesting and highly prestigious venture. The respected panel of participants, moreover, would also be introduced to the newest Marxist approaches for the philosophical interpretation of quantum physics. The Marxist position would be advanced by physical chemist (and later GDR dissident) Robert Havemann, who would give a paper there; one official even boasted that Havemann's lecture might "become a world sensation."[32]

In preparation for the colloquium, Havemann traveled to Göttingen to speak with Heisenberg. SED officials, however, had not sent him there just to wax theoretical with Heisenberg; they were interested in concrete information about the attitude of Heisenberg and other West German physicists towards the nuclear arms race; and their general political positions on domestic and foreign affairs. Most importantly, as a detailed report of the encounter showed (which found its way up to Ulbricht's desk), Havemann pressed Heisenberg for concrete information about his planned stay in the GDR.[33] Havemann's inquiry had been given urgency by the advance word that Heisenberg would accept an invitation from church superintendent Jakob to lecture at the Evangelic Academy in Cottbus (a little town south of Berlin). This serious development merited the immediate attention of the Politburo, which asked that Professor Heisenberg be convinced "to reject the church's invitation to lecture in Cottbus."[34] Some party officials suggested telling Heisenberg—in no uncertain terms—that a Cottbus appearance would be seen "as a political counterpoise . . . He must not talk in the Friedensrat (Peace Council), that being a political organization."[35] An outright prohibition of Heisenberg's invitation to speak at the Evangelic Academy, however, was never issued. Such drastic measures would have endangered Heisenberg's visit, the prestige of East German physics, and the upcoming Planck centennial. Moreover, Havemann's Göttingen report had no new ammunition that might be used against Heisenberg: "Heisenberg is not a man of church and Christian religion or education, but rather tends to materialism in his opinions."[36] ZK officers instructed local authorities to be "cooperative" and, in a kind of "political correctness," revoked a ban that local authorities had placed on speaking space for Heisenberg. A generous room for between 150 and two hundred people was allotted with the hope that "if the meeting takes place outside the church and is not an event of the church, the intention of the Evangelist academies [of] using Heisenberg for their purpose could be crossed out."[37] Aware of the explosive nature of his visit and the possibility of being politically exploited, Heisenberg had asked Hertz to reiterate his desire to stay aloof from state affairs. "He does not want to be welcomed by government representatives or political personalities," Hertz reported bluntly. "If this is not guaranteed he would rather stay home."[38]

[32] See Dieter Hoffmann, "Robert Havemann: Antifascist, Communist, Dissident," in *Science under Socialism,* ed. Kristie Macrakis and Dieter Hoffmann (Cambridge, Mass.: Harvard University Press, 1999), pp. 269–85.

[33] SAPMO IV 2/2.024/1; IV 2/11/v. 4920, p. 209–21.

[34] SAPMO J IV 2/2/587.

[35] SAPMO J IV 2/2.024/25, p. 25.

[36] SAPMO J IV 2/2.024/1, p. 16.

[37] Ibid., p. 18.

[38] SAPMO IV 2/2.024/25, p. 39.

In April, one coordinating meeting followed another as the ZK apparatus maintained its efforts to influence the outcome of the Planck celebrations.[39] Even the Politburo of the SED attended to the details of the jubilee, giving final approval to the Planck program on 1 April. Party groups were formed for the Planck celebration and for the Leipzig physicists' conference in accordance with the dogma of the "leading role of the party." Such groups not only took on organizational problems like lodging, travel, and receptions, but also assumed responsibility for a general state of political and ideological vigilance. Three weeks later, Secretary Hager issued a final report on the centennial. All involved could by now sense that the jubilee was not just for physicists anymore, even though party business had been conducted behind the scenes. Laue, for his part, made one last strong appeal before the Planck committee of the German Academy of Sciences, emphasizing "how important it is to ensure the political neutrality of the events on both sides. I attach great importance to this aspect of the agreement, for the sake of this event and future ones like it . . . We've done everything to maintain a purely scientific character on our part."[40]

More than just supervising and organizing the Planck conference, the Politiburo worked to move Max Planck and his scientific achievement into the "right light" and exploit them as propaganda. The Politburo composed an "address of the Central Committee of the SED," which provided the jubilee (and the conference) with a Marxist-Leninist exegesis. The ZK address not only acknowledged Planck's status as "one of the greatest natural scientists" who opened "the world of atoms and the atomic nucleus" by discovering the "quantum of energy," it also attempted to give Planck's philosophical and political views a materialistic spin. Planck's antipositivism and his opinion "that the outside world is something totally independent and absolute" became an entry point for dialectical materialism and the Leninist reflection theory, although Planck had never found his way ideologically to modern materialism. Likewise, Planck's Christianity was glossed over as a belief in "main administration Eternal Truth," pantheism, and a nonpersonal God. Finally, leaving no room for doubt, the address concluded, "Only the working class, which has built up socialism and defended world peace, has the right to celebrate the great natural scientist Max Planck. The bourgeoisie has forfeited its right to the pioneers of science . . . What Planck and, with him, a generation of younger physicists have created, capitalism can no longer assimilate."

This address received full-page coverage on 23 April—Planck's birthday—in the main SED daily newspaper, *Neues Deutschland.*[41] Other newspapers and television and radio stations across the GDR had also paid tribute to Planck and called attention to the great festivities planned in Berlin, commencing 24 April with an official ceremony sponsored by the German Academy of Sciences in the State Opera on Unter den Linden. (A more solemn and prestigious venue could hardly have been found in the then mostly destroyed city of East Berlin.) Numerous scientists of considerable standing, former students, and émigré colleagues of Max Planck took seats on the platform and in the parquet, with the most prominent among them comprising a veritable who's who of modern physics: Max von Laue, Gustav Hertz, Otto Hahn, Werner Heisenberg, Paul A. M. Dirac, Max Born, Lise Meitner, James Franck,

[39] Ibid., p. 21.
[40] Max von Laue to German Academy of Sciences, 21 April 1958, SAPMO IV 2/2.024/25, p. 147.
[41] *Neues Deutschland,* 23 April 1958, p. 3.

Victor Weisskopf, and Peter Pringsheim. Large delegations from socialist "brother countries" (naturally led by the Soviet Union) also filed onto the floor; for instance, Abram Joffe, Dmitri Iwanenko, and Nikolai Bogoljubow from Leningrad/Moscow and Lajos Janossy from Hungary. The remarkable assembly represented one of the foremost conferences of physicists of the post-World War II era.

The morning ceremony was inaugurated by the president of the German Academy of Sciences, physical chemist Max Volmer, who implored scholars to keep their "international relations" free from "politics." Volmer's short opening address was certainly more than a plea for scientific cooperation in the midst of the cold war; it also served to rebuke the East German government for its veiled attempts to interfere with the organization of the event. One GDR official quipped in a report to SED Secretary General Walter Ulbricht,

> President Volmer's introductory and closing words have been censured by the academy party and nonparty members alike. . . . Volmer hadn't shown this speech to anybody beforehand, and in the last couple of days he was carrying around two manuscripts. Academy members are indignant that Volmer didn't fulfill his agreement to share his remarks prior to speaking. Some think his remarks sprang from heat of the moment.[42]

Most Western scientists were astonished to find high GDR government representatives at the centennial. Walter Ulbricht and Prime Minister Otto Grotewohl filled the opera's grand loge, accompanied by Hager, the Soviet ambassador, and some leading scientists from the GDR (Figure 1). On the rostrum, the chairpersons of the ceremony sat directly opposite on the stage—as von Laue put it "in front of the iron curtain, I mean it literally" (Figure 2).[43] Hans Frühauf's speech, which was rated as a political address, dealt with Planck's long-term role as secretary of the academy.[44] Planck's former student von Laue, however, returned the spotlight to Planck's scientific work.[45] Finally, Otto Hahn, president of the Max Planck Society, closed the ceremony by presenting a bust of Planck to the Physical Society in the GDR.

Afterwards, the Physical Society sponsored a reception at the renovated Magnus House, during which Mayor Friedrich Ebert turned over the keys to the house to the society. The head of the Soviet delegation, A. F. Joffe, returned to the society volumes from Planck's library that had been taken by the Red Army in 1945. Subsequent to this official gesture, Lise Meitner commented on Planck's personality. The East Berlin festivities came to an end with a production of the opera *Iphigenis in Aulis* and an evening reception in the State Opera.

The following day, 25 April, the Planck honors continued in Berlin. The venue for the conference was also a place with high prestige and political symbolism: the newly built Congress Hall in Berlin Tiergarten. The Congress Hall stood proudly as a symbol of German-American friendship and of the city's tie to the West, and it was surely made available to the Physical Society rent free.[46] Werner Heisenberg, Gustav Hertz, and Wilhelm Westphal spoke on this occasion to an almost exclusively

[42] SAPMO IV 2/2.024/25, p. 98.

[43] Max von Laue to Otto Hahn, 8 January 1958, MPGA-L III/50/1537, p. 5.

[44] Hans Frühauf, "Max Planck als beständiger Sekretar," *Mitteilungen der deutschen Akademie der Wissenschaften zu Berlin,* 1958, Heft 4–5.

[45] Max von Laue, "Zu Plancks 100: Geburtstag," in Max von Laue, *Vorträge und Reden,* vol. 3 (Braunschweig: Vieweg und Sohn, 1961), pp. 257–62.

[46] Max von Laue to Otto Suhr, 21 June 1957, MPGA-L II/50/1535, p. 27.

***Figure 1.*** *View of the Opera's grand loge. Sitting in the front row (from left to right): Walter Ulbricht (secretary general of the SED) and Otto Grotewohl (prime minister); behind them (from left to right) Max Steenbeck; unknown person; Michail Perwuchin (Soviet ambassador), Kurt Hager, and Peter Adolf Thiessen.*

***Figure 2.*** *On the Opera's stage, the chairpersons of the ceremony (from left to right): Max von Laue, Walter Friedrich, Lise Meitner, Max Volmer, and Otto Hahn.*

***Figure 3.*** *Reception given by Theodor Heuss, FRG's president, for the guests of honor at the Planck ceremony (from left to right): Otto Warburg, Otto Hahn, Nicolai Hartmann, Lise Meitner, Werner Heisenberg, Theodor Heuss, Willy Brandt, and Max von Laue.*

"scientific" audience. But that afternoon, West German President Theodor Heuss and West Berlin Mayor Willy Brandt received prominent conference participants at Castle Bellevue; this event was reported by the Western press in great detail (Figure 3).

The legacy of the Planck centennial is thus surprisingly mixed. On the one hand, the event provides a clear study of how GDR officials tried to exploit Planck to heighten the international prestige of East German science and the reputation of their state. On the other hand, the scientific excellence and aura of this conference, as well as the rebuke of party policy by Volmer and others, point to the limits of party power. Despite SED policy, many conferees saw the Planck centennial as a victory; as a sign that, after years of fascism, war, and terror, scientific discussion and dialogue beyond the trenches of the cold war would be possible. As one newspaper report put it, the conference gave hope that scientists might work "at least to counteract the intellectual separation, if not the political." For GDR physicists and especially for students who attended, the ceremony in the State Opera was a formative exposure to the internationality of science, providing an opportunity to meet the heros of the field and break the chains of political isolation. Later events honoring Alexander von Humboldt (1960), Albert Einstein (1979), Leonhard Euler (1983), and others can be constructively viewed in this context. True, these events increasingly served the political function of bolstering the legitimacy of GDR science and policy. But, especially after the Berlin Wall went up in 1961, such international conferences and jubilees also provided Eastern scientists with an outlet for uncensored contact with Western colleagues, chances for direct comparisons of research, and first-hand experience evaluating international standards.

# *Part II. Commemorating Scientific Institutions: The Re/Production Sites of Scientific Progress*

# The Tercentenary of Harvard University in 1936

## The Scientific Dimension

*By Clark A. Elliott**

### I. HISTORICAL INTRODUCTION

CONTINUITY AND LONGEVITY are important features of many institutions. This lends a quality to their commemorative events that is not present in retrospective observances of individual lives or of momentous but finite events. A further distinction is the inherent self-celebration that characterizes institutional observances; they ordinarily are organized and promoted by the institution itself. If the celebrants of such anniversaries are able to achieve any level of faith, they can assume that the institution will continue for an indefinite time. Consequently, emphasis on the institution's future role and development is an important, perhaps overshadowing, aspect of the observances.

Harvard University is one of the nation's most enduring institutions and is its oldest in terms of organized support for the promotion of science. An examination of Harvard's anniversary observances, especially the three-hundredth in 1936, reveals how the celebrations have emphasized and served various historical, institutional, and societal goals and indicates the place of science within the context of larger commemorative events.

In 1636, only six years after Governor John Winthrop's Puritan party took up residence in the Massachusetts Bay Colony, a vote was taken whereby "The [Great and General] Court agreed to give 400*L* towards a schoale or colledge, whearof 200*L* to bee paid the next yeare, and 200*L* when the worke is finished, and the next Court to appoint wheare and what building." This legislative act for the founding of a college was not a small commitment on the part of the colonial government; it represented about a quarter of the tax income in that year.[1] When John Harvard, a young minister and graduate of Emmanuel College, Cambridge, died in September 1638—the same year that the college began operation—he left his library of four hundred volumes and half of his estate to the new institution. In 1639, it was named Harvard College in honor of its early private benefactor.[2]

Historian Michael Kammen has pointed to the assertion by some that "the manip-

*105 Beech Street, no. 2, Belmont, MA 02478.

[1] Samuel Eliot Morison, *The Founding of Harvard College* (Cambridge, Mass.: Harvard University Press, 1935), pp. 168–69.

[2] Samuel Eliot Morison, *Three Centuries of Harvard, 1636–1936* (Cambridge, Mass.: Harvard University Press, c. 1936), pp. 8–9.

ulation of collective memory and the invention of tradition are markedly cultural manifestations," and that in the United States these did not emerge until late in the eighteenth century and did not become a significant force until after 1825.[3] This assessment is verified with regard to Harvard. There is no evidence that any notice whatsoever was taken of the centennial of the college in 1736. President Benjamin Wadsworth's journal entries for that year make no reference to an observance of the anniversary and, in fact, express no awareness of it.[4]

In the one hundred years between 1736 and 1836, historical consciousness changed drastically. On 8 September 1836, a major public observance of two hundred years of Harvard history took place. President Josiah Quincy was the major speaker in the ceremonies, which took place in the nearby Congregational Church in Cambridge. His two-hour address reviewed the history of the institution and became the foundation for his two-volume work published in 1840.[5] Such an extensive historical engagement by the university's president is a telling factor in appraising the place that history had taken in public and institutional life.

The ceremonies in the church were followed by a dinner organized by the alumni, which met in Harvard Yard under a pavilion covered with white canvas that extended nearly eighteen thousand square feet. More than 1,500 people were in attendance.[6] Massachusetts Governor Edward Everett, himself a future president of the university, gave the dinner's opening address. (Everett would go on, when president, to play a crucial role in establishing the Lawrence Scientific School.) In his bicentennial speech, the great orator imitated the ancient historians, putting words in the mouths of deceased persons that they might have spoken. He spoke for colonial Deputy Governor John Winthrop, imagining what Winthrop might have said in arguing for the legislature's founding and initial support of a college in 1636.

Giving words to a historical figure is a curious rhetorical device in itself, but it is even more curious in the way it reverberates through a study of anniversary celebrations. Everett reflected from his perspective in 1836 back to 1636 and, in our rereading, his words reflect ahead to our time. We thus engage in a mutual effort, though from different historical vantage points, to understand how Winthrop might have argued. Winthrop saw a modest goal for the college in his own time, but a bright future. For science, Winthrop predicted a time "when libraries and cabinets will open their treasures in these precincts, now scarcely safe from the beasts of the forest; when Nature, tortured in our laboratories, shall confess her hidden mysteries; when, from the towers of our Academy, the optic tube, lately contrived by the Florentine philosopher, shall search out the yet undiscovered secrets of the deepest heavens."[7] Such words no doubt spoke as well for Everett's own unrealized goals for science in the college.

[3] Michael Kammen, *Mystic Chords of Memory: The Transformation of Tradition in American Culture* (New York: Alfred A. Knopf, 1991), p. 4.

[4] *Harvard College Records III: Benjamin Wadsworth's Book,* vol. 31 of Colonial Society of Massachusetts Publications (Boston: Colonial Society of Massachusetts, 1935), pp. 506–7. See also Samuel Eliot Morison, "The Bicentennial of 1836," *Boston Herald,* 13 September 1936, special magazine section on Harvard tercentenary, p. 12.

[5] Josiah Quincy, *The History of Harvard University* (Cambridge, Mass.: John Owen, 1840). Quincy included an extended account of the bicentennial celebration, complete with the speeches of participants; see vol. II, pp. 639–708.

[6] *Ibid.,* vol. II, pp. 645, 651.

[7] *Ibid.,* vol. II, p. 655, quotation on p. 656.

Between Everett's remarks and the toasts and responses from a number of others, the event continued until eight o'clock at night. The toasts were made to various personages connected with the university, the professional schools, the Supreme Court of Massachusetts, former governors of the Commonwealth, several states, ideals such as liberty (to which Senator Daniel Webster replied), other colleges, and to other topics. Though the concerns and special interests of the university were discussed, the alumni dinner's political dimension now seems most evident. This undoubtedly reflected the character of many who were called upon to speak, including Josiah Quincy, who had served in Congress and as mayor of Boston before becoming university president. Science occasionally got passing reference, but it was no one's primary interest.

In 1886, Harvard's 250th anniversary was observed in ceremonial style. It is worth examining this event for the clues it yields in comparison with the 1836 and 1936 celebrations. In reading the account of the 1886 observance, it appears that even the presence of United States President Grover Cleveland could not turn the alumni gathering away from its predominantly literary character.[8] The principal speakers at the Monday, 8 November 1886 ceremonies were James Russell Lowell and Oliver Wendell Holmes, orator and poet of the day, respectively. Lowell, in envisioning the ideal of what is taught, observed in passing, "Give us science too; but give first of all, and last of all, the science that ennobles life and makes it generous." That is essentially all he had to say on the subject of science, although he did not intend to be exclusive but stood there "as a man of letters, and as a man of letters I must speak."[9] Sixteen of the forty-two honorary degrees went to scientists.[10]

At the dinner that followed the 8 November ceremonies, science did have its rhetorical moment in reply to a toast to "The advancement of Science." The respondent was Alexander Agassiz, zoologist and oceanographer, financial supporter and one-time head of Harvard's Museum of Comparative Zoology, and a member of the Harvard Corporation. He recognized the recent growth in science at the university, for "in the last fifty years—the half-century we celebrate today—the greater number of the scientific departments of Harvard have sprung into existence." Dwelling on the importance of organizational factors in assessing the future of science, and eying the prospects for American science within the world context, Agassiz concluded,

[8] See *A Record of the Commemoration, November Fifth to Eighth, 1886, on the Two Hundred and Fiftieth Anniversary of the Founding of Harvard College* (Cambridge, Mass.: John Wilson and Son, University Press, 1887). The report was prepared by the college librarian, Justin Winsor.

[9] *Ibid.*, p. 217.

[10] *Ibid.*, pp. 38–41. Among the foreign scientists receiving honorary degrees were mathematician and Semitic scholar Charles Taylor, the delegate from the University of Cambridge; Sir Lyon Playfair, delegate from the University of Edinburgh; and Michel Eugène Chevreul, French chemist who, at age 100, was not able to attend. The thirteen American recipients were distinguished but did not include anyone in physics or chemistry; heads of important scientific institutions were particularly evident. Biologists and earth scientists predominated and included Spencer Fullerton Baird (zoologist and secretary of the Smithsonian Institution) and John Wesley Powell (director of the United States Geological Survey). Three recipients were from Yale: mineralogist George Jarvis Brush (who also was director of its Sheffield Scientific School), geologist and mineralogist James Dwight Dana, and paleontologist and master collector Charles Othniel Marsh. Paleontologist James Hall (director of the New York State Museum) and Joseph Leidy, paleontologist, zoologist, and delegate of the University of Pennsylvania, also received honorary degrees. Astronomers Asaph Hall (Naval Observatory) and Samuel Pierpont Langley (director of the Allegheny Observatory); army engineer Colonel Henry Larcom Abbot; physician Silas Weir Mitchell; surgeon and head of the surgeon general's library John Shaw Billings; and science patron Jonathan Ingersoll Bowditch were also honored.

"Borrowing what is best from each of these [European] examples, but adapting their methods to our own national conditions, so different from those of Europe, this, the oldest University of the land, may now challenge her transatlantic sisters to a friendly rivalry in the development of the highest scientific culture."[11]

Though too-briefly reviewed here, the contrasts in tone and emphasis between the 1836 and 1886 anniversary celebrations are instructive. In the earlier instance, statesmanship was a dominant theme and had high value. By 1886, a native literature had developed and the poetic imagination had found a place in the values of the university and in public life. From our historical perspective, 1886 was a poet's anniversary. At the same time, science did have a place, if not a prominent one, and in recognizing the place of science in the university, the occasion made it possible to acknowledge and promote the elevation of Harvard among the universities of the world. These were themes that would become foundational to the observances of Harvard's origins fifty years later.

## II. THE HARVARD UNIVERSITY TERCENTENARY CELEBRATION

The planning for and celebration of Harvard's three-hundredth anniversary came in the midst of national and international crises of worrisome and seemingly intractable dimensions. The Great Depression had shaken faith in the established order, but the rapid advance of fascism in Europe was the more salient factor for the tercentenary. Jewish scholars and European universities were under attack in Germany and elsewhere; freedom of thought was in peril; and these were issues that spoke to the primary interests and operating credo of academic institutions. It is a signal of Harvard's development since 1886 that its three-hundredth birthday celebration was cast and played out on this world stage.

As will become evident in the following account, the tercentenary offered an opportunity to address and promote multiple interests. It is inherent to such occasions that the commemorative or historical aspect often takes second place to other interests. Although the event was precipitated by the calendar, more than the founding of the university was at the fore.

Formal planning for the tercentenary was underway as of 1930 and as early as September 1932, James B. Conant (who became the university's president the next year) was involved in the planning structure. Jerome D. Greene was appointed director of the anniversary observances by Conant in 1934. Greene was a member of the Harvard College class of 1896, former secretary to the Harvard Corporation (the chief governing board) under President Charles W. Eliot, a first executive of the Rockefeller Foundation, and was later in the investment business, where he lost most of his money in the early years of the depression. His appointment in 1934 also renamed him as secretary to the Harvard Corporation. Greene later reported that before he became tercentenary director, it had been determined that "the emphasis of the Celebration should be laid on exhibiting the resources of the University of today rather than on the retrospective aspects of the anniversary."[12] Nonetheless, Samuel

[11] *Ibid.,* pp. 313–15.

[12] *The Tercentenary of Harvard College: A Chronicle of the Tercentenary Year 1935–1936* (Cambridge, Mass.: Harvard University Press, 1937), pp. 5–7, 381; quotation on p. 6. This report on the tercentenary was prepared by director Jerome Greene. See also Records of Tercentenary Celebration

Eliot Morison, the official historian of the tercentenary, wrote three fully documented volumes on Harvard in the seventeenth century and edited a volume on the college in the period 1869–1929.[13] In a private capacity, he also wrote *Three Centuries of Harvard* "in order that Harvard graduates and others may have something of this scope to read in 1936."[14] In effect, Morison recognized that focus and interest in the tercentenary year went beyond the founding. It was on the entire three-hundred-year history of the institution.

In addition to contemplating salient contemporary concerns and the sweep of Harvard's past, a forward-looking stance also came forth frequently in the observances. The final event of the tercentenary was the meeting of the Alumni Association, over which President Emeritus A. Lawrence Lowell presided. At its conclusion, Conant moved that the alumni meeting be adjourned to the year 2036, to which Lowell added, "In spite of the condition of many things in the world, I have confidence in the future. Those of you, therefore, who believe that the world will exist 100 years hence and that universities will then be faithful to their great purpose will say 'aye', contrary-minded 'no'. The vote was unanimous."[15]

As the focus was both local and international, and on the past, present, and future, the choice of a specific date for the celebration (as recorded by tercentenary director Greene) was dictated by other than strictly historical considerations. Early in the planning, it was decided that the culminating public ceremonies should include 18 September, the date on which the General Court session had begun in 1636. Part of the reason for choosing this date was an assumption that the weather would be more

Office, Minutes and Papers of the Tercentenary Committee 1930–1936, Harvard University Archives, Cambridge, Mass., UAV 827.2 (hereafter cited as Tercentenary Records); see note by Henry James and report on meeting of 27 September 1932 (1932 folder) for Conant's early involvement. On Greene, see *Dictionary of American Biography Supplement Six 1956–1960,* ed. John A. Garraty (New York: Scribner's, 1980). At the time of his retirement in 1943, Greene was secretary to both the Harvard Corporation and to the Harvard Board of Overseers.

[13] Morison, *Founding of Harvard College* (cit. n. 1); Samuel Eliot Morison, *Harvard College in the Seventeenth Century,* 2 vols. (Cambridge, Mass.: Harvard University Press, 1936); *The Development of Harvard University since the Inauguration of President Eliot, 1869–1929,* ed. Samuel Eliot Morison (Cambridge, Mass.: Harvard University Press, 1930). Morison was appointed tercentenary historian in 1924; *Tercentenary of Harvard College* (cit. n. 12), p. 5. George Sarton reviewed the Morison volumes in *Isis,* September 1936, *25,* n. 2:513–20, the writing of which he described as "the fundamental requirement" of the tercentenary celebrations. But Sarton did not relate the volumes or the anniversary to the history of science much further than the curious observation that slow progress in expanding the interests of most people beyond the local or personal will continue to favor interest in provincially focused or antiquarian societies above a society for the history of science. For a man who dedicated his life to promoting a broad and integrated view of the history of science and civilization, his comments seem curiously cynical in light of the idealistic hopes for a world community of scholarship that characterized the academic aspects of the tercentenary celebration.

[14] Morison, *Three Centuries of Harvard* (cit. n. 2), preface. See also Max Hall, *Harvard University Press: A History* (Cambridge, Mass., and London: Harvard University Press, 1986), pp. 73–4. In 1934, the celebration planning committee voted to recommend to the corporation that Morison undertake a one-volume history that should take priority over his other tercentenary writings; see J. B. Conant Presidential Records, 1934–1935, folder: Tercentenary: Central Committee, Minutes of meeting 30 April 1934, Harvard University Archives, Cambridge, Mass. (hereafter cited as Conant Records).

[15] *Harvard Alumni Bulletin,* 30 September 1936, *39,* n. 1:48. Also quoted in Richard Norton Smith, *The Harvard Century: The Making of a University to a Nation* (New York: Simon & Schuster, 1986), p. 131, which was produced at the time of Harvard's 350th anniversary. A similar resolution to meet a century hence concluded the anniversary celebrations in 1836; see Quincy, *History of Harvard* (cit. n. 5), vol. II, pp. 706–7.

favorable in September, but perhaps more importantly, the actual date of founding, 7 November, would fall during the week of national presidential elections.[16]

Presidential politics, in spite of this preemptive attempt, did not stay out of the celebration entirely. Franklin D. Roosevelt, who was running for his second term in 1936, was a Harvard graduate; he had been invited by Conant to speak at the alumni meeting on the afternoon of the last day of the celebration.[17] Furthermore, FDR was a member (and honorary chair) of the United States Harvard University Tercentenary Commission, created by Congress in April 1936 as a vehicle for the federal government's invited participation in the celebration.[18] Conant was sympathetic to the New Deal, but his predecessor A. Lawrence Lowell was scandalized by it. When Lowell was chosen as "President of the Day" to preside over the tercentenary alumni meeting, he initially refused to participate if Roosevelt was to speak. He finally was persuaded to play his part, but before the event Lowell and Roosevelt exchanged correspondence in which Lowell attempted to dictate the character of FDR's remarks at the alumni gathering. In his memoirs, Conant explained Lowell's actions as being those of a person "placed in a position of having to introduce a man whom he despised to an audience which he loved."[19]

International politics, as well, cast a shadow over the planning. In 1934, Conant had a chance to take a stand in support of the besieged German universities when Ernst "Putzi" Hanfstaengl, Harvard class of 1909 and associate of Adolf Hitler, offered the university one thousand dollars for a German scholarship. (This was in connection with Hanfstaengl's presence for the twenty-fifth anniversary of his Harvard College class.) Conant and the Harvard Corporation rejected the offer in the face of criticism from the *Harvard Crimson.*[20] Later, an invitation to participate in Ruprecht-Karl-Universität Heidelberg's 550th anniversary ceremonies, in June 1936, came closer in time to Harvard's tercentenary. Universities in Great Britain and Europe refused the invitation. Harvard, nonetheless, decided to send a delegate and underscored the theme of academic unity that would characterize its Cambridge tercentenary. But the step was not taken without fierce criticism. Along with high-sounding principles, President Conant had to confess in private that the decision was also motivated by the desire to avoid a possible German boycott of Harvard's celebration. The observances at Heidelberg, in fact, substantially reflected the

[16] *Tercentenary of Harvard College* (cit. n. 12), p. 7. James B. Conant, *My Several Lives: Memoirs of a Social Inventor* (New York: Harper & Row, 1970), p. 149.

[17] Conant, *My Several Lives* (cit. n. 16), p. 153.

[18] *Tercentenary of Harvard College* (cit. n. 12), pp. 15–18, 420. The commission was formed after the university invited the national government to name delegates to the tercentenary observances; its functions apparently were limited to that role. The official history of the tercentenary, in mentioning the commission, was careful to point out that the general organization of the event was directed to the association between Harvard and other academic and learned bodies and individuals, not governments. Exceptions were made for the Commonwealth of Massachusetts, the Cities of Boston and Cambridge, and the United States government. The rationale given for inviting federal government representation was observance of the three-hundredth anniversary of the establishment of higher education in the country, rather than celebration of a particular institution. This theme appeared in other contexts in the observances; for example, at the reception of delegates on 16 September that opened the three primary days of celebration. See *Harvard Alumni Bulletin,* 30 September 1936, *39,* n. 1:4–5.

[19] James G. Hershberg, *James B. Conant: Harvard to Hiroshima and the Making of the Nuclear Age* (New York: Alfred A. Knopf, 1993), pp. 90–1. Conant, *My Several Lives* (cit. n. 16), pp. 152–3 (quotation on p. 153). Smith, *Harvard Century* (cit. n. 15), p. 127.

[20] Smith, *Harvard Century* (cit. n. 15), pp. 117–19.

crass propagandistic intentions of the Nazi regime, but Conant was relieved that the German officials did not exploit the view that Harvard's presence suggested complicity.[21] The following year, after Harvard's own observances were over, the university concluded that it would not be able to send a delegate to the bicentennial ceremonies for the Georg-August-Universität zu Göttingen.[22]

Though it is apparent that planning for the tercentenary involved compromise between ideals and pragmatic choices, director Jerome Greene, in his official account of the celebration, highlighted "the dominant note of the whole [tercentenary] occasion, namely, that of a community of interest in the upholding of scholarship, free and disinterested, in a world harassed by social and political upheavals—a note of confidence and courage."[23] In effect, Greene underscored Conant's professed desire to put international academic unity above other, more worldly political concerns.

The Harvard tercentenary was a truly international event; delegates were sent by universities and learned societies from all parts of the world. In the official report on the celebration, the institutions represented and their delegates are listed by processional number, with the oldest institution first. That place went to Al-Azhar University (Cairo), founded in 970. Second was the Università degli Studi di Bologna and third the Université de Paris. In all, about 160 foreign institutions were represented; including American institutions, there were over 500.[24]

## III. OVERVIEW AND TEXTURE OF THE TERCENTENARY CELEBRATION

The tercentenary entailed a number of events that addressed the multiple interests and expectations motivating the celebration. A Tercentenary Conference of Arts and Sciences preceding three days of open celebration was the primary academic or scholarly component of the observances. During the festive climax, from Wednesday, 16 September to Friday, 18 September, the days (and nights) were filled with pageantry, receptions and dinners, concerts, religious services, fireworks, and other observances that recognized the American and foreign delegates and alumni organizations. These events included addresses or greetings by, among others, the university president and other officers, students, the master of Emmanuel College, Cambridge (John Harvard's alma mater), and officers of other academic institutions, Massachusetts's governor, and the president of the United States. The large public ceremonies were held in the open area between the Widener Library and Memorial Church, in Harvard Yard, designated the Tercentenary Theatre and seating some 17,000 people.[25] A number of displays were produced around the university, forming a "Great Exhibition: one which, while it scarcely follows the World's Fair pattern, could not be duplicated outside the boundaries of a great university," as the *Tercentenary Gazette* observed.[26]

Among the aspects of the celebration that relate to the sciences (in addition to the

[21] Hershberg, *James B. Conant* (cit. n. 19), pp. 96–7. Hershberg alludes to Conant's concern that a German reaction against Harvard might result in "publicity interfering with the planned fund-raising and self-congratulations."

[22] William M. Tuttle, Jr., "James B. Conant, Pressure Groups, and the National Defense, 1933–1945" (Ph.D. dissertation, University of Wisconsin, 1967), pp. 78–9.

[23] *Tercentenary of Harvard College* (cit. n. 12), p. vii.

[24] *Ibid.*, pp. 394–419 (lists of institutions and delegates).

[25] See the summary report in the *Harvard Alumni Bulletin,* 30 September 1936, *39,* n. 1 and the more extended account in *Tercentenary of Harvard College* (cit. n. 12).

[26] *Harvard University Tercentenary Gazette,* 3 July 1936, *3:*1.

tercentenary conference) were the first International Conference on Soil Mechanics and Foundation Engineering (22–26 June 1936), and a special tercentenary session of the Harvard Summer School that included notable foreign scholars.[27] There were tercentenary sessions at the Harvard Medical School (14–15 September), Dental School (15 September), and School of Public Health (24–29 August). As part of the American Astronomical Society meeting held in tandem with Harvard's tercentenary conference, one morning was set aside for the presentation of twenty-eight papers by the Harvard College Observatory staff.[28] Radcliffe College mounted what must have been an extensive exhibit on women in science, featuring contemporary women but apparently incorporating some historical aspects as well. The exhibit was in Radcliffe's Byerly Hall, which opened its laboratories to visitors, and included publications and photographs.[29] But this was virtually the only reference to women in science in connection with the tercentenary.

The substantive aspects of the occasion were, of course, represented in its formal programs and the published books and papers that resulted. The texture was in the level of detail that reflected the work of creating the event and the experience of being part of it. It is difficult to imagine the amount of clerical and other work that went into the tercentenary celebration, but a mere consideration of the records that have survived in the university archives will give an idea. To consider these records—which document repetitive detail, careful planning, diplomatic negotiation, and the process of translating conflicting values and ideas into a viable and appropriate program of events—is to enter ethnographically into a community at work. Some 160 containers of records from the Tercentenary Celebration Office alone have survived, and they relate to all aspects of the event, large and small: e.g., the choice of conference participants and attendees; arrangements for transport and housing of participants and their families; publications of conference papers; issuance of passes; seating charts; exhibition catalogues; publication of a handbook on the university for visitors; design and preparation of medals; news releases and radio broadcasts; hundreds of printed items such as programs, flyers, tickets, and invitations; arrangements for official delegates; etc., etc.

## IV. TERCENTENARY CONFERENCE OF ARTS AND SCIENCES

On the one hand, the tercentenary was spread on the broadest canvas possible. But it also served local institutional interests, and the ways in which these missions fit together is intriguing but not easy to articulate. This dual nature was especially true of Conant's early intentions as president for the development of the university, and the process of planning and carrying out the Tercentenary Conference of Arts and Sciences. Director Jerome Greene said of the conference, which was held at Harvard between 31 August and 12 September 1936, that it was "The most substantial and distinctive feature of the Tercentenary Celebration."[30] The conference was, in fact,

[27] This description is based largely on the *Harvard University Tercentenary Gazette,* published in eight issues, 12 June–11 September 1936. There also are accounts or descriptions of events in *Tercentenary of Harvard College* (cit. n. 12).

[28] *Scientific Monthly,* October 1936, *43:*376.

[29] *Tercentenary of Harvard College* (cit. n. 12), p. 29. Also, unpublished finding aid, "Radcliffe 'Women in Science' Exhibit, 1936. Papers, 1935–1940," Radcliffe College Archives, Cambridge, Mass., B-6,

[30] *Tercentenary of Harvard College* (cit. n. 12), p. 87.

a major international scholarly event, with leading academics from all parts of the world contributing. It was carried out before an audience of about 2,400 American and Canadian scholars, with some lectures open to the public.[31] The conference, therefore, is the place to look for the special relevance of the tercentenary to the history of science.

Several threads came together in planning for the conference. In the first page of his first annual report as president, covering the academic year 1933–1934, Conant observed that "The administration of a great university must endeavor to find methods of counteracting the centrifugal forces which tend to separate our faculties into an ever-increasing number of subdivisions. [But] There must be no hint of regimentation, no blueprint plan . . . " With these words so early in his term, Conant identified a leading theme that played throughout the tercentenary: that of the individual and cooperation; the particular and the universal. His immediate plan for the institution, as outlined in this annual report, was to appoint advisory research committees in "each of the four major divisions of knowledge: the Humanities, the Social Sciences, the Biological Sciences and the Physical Sciences." The members of the committees were from the various faculties and (the president insisted) had none but advisory functions. As a concrete suggestion along these same lines, Conant called for the institution and endowment of "university professors," whose teaching and research activities were not confined to specific academic departments.[32]

When Conant was called to report on the 1934–1935 academic year, he featured the upcoming tercentenary and observed that "We do well to salute this country's distinguished past, but clearly our most important task is to face the future." His vision for the future was to plan for faculty and for students, a vision actualized in a Three Hundredth Anniversary Fund enacted by the governing boards. Its twin goals were to create university professorships and to establish what Conant called Harvard National Scholarships. The campaign had been devised over a period of two years, with the help of a special faculty committee.[33] Though Conant stated that the tercentenary fund had not set a dollar goal, it is obvious from the records of the president's office that considerable work had been done to begin the campaign. In one compilation of the aggregate needs of the university departments, the total came to $87 million.[34] Such a figure demonstrated the dimension of the overall problem and the poignancy of Conant's plan to concentrate on the larger picture of coordination and cooperation. The President's Committee on Special Research, which reported in mid-1934, took Conant's intent for the university professorships seriously. The committee's report emphasized the point that the hoped-for unity within the university would be based on the existing strength of its component parts.[35]

At the tercentenary Alumni Association meeting on the afternoon of 18 September

[31] *Ibid.*, pp. 9–11.

[32] "President's Report 1933–34," *Official Register of Harvard University*, 18 February 1935, *32*, n. 3:5–7.

[33] "President's Report 1934–35," *Official Register of Harvard University*, 29 February 1936, *33*, n. 4:5–7.

[34] Conant Records 1934–1935, folder: Tercentenary Fund; also see folder: Tercentenary: Special Research Committee, which includes a number of signed statements by faculty members on work and needs of various parts of the university.

[35] Conant Records 1935–1936, folder: Tercentenary Fund: Special Gifts Committee, printed confidential document, "Extracts of the Report to the President from the Committee on Special Research June 1934."

1936, the last day of the celebration, Conant reported on the anniversary fund. A total of $2,814,971 had been received. Of this amount, about $1 million was for the Harvard National Scholarships and a little more than $500,000 was for the university professorships. In addition, about $775,000 was unrestricted, and slightly less than a half million was for "the encouragement of work in the physical sciences."[36] The funds for the university professorships must have been somewhat of a disappointment to Conant, who reported to the corporation in October 1936 that the investment income from $500,000 would be required to support each such position.[37] In fact, in June 1936, Conant felt the ground shifting within the university, where he was under pressure to address particular departmental needs at the expense of university-wide considerations. He feared then that the scheme of university professorships might have little chance of success.[38] As Conant insisted, the game of coordination and individuation was a matter of keeping a difficult balance. When he looked back on the Tercentenary Conference of Arts and Sciences for his annual report for 1935–1936, he observed that "The joint labors of these scholars proved once again that it is because of specialization that knowledge advances, not in spite of it; and that cross fertilization of ideas is possible only when new ideas arise through the intense cultivation of special fields."[39]

The exact moment when the idea for the tercentenary conference arose in the planning is not entirely clear. What is clear is that it represented the resolve to make the three-hundredth anniversary celebration an academic and not a political or literary event. A report dated 1 September 1934 set aside the period from 17 August to 12 September 1936 for the meetings of learned societies and special symposia.[40] In fact, several national mathematical organizations and the American Astronomical Society held meetings in tandem with the tercentenary, and selected sections of their programs were integrated with the Tercentenary Conference of Arts and Sciences.[41]

The university more than paid lip service to the academic and educational aspects of the celebration. In a memorandum concerning the budget for the tercentenary, prepared for the Harvard Corporation book of understandings (meeting of 5 November 1934), the total projected budget for the celebration was set at $150,000, of which $90,000 was intended for educational purposes, including colloquia. The remaining $60,000 was for ceremonies and festivities.[42]

The planning for the celebration began to take real form by late fall and winter of 1934. Members of the faculty were involved at an early date. In October of 1934, the president had meetings with four small committees, each representing one of the four traditional areas of knowledge. (These were the advisory committees that Conant had appointed relative to his plans for university coordination.) The committee

[36] *Tercentenary of Harvard College* (cit. n. 12), p. 228.

[37] Conant to Corporation, 18 October 1936, Conant Records 1935–1936, folder: Tercentenary Fund: Special Gifts Committee.

[38] Conant to Henry James, 20 June 1936, Conant Records 1935–1936, folder: Tercentenary Fund: Special Gifts Committee.

[39] "President's Report 1935–36," *Official Register of Harvard University,* 22 March 1937, *34,* n. 11:6.

[40] Conant Records 1934–1935, folder (item): "1636 Harvard Tercentenary 1936 A Sketch of the Celebration First Draft September 1, 1934."

[41] *Tercentenary of Harvard College* (cit. n. 12), p. 88.

[42] Conant Records 1934–1935, folder: Tercentenary Celebration: Greene, "Memorandum for the Book of Understandings."

on biological sciences included Thomas Barbour, Walter B. Cannon, Lawrence J. Henderson, Alfred C. Redfield, and Hans Zinsser. The physical sciences advisory committee included George D. Birkhoff, Percy W. Bridgman, Elmer P. Kohler, Donald H. McLaughlin, George W. Pierce, and Harlow Shapley. There also were committees on social sciences and on the humanities (the latter including George Sarton and Alfred North Whitehead).[43]

When the advisory committees each met with Conant in October 1934, they had a standardized but mixed agenda. The president introduced the concept of the committees as relating to general concerns for knowledge and scholarship free of departmental considerations, emphasizing instead the overall needs of the university. The way to attack the problem, however, began with the compilation of notable names in the general area of knowledge of each committee. The intention was, thus, to serve the coordination needs of the university while also generating lists of individuals who might be invited to take part in the tercentenary. (Jerome Greene, as tercentenary director and secretary to the Harvard Corporation, was a participant in these meetings.) Following these exercises, the meetings turned to the president's agenda of university professorships and research funds.[44]

A joint meeting of all four advisory committees took place on 22 November 1934 and there plans for the conference began to take shape. In the discussion, Alfred North Whitehead suggested that some of the symposia be comparative between 1636 and 1936, but Harlow Shapley explained that the exhibits committee, another component of the celebration, had shown little interest in an historical approach. Discussion of symposia then were referred back to the four advisory committees, while an executive committee was formed by the president, with Greene as chair. This executive committee included Edwin F. Gay for social sciences (later replaced by Edwin Bidwell Wilson), Harlow Shapley for physical sciences, Lawrence J. Henderson for biological sciences, and Arthur Darby Nock for the humanities.[45] The advisory committees met and passed their suggestions to the executive committee, which considered them at a meeting on 19 December 1934. The committee on biological sciences recommended a symposium on "Factors Determining Human Behavior," while social sciences and humanities suggested "Authority and the Individual" and "Independence, Convergence, and Borrowing in Institutions, Thought and Art."[46] These symposia eventually became part of the tercentenary conference.

On 14 December 1934, the Harvard University News Office sent out the first public release of the tercentenary schedule, including an announcement of the Tercentenary Conference of Arts and Sciences. It also included a statement by Greene on the purposes of the celebration, of which there were three: first (but not of greatest importance) was festive remembrance, and second was historical reflection, which also was not the central purpose. The primary place was given to the goal of contributing to education and learning through a special session of the Harvard Summer

[43] Tercentenary Records, Conference of Arts and Sciences, folder: Advisory Committee: General, UAV 827.114, box 26.

[44] *Ibid.*

[45] *Ibid.* Also, *Tercentenary of Harvard College* (cit. n. 12), p. 383.

[46] Tercentenary Records, Conference of Arts and Sciences, Minutes of Executive Committee, 19 December 1934, UAV 827.114, box 27.

School, exhibits, meetings of learned societies, and by bringing together scholars in the symposia preceding the concluding ceremonies.[47]

Plans for the conference, as formulated by the executive committee, operated on two principles, undoubtedly reflecting the views of the four advisory committees. For mathematics and the physical sciences (and less universally in the biological sciences) invited participants were asked to contribute a paper on a topic of their own choosing, and the various contributions were later arranged in a logical way on the program. The three collaborative symposia (named earlier) were devised to focus some of the biologists and all of the humanists and social scientists on large areas of common concern. The executive committee considered these integrated symposia to be the most significant of the conference's offerings.[48]

After extensive consultation within the Harvard faculty about prospective participants in the conference, invitations were sent out by the president on 5 February 1935.[49] Since Harvard was to serve as host to the conference, no university faculty members were asked to participate. In presenting the list to the president in early January, Greene explained that all the individuals on the list were worthy of receiving an honorary degree and were still actively working in their fields. All who were invited and accepted did receive an honorary degree (if Harvard had not so honored them previously) at the concluding ceremonies on 18 September. None other than these scholars received honorary degrees as part of the tercentenary, again underscoring the academic character of the celebration. Sixty-two degrees were conferred, of which thirty-four were doctors of science (see Table 1).[50]

[47] Conant Records 1935–1936, folder: Tercentenary: Director J. D. Greene, news release no. 1934–332.

[48] Greene to Conant, 31 May 1935, Conant Records 1934–1935, folder: Tercentenary Celebration: Greene.

[49] Tercentenary Records, Drafts of the Tercentenary Plans 1934, Suggested Participants 1935; this file series includes List of Scholars to be Invited (president's copy) with letter from Greene to Conant, 3 January 1935 that outlines the process of selection, UAV 827.10.19. Also see Tercentenary Records, Conference of Arts and Sciences, participants files, UAV 827.114, boxes 14–21.

[50] *Tercentenary of Harvard College* (cit. n. 12), pp. 8–9, 217–21 (list of honorary degree recipients), pp. 457–72 (tercentenary conference program). The scientists honored as participants in the Conference of Arts and Sciences were: Edgar Douglas Adrian (physiology, University of Cambridge); Edward Battersby Bailey (geology, University of Glasgow); Sir Joseph Barcroft (physiology, University of Cambridge); Friedrich Bergius (chemistry, Heidelberg); Norman Levi Bowen (petrology, Carnegie Institution of Washington); Rudolf Carnap (philosophy, Deutsche Universität, Prague); Élie Joseph Cartan (mathematics, Université de Paris); James Bertram Collip (biochemistry, McGill University); Arthur Holly Compton (physics, University of Chicago); Peter Debye (physics, Universität Leipzig); Leonard Eugene Dickson (mathematics, University of Chicago); Sir Arthur Stanley Eddington (astronomy, University of Cambridge); Hans Fischer (chemistry, Technische Hochschule, Munich); Ronald Aylmer Fisher (eugenics, University of London); Corrado Gini (statistics and sociology, Università degli Studi di Roma 'La Sapienza'); Godfrey Harold Hardy (mathematics, University of Cambridge); Ross Granville Harrison (biology, Yale University); Johan Hjort (marine biology, Universitet i Oslo); Sir Frederick Gowland Hopkins (biochemistry, University of Cambridge); Bernardo Alberto Houssay (physiology, Universidad de Buenos-Aires); Pierre Janet (psychology, Collège de France); Carl G. Jung (analytic psychology, Technische Hochschule, Zurich); August Krogh (zoophysiology, Københavns Universitet); Karl Landsteiner (physiology and medicine, Rockefeller Institute for Medical Research); Andrew Cowper Lawson (geology, University of California); Tullio Levi-Civita (rational mechanics, Università degli Studi di Roma 'La Sapienza'); Bronislaw Malinowski (anthropology, University of London); John Howard Northrop (biochemistry, Rockefeller Institute for Medical Research); Antonie Pannekoek (astronomy, Universiteit van Amsterdam); Leopold Ruzicka (chemistry, Technische Hochschule, Zurich); Kiyoshi Shiga (medicine, Kitasota Institute, Tokyo); Filippo Silvestri (zoology, Regia Scuola Superiore de Agricultura, Portici, Italy); Hans Spemann (physiology and medicine, Albert-Ludwigs-Universität Freiburg); The Svedberg (chemistry, Uppsala Universitet). Scientists (or affiliates of the scientific community) who took part but who already held honorary

**Table 1. Honorary Degree Recipients Arranged by the Tercentenary Conference of Arts and Sciences Symposium in Which They Participated[1]**

| "Factors Determining Human Behavior" | "Authority and the Individual" | "Physical Sciences" | "Biological Sciences" |
|---|---|---|---|
| Edgar D. Adrian[2] | Corrado Gini | Edward B. Bailey | Joseph Barcroft |
| Rudolf Carnap | John Dewey[3] | Friedrich Bergius | Ross G. Harrison |
| James B. Collip | | Norman L. Bowen | Johan Hjort |
| Pierre Janet | | Rudolf Carnap | Frederick G. Hopkins |
| Carl G. Jung | | Elie J. Cartan | Bernardo A. Houssay |
| Bronislaw Malinowski | | Arthur H. Compton | August Krogh |
| Jean Piaget | | Peter Debye | Karl Landsteiner |
| | | Leonard E. Dickson | Elmer D. Merrill[3] |
| | | Arthur S. Eddington | John H. Northrop |
| | | Hans Fischer | Kiyoshi Shiga |
| | | Ronald A. Fisher | Filippo Silvestri |
| | | Godfrey H. Hardy | Hans Spemann |
| | | Frank B. Jewett[3] | The Svedberg |
| | | Andrew C. Lawson | |
| | | Tullio Levi-Civita | |
| | | Robert A. Millikan[3] | |
| | | Antonie Pannekoek | |
| | | Henry N. Russell[3] | |
| | | Leopold Ruzicka | |
| | | William B. Scott[3] | |

[1]No scientists participated in the symposium on "Independence, Convergence, and Borrowing in Institutions, Thought, and Art."

[2]For brief biographical identifications of the listed scientists, see cit. n. 50.

[3]Previously received an honorary degree from Harvard.

In an undertaking as complex as the Conference of Arts and Sciences, various problems arose and readjustments were inevitably required. The theoretical physicists suffered a particularly high rate of attrition. Niels Bohr, Albert Einstein, and Werner Heisenberg all agreed to participate but eventually were not able to attend. Heisenberg wrote to Conant on 5 July 1936 because he wanted the president to know that there was no political component to his withdrawal, as he had previously received permission (apparently from the German government) to take part. His reasons included a need to defend theoretical physics from a recent attack in the press, and an obligation to perform military service. Einstein had to withdraw because of his wife's illness but this was not without controversy, since a news story circulated that he had refused to take part in a program that included scientists from

degrees from Harvard included: John Dewey (philosophy, Columbia University); Frank Baldwin Jewett (president, Bell Telephone Laboratories); Elmer Drew Merrill (botany, Harvard University; he was director of the New York Botanical Garden when chosen as a conference speaker); Robert A. Millikan (physics, California Institute of Technology); Henry Norris Russell (astronomy, Princeton University); William B. Scott (geology, Princeton University). Jean Piaget (history of scientific thought, Université de Genève) received the doctor of letters.

Nazi Germany. Among the scientists who were invited but could not attend was physiologist Ivan Pavlov. In his mid-80s when invited, Pavlov died before the date of the tercentenary.[51]

The conference planners encountered the Nazi problem directly when they invited zoologist Hans Spemann. The German, through Harvard zoologist George H. Parker, had made the request that Conant and Harvard curb their anti-Nazi attitude. Jerome Greene replied to Spemann through Parker that this was quite impossible, and reiterated the university's stand: Harvard had appointed a delegate to and invited one from Heidelberg, but it had stated reasons (which he quotes) for declining the Hanfstaengl gift. Greene also explained Harvard's overriding concern with the fate of Germany's universities and not with its form of government; in fact, no foreign governments had been asked to take part in the tercentenary, only universities and learned societies, and this stance had been communicated to the German embassy in Washington.[52]

The case of Carl Jung, a participant in the tercentenary conference, also revealed the problems encountered in organizing an international conference in an atmosphere of extreme divisiveness and danger. Greene was faced with a concerned correspondent, New York lawyer Maurice Leon, who argued that Jung supported or sympathized with the Nazis and that Harvard's involvement with him, especially the conferral of an honorary degree, would appear to endorse Jung's supposed complicity. Greene replied to Leon on 31 August 1936, the very day the Conference of Arts and Sciences opened. He explained that the case of Jung had been carefully considered by the university and that, as tercentenary director, he accepted the faculty's view that Jung had an important contribution to make to the symposium on "Factors Determining Human Behavior." Nonetheless, Greene appears to say that, for him, the case against Jung produced some ethical considerations and concerns about Jung's objectivity as a scientist. Finally, Greene fell back on the defense that Jung's point of view would have to withstand the critique of his peers.[53]

The Conference of Arts and Sciences impressed by its very ambition, and was referred to as "the greatest single assemblage of the scholars of the world since the Middle Ages."[54] The richness of the program is difficult to convey but, as indicated earlier, it included three integrated symposia, individually submitted papers by physical scientists and biologists, and aspects of the meetings of mathematical and astronomical societies. The symposium on "Factors Determining Human Behavior" was the most broadly integrated: Edgar Douglas Adrian (physiology, University of Cambridge) spoke on "The Nervous System"; James B. Collip (biochemistry, McGill University) on "Hormones"; Jean Piaget (child psychology and the history of scientific thought, Université de Genève) on intellectual evolution; Carl Jung (ana-

[51] For correspondence regarding the invitees who did not participate, see Tercentenary Records, Conference of Arts and Sciences, folders: Participants Dropped, UAV 827.114, box 13. For Einstein, see the same series, folders: Participants, box 16.

[52] Greene to Conant, 2 July 1936, enclosing copy of letter from Greene to G. H. Parker, 30 June 1936, Tercentenary Records, Conference of Arts and Sciences, participants files, Spemann, UAV 827.114, box 20. Other documents in the folder indicate that Parker sent Greene's letter to Spemann.

[53] Tercentenary Records, Conference of Arts and Sciences, participants files, Jung, UAV 827.114, box 17.

[54] Joseph Barnes, "Harvard Tercentenary—The Greatest Scholars View a Darkened Universe," *Current History,* November 1936, *45:*73.

lytic psychology, Technische Hochschule, Zurich) on psychological factors; Pierre Janet (psychology, Collège de France) on psychological strength and weakness in mental diseases; Rudolf Carnap (philosophy, Deutsche Universität, Prague) on logic; A. Lawrence Lowell (Harvard president emeritus) on "An Example from the Evidence of History"; and Bronislaw Malinowski (anthropology, University of London) on "Culture as a Determinant of Behavior." The other two symposia, "Authority and the Individual" and "Independence, Convergence, and Borrowing in Institutions, Thought and Art," included scholars from the humanities and the social sciences (including economics, history, law, and sociology). All three symposia were intended to address, from an interdisciplinary view, large problems confronting humanity and the international community. Three volumes containing papers from the symposia were published by Harvard University Press in 1937 under the titles of the symposia.

It is to some degree ironic that, despite the efforts of Conant (a leading organic chemist before becoming president in 1933) to promote crossdisciplinary work within the university, the scientist participants in the tercentenary conference proceeded as individuals with their own research agendas. The organizers, nonetheless, attempted to gather the papers into large thematic groups. But in publication, the scientists did not follow the route of cooperation or integration; their papers appeared individually in various scientific journals. However, the Tercentenary Celebration Office gathered reprint copies of those it could and issued them in a bound volume, of which fifty copies were prepared.[55]

Although a good part of the tercentenary conference was of a technical nature, it received extensive press coverage.[56] This was no accident; a press headquarters had been established in Grays Hall in Harvard Yard. The *Harvard Crimson* described the activity there, reporting that "The heads of the science departments of the three big wire services, the AP, the UP, and the INS, as well as three or four men from each Boston paper and from several other out of town papers, were also present. Science Service, an organization specializing in the gathering of all scientific news, sent a large fraction of its whole staff." Twice a day during the conference, a news conference was held at which a younger member of the Harvard faculty was on hand to help reporters understand the papers that were presented. Occasionally, the authors of the papers would attend the news conference as well.[57] In addition to press coverage, the World Wide Broadcasting Foundation of Boston (WIXAL) broadcast forty-five of the conference papers; reports from listeners were received from as far away as Great Britain and New Zealand. Jerome Greene's report on the tercentenary claimed the broadcast of the conference "established a new record in the use of the radio in making available by short-wave transmission learned discourses ordinarily regarded as unavailable for popular entertainment."[58]

[55] For the tercentenary conference program, which also indicates where the papers were published, see *Tercentenary of Harvard College* (cit. n. 12), pp. 457–72; also, *Scientific Papers Contributed to the Tercentenary Conference of Arts and Sciences August 31–September 12, 1936, Other than Those Related to the Three Collaborative Symposia Published By the University* (Cambridge, Mass.: Harvard University, 1938). Abstracts of the papers were printed by Ginn and Company, 1936, as a supplement to their *What the Colleges Are Doing* (see Harvard University Archives, HUA 936.33).

[56] See, for example, Scrapbooks of Clippings from the Tercentenary in the Harvard University Archives, HUA 936.80, especially vol. 1.

[57] *Harvard Crimson,* 16 September 1936, *107,* n. 1:10.

[58] *Tercentenary of Harvard College* (cit. n. 12), pp. 42–3.

An impromptu session at the end of the conference brought together four participants—philosopher John Dewey, French historian Etienne Gilson, anthropologist Bronislaw Malinowski, and Chinese philosopher and historian Hu Shih—in a discussion of the conference's expected outcome and prospects. Though Hu demurred, the other three endorsed the idea of what one newspaper headline called a "Supreme Court of Learning"—that is, some type of permanent organization that would channel scholarly knowledge toward solving world problems.[59] Others, however, were less optimistic about the outcome while lauding the event itself. A commentator in *The Nation* observed, "To the surprise of almost no one, the ambitious plan of Harvard University on its three-hundredth birthday to synthesize the specialized branches of modern scholarship into a unified and coherent system of thought and belief progressed no farther than the titles of the symposia . . . " But the same author, a 1927 graduate of Harvard who covered the tercentenary for the *New York Herald Tribune,* seemed to find a degree of unity in the speakers' methodology when he wrote, "With honorary degrees from Harvard awaiting them at the end of the conference, many of the invited guests in the liberal arts rivaled their colleagues in the physical and biological sciences in the toughness of their empiricism."[60]

## V. THE TERCENTENARY AND HISTORY

As indicated earlier, the planners were ambivalent about the place of history in the tercentenary, and it mainly served as a stepping stone to contemporary concerns and future directions. The Conference of Arts and Sciences most clearly demonstrated the broader—both international and academic—intent of the planners. In the three festive Tercentenary Days in mid-September, an historical consciousness did come to the fore. At the opening ceremonies on 16 September, where nearly 550 delegates were formally greeted, Conant placed the event in the context of the academic history of the United States, and suggested that Harvard was the link to the foreign universities represented.

The second of the Tercentenary Days was especially devoted to the alumni, and was highlighted by a meeting of the Associated Harvard Clubs in the outdoor Tercentenary Theatre. On that occasion, Conant observed that "the Tercentenary celebrations are insidious in their effect, and all of us who have been closely connected with the plans have perhaps become too historically minded," but he took the moment to emphasize that the meeting of the various Harvard clubs—with local organizations in all parts of the United States as well as in Europe, Asia, and elsewhere—indicated that the university was as well extended geographically as it was temporally. This fact would have been of particular interest to the president, whose national scholarships program was intended to extend the range of the student body beyond its traditional New England base.[61] Conant's remarks then turned back to history, underscoring the range of intellectual development over the three hundred years of Harvard's existence by demonstrating the state of scientific thought at its founding and in 1836

[59] *Scientific Monthly,* November 1936, *43:*490. Scrapbooks of Clippings from the Tercentenary (cit. n. 56), vol. 1, p. 96, clipping (apparently) from *Boston Herald,* 14 September 1936.
[60] Joseph Barnes, "Harvard's United Front," *The Nation,* 26 September 1936, *143:*355–6.
[61] Tuttle, "James B. Conant" (cit. n. 22), p. 36.

***Figure 1.*** *A view of the Tercentenary Theatre in Harvard Yard, from the west, during the 18 September final ceremonies, held under clouds and rain. The platform is on the left and a portion of the audience on the right; in all, some 17,000 were in attendance.*

by way of contrast to 1936. He concluded by reading from letters from the alumni prepared in 1836 and by sealing a similar package to be opened in 2036.

The final climatic day of the tercentenary celebration, 18 September, was well-planned but disrupted by the weather, which forced the afternoon Alumni Association meeting indoors. The morning events, which were the most ceremonious of all, were held in the Tercentenary Theatre though partly under falling rain (see Figure 1). It was the anniversary date of the founding of the college, but the official published account would closely link that last day to the preceding Conference of Arts and Sciences, arguing that "Much of what was said and done on the last day derived its meaning and inspiration from that great assemblage of scholars." The 18 September ceremonies included a salutatory oration (in Latin), an address on the founding of Harvard College by historian Samuel Eliot Morison, official greetings from the Massachusetts governor (the notorious James Michael Curley), a reading by England's poet laureate John Masefield, an oration by President Conant, and the conferral of honorary degrees (see Figure 2).[62]

Though the tercentenary produced a sea of words, both scholarly and popularized, some considered Conant's oration at the festive morning ceremonies on 18 September as the highlight of the Celebration (see Figure 3). Conant had devoted much

[62] The foregoing summary of the Tercentenary Days is taken chiefly from *Tercentenary of Harvard College* (cit. n. 12), especially pp. 98–101, 117–18, 120–2, 187, 440–1; quotations on pp. 117, 187.

***Figure 2.*** *The doctor of science diploma is presented to mathematician Tullio Levi-Civita. Some of the other honorary degree recipients are seated in the first and second rows, in front of academic delegates and professors.*

time and thought to this oration; one historian characterized it as a delayed inaugural address.[63] The president addressed the issue of history, pointing, on the one hand, to "The future of the university tradition in America—that is the problem that must concern all of us who are assembled here today." If the future was the focus, however, Conant showed that he also took history seriously, even if it served him a purpose. "What is this tradition; indeed, what is a university?" he asked. "Like any living thing, an academic institution is comprehensible only in terms of its history." In examining the university tradition, Conant found "four main streams" that he identified as "the cultivation of learning for its own sake," liberal arts education, education for the professions, and student life. The goal was to maintain a representation and balance among these four aspects of the tradition.[64]

The Tercentenary evoked more than academic self-examination. Commercial organizations were ready to attach their interests to Harvard's own. One interesting example is the generally tasteful advertisements that appeared in a special tercentenary magazine issue from the *Boston Herald;* banks, life insurance companies, railroads and other transportation companies, and manufacturers presented aspects of their own histories as a way of appropriating the opportunity offered by the Harvard

[63] Conant's formal installation as president, bypassing pageantry, resurrected an early eighteenth-century ceremony that took only fourteen minutes. Hershberg, *James B. Conant* (cit. n. 19), pp. 77, 98. Characterization as inaugural in Smith, *Harvard Century* (cit. n. 15), p. 128.

[64] *Tercentenary of Harvard College* (cit. n. 12), pp. 208–9.

***Figure 3.*** *Harvard President James B. Conant addressing the celebrants in the Tercentenary Theatre.*

celebration. NBC and the Radio Corporation of America claimed that their ten-year history was time enough to have "earned [them] a place comparable to that of a great university as a creative force in education."[65] The tercentenary planners faced somewhat less benign concerns from auxiliary groups; one possibility made them worry that the event would resemble a world's fair, which conjured up the idea not of education but of a circus.[66] Conant and others did not see the tercentenary as being anything like a world's fair, but it was placed between two of the largest and most successful world expositions (Chicago's in 1933–1934 and New York's in 1939–1940), and there were shared elements. All three events used an historical anniversary as a reason to organize; the Chicago Century-of-Progress fair commemorated the city's founding while the New York fair celebrated 150 years since the inauguration of George Washington. In each case, however, the place of history became equivocal—the eye was on the present and the kind of world the organizers imagined for the future.[67] Both the world's fairs were strongly oriented toward science and technology as means to achieve the future they envisioned.[68] As this ac-

[65] *Boston Herald,* tercentenary magazine issue, 13 September 1936.

[66] See Conant to Henry James, 10 May 1934, Conant Records 1934–1935, folder: Tercentenary Central Committee.

[67] *Historical Dictionary of World's Fairs and Expositions, 1851–1988,* ed. John E. Findling (New York: Greenwood Press, 1990), pp. 266, 293–4.

[68] Robert W. Rydell, *World of Fairs: The Century-of-Progress Expositions* (Chicago and London: University of Chicago Press, 1993), e.g., p. 93.

count indicates, science also had an important place in the Harvard celebrations and especially in the Conference of Arts and Sciences. The tercentenary, however, featured the systematic study of several areas of knowledge. Furthermore, the organized symposia—broad-based, interdisciplinary, and thematically reflecting an overarching concern for the state of the world—involved the humanities and social sciences more than the natural or physical sciences.[69]

A revival of interest in American colonial history began in the 1920s, although at the same time, the Puritan heritage was heavily criticized.[70] During the 1920s and 1930s, as Michael Kammen notes, more and more local and regional anniversaries were observed. And in New England, at least, there was an increasingly positive interest in Puritan values.[71] The Harvard tercentenary can be seen as part of this phenomenon. For example, historian Bernard DeVoto took a sympathetic look at Harvard from the Puritan perspective in a 1936 article in *Harper's Monthly Magazine.* He was aware of the many perceived negative features of the Puritans and their progeny, yet observed that "the establishment and maintenance of one of the world's great universities is a flowering of the Puritan culture, its finest flowering." Central to DeVoto's argument was his identification of Puritanism as the historic source of Harvard's liberalism and focus on the individual.[72]

Given the direction that the celebration took, however, and the importance given to the Conference of Arts and Sciences, local history had to compete with a longer view. To some degree, President Conant, in his tercentenary oration, recognized

[69] The overall tenor of the tercentenary reflected a concern to maintain the independence of academic institutions from the political currents of the 1930s. Coupled with the scholarly character of the conference (unlike the world's fairs), science in the Harvard context appeared less as an agent for the betterment of society than as a community seeking unity and insularity for its own functioning. But see the discussion of A. N. Whitehead's commentary on the tercentenary in this article.

[70] Rydell, *World of Fairs* (cit. n. 68), p. 74.

[71] Kammen, *Mystic Chords of Memory* (cit. n. 3), pp. 375–8, 391–2.

[72] Bernard DeVoto, "A Puritan Tercentenary," *Harper's Monthly Magazine,* September 1936, *173:*445, 447.

During the planning stages for the tercentenary, Robert K. Merton was a graduate student at Harvard writing his dissertation on science and Puritanism, which was completed in December 1935. Early in this project I speculated that Merton's Puritanism and science thesis may have been influenced by the attention given at Harvard during the 1930s to its own Puritan roots, although Merton made no references to it in his published reminiscences for the years 1933–1935. See Robert K. Merton, "Preface:1970," in *Science, Technology and Society in Seventeenth Century England* (New York: Howard Fertig, 1970), pp. vii–xxix; and *idem.,* "George Sarton: Episodic Recollections of an Unruly Apprentice," *Isis,* December 1985, *76:*470–86. In a telephone interview (8 January 1998) with Professor Merton, it became apparent that timing of his work's inception and the public planning of Harvard's celebration would have made such an influence unlikely (the first press release of the tercentenary schedule came on 14 December 1934). Also see the documentary evidence in *idem.,* "The Sorokin-Merton Correspondence on 'Puritanism, Pietism and Science,' 1933–34," *Science in Context,* 1989, *3,* n. 1:291–8 (reprint received from Professor Merton). The more subtle question remains, of course, regarding what influences might have come from the general reassessment of Puritanism in America in the 1930s. In this regard, *idem.,* "STS: Foreshadowings of an Evolving Research Program in the Sociology of Science," in *Puritanism and the Rise of Modern Science: The Merton Thesis,* ed. I. Bernard Cohen (New Brunswick and London: Rutgers University Press, 1990) is relevant. Drawing on an earlier paper by I. B. Cohen on the topic, Professor Merton there discusses and affirms the origins of the Puritanism and science thesis in the mid-1930s as an example of independent multiple discovery (pp. 352–6) (reference provided by Professor Merton). There is this evidence, therefore, that the reevaluation of Puritanism was an idea with some currency among those concerned with the history of science. The place of Harvard and its tercentenary celebration in this phenomenon is an intriguing one.

the long-term historical context within which Harvard's development fit. And the inherent conflict between academic specialization and generalization that the conference addressed had its counterpart in the more general conflicts between particularism and the values of universalism. Etienne Gilson, professor of philosophy at the Collège de France, participated in the symposium on "Independence, Convergence, and Borrowing" and took his task seriously, not only presenting a scholarly paper but asking how his topic related to current needs in a divided world. He spoke on "Medieval Universalism and Its Present Value," which related the tercentenary celebration not to three hundred years of history but to a millennium. Gilson's argument concerned "the deeply rooted medieval conviction that though the various expressions of truth unavoidably bear the mark of their local origins, truth itself, both in the speculative and in the practical order, is not true for a certain civilization, nor for a certain nation, but belongs to mankind as a whole." As part of his exposition, Gilson showed that teachers in Paris in the thirteenth century came from all parts of Europe, and he wanted that point to be clear to the nations of his own time.[73] His argument was ultimately both historical and philosophical.

An unnamed commentator in *The Nation* considered Gilson's paper, presented as an evening public lecture, to be "the keynote of the meetings." This "realization of the need for an intellectual unity as stable and as universal as that which Christendom enjoyed in the Middle Ages" was recognized as a focal and repeated point at the conference.[74] In a magazine article on the tercentenary, Alfred North Whitehead (who chaired the conference session at which Gilson presented his paper) used his philosophical background to argue for a greater union of thought and action in the modern university, pointing (in part) to the medieval institutions as a source and example. In closing, Whitehead stated, "To-day Harvard is the greatest of existing cultural institutions. . . . Will Harvard rise to its opportunity, and in the modern world repeat the brilliant leadership of medieval Paris?"[75] With the importance of medievalism at the tercentenary, history came into play not so much as a projection of Harvard onto the world, but out of overarching concern for an academic unity that was slipping into an abyss of international and intellectual chaos. None could deny that the Harvard anniversary gave that concern a universally visible platform.

## VI. CONCLUSION

The tercentenary operated on and was experienced at many levels. It must have permeated the Harvard community over a period of time and influenced the thinking or attitudes of its members in varying ways. Furthermore, nearly 67,000 individuals

[73] Etienne Gilson, "Medieval Universalism and Its Present Value," in *Independence, Convergence, and Borrowing in Institutions, Thought, and Art* (Cambridge, Mass.: Harvard University Press, 1937), pp. 194, 198–9.

[74] "'The Truth Shall Make You Free'," *The Nation,* 12 September 1936, *143:*214 (quote on "keynote"). *Tercentenary of Harvard College* (cit. n. 12), p. 462. Barnes, "Harvard's United Front" (cit. n. 60), p. 356 (quote on medieval unity). (It is possible that Barnes also wrote the first quote; both were published in *The Nation.*)

[75] Alfred North Whitehead, "Harvard: The Future," *The Atlantic Monthly,* September 1936, *158:*269, 270. Robert M. Hutchins, in *The Atlantic Monthly,* November 1936, *158:*582–8 countered Whitehead's plea for a greater practical educational component in the universities.

from all parts of the world registered as visitors to the university during the tercentenary period.[76] Each of the participants and observers carried away some impression, and in this sense the celebration projected into the future both an historical and a contemporary image—no matter how small, unfocused, or short-term it might have been in any given instance.

Some individuals did grasp the larger meaning at which Jerome Greene and the other organizers had aimed. Among the many commentaries was one in *The Christian Student* by William Pearson Tolley, who at the time was president of Allegheny College. "There is something seminal and symbolic about these meetings at Harvard," Tolley wrote. "They suggest that the world already physically one can be unified spiritually by these same forces of light. They suggest that the achievements in the field of physical and biological sciences can be duplicated in the field of social sciences; they call forth the hope that the next advancement of learning will see the fruits of philosophy as well as science, a concern for values as well as fact."[77] Tercentenary director Greene, a child of missionary parents, particularly liked Tolley's commentary because it captured the greater, spiritual significance of the event.[78] Greene revealed similar sentiments in a closing letter to Conant on 22 September 1936, in which he reflected, "In the welter of staff details, the preparation of programmes, tickets, seating arrangements, and hospitalities, it is not easy to discover anything very spiritual or idealistic; but the fact is that the execution of these mechanical details offers the indispensable framework for things that are important but which otherwise could not find expression. That is the only thing that makes ritual and etiquette important: they make it possible for the things that are worth while to have attention."[79]

Grateful though he undoubtedly was to Greene, the president must have, above all, felt relief with the completion of the tercentenary. In the immediate aftermath of the celebration, Conant expressed apprehension as well about its legacy, indicating to one correspondent that the less said about the tercentenary, and especially within the university, the better.[80] Five days after the ceremonies concluded, Conant left for a vacation trip to England.[81]

When he wrote his annual report for the year, however, Conant considered again how history could be put to work to address some of the problems of the academic community. He thought that the public's, as well as the scholar's, understanding of universities would be promoted if they knew more of academic history. He had more in mind than the history of particular institutions. Conant mentioned Harvard's implementation that year of a doctoral program and an undergraduate concentration in the history of science and learning. In the context of his argument, study of the

[76] [David McCord], *Notes on the Harvard Tercentenary* (Cambridge, Mass.: Harvard University Press, 1936), p. 7.

[77] William Pearson Tolley, "The Tercentenary of the Mother of American Colleges," *The Christian Student,* May 1937, *38,* n. 2:4.

[78] Greene to William Pearson Tolley, 29 March 1937, in response to manuscript sent in advance of publication, Tercentenary Records, Greene's Files, folder: Archives (Supplementary), UAV 827.5, box 3.

[79] Greene to Conant, 22 September 1936, Conant Records 1935–1936, folder: Tercentenary: Director J. D. Greene. Quoted by permission of the Harvard University Archives.

[80] Conant to Professor E. J. Cohn, 21 September 1936, Conant Records 1935–1936, folder: Tercentenary Publicity Committee.

[81] *Science,* n. s., 2 October 1936, *84:*308.

history of science and scholarship was a way to regain the much-desired unity of academia that had slipped away during the course of development of the modern university.[82] It was a way to face the future through knowledge of the past. This creative tension is one that many must have contemplated, if only momentarily, during the international academic event that had just concluded in Cambridge.[83]

[82] "President's Report 1935–36" (cit. n. 39), pp. 7–8. Although study of the history of science had begun at the university, initial development was slow and not vigorously supported. George Sarton's courses actually declined in enrollment in the late 1930s before undergoing substantial growth in the 1940s. In spite of his larger vision for the history of science and learning, Conant was reluctant to prepare young students for a field that did not offer hopes of employment. See Arnold Thackray and Robert K. Merton, "On Discipline Building: The Paradoxes of George Sarton," *Isis,* December 1972, *63:*489–90; and I. Bernard Cohen, "A Harvard Education," *Isis,* March 1984, *75:*16–17.

[83] In his memoirs, Conant confessed to an ultimate disillusionment with the values of history as a basis for cultural unity. See Conant, *My Several Lives* (cit. n. 16), p. 366.

# The *Enola Gay* Affair: What Evidence Counts When We Commemorate Historical Events?

***By Stanley Goldberg****

AS ONE WHO SERVED ON THE ADVISORY BOARD for the Smithsonian's upcoming exhibit commemorating the fiftieth anniversary of the atomic bombings of Hiroshima and Nagasaki, I was astounded and flabbergasted at the public uproar engendered by the proposed script of the exhibit. Much of it appeared in the pages of the *Washington Post.* There, the Air Force Association, columnists Charles Krauthammer and Chalmer Roberts, the *Post's* editorial board, and others lambasted National Air and Space Museum (NASM) director Martin Harwit and the exhibit curators for being, at worst, silly and ignorant and, at best, biased and without objectivity. Twenty-four congressmen signed a letter to a sympathetic and lame-duck Smithsonian Secretary Robert McCormick Adams complaining that the exhibit lacked balance and objectivity. According to one report, the congressmen charged curators with "anti-American" bias. Judging by the *Post's* letters to the editor, the newspaper must have heard from hundreds of World War II veterans, most of whom had been prepared for an invasion of the Japanese mainland at the end of the war. Almost all of these veterans find it unthinkable that, in the absence of the atomic bomb, the Japanese government might have surrendered without an invasion.

Save for Harwit's defense, if defense it can be called (in the *Washington Post,* 7 August 1994, Harwit seemed to suggest that the dispute was a matter of differing perspectives in evaluating the same evidence), much of the dispute progressed without one word from those of us who have spent years studying the military, political, and social factors at work at the end of World War II. Apparently, most of my colleagues have not only ducked into their foxholes, they have pulled the lids shut. That is too bad, because it is obvious to me that an enormous gulf has developed between public perceptions of the end of World War II and the histories of the period that have been produced over the last twenty years—histories that are enriched by the

*Editors' note: Stanley Goldberg died before he could revise for publication the paper that he presented at the April 1995 Boston University symposium on "The Construction of Scientific Memory." The editors have chosen to undertake only a minimal revision of that paper, making chiefly stylistic changes, so that it reflects both Stanley's perspective and the moment of its composition. The factual accuracy of the paper, insofar as possible, has been checked and confirmed. It did not include footnote references, but a bibliographic note has been appended by the editors. The opinions expressed, given as Stanley wrote them, are based in part on his personal experience. That included his research for many years on a biography of General Leslie R. Groves (military officer in charge of the Manhattan Project that developed the atomic bomb) and his service on the advisory board for the National Air and Space Museum's *Enola Gay* exhibit.

 0369-7827/99/1401-0010$02.00

great quantity of hitherto classified material that has become available during the last two decades. Yet in spite of this wealth of evidence, Martin Harwit and the exhibit curators capitulated to many of the demands of the Air Force Association and the American Legion. Perhaps the greatest irony of all was the fact that the exhibit was finally cancelled as a result of a dispute between Harwit and the representatives of the American Legion concerning how many casualties there would have been had there been an invasion of Japan—that is, a dispute about events that never happened!

Many of the critics of the original *Enola Gay* exhibit script snort with disdain at what they term "revisionist" history, by which they mean, apparently, accounts written by those who come along after the fact that draw different conclusions than those ascribed to the events at the time they were newsworthy. These critics deem such histories to be exercises in abstraction and this, apparently, makes them suspect.

All history is by its very nature abstract, and this includes eyewitness chronicles. And all history is revisionist history. The new pictures that we historians have been drawing of certain specific World War II milestones are not the *result* of "scholarly abstraction" (*Washington Post,* 26 September 1994) or "intellectual sophistication" (*Post* editorial, 14 August 1994). They arise largely because hitherto unavailable *evidence* contained in recently declassified documents has forced us to reconsider earlier histories, many of them so-called official histories. (Unfortunately, the National Archives is required by law to withhold millions of pages of still-classified material from World War II.)

Anyone writing a serious history of the circumstances surrounding the surrender of Japan has to grapple with the following information:

- Because the United States had long since broken the Japanese diplomatic code, it was known, as early as May 1945, at the highest levels of the American government, that a significant faction of the Japanese leadership was in favor of surrender. It was also known that the key concession that might have given this faction the upper hand was that the Japanese be allowed to retain the emperor. We insisted on unconditional surrender. Furthermore, at the time the atomic bomb was dropped on Hiroshima, negotiations with the Japanese were being undertaken through diplomatic channels in Switzerland. It is of more than passing interest that after surrender, we did allow the Japanese to retain the position of emperor.

- In the spring of 1945, there was intense interservice rivalry as to how the war could be won. By 1 June, the United States Army air forces had fire-bombed over thirty Japanese cities. General Curtis LeMay, commander of the 20th Bomber Command, reported that he expected to run out of targets by September and he did not "see very much war after that." The United States Navy, on the other hand, was convinced that its blockade of the Japanese islands would, by itself, bring the Japanese to their knees. Meanwhile, the army was planning an invasion of the islands.

- In planning the invasion, the army's Operations Division of the General Staff (OPD) estimated between thirty thousand and fifty thousand casualties (i.e., about ten thousand deaths).[1] Contrary to what the Air Force Association has recently suggested, OPD was

[1] Editors' note: The issue of expected invasion casualties is complex and controversial. It should be noted, however, that the figures given here are in line with those projected in mid-June 1945 for

well aware of Japanese fanaticism. The United States had had ample experience with it at such places as Guadalcanal, Saipan, Tarawa, Luzon, Iwo Jima, and Okinawa, to name but a few. Forty thousand casualties are not to be sneezed at, but, as historian Barton Bernstein and others have shown, higher estimates of several hundred thousand to a million were unstudied, self-serving inventions, most of them created by Secretary of War Henry Stimson and his staff after the war as part of a campaign to justify the use of the atomic bomb.

• As the war progressed, there was a great deal of unease about and displeasure with the Manhattan Project. The powerful Senate Special Committee investigating the National Defense Program, headed by Harry Truman, was repeatedly frustrated when it tried to get any information on why $750 million was being spent on the Hanford reservation in the desert of south-central Washington or why a billion dollars (to translate these expenditures into 1994 dollars, multiply by 10) was being invested in building and operating the facilities at Oak Ridge, Tennessee. In fact, the committee did not discover that the two installations were part of the same project until after 6 August 1945. All during the war, manufacturers whose products were essential to the effort were astonished to be told that their priority claims to vital raw materials would have to take a back seat to the Manhattan Project, a project which seemed to require a great deal but produced nothing. Workers at Oak Ridge, not allowed to know the purpose of their work, described the facility as a place where they "built the front end of horses which were then shipped to Washington for final assembly.") Yet the Manhattan Project's priority status put it above bullets and other munitions, above the landing craft program, above the rubber program, above high-octane gasoline, above everything. In the fall of 1944, the manager of a refinery was told that he would have to suspend the manufacture of high octane for military aircraft for two weeks in order to fill an order for the Manhattan Project. When all his efforts to find out anything about the Manhattan Project were blunted, he made one last inquiry, in disgust. "Just tell me," he said, "is it part of the war?" Time and again, Secretary of War Henry Stimson and Assistant Secretary of War Robert Patterson had to remind the backsliding War Manpower Commission and skeptical War Production Board that the Manhattan Project was "the most important project of the war." The situation became so fractious that on 3 March 1945, James Byrnes, the retiring head of the Office of War Mobilization, wrote a troubled memo to President Franklin D. Roosevelt. Byrnes told Roosevelt that he knew very little about the nature of the Manhattan Project but that it had already spent about $2 billion without issuing a product. It was probably true, Byrnes went on, that they could continue to hold Congress at bay for the rest of the war; "However, if the project proves to be a failure, it will then be the subject of relentless investigation and criticism." Byrnes begged Roosevelt to protect himself and the Democratic Party by getting the opinion of a totally independent commission of scientists—a proposal that was instantly rejected by Secretary of War Stimson.

• General Leslie R. Groves, the military head of the Manhattan Project, had a reputation for getting things done. The civilian and military leadership to whom he answered had given him *carte blanche*. He was given whatever priority he demanded, whatever funds

---

the first thirty days of the Kyushu invasion. Consideration of subsequent military operations and other factors led to estimates that were higher than those given here.

he required. Questions were never asked; there was no upper limit. Under those circumstances he could and he did apply intense pressure to project scientists and civilian contractors: first, to build the required enormous installations, and then, beginning in January 1945, to manufacture nuclear fuel at rates which at first seemed impossible to those doing the work. He made sure that nothing distracted the designers of the first bombs from getting a *workable* (as opposed to the most elegant) design in the shortest time possible. And once the bombs were delivered to the island of Tinian, from where they were to be dispatched to Japan, he relentlessly drove scientists and technicians who were responsible for assembling and loading the bombs to the point of speeding up the second atomic attack (on Nagasaki) by two days. After the war, in response to the congratulations of a close friend of his, Groves wrote, "I had to do some good hard talking at times. One thing is certain—we will never have the greatest congressional investigation of all time."

• When Secretary of War Stimson formed the Interim Committee which, among other things, was to recommend to President Harry Truman how the atomic bomb should be used against Japan, he asked Truman to name a personal representative to the committee. Truman named James Byrnes. Byrnes was already on record as being concerned about the political consequences of a "failed" Manhattan Project. Interim Committee meeting minutes show that Byrnes was also intent on using the bomb to limit Russian involvement in the Pacific war and as a way of warning the Russians about postwar adventurism. There can be little question that Byrnes, who, it was known, soon would become Truman's secretary of state, had the president's ear more than anyone else in the early days of the Truman administration.

These are not dry abstractions. The only thing musty about this information is the paper on which it is documented. Any account of the conditions surrounding the use of the atomic bomb and the Japanese decision to surrender has to deal with these facts. The Air Force Association's simplistic account, in which dropping the bomb was motivated by nothing more than a desire to bring a quick end to the war so as to result in a net savings of lives, can't begin to cope with the complexity of the evidence.

Later histories of a set of events often differ from immediate accounts because the emotional surge surrounding the events dissipates over time. As that process goes on, some of those who lived through events are unwilling and often unable to let go of the strong emotions of an earlier time. For example, some Civil War veterans to their death damned "the Yankees" or despised "the Rebs"; some veterans of World War II refuse, to this day, to purchase cars designed or manufactured in Japan; some consumers recalling Cesar Chavez's crusade for migrant farm workers still refuse to purchase grapes. Such strong emotional responses are understandable, but they are almost always unreliable guideposts for writing history. The very fact that one is a participant means that, at best, he or she knows only part of the story. It is often the case that assumptions about the whole made by participants on the basis of limited experience and anecdotal information are far off the mark. This does not invalidate an individual's experiences; it does not devalue a person's emotional reactions to the unfolding juggernaut of a set of events. Those events, however, are but part of one scene in a multi-act drama in which many of the scenes simultaneously play on a stage that can be as large as the earth. Comparing a history of a global war based on an individual's personal response to local conditions with a thoroughly researched history based on both documentary and wide-ranging oral evidence is to compare apples and oranges.

This chasm is nowhere better illustrated than in *Washington Post* reporter Ken Ringle's analysis of the *Enola Gay* dispute (26 September 1994, p. 1). Ringle characterizes it as a collision over two views of history. One of these Ringle describes as "old history, a scholarly abstraction composed of archival records, argumentative books, and . . . fading images on black and white film." The other is, apparently, the history as perceived by those whose lives were directly touched by the war—especially those who had fought the Japanese (some of whom became prisoners of war), as well as those who were told that they were being prepared for the invasion of Japan. Ringle's interview subject Grayford Payne, interned in a Japanese prisoner-of-war camp for over three years, understandably associates the end of the war and his freedom with the dropping of the atomic bomb. He recalls that the commander of the prison had posted an order from Premier Hideki Tojo that the moment an American soldier set foot on the Japanese mainland, all prisoners of war were to be shot. It is fixed in his mind that the atomic bomb saved him from that fate. But the fact of the matter is that (fortunately) when American soldiers did set foot on Japanese soil (a radiation survey team left San Francisco for Japan on 14 August), Payne was not executed. His reprieve had less to do with the atomic bomb than with the fact that Tojo was forced to resign on 18 July, three weeks before the first atomic bomb was dropped.[2] Whether or not an American prisoner was executed at the end of the war depended on the mind-set of the prison's local commander. At the time, Payne could not have known that. It is understandable that he would not have bothered to check it out later. But it is unforgivable that the *Post* reporter and those who supervise his stories did not verify their facts. In the interests of making the Smithsonian historians appear totally out of touch with reality, the *Post* staff was willing to equate idiosyncratic impressions of an event with carefully researched histories of the same events. Paradoxically, it is the impressionistic memories of individuals that are akin to "fading images on black and white film."

Critics of the exhibit script charge that its creators lacked objectivity. The charge is unwarranted. There is a great deal of confusion in our culture about objectivity. Many people seem to believe that attaining objectivity is a matter of individual will and personal psychology. This fable has been reinforced by Hollywood portrayals of great scientists such as Louis Pasteur, Marie Curie, and George Washington Carver. As the great scientist puts on his or her lab coat (it is almost always white), the mantle of objectivity descends and all preconceptions, all biases, all self-interest are wiped from the mind, like chalk from a blackboard. The noble investigator is now prepared to settle for nothing but the naked truth in his or her quest for the secrets of nature. Such a picture, while popular, is a fairy tale. So, too, is the myth promulgated by the press that serious, professional newspeople are by habit and training able to clear their minds of all prejudices and prejudgments, thereby allowing them to report only the "facts." All of us—reporters and historians,

[2] Editors' note: This is an error. In fact, Tojo resigned as prime minister in July 1944 (the year before these events). The confusion began with the *Post* story of 26 September 1994, in which Payne was quoted as stating that Tojo's order was posted in June 1945. The reporter, Ringle, accepted the statement, merely adding an identifying "[Japanese premier]" before Payne's reference to Hideki Tojo.

scientists and artists, policemen and engineers, laborers and congressmen, engineers and poets, men and women, white and black, yellow and brown, old and young—carry around considerable psychological baggage (much of it unconscious and deeply imbedded in our psyches) that colors everything we see, everything we believe, and everything we do. Any scientist worth his or her salt has clear expectations of how the experiment will come out. Any good historian starting a fresh project does so with a clear set of expectations. Any alert policeman on a crime scene has hunches. This does not mean that we are not often surprised, but it does mean that we are all predisposed to certain kinds of outcomes and disinclined to some kinds of conclusions. Living a life is never a matter of logic.

This is not to say that we don't do the best we can to strike a balanced approach. No matter how hard we try to get out from under, however, our outlook on the world is always colored by an overlay of shrouded preconceptions—the result of idiosyncratic, familial, regional, and national influences. Even so, objectivity about the past, about the nature of the physical world, about day-to-day events is achievable. While objectivity cannot be a matter of personal will and discipline, it can be attained through a social process—one which requires forums for the free flow and exchange of ideas for its successful implementation. For example, the heart of the Scientific Revolution of the seventeenth century was not any key discovery or any one set of experiments. It was the decision on the part of those participating to insist on public disclosure, not just of results, but of techniques and, if demanded, of the raw data on which conclusions were based. That which had hitherto been secret and private, as for example in the study of alchemy, became open and public. In this way, others of different persuasions can independently check all aspects of the work in question. It is out of the ensuing deliberations that consensus emerges—not so much a consensus of interpretation, but of the evidence on which interpretation is to be based. Similar processes were adopted by the medical and biological sciences and by the social sciences, including history. Similarly, the press not only relies on an open forum in which newspapers and broadcast stations compete, it also has internal processes by which a reporter is often required by editors to produce one or more independent corroborations that an event occurred or a public official made a particular statement.

In this case, however, who is to corroborate the *real* reason the Japanese decided to surrender? Who is to confirm the *real* reason the atomic bomb was dropped? The Air Force Association and the American Legion seem sure that they know, and the *Post's* editorial writers and some of its reporters appear to endorse that position: that the bomb was dropped to bring the war to a quick end and thereby effect a net saving of lives.

Among historians there is no such agreement. On the basis of documentary and oral evidence, several reasons are being proposed:

1. **Momentum.** A $2 billion, three-year head of steam had been built up and no one said, "No, it is not needed."

2. **Protection of personal reputations.** Should the bomb not be used, the political, scientific, and military leadership, which all during the war had pushed the project to the top of military priorities, would have had to face an outraged Congress and an irate public.

3. **Personal ambition.** Individual scientific leaders and military commanders saw the project as a route for their own personal advancement.

4. **International diplomacy.** The atomic bomb would shorten the war and thereby minimize Soviet involvement in the Pacific theater. At the same time, it would serve notice to the Soviets with regard to Soviet designs on Western Europe.

5. **Humaneness.** Use of the bomb would shorten the war and effect a net saving of lives.

Most of us agree on the evidence; we disagree and debate each other on how that evidence should be interpreted. My own view of the situation is that one can find individuals who participated in the decision who evince at least one or more of these motivations. As a colleague recently remarked, "The decision to use the bomb was over-determined." In other words, there was no one single reason.

The NASM advisory committee, which met in February 1994 to consider the first draft of the *Enola Gay* exhibit script, was unanimously supportive of the effort. We all knew that some labels would require rewriting. Each of us provided a unique perspective—things to avoid, things to reinforce—but we all (and this included United States Air Force Historian Richard Hallion) voiced approval of the general tone and plan of the exhibit. The stories on the exhibit that subsequently appeared in the press illustrate how easy it is to give a misleading picture of a piece of writing by highlighting a few snippets out of context.

The original script for the *Enola Gay* exhibit made an honest attempt to reflect the complexity of the decision to drop the bomb without overwhelming the visitor. It is not an easy job. Unfortunately, because the museum's administration allowed the Air Force Association and the American Legion to eviscerate the exhibit, much of the complexity of the real-life drama in which the atomic bomb was dropped on two Japanese cities will be kept from museum visitors.

The first draft of the exhibit script, which I received in January 1994, was quite a tome—upward of 750 pages of text and pictures of artifacts, of sites and action, and of original documents. Rather than retelling the whole story of the Pacific war, which would have been impossible in the space allotted, the curators had chosen to link the atomic bombing of Hiroshima and Nagasaki to the evolution of the strategic bombing campaign of the American 20th Bomber Command which, beginning in March 1945, had been systematically incinerating one Japanese city after another with carefully designed fire-bomb raids. Rather than having to use the hundreds of planes and thousands of tons of high explosives that these attacks required, the air force could now achieve the same effect with one plane and one atomic bomb. This theme was not original with the exhibit curators; it has been well developed in the literature of World War II, for example, in Michael S. Sherry's *The Rise of American Air Power: The Creation of Armageddon* (New Haven, Conn.: Yale University Press, 1987).

The flow of the exhibit was quite logical. After some references to the start of the war and Pearl Harbor, the exhibit began with V-E Day (8 May 1945, marking victory for the Allies in Europe) and concentrated on the battles for Iwo Jima and Okinawa, which together were almost more costly in American casualties than were the first three years of the Pacific war. It then moved to a discussion of the decision to drop the bomb. I was quite taken aback by the sophistication of the script. As is the case with most decisions made from within complex organizations, the decision to use the bomb had come about as a result of a very complicated set of circumstances and relationships. The curators, I thought, had handled that part of the script quite well and had done a fine job of making the subject accessible to a general audience—no easy task. The story then moved to an exploration of the "miracle" of the design and production of the B-29. This was followed by a section on the training of the crews chosen to deliver the atomic bomb—the 509th Composite Group—and the construction of the 509th's facilities on the island of Tinian in the Marianas. Tinian was home to a large segment of the 20th Air Force, which had been raiding Japan daily for almost a year. After describing in words, photographs, and artifacts the scene at and near ground zero in Hiroshima and Nagasaki, the exhibit ended with a section on the legacy of the use of the atomic bomb: an abrupt halt to the war and, as predicted by some of the scientists and administrators who had seen the project to completion, a fierce, competitive nuclear arms race between the United States and the Soviet Union that would dominate international politics for the next four decades. Even now the world is threatened by the existence of weapons capable of unimaginable destruction.

I was astonished when the NASM administration conceded to the American Legion and the Air Force Association and forced the exhibit's curators to negotiate its content and the wording of individual labels. I was thunderstruck when I learned that twenty-eight members of Congress had signed a letter to the secretary of the Smithsonian denouncing the exhibit and urging him to intercede; and that the museum director had been visited in his office by twenty-one lawmakers, some of whom wanted to know why the curators were so un-American. At this point (7 September 1994), I informed NASM's director that I was resigning from the advisory board. What upset me most was that the museum administration had exposed the curators to the *direct* pressure of organizations such as the Air Force Association and Congress. The fact that a significant portion of the funds for NASM comes from public sources no more entitles Congress or anyone else to dictate substantive conclusions to curators than does the fact that public monies support other kinds of research and writing projects. That kind of thought control should have no place in a government committed to democratic ideas. It suggests reexamining the qualifications of the twenty-one members of Congress who stormed the National Air and Space Museum and thought it appropriate to micromanage the content of the exhibit.

To add insult to injury, the information the congressmen, the *Washington Post,* and the general public had received from a campaign led by the Air Force Association and the American Legion was largely based on taking exhibit labels and portions of labels out of context. This had resulted, intentionally or not, in serious distortions. Take, for example, the following passage, which was often quoted by the American Legion and the Air Force Association and which the *Post* used as evidence of the lack of balance in the exhibit:

> For most Americans, this war was fundamentally different than the one waged against Germany and Italy—it was a war of vengeance. For most Japanese, it was a war to defend their unique culture against Western imperialism.

The full label, a main unit label near the beginning of the exhibit, was entitled "A Fight to the Finish." Here is the full text of that label:

> In 1931 the Japanese Army occupied Manchuria; six years later it invaded the rest of China. From 1937 to 1945, the Japanese Empire would be constantly at war.
>
> Japanese expansionism was marked by naked aggression and extreme brutality. The slaughter of tens of thousands of Chinese in Nanking in 1937 shocked the world. Atroci-

> ties by Japanese troops included brutal mistreatment of civilians, forced laborers and prisoners of war, and biological experiments on human victims.
>
> In December 1941, Japan attacked U.S. bases at Pearl Harbor, Hawaii, and launched other surprise assaults against Allied territories in the Pacific. Thus began a wider conflict marked by extreme bitterness. For most Americans, this war was fundamentally different than one waged against Germany and Italy—it was a war of vengeance. For most Japanese, it was a war to defend their unique culture against Western imperialism. As the war approached its end in 1945, it appeared to both sides that it was a fight to the finish.

Rather than being absurd, the questioned sentences required some rewriting. The substance contained in the label is all well documented and gives the lie to those who accused the curators of distorting history or pandering to the Japanese.

Many of the critics of the *Enola Gay* exhibit believe that the contents of the labels in the exhibit should be limited to artifact identification. For example, Paul Tibbets, the man who piloted the *Enola Gay* on its run to Hiroshima, believes that history would be best served by such an approach. A *Post* editorial writer, Amy E. Schwartz, agrees. In Schwartz's view, when there is no historical consensus

> you're left by default with the same old authority that seeps from the presence of the real object. Keeping silent in its presence may paradoxically turn out to be the best way of saying the unsayable.

There is a longstanding debate among museum curators over this issue. On one side, curators who believe that artifacts speak for themselves cite the traditional role of labels in art museums. Usually, the only information in such labels is the name of the artist, the year the work was completed, and, sometimes, the work's title.

Those of us on the other side acknowledge that labels on works of art usually contain a minimum of information. When an individual contemplates an *objet d'art,* however, he or she brings personal experience into play and the range of allowable interpretations is limitless, regardless of how realistic the portrayal. It is hard to imagine what "seeps" from the *Enola Gay* that, together with the experiences of the average viewer, would provide insight into the meaning of the *Enola Gay* as an *objet d'art.*

The truth of the matter is that when art museums mount historical exhibits, great varieties of artifacts are introduced in support of the thesis and the labeling becomes quite expansive. My favorite example is the exhibit *Raphael in America* at Washington, D.C.'s National Gallery of Art in 1982. Even though there are only three original Raphaels in this country, the exhibit required over five thousand square feet of exhibit space and included movies, an exegesis on why Raphael is such an important figure in the history of art, another on why there was informal competition among American arts patrons to be the first to import a Raphael, and notes on how one can distinguish the genuine item from a counterfeit. In a history exhibit, the artifacts serve the role of providing a context for the labels, not vice versa. A history exhibit is the spatial and visual equivalent of a monograph. And there is no *a priori* reason why it cannot be both commemorative and analytical. In fact, I believe the *Enola Gay* exhibit as proposed admirably served both functions.

On 30 January 1995, newly appointed Smithsonian Secretary I. Michael Heyman announced the cancellation of the original *Enola Gay* exhibit. In its place, he said, would be a very simple display of the *Enola Gay* fuselage and perhaps some videotaped interviews with surviving crew members. Many of my colleagues, whom I had joined in protesting the initial censoring of the exhibit, expressed relief that a show so butchered had been withdrawn. While I understand how they must feel, I do not share in their relief. Even in its damaged state, the exhibit would have brought a challenging and heuristic portrayal of the destruction of Hiroshima and Nagasaki, the end of the era that had preceded it, and the portents of the era that ensued.

Since the closing of the original exhibit was announced, I have had several opportunities to discuss the issue privately with members of the American Legion team. It is now clear to me that the motives and concerns of the American Legion representatives and those of the Air Force Association need to be distinguished.

Air Force Association press releases, as well as the remarks of individual members, in calling for a balanced exhibit also express preference for labels without historical context. This harkens back to the days when the Air and Space Museum was directed by an air force veteran. At that time, virtually no context was provided for most of the artifacts on display. For example, rather than the extensive labeling now on the V2 rocket explaining when and where it was built and its purpose, the label used to say something like "V2—first rocket into space." I believe that behind the Air Force Association's complaints about historical labels they saw, in the controversy over the *Enola Gay* exhibit, a wedge that might be exploited to embarrass the Smithsonian Institution into replacing the current NASM director with someone whose ideas of a good museum are more compatible with those of the association. They might have even seen this as an opportunity for recapturing "their" museum. For the Air Force Association, the issue was power.

For the American Legion, the issue was much different. For them, the exhibit contradicted their personal memories and appeared to slight their contributions to the victory over Japan in World War II. For the American Legion, the issue was sentiment.

It is not the Smithsonian that is the greatest loser in this tug-of-war. The biggest loser, unfortunately, is the public. It has been denied the opportunity to assess for itself some of the perspectives on the end of World War II that are now emerging from intense study of the documented views and actions of those who shaped the course of events in the Pacific during July and August, 1945. It is true that the fiftieth anniversaries of the atomic bombing of Hiroshima and Nagasaki witnessed an explosion of articles, books, and television specials about the first and only uses of atomic bombs in warfare, each presenting a unique interpretation of the evidence concerning the motives behind the decisions of our leaders and the actions those decisions precipitated. Unfortunately, if you want to see an interpretation accompanied by actual artifacts that bear on the story, other than the *Enola Gay* itself, you will have to make the trek to Nagasaki and Hiroshima.

## BIBLIOGRAPHIC NOTE

A number of publications have appeared in the aftermath of the *Enola Gay* controversy and the pared-down exhibit that opened in the summer of 1995. Readers who wish to pursue this topic further and to compare Dr. Goldberg's views with those of other scholars and commentators may find the following of particular interest: *Judgement at the Smithsonian: The Bombing of Hiroshima and Nagasaki,* ed. Philip

Nobile (New York: Marlowe & Company, [1995]), which includes the entire text of the original exhibit script; "History and the Public: What Can We Handle?: A Round Table about History after the *Enola Gay* Controversy," *Journal of American History, December* 1995, *82,* n.3; Martin Harwit, *An Exhibit Denied: Lobbying the History of the Enola Gay* (New York: Springer-Verlag, 1996); *History War: The Enola Gay and Other Battles for the American Past,* eds. Edward T. Linenthal and Tom Engelhardt (New York: Metropolitan Books, 1996); *Hiroshima's Shadow: Writings on the Denial of History,* eds. Kai Bird and Lawrence Lifschultz (Stony Creek, Conn.: The Pamphleteers' Press, 1998); "Special Section: The Last Act [commentaries and reviews relating to the *Enola Gay* controversy]," *Technology and Culture,* July 1998, *39,* n.3.

Stanley Goldberg contributed two pieces to Bird and Lifschultz, *Hiroshima's Shadow:* "Racing to the Finish," pp. 119–29, on Leslie Groves and the building of the bomb; and "Smithsonian Suffers Legionnaires' Disease," pp. 355–63, written in the vein of the paper published here.

# The Golden Jubilees of Lawrence Berkeley and Los Alamos National Laboratories

*By Robert W. Seidel**

> " 'It's a poor sort of memory that only works backwards,' the Queen remarked."
>
> —Lewis Carroll, 1832–1898

J. Robert Oppenheimer, in his *Reflections on the Resonances of Physics History,* expressed the hope that studies of the history of science might bring some coherence to the general intellectual and cultural life of our time. "We are so engulfed by the changes, the massiveness, the ferocity, the brashness, the virtuosity, the confusion of the current scene in physics, that we do not understand it very well, and it may not be possible for us to understand it," he wrote.[1] Applying the "wisdom and insight of history," Oppenheimer had long hoped, could elucidate how science had changed the conditions of man's life, even though he recognized the very great difficulty of doing so without error.[2]

The fiftieth anniversaries of the Lawrence Berkeley Laboratory (LBL) and Los Alamos National Laboratory (LANL) featured a variety of reflections on their pasts: public ceremonies, formal and informal histories, secret celebrations, banquets, and exhibitions. Comparing the labs' collective memories with the research of historians reveals not only the distinct perspectives of the labs' observers, but also the different lenses through which they viewed the past.

Commemorations are "solemnifications of the memories of events" and secular rites, with a number of similarities to their religious and sacred counterparts.[3] They are used to suggest permanence and legitimacy, to mark changes in the life of an institution, to declare the value of the past for the future, to explain that value in terms of internal efficacy, and to influence the world.[4] Anthropologists have paid a great deal of attention to such rites. One saw them as "a by rote performance of what

*Charles Babbage Institute, Center for the History of Computing, 117 Pleasant Street, University of Minnesota, Minneapolis, MN 55455.

[1] J. Robert Oppenheimer, *Reflections on the Resonances of Physics History* (New York: American Institute of Physics, 1962), pp. 5, 7.

[2] J. Robert Oppenheimer, *Science and the Common Understanding* (New York: Simon & Schuster, 1952), p. 3.

[3] The definition of "commemoration" is from *The Compact Edition of the Oxford English Dictionary,* 2 vols. (Oxford: Oxford University Press, 1971), vol. I, p. 672.

[4] Sally F. Moore and Barbara G. Myerhoff, *Secular Ritual* (Amsterdam: Van Gorcum, 1977), pp. 3–15.

once had meaning—expressions of genuine, sincere emotions. . . . No more. In fact, some authors in the public sector have come to feel that administration is itself becoming a rite or ceremony."[5]

Historians, on the other hand, seek the coherence of which Oppenheimer spoke by placing past events in larger contexts and seeking deeper meanings than those that may be of interest to commemoration participants. Unlike them, they cannot discard inconvenient facts in their attempts to interpret the past, and yet they must select among those facts in order to make their reconstruction intelligible. As a professional historian, I was in a unique position to observe the interactions between history and memory, formal and public history, and men of knowledge and men of power that were entailed in planning and executing the public rituals to commemorate the laboratories' fiftieth anniversaries.[6] My reflections on them have some of the quality of the participants' view, which is predicated on action, as well as of the historian's, which is predicated on understanding.[7] In the case of LBL's celebration, I was drafted; in the case of LANL's, I volunteered.[8] In the following pages, I contrast the participants' and the historians' views, and their purposes, as expressed in these ceremonies.

## I. WAR AND REMEMBRANCE

LBL played a central role in World War II. (At the time, it was called the University of California Radiation Laboratory.) Its scientists discovered plutonium (Pu) and perfected the Calutron electromagnetic separation technique to isolate uranium-235 (U235), nuclear explosives used in the atomic bombs dropped on Hiroshima and Nagasaki. Los Alamos National Laboratory (then called Project Y), created during the war, designed those weapons and was popularly recognized as the home of the atomic bomb. Both laboratories grew rapidly during the war and engaged in research, development, testing, evaluation, and operation of their products, which revolutionized the nature of warfare.

[5] Anthony G. White, *Administrative Rituals, Rites, and Ceremonies: A Selected Bibliography,* in *Vance Bibliographies: Public Administrative Series,* # P-2767 (Monticello, Ill.: Vance Bibliographies, 1989), p. 2.

[6] John Heilbron, Robert W. Seidel, and Bruce R. Wheaton, *Lawrence and his Laboratory* (Berkeley: Office for History of Science and Technology, 1981); also published in *LBL Newsmagazine,* Fall 1991, *6,* n. 3:1–106; and available on line at http://imglib.lbl.gov/LBNL_Res_Revs/RR_online/81F/81fcontents.html. A multi-volume history of LBL is in process, of which the first volume has appeared: John Heilbron and Robert W. Seidel, *Lawrence and His Laboratory,* vol. I of *A History of the Lawrence Berkeley Laboratory* (Berkeley: University of California Press, 1989). *Critical Assembly: A Technical History of Los Alamos during the Oppenheimer Years 1943–1945* (New York: Cambridge University Press, 1993) was written by Lillian Hoddeson, Paul W. Henriksen, Roger A. Meade, and Catherine Westfall with the contributions from Gordon Baym, Richard Hewlett, Alison Kerr, Robert Penneman, Leslie Redman, and Robert W. Seidel. In addition, I composed a popular history, *Los Alamos and the Making of the Atomic Bomb* (Los Alamos: Otowi Station, 1995) as a series of columns for the *LANL Newsbulletin.*

[7] Edward Hallett Carr, *What is History?* (New York: Alfred A. Knopf, 1964); R. G. Collingwood, *The Idea of History* (New York: Galaxy, 1956); Murray Murphey, *Philosophical Foundations of Historical Knowledge* (New York: State University of New York Press, 1994); Michael Mandelbaum, *The Problem of Historical Knowledge: An Answer to Relativism* (New York: Liverright, 1938; reprint, Harper Torchbooks, 1967).

[8] From 1980 to 1982, I was employed by the University of California's Office for History of Science and Technology, which operated under contract with the Department of Energy through the Lawrence Berkeley Laboratory to prepare its history, and from 1985 to 1994, I was employed by the Los Alamos National Laboratory as a museum administrator, project leader, and senior policy analyst.

The cold war saw the development of the hydrogen bomb and all of the other nuclear weapons in the stockpile at Los Alamos and at the Livermore branch of the Lawrence Berkeley Laboratory. As an historian, I saw political conflicts shaping the institutional fate of the laboratories, if not all of the science and technology they produced.[9] But participants in the fiftieth anniversary celebrations held another view. Scientists were much more aware of the two labs' achievements in science and technology, which had led to nine Nobel Prizes, the discovery of several new elements and many subatomic particles, the development of several types of particle accelerators, and many other advancements in physics, chemistry, biology, and medicine.[10] As most scientists do when playing at history, they saw a stream of accomplishments leading logically and inevitably to each other, and ignored the bureaucratic, political, and economic contexts, the tragedies and the ironies of their past. However, when they turned to the future, these contexts became all important.

The anniversaries occurred at a time when the laboratories' future was in doubt. By the time of its fiftieth in 1981, LBL's leadership in high-energy physics had been usurped by other laboratories. No clear mission had replaced founder Ernest Lawrence's vision. Twelve years later, when Los Alamos celebrated its golden anniversary, the end of the cold war had cooled the passions that drove the nuclear arms race, as well as the space race and the international competition in high-energy physics, which had served as a sort of *macht ersatz* as the *macht* of nuclear weapons was found to be unusable. Reduced budgets and increasing regulation curbed Los Alamos's growth after 1987. Federal reviews of the labs had failed to produce new missions to replace nuclear weapons R & D, control the erosion of their support, and make the best use of their scientific and technical capabilities.

This search for new bottles in which to pour old wine was one motive for historical repackaging. Capabilities, in order to be understood, must be explained, especially when the nature of their application is unclear outside the original realm in which they arose. Although research spin-offs were obvious to scientists, the policy makers who would shape the future of the laboratories had to be told about them.

At the same time, a growing number of historians had adopted an increasingly critical moral stance towards cold war science and technology, as well as towards the institutions that conducted it. For example, the fiftieth anniversary of the bombing of Hiroshima and Nagasaki occasioned a critical evaluation of the use of nuclear weapons in World War II that stirred the historical profession as few other topics have done. While the weapons labs of the Department of Energy (DOE) saw the fall of the Berlin Wall and the Soviet Union as their triumph, historians saw the laboratories as having perpetuated both World War II and the cold war.[11]

[9] John Heilbron, Robert W. Seidel, and Bruce R. Wheaton, "Science and Society, 1931–1981," *CERN Courier,* 1981, *221:*335–46; Robert W. Seidel, "Accelerating Science: The Postwar Transformation of the Lawrence Radiation Laboratory," *Historical Studies in the Physical Sciences,* 1983, *13:*375–400; *idem.,* "A Home for Big Science: The AEC and its Laboratory System," *Hist. Stud. Phys. Biol. Sci.,* 1986, *16:*135–75.

[10] Nobel Prizes were awarded to Ernest Lawrence in 1939 for the cyclotron; Edward M. McMillan and Glenn T. Seaborg in 1951 for new elements; Emilio Segrè and Owen Chamberlain in 1959 for the antiproton; Donald A. Glaser for the bubble chamber in 1960; Melvin Calvin for the chemistry of photosynthesis in 1961; Luis W. Alvarez for new particles in 1968; and Yuan T. Lee for dynamics of elementary chemical processes in 1986.

[11] Gar Alperowitz, *Atomic Diplomacy: Hiroshima and Potsdam: The Use of the Atomic Bomb and the American Confrontation with Soviet Power* (New York: Simon & Schuster, 1965; expanded and updated, New York: Penguin, 1985; rev. ed., New York: Alfred A. Knopf, 1995); *The Atomic Bomb:*

**II. THE FACE OF DOE LABS**

LBL, the oldest of the nine general-purpose labs now supported by the DOE, developed the cyclotron and other particle accelerators for research in nuclear physics, chemistry, biology, and medicine between its founding in 1931 and the onset of World War II.[12] After the wartime mobilization, the laboratory resumed accelerator development and nuclear science research. It built a synchrocyclotron, an electron synchrotron and the modern linear accelerator, and resumed the manufacture of pions, antiprotons, and many other subatomic particles. Although its leadership in high-energy physics lapsed in the 1960s, LBL successfully mated the Heavy-Ion Linear Accelerator (HILAC) and the Bevatron into the Bevalac and developed advanced particle detectors for the Stanford Linear Accelerator Center (SLAC) in the 1970s; in the 1980s, it helped design the Superconducting Super Collider.[13] Its multidisciplinary research embraced nuclear medicine, materials science, biodynamics, computer science, and many other fields.

The defense work done at LBL after World War II was sequestered primarily at its Livermore branch, which was formed by Lawrence, Edward Teller, and Luis Alvarez in response to the Soviet atomic bomb. Livermore rivaled Los Alamos in the design of nuclear weapons.[14] Like the picture of Dorian Gray, the Livermore branch revealed far more about the Faustian bargain between the Atomic Energy Commission (AEC) and the University of California than did the Berkeley branch, with its pacific, productive, and public expression of fundamental science. Livermore remained an arm of the Radiation Laboratory until 1971 (both were renamed in memory of Lawrence in 1959), relying on the Berkeley lab for manpower, administration, and direction.

Los Alamos is popularly and historically associated with the weapons that were dropped on Hiroshima and Nagasaki in World War II. The romantic aspects of the wartime effort, which involved eminent physicists such as Hans Bethe, Niels Bohr, Enrico Fermi, Otto Frisch, and Los Alamos's charismatic and tragic director J. Robert Oppenheimer, have been indelibly etched in the collective memory. The postwar accomplishments of the laboratory are much less well known, although the weapons it subsequently produced were far more powerful, sophisticated, and deadly. These

---

*The Critical Issues,* ed. Barton J. Bernstein (Boston: Little, Brown, 1976); Stuart W. Leslie, *The Cold War and American Science: The Military-Industrial-Academic Complex at MIT and Stanford* (New York: Columbia, 1993); Paul Forman, "Behind Quantum Electronics: National Security as Basis for Physical Research in the United States, 1940–1960," *Hist. Stud. Phys. Biol. Sci.,* 1987, *18:*149–229; *idem.,* "Inventing the Master in Postwar America," *Osiris,* 1992, *7:*105–34; *idem.,* "'Swords into Ploughshares': Breaking New Ground with Radar Hardware and Technique in Physical Research after World War II," *Reviews of Modern Physics,* 1995, *67:*397–455; *idem.,* "Independence, Not Transcendence, for the Historian of Science," *Isis,* 1991, *82:*71–86. For the *Enola Gay* exhibit, see Martin Harwit, *An Exhibit Denied: Lobbying the History of Enola Gay* (New York: Copernicus, 1996); Philip Nobile and Barton Bernstein, *Judgment at the Smithsonian* (New York: Marlowe, 1995). Other historians had thrown cold water on the spin-off theory, e.g., Mary Kaldor, *The Baroque Arsenal* (New York: Hill and Wang, 1981).

[12] LBL was known as the University of California Radiation Laboratory from 1931 to 1959, the Lawrence Radiation Laboratory from 1959 to 1971, and the Lawrence Berkeley Laboratory from 1971 to 1995, when it was renamed the Lawrence Berkeley National Laboratory.

[13] See Heilbron and Seidel, *Lawrence and his Laboratory* (cit. n. 6); Robert N. Kahn, "Update on the Superconducting Super Collider," *Physics Today,* 1988, *41:*S34–S35.

[14] See, *inter alia,* Herbert F. York, *The Advisors: Oppenheimer, Teller, and the Superbomb* (San Francisco: W. H. Freeman, 1976; reprint Stanford: Stanford University Press, 1989).

include the hydrogen bomb and most of the weapons in the cold war nuclear stockpile.

Los Alamos Scientific Laboratory did not disband at the end of the war, as Oppenheimer and others had planned. The Manhattan Engineer District (MED) underwrote fundamental research at Los Alamos until the AEC was created in 1947. In order to recruit and retain scientific manpower to replace the wartime staff, the MED and AEC authorized the construction of accelerators, experimental reactors, and a new laboratory complex that made Los Alamos the largest physics laboratory in the nation, if not the world.

In the 1950s Frederick Reines and Clyde Cowan detected the neutrino, for which, much later, Reines won the Nobel Prize, the first work in a weapons laboratory to be so recognized. The laboratory also investigated controlled thermonuclear reactions, nuclear-powered rocket engines, and advanced fission reactors; built and bought the highest-performance computers for both weapons design and scientific computing; and designed and built the world's largest proton linear accelerator in the 1960s. President Ronald Reagan's restocking of the nuclear arsenal and Strategic Defense Initiative (SDI) in the 1980s reinvigorated strategic weapons programs at the laboratory. As its fiftieth anniversary approached in 1993, however, the end of the cold war threatened its future.[15]

## III. GOLDEN ANNIVERSARIES

People bundle the past in packages called "anniversaries." A cult and an industry have grown up to celebrate them both in Europe and in the United States, although cultural analysis suggests anniversaries have different meanings in these regions.[16] Americans more often commemorate the anniversaries of events—the bicentennial of the American Revolution, the founding of organizations, the five hundredth anniversary of the discovery of the New World—than they do the births or deaths of individual luminaries.[17]

Anniversaries in physics also tend to celebrate events—e.g., the fiftieth anniversaries of the discoveries of the neutron, of beta decay, the meson, Bose-Einstein statistics, the invention of particle accelerators—although the Nobel Prize, a Swedish innovation, tends to celebrate individuals or small teams. The work of large laboratories in constructing massive tools like particle accelerators tends to be recognized in terms of significant discoveries made by these devices.[18]

The anniversary of a laboratory is a recognition of collective accomplishments.

[15] Scott McCartney, "Lab Partners: With Cold War Over, Los Alamos Seeks New Way of Doing Business," *Wall Street Journal,* 15 July 1993; sec. A, p. 1; William J. Broad, "Science Times: Former Foes Team Up," *New York Times,* 4 May 1993, sec. 6, p. 5; Jonathan Weber, "To Trade Wars from Star Wars?" *Los Angeles Times,* 19 July 1992, sec. A, p. 1.

[16] William M. Johnston, *Celebrations: The Cult of Anniversaries in Europe and the United States Today* (New Brunswick, N.J., and London: Transaction Publishers, 1991).

[17] *Ibid.* In physics, celebrations have included the centennial of Albert Einstein's birth in Belgium and that of Vladimir Zworykin and Abraham Ioffe in Russia. Americans tend to celebrate events in an individual's life like Henry Rowland's introduction of the concave diffraction grating, the seventy-fifth anniversary of general relativity, etc.

[18] For an illuminating counterexample, see the memorial volume on the occasion of the decommissioning of the Zero Gradient Synchrotron at Argonne National Laboratory: *History of the ZGS,* eds. Alan D. Krisch, Lazarus G. Ratner, and Joanne S. Day (New York: American Institute of Physics, 1980).

In a multidisciplinary laboratory, pains are taken to highlight achievements that result from the synergy of practitioners of different disciplines. The laboratory seeks to exalt the team rather than its leader, the laboratory rather than its director, and the collective rather than the individual accomplishment.

LBL's fundamental activity had been in high-energy physics. The award of the 200-GEV accelerator that LBL designed to Fermilab in 1966 by the AEC signaled the end of the laboratory's leadership in this field. Although LBL subsequently participated in experiments at SLAC and other high-energy physics laboratories, and helped design the positron-electron proton storage ring, the Time Projection Chamber, the Superconducting Super Collider, and other accelerators and detectors for other sites, it no longer managed a major facility in the field by the time of its fiftieth anniversary in 1981. The Bevatron had been transformed to a medium-energy physics machine in 1974, and no other high-energy machine replaced it.

As LANL approached its fiftieth anniversary in 1993, President George Bush signed the Strategic Arms Reduction Treaty, reducing nuclear weapon stockpiles to six thousand "accountable" warheads, and announced additional unilateral cuts in the nuclear weapon arsenal. Secretary of Energy James Watkins testified before the Senate Armed Services Committee that for the first time since 1945, the United States was not building any nuclear weapons. Congress voted to impose a moratorium on nuclear weapons testing. Adding insult to injury, Watkins dispatched "Tiger Teams" to the labs in 1990 to search out violations of environmental and safety regulations. His successor, Hazel O'Leary, launched "Openness" initiatives to uncover victims of medical experiments conducted at Los Alamos and LBL that had used plutonium and other radioactive substances.[19]

From an historical point of view, these changes in federal policy reflected changes in the underlying political economy of science in the postwar era. Like the nation itself, American science grew prosperous in the 1950s and 1960s. Its share of the federal budget increased when its technological applications, especially in defense, were sought after and successful. It decreased when, as in the economic dislocations of the 1970s and 1980s, other national needs seemed more urgent.[20] Similarly, when the passions of hot and cold war had faded from memory, radiation safety measures that had seemed legitimate protective measures at the time appeared to be gross violations of medical ethics, and it took several years to recognize the historical difference in the ethical standards involved.[21]

## IV. LABORATORY CULTURES

The fiftieth anniversary celebrations cannot be understood without taking these historical contexts into account. The labs sought to present a useful past, not only to appeal to patrons, but also to cement their own laboratory cultures. These cultures

[19] United States Advisory Committee on Human Radiation Experiments, *Final Report of the Advisory Committee on Human Radiation Experiments* (New York: Oxford University Press, 1996).

[20] I presented my views on the effects of national policy shifts on science in "Science Policy and the Role of the National Laboratories," *Los Alamos Science,* 1993, *21:*218–26, to which the chief participant in the celebrations, Los Alamos National Laboratory Director Siegfried S. Hecker, responded with "Los Alamos: Beginning the Second Fifty Years," *ibid.,* pp. 227–37.

[21] United States Advisory Committee, *Final Report* (cit. n. 19).

resist attempts at analysis by anthropologists and political scientists.[22] One difficulty arises from the fact that, as corporate entities, the laboratories define themselves not only in terms of their missions but also in opposition to each other; they cannot be understood in isolation. Anniversaries reinforce values, create a shared sense of mission, and engender beliefs and behavioral norms.[23] Committees of scientists at both laboratories selected speakers who could serve as role models for the laboratory, place its achievements in the best possible light, send a message to laboratory management, and compete for distinction with their rivals. When LANL's director learned that Oak Ridge National Laboratory planned to hold its fiftieth anniversary celebration before Los Alamos's, he chose to launch the commemoration on the fiftieth anniversary of the army's condemnation of the Los Alamos Ranch School in 1942, instead of the fiftieth anniversary of the lab's first meeting in April 1943. He invited the secretary of energy, the governor of New Mexico, and other notables to a banquet on 7 December, preempting Oak Ridge's attempt to present itself as the first laboratory of the DOE.[24]

Sally Moore and Barbara Myerhoff have argued that "a collective ceremony is a dramatic occasion, a complex type of symbolic behavior that usually has a stated purpose, but one that invariably alludes to more than it says and has many meanings at once."[25] For example, commemorations can reorganize and help create social arrangements and modes of thought.[26] Modern secular rites, e.g., trade fairs and international exhibitions, display their products to facilitate marketing.[27] National laboratories use anniversaries for similar purposes.

## V. THE ROLE OF HISTORY: CELEBRATION OR CEREBRATION?

Given the variety of motives involved in the celebration of a laboratory's anniversary, the past can be perceived as either an anchor or as a sail, invoked either to justify or to undermine shared values and to serve conservative or radical ends.[28] For those who see change as desirable, cutting the anchor chain and laying on more sail may seem the best policy.

At its forty-fifth anniversary in 1976, LBL explicitly turned its back on "a tendency to look back to the good old days [and] overemphasize our glorious history," and adopted the theme "Looking Ahead to the Next Half-Century." At the public symposium held to mark the occasion, Ed McMillan, the laboratory's only living

[22] See Hugh Gusterson, *Nuclear Rites: A Weapons Laboratory at the End of the Cold War* (Berkeley: University of California Press, 1996); Debra Rosenthal, *At the Heart of the Bomb: The Dangerous Allure of Weapons Work* (Reading, Mass.: Addison-Wesley Publishing, 1990); Sharon Traweek, *Beamtimes and Lifetimes: The World of High Energy Physicists* (Cambridge: Cambridge University Press, 1988; reprint 1992).

[23] Corporations use a combination of rewards and disincentives to promote an increasingly impersonal culture. See Terrence C. Deal and Allan A. Kennedy, *Corporate Cultures: The Rites and Rituals of Corporate Life* (Reading, Mass.: Addison-Wesley Publishing, 1982).

[24] Robert W. Seidel to Siegfried S. Hecker, 23 October 1992, memorandum n. CNSS-75-RWS, Los Alamos National Laboratory Archives (hereafter cited as LANL Archives).

[25] Moore and Myerhoff, *Secular Ritual* (cit. n. 4), p. 5.

[26] *Ibid.*

[27] Burton Benedict *et al., The Anthropology of World's Fairs: San Francisco's Panama Pacific International Exposition of 1915* (London and Berkeley: Scolar Press, 1983), p. 15.

[28] See, e.g., Nicola Gallerano, "History and the Public Use of History," in *The Social Responsibility of the Historian,* ed. François Bédarida (Providence, R.I., and Oxford: Berghahn Books, 1994), pp. 85–102.

former director, was allowed to speak on the "Early Days." Other speakers focused on "scientific and public policy issues."[29]

The editor of the *LBL Newsmagazine* asked, "What meaning can an event like this 45th anniversary observance have to an institution like the Laboratory?" and answered,

> The same meaning that a birthday has to an individual—a time to remind ourselves, and our friends, of where we came from and where we're going. As individuals, we all like to think that each of our actions comes from the whole person, not just that fragmented part of us—worker, boss, scientist, chairperson, parent, or whatever—that the outside world may be noticing at any given moment. So it's well to remember that an institution like the Laboratory has a life too—with a past that got us to where we are today, a present that we're maybe too involved in to understand in perspective, and a future that, like anybody else, we're looking forward to with mixed anticipation and uncertainty.[30]

On two pages of the newsmagazine devoted to the past, Nobel laureates Luis Alvarez, Melvin Calvin, Owen Chamberlain, Emilio Segrè, and Glenn Seaborg shared their "most exciting moments in research" and McMillan contributed an account of "How LBL Began." The rest of the issue described contemporary research.[31] When history was discussed, events of long past were the primary focus: the invention of the cyclotron in 1930, the (unofficial) founding of the laboratory in 1931, and discoveries in physics in the 1930s that had led to Lawrence's Nobel Prize. The founder's myth was cultivated at the expense of later historical events that had shaped the growth of the laboratory.

Professional historians of science were invited to prepare formal histories for both laboratories' fiftieth anniversaries LBL's History Committee was set up in 1979 to plan for the celebration. Emilio Sergè, one of its members, suggested that a full history of the laboratory be written. The committee also wanted a briefer history prepared for the occasion. Both were contracted to the Office for History of Science at the University of California.

At LBL, the laboratory director asked that the brief history stop with the death of Ernest Lawrence in 1958. Although it did not, therefore, touch upon more recent sore spots, the brief history nevertheless included episodes like Lawrence's "deuteron-disintegration hypothesis" announced at the Solvay Congress of 1933, the wartime calutrons for uranium isotope separation for the atomic bomb, the Materials Testing Accelerator built to produce fissile materials, and the creation of the Livermore branch. All of these were less-than-flattering reflections. The editor of the *LBL Newsmagazine* remarked, "I can't believe we're publishing this," and the sub-

[29] Judith Goldhaber to Louis Robinson, memorandum re 45th Anniversary Event, 29 July 1976, LBL. Director Andrew M. Sessler, "Viewpoint: LBL in a World of Change," *LBL Newsmagazine,* Summer 1976, *1,* n. 1:3, expressed the need for a fresh start in the wake of the establishment of a new patron, the Energy Research and Development Administration. Judith Goldhaber, "In This Issue," *LBL Newsmagazine,* Fall 1976, *1,* n. 2:1; *Symposium Agenda, 45th Anniversary,* Lawrence Berkeley Laboratory, 30 October 1976.

[30] Goldhaber, "In this Issue" (cit. n. 29).

[31] "Dialogue: Your Most Exciting Moment in Research?" and Ed McMillan, "Viewpoint: How LBL Began," *LBL Newsmagazine,* Fall 1976, *1,* n. 2:2–3.

title she gave the issue, "A Historian's View," made clear whose voice was being heard.[32]

The director's four-sentence summary of this history, which opened LBL's public anniversary symposium, conceded certain "essential characteristics" of the lab observed by the historians: a major concern with scientific research at the frontiers of knowledge, a multidisciplinary approach, and large and complex experimental facilities. These qualities, he thought, were useful for meeting "national technological needs and for solving problems ranging from the energy issues facing our nation to human health and suffering," while the laboratory's "strong commitment to the training of students and the advanced training of graduates" supplied manpower to the DOE. The multidisciplinary nature of the laboratory suited it for applied as well as fundamental research.[33] This précis neatly avoided opening old wounds and preserved a peaceful visage.

In its planning for the symposium, the LBL program committee "agreed that . . . all the speakers should be distinguished invited guests." Of the laboratory's Nobelists, only Alvarez, a last-minute stand-in for Philip Handler, spoke. His subject, the asteroid hypothesis for the extinction of dinosaurs, had "nothing to do with any of the programs at the Laboratory," he claimed, although "Ernest Lawrence would really approve of this because one of the things that he did . . . was to bring in people from other disciplines and let them share the resources of the Laboratory . . . and I'm sure he would have enjoyed seeing what the Laboratory has done by sharing its resources."[34]

Most of the rest of the symposium speakers also recognized laboratory contributions to their work. Philip Abelson, who had codiscovered neptunium with McMillan before the war at the Radiation Laboratory, pointed out its "great contributions to the country" during World War II.[35]

Steven Weinberg, who had been a research associate at the Radiation Laboratory from 1959 to 1969, acknowledged LBL contributions to his formulation of the "standard model"—the modern unified theory of electromagnetic, weak, and strong interactions for which he had recently won the Nobel Prize in physics.[36] David Kuhl of the University of California, Los Angeles paid tribute to the development of cyclotron-produced radioactive tracers and positron emission computed tomography at LBL as examples of "wonderful benefits from physics and chemistry labs [that] have given us the ability to visualize the gross structure of the brain of a living patient with a facility that not too many years ago was only possible at the autopsy table."[37]

John Adams of CERN praised the invention of the cyclotron and "the foundation

[32] John Heilbron, Robert W. Seidel and Bruce R. Wheaton, "A Historian's View of The Lawrence Years," special issue of the *LBL Newsmagazine,* Fall 1991, *6,* n. 3:1–106.

[33] David Shirley, "Opening Remarks by the Director of Lawrence Berkeley Laboratory," *Proceedings: 50th Anniversary LBL Symposium,* report n. LBL-13613 (Berkeley: Lawrence Berkeley Laboratory, 1981), p.v.

[34] Luis W. Alvarez, "Asteroids and Dinosaurs," *Proceedings: LBL Symposium* (cit. n. 33), pp. 4–5. See p. 7.

[35] Philip H. Abelson, "Energy and Electronics in a Changing World," *Proceedings: LBL Symposium* (cit. n. 33), pp. 49–65.

[36] Steven Weinberg, "The Ultimate Structure of Matter," *Proceedings: LBL Symposium* (cit. n. 33), pp. 78–9, 82.

[37] David Kuhl, "From Science Laboratory to Hospital: New Imaging Instruments," *Proceedings: LBL Symposium* (cit. n. 33), pp. 90–100, quotation on p. 100.

of this great laboratory." His review of accelerator development, however, acknowledged the influence nuclear physicists acquired in building the atomic bomb, "which gave them access to government funds in place of private donations," and concluded that

> since Berkeley was set up in the aftermath of a major economic depression and CERN after a World War, it might almost be concluded that such ventures need desperate situations for their birth. So if the physicists are convinced that wider international collaboration is essential for the future of their research and if the governments of the regions of the world want to show that they can do something together and are looking for a subject which does not have commercial or military complications, then joint action on an inter-regional basis may well succeed. It may even release more money for this research. . . . the conjunction of circumstances which took place after the Second World War and led to the rapid expansion of this research is unlikely to repeat itself—at least I sincerely hope not. Surely not even the most desperate physicists would wish for another war just to increase his budget."[38]

Reflections on war were as welcome as Banquo's ghost at the banquet that followed. The DOE program managers for high energy and nuclear physics and George Keyworth, the president's science advisor, heard instead brief reminiscences of Ernest Lawrence and of the Radiation Laboratory in the 1930s by Lawrence's widow, Molly, and by his prewar protégé, Robert Wilson, then director of Fermilab. Wilson cast the anniversary as "a time to renew the Berkeley ambience, the memories of Ernest Lawrence, and of the inspiration that he provided for so many of us" in order to respond to a crisis: "we have seen our funds cut back . . . perhaps by a factor of 2 from the late 1960s," he lamented, but

> Ernest would [not] stand around wringing his hands . . . there's a lot of money and resources that we do have, and he would pause and he would think about what he wanted to do, or he thought ought to be done, take a long view and he would start to do it with the resources and the funds at hand. . . . not only would he be making innovations himself and inspiring young people to be making innovations also . . . but he would also be participating in the fund-raising experience, and with no holds barred and [in] every direction that he could. And that I think is something we also should emulate. Doing our best to see that the funds are adequate would be maintaining the ingredient of our culture, the ingredient of the advances of technology that are so badly needed by this country.[39]

In building the great accelerator laboratory that had displaced LBL from its leadership in high-energy physics, Wilson had consciously followed such a philosophy.[40] After Lawrence's death in 1959, however, LBL had become a web without a spider. Wilson, who claimed to be closer to Lawrence than were those who had remained at the Radiation Laboratory, had reaped the harvest of opportunities in high-energy physics at Fermilab.

In his keynote address, "National Science Policy and its Relationship to the Needs and Capabilities of the Nation Today," George Keyworth stressed fiscal responsibil-

[38] John Adams, "The Evolution of a Big Science," *Proceedings: LBL Symposium* (cit. n. 33), pp. 110, 129–30.

[39] Robert Wilson, *Proceedings: LBL Symposium* (cit. n. 33), p. 140.

[40] See, *inter alia,* Catherine Westfall and Lillian Hoddeson, "Thinking Small in Big Science: The Founding of Fermilab, 1960–1972," *Technology and Culture,* 1997, *37:*457–92.

ity, increased productivity, excellence, and pertinence as criteria for federal support in "areas where the U.S. is, and shall remain, a leader," e.g., space, biology, chemistry, computers, aviation, and materials. LBL, "unique in its primary missions of basic research combined with its location on the campus of one of the world's most esteemed academic and scientific institutions" would enjoy "opportunities for a special future," Keyworth promised.[41]

Although LBL suffered a 12 percent budget cut the month following the anniversary, within a year the Reagan administration increased funding for basic research as well as for a major new mission—of the kind that Lawrence and the Radiation Laboratory had seized when creating the materials testing accelerator in 1950 and a second weapons laboratory at Livermore in 1952. At the Reagan banquet for science and technology, however, it was not LBL, but its errant sibling, that was to feast.

## VI. THE PRODIGAL SON

The Lawrence Livermore National Laboratory, a branch of LBL until 1971, went unmentioned and unrecognized in LBL's fiftieth anniversary festivities. Molly Lawrence launched a campaign to strip her late husband's name from the Livermore laboratory a few months later, dismayed at Livermore's participation in "President Reagan's tragically ill-advised nuclear-weapons buildup," one of the "special opportunities" offered to the DOE weapons labs in the 1980s.[42]

The historians recognized the importance of Livermore in the history of its parent institution, and we asked Edward Teller, among others, to comment on the history of Los Alamos and Livermore during a lecture series we had organized for the occasion. Instead, Teller described a novel idea he had for a nuclear-powered defensive weapon in space. He was able to sell this idea a little later to Keyworth and Reagan as part of SDI. War, weapons, and wizardry had more appeal for Reagan than the basic scientific research LBL was willing to undertake.[43]

Like its biblical equivalent, LBL refused to join in the banquet for Lawrence's prodigal son. Although it had long cultivated the fields of fundamental research that supported the applied R & D at Los Alamos and Livermore, to which it had also contributed not a few of its scientists, LBL was unwilling to accept a role in SDI. This was true despite the fact that the intellectual origins of SDI particle-beam technologies can be traced back to Luis Alvarez's idea that an electron accelerator might pre-detonate a falling atomic bomb.[44]

Why was LBL's martial past so readily discarded? The use of selective memory to create ambiguity, ambivalence, and a rationale for our personal choices is commonplace enough. The participants in the 50th Anniversary Celebration of the Lawrence Berkeley Laboratory collectively excluded from their commemorations what

[41] George A. Keyworth, *Proceedings: LBL Symposium* (cit. n. 33), pp. 150–59.

[42] Barbara A. Serrano, "Mrs. Lawrence Unhappy with Nuclear Laboratories," *The Daily Californian* (University of California, Berkeley), 8 April 1982, p. 13; Elizabeth Mehren, "Nuclear Scientist's Widow Reconsiders Role of Lab," *Los Angeles Times,* 23 April 1982, sec. V, pp. 1, 26; Molly B. Lawrence, "Rename Livermore," *Physics Today,* October 1986, *39*:9, 11, 13.

[43] William J. Broad, *Teller's War: The Top Secret Story Behind the Star Wars Deception* (New York: Simon & Schuster, 1992), pp. 114, 118–19, 224–5, 305–6; Martin Baucom, *The Origins of SDI, 1944–1983* (Lawrence, Ks.: University Press of Kansas, 1992), pp. 134, 148–9.

[44] "Military Applications of Particle Accelerators: Report by the Director of Research," 23 February 1952, Department of Energy Archives, RG 326, Box 1305.

the historians considered the defining transformations wrought by World War II and the cold war. As might be expected, they also limned the past in hues of success, heroism, and wealth, and elevated the founder to mythical status. What is more, they explicitly distanced themselves from their past in order to suggest a new future for the laboratory. Their selective use of the past, however, excluded them from some of the opportunities available, including those of the type Lawrence had seized upon with alacrity at the beginning of the cold war.

## VII. LOS ALAMOS: JOINT TASK FORCE 50

To complain of the age we live in, to murmur at the present possessors of power, to lament the past, to conceive extravagant hopes of the future, are the common dispositions of the greatest part of mankind."[45]

In the era of atmospheric testing, flotillas of naval vessels bearing LBL and Los Alamos technical staff detonated hundreds of nuclear devices in the Pacific. A veteran of these exercises was chosen to lead the planning for Los Alamos's fiftieth. He identified constituencies of the laboratory and assigned committees to target them. His weapons included rock concerts, laboratory picnics, a commemorative stamp, two days off for all laboratory employees, several seminar series, a popular history, local celebrations in all of the northern New Mexico towns and cities in which the laboratory's employees lived, a traveling exhibit, a video, and an open house, embracing even secret laboratory facilities. The town of Los Alamos would provide sweatshirts, coffee cups, commemorative coins, calendars, and a sculpture garden around Ashley Pond. The laboratory's theme for the celebration, "Partners in Science Making a Difference," emphasized cooperation with its constituencies. The intent was to market the laboratory as widely as possible, to sell multidisciplinarity, "world-class," pure and applied science and engineering and the capability to solve technical, civil, and military problems that no other laboratory was prepared to tackle.

The DOE's director of public affairs, however, decided that there should be "no birthday for the bomb" at Los Alamos. Reluctantly, she agreed to permit a "small, local celebration" costing less than $30,000 to mark the fiftieth anniversary of the founding of the laboratory. Her office killed the stamp proposal by submitting a mushroom cloud design with it to the secretary of energy.[46] It seemed that Joint Task Force 50 was sunk.

The fleet was salvaged by redefining its ships as merchantmen. For example, the seminar series became part of the laboratory's programmatic activities, with funding from operating budgets. A special fiftieth anniversary edition of *Los Alamos Science* replaced the short history.[47] By reducing the scale of activity and paying for most of it out of its programmatic funds, the laboratory won approval of the DOE for most

[45] Edmund Burke, *Thoughts on the Cause of the Present Discontents,* 5th ed. (London: Printed for J. Dodsley, 1775, c. 1770).

[46] These recollections are drawn from the author's notes of the planning sessions for the celebration. I was transferred to the Center for National Security Studies to assist in planning. I traveled to Washington on at least one occasion with John Hopkins to discuss the events described here with senior DOE officials.

[47] "The Laboratory's 50th Anniversary," *Los Alamos Science,* 1993, vol. 21.

of its plans. The celebration became a "commemoration": to "celebrate" the founding of the laboratory might appear as a "birthday for the bomb." The open house became another Family Day, like those the laboratory had conducted every fifth year for most of its history. The idea of opening all facilities to the public was dropped.

Unlike the modest one-day ceremony that had marked the fiftieth anniversary at LBL, Los Alamos's activities extended over a year and showcased laboratory work in many fields. Special exhibits and displays, including a vast array of nuclear testing instrumentation, were erected for the occasion. Plans to mount a display of a multistory test rack, used for underground testing, were, however, abandoned as unseemly after a number of laboratory personnel were laid off due to programmatic budget cuts.

On 1 April 1993, fifty years after the first planning meetings were held at Los Alamos, a four-day classified series was inaugurated by Robert Serber, whose lectures—now published as *The Los Alamos Primer*—had originally launched the laboratory.[48] The theme of the series was "creativity": the laboratory's responses to past challenges in the weapons program presaged a creative response to the challenges it then faced with the end of the nuclear arms buildup. Although he did not disagree with the theme, the laboratory's associate director for nuclear weapons technology called for a more forward-looking program, coupling specific speculations on the future with reviews of the past. The series was, with the exception of the Serber talk and a round-table on the first nuclear test at Trinity, closed to the public, so their reaction to it was minimized. It satisfied the working weaponeers, whose secret arts go uncelebrated, and who delighted to commune with past luminaries like Teller, John Wheeler, and Serber.[49]

The public seminar series involved protracted negotiations between a committee of scientists who sought to display the fruits of laboratory efforts in science and technology, and the administration, who sought to entice policy-makers from the Bush administration and, after November 1992, the new Clinton administration, to participate in the commemoration.[50] Laboratory scientists were invited to participate by the director in an effort looking towards the future, but using the past:

> The goal of the Seminar Series is to present an overview of the science and technology activities at the Laboratory emphasizing those scientific strengths that can help meet and define future requirements and directions. These activities are diverse in character. Many are rooted in the past and arose as part of the broad base of support of the defense mission of the laboratory. The end of the Cold War has not only altered the defense needs of the nation, but has also opened up new opportunities for the Laboratory to serve the country. Advances in biology, new approaches to computing, and properties of materials and their applications to health, environment, global monitoring and change, industrial competitiveness, space exploration and astrophysics are examples of areas to which our scientific experience can make significant contributions. At the same time, we must remain committed to meeting the nation's defense needs in a rapidly changing world situation.[51]

[48] Robert Serber, *The Los Alamos Primer* (Berkeley: University of California Press, 1992).
[49] ADNWT Seminar Planning Meetings, 23 November and 22 December 1992, in the author's files.
[50] *50th Anniversary Seminar Series* (Los Alamos: Los Alamos National Laboratory, 1993), video recording.
[51] S. S. Hecker, memorandum to laboratory staff invited to speak in the seminar series, 17 September 1992, LANL Archives.

The summoning of the oracles was difficult. Because of the transition between the Bush and Clinton administrations, the first list of policy-makers to be invited did not even appear until February 1993, by which time many of the most desirable were unavailable. Several rounds of invitations were sent out before enough faces to fill the panels could be recruited.

The seminar series celebrated the science that had sprung from the ashes of nuclear weapons: accelerator ideas born at Los Alamos; the early development of computing; the detection of the neutrino; explosives as objects of scientific interest; the discovery of elements 109 and 110; and Los Alamos's contributions to fusion. All sought "to show how the Laboratory's scientific base, rooted in its past experience, adds significantly to the nation's ability to solve complex problems."[52]

Speakers were selected "because of their knowledge of their subjects, communication skills, and ability to capture in an imaginative and forthright manner the spirit of a Laboratory committed to meeting the challenges of the future."[53] They were brought from outside the laboratory whenever possible, on the principle that it is better to have someone else sing your praises. Their talks were intended to be instructive to the political patrons of the laboratory, as well as to the director and the audience. For example, New Mexico Senator Jeff Bingaman, chaired the session "Energy and Technology for the Future," where laboratory work benefiting the New Mexico petroleum industry was celebrated by, among others, the president of the Murphy Operating Company. The acting assistant secretary for DOE Defense Programs, the laboratory's chief patron, presided over a session on national security.

A banquet honoring Manhattan Project pioneers, local municipalities, former directors, and the DOE's director of energy and research, as well as leaders of the Russian nuclear weapons labs who had brought gifts commemorating the end of the cold war, was more popular. Hopes were expressed for continued cooperation between the nuclear weapons labs of both countries in dealing with the nuclear danger, and commemorative coins and laboratory liquor were distributed in an ersatz banquet hall at the Lodge, the city's oldest and most distinguished architectural monument, to which a tent had been affixed for the occasion.[54]

The annual Manhattan Engineer District reunion a month later was a more popular celebration of the past. Richard Rhodes, author of *The Making of the Atomic Bomb,* was the guest speaker, and his comparison of Los Alamos weapons makers with the public health crusaders of the nineteenth century offered a congenial historical interpretation for his audience:

> So now, fifty years later, fifty years after this city on a hill opened its doors, when we know what the consequences are, I believe the world owes you, and those of your colleagues who are gone now and no longer among us, an immense debt of gratitude. . . . With courage, with vision, and with intelligence, braving the most cruel part of reality, you started us down the road toward removing a terrible scourge from the earth. Knowledge is itself the basis of civilization, and you helped to civilize us.[55]

[52] Jim Louck to Darleane Hoffman, 22 February 1993, LANL Archives. A similar instruction was sent to all participants.

[53] Hecker to Seminar Speakers, 17 September 1992, LANL Archives.

[54] *Los Alamos National Laboratory 50th Anniversary Seminar Series Dinner,* 15 April 1993, LANL Archives.

[55] Richard Rhodes, "A Different Country," presented at Los Alamos National Laboratory 50th Anniversary, 10 June 1993, video recording. Also quoted in S. S. Hecker, "Science Looks Toward the Future," *Los Alamos National Laboratory Research Highlights 1993,* April 1994, Los Alamos Labo-

Joining Rhodes in the celebration was Edward Teller, who attracted a standing-room-only audience to his laboratory colloquium, which looked forward to another fifty years of nuclear weapons development.[56]

## VIII. PAST AND FUTURE

"Postmodern sensibility expresses itself not least through the cult of anniversaries. . . . Postmoderns commemorate what they no longer wish to imitate."[57]

Both labs struggled with their histories in these celebrations. By implicitly denying the legacy from the labs' wartime and postwar preparations for war, the celebrations publicly ignored the underpinnings of their existence during those eras. The past, placed in the hands of historians, was buried between the covers of their books, with the proper warning labels attached.

The laboratories interpreted their past polemically in order to insure their survival. Inconvenient facts, like nuclear weapons or wars, were either ignored or placed in secret ceremonies. Outside the fences at Los Alamos, however, these facts were selected for celebration. In *Behind Tall Fences: A Collection of Stories and Experiences About Los Alamos,* the J. Robert Oppenheimer Memorial Committee collected a number of anecdotal recollections, avoiding "issues":

> There is no congruity of theme, or discussion of issues of morality and ethics, no messages of doom, or hope, on that which can be read into the expressions and actions of individuals dedicated to meeting the day-to-day challenges presented by the task at hand—the building of the bomb. It is hoped, however, that the reader may sense from these representative stories, written by working scientists, the deep commitment, sensitivity, and humanity of all those involved: scientists, engineers, technicians, secretaries, families, and supporting military personnel.[58]

## IX. CONCLUSION

James Bryant Conant and others created the history of science at the end of World War II to celebrate the distant accomplishments of science and divert attention from the more recent and horrendous demonstrations of its power. Those of us who, as historians of science, have depended upon cold-war institutions for our support should not be surprised to find that the history of science is such an institution. Like the practitioners of high-energy physics and of nuclear weapons design, development, and testing, historians of science have been in denial about the larger historical context of their subject, preferring to believe that its support was contingent on its intellectual interest, which they often shared, rather than its applications, which they mostly ignored.

---

ratory Publication 94–20. Rhodes also presented a talk, "Trademark Los Alamos: The First Soviet Bomb," Director's Colloquium, 10 June 1993 (Los Alamos: Los Alamos National Laboratory, 1993), video recording which detailed how the first Soviet bomb was a copy of the Trinity device based upon Fuch's intelligence.

[56] Edward Teller, "Nuclear Weapons & International Relations at the 100th Anniversary at LANL," Director's Colloquium, 8 June 1993 (Los Alamos: Los Alamos National Laboratory, 1993), video recording.

[57] Johnston, *Celebrations* (cit. n. 16), p. 6.

[58] J. Robert Oppenheimer Memorial Committee, *Behind Tall Fences: A Collection of Stories and Experiences About Los Alamos* (Los Alamos: J. Robert Oppenheimer Memorial Committee, 1994).

Just as those bearing the legacy of World War II and the cold war diverged into different camps of weaponeers and basic researchers, so the history of science has diverged from other social science and humanities disciplines in its focus on a redemptive agenda for science. The adoption of an internalist orientation by traditional historians of science, and their disinclination to investigate the technologies and polities associated with nuclear science, have left the field open to others in science studies, sociology, and political science, who may not appreciate the redemptive virtues of modern science. As I have pointed out before, most of the historians who have analyzed defense research and development came from different traditions than our own.[59]

The reaction is in full swing: postmodern historiography denies the historian's privileged perspective on the past, and would permit the polemical narratives I have described here equal status with the products of the historian's "methodological" approach. As was the historian in the midst of these anniversaries' celebrants, so we may all be witnesses to the past, our interpretations differentiated only by our interests or by the futures we covet. The shaping of the past that is so evident in the fiftieth anniversary celebrations of the Los Alamos and Lawrence Berkeley Laboratories may rather serve to indicate how successful historians of science have become at penetrating the veils of interest and affect in their own accounts.[60] If the past is disputed, ought not historians strive to weigh rather than inflate partisan claims, and thereby to produce a balanced, if not "objective" view?

[59] Robert W. Seidel, "Clio and the Complex: Recent Historiography of Science and National Security," *Proceedings of the American Philosophical Society,* 1990, *134:*420–41.

[60] For a review of the historiographical issues from a postmodern perspective, see Keith Jenkins, *On What is History* (London and New York: Routledge, 1995).

# Commemorative Practices at CERN

## Between Physicists' Memories and Historians' Narratives

*By Dominique Pestre**

### I. SCIENCE AND HISTORY

SINCE SCIENTISTS ARE OFTEN DESCRIBED as being only concerned with future developments and as living in a world without history, let us not forget that they often show a marked interest in the history of their discipline, and have always done so. By way of example, just consider the universities and *sociétés savantes* over the last three centuries and the historical texts that have accompanied their nominations, receptions, and other commemorative ceremonies. At the Académie des Sciences in Paris, one of the tasks of the *secrétaire perpétuel* is to write historical "notices" whose quality and accuracy are indicative of methodical research carried out with diligence. Two excellent examples are provided by the (published) speeches of Emile Picard and Louis de Broglie, both *secrétaires perpétuels* for the academy's division of mathematical and physical sciences from 1917 to 1975.[1]

History, however, does not only manifest itself in commemorations. It can be seen in daily scientific work, when physicists describe their links with their predecessors, and when they expose the chronological development of the problems they address. Like intellectuals and academics in all fields, at the beginning of their books or articles physicists tend to create a frame in which to insert their own research program. They tend to build a narrative; to draw a coherent itinerary through the past—an itinerary which links their own questions and solutions to the ones previously debated, and currently accepted, by at least part of the community. When a physicist writes a synthetic article for *Annual Reviews* or for *Advances in ( . . . ) Physics* (an extremely widespread exercise since World War II due to the exponential development of the field), he or she often offers a complete intellectual history of the problems being addressed. The physicist tells how theories and models have been replaced over the years, each time enabling the solution of a pending difficulty or a contradiction. To take a random example, in an article of this kind published in 1982

*Centre Alexandre Koyré, Paris, Muséum National d'Histoire Naturelle, Pavillon Chevreul, 57 rue Cuvier, 75231 Paris Cedex 05, France.

I would particularly like to thank Jean-Paul Gaudillière, John Krige, and Simon Schaffer for more than useful comments on various versions of this text.

[1] Complete references are in Dominique Pestre, *Physique et physiciens en France, 1918–1940* (Paris: Éditions des Archives Contemporaines, 1984), p. 338.

on "changes in electric properties induced by irradiating certain MOS compounds," we find a ten-page section entitled "Brief History and Introduction" that goes back to the atomic models proposed by Ernest Rutherford in 1911 and by Niels Bohr in 1913 (first paragraph), to the quantum solutions suggested by Felix Bloch and A. H. Wilson in 1928 (second and third paragraphs), and so on.[2]

More elaborate, self-proclaimed historical studies of the discipline are also regularly published in important scientific journals.[3] Usually written by influential personalities, they propose a conceptual and technical history of the field's development that draws heavily on the versions then commonly accepted among physicists. On some occasions, these historical works are meant to remind readers of abandoned research directions and to reestablish approaches or authors that have been invalidated or forgotten. When there is a significant paradigmatic change in the discipline (as is currently the case with chaos theories and mild physics), or when a minority group tries to reverse an intellectual trend (as happened more than once with quantum mechanics), long-forgotten people, books, and historical events are revived—to be used as tools in future debates and research.

As one may guess, these historical narratives do not always correspond to the canons of academic history. If I had to define them—even while recognizing that they differ widely according to contexts and audiences—I would first say that one aim is often to celebrate past successes, to show their importance, and to measure the distance covered. Sometimes this aim is served by an heroic form of narrative telling of the endless quest for truth that animates scientists—but this is not necessarily the case (it might be less hagiographic). In other, and in fact currently rarer cases, scientists are introduced as moral characters, embodying abnegation, probity, modesty, and hard work. Usually, the reported advances are not described as being the end result of controversies involving a whole range of practices and conceptualizations, or as being the result of complex social and material debates about what must be considered a "fact," but as having imposed themselves through the evidence of the experiment or of the mathematical demonstration—as having been immediately compelling and unavoidable. Because the true nature of things is now known—it is tantamount to today's science's descriptions and categories—these narratives do not hesitate to select from the productions of the past and spin a thread of discoveries that leads, often without disturbance, to what we now know. Retrospectively chosen, the episodes are judged more according to our standards of proof than according to those reigning when they occurred. In this sense, these stories alter the ways of thinking that they are reporting; they read, reorganize, and reinterpret past intellectual achievements without avoiding what professional historians call "anachronism."[4]

Another feature of these works is that they often tend to take a stand in current debates among scientists; they often aim to arbitrate priority disputes. As they try to establish for posterity what the most important discoveries were and by whom they

[2] Article published in *Advances in Electronics and Electron Physics,* 1982, *58:*1–79.

[3] As examples, two long such articles published in 1980 are: Léon Van Hove and Maurice Jacob, "Highlights of 25 Years of Physics at CERN," in *Physics Reports: A Review Section of Physics Letters,* and, for solid state physics, Pierre Grivet, "Sixty Years of Electronics," *Advances in Electronics and Electron Physics,* 1982, *58:*1–73. Jacob and Van Hove are former directors of the theory division at CERN. Grivet held the first chair in electronics at the Sorbonne.

[4] A classic book on these questions is Paul Veyne, *Comment on écrit l'histoire* (Paris: Seuil, 1971).

were made—thus drawing up a double list of the most well-known facts and of the names to be associated with them—they are easily trapped in endless confrontations. These narratives give evidence for some and against others; they provide evidence for the judgment of History—and, more prosaically, for those who decide on promotions and rewards like Nobel Prizes. For a "discovery" never appears full-blown but is put together contradictorily and retrospectively by the actors. It entails an *a posteriori* choice of events and is a reading of what was said and done in the light of a later state of science. It is always a post-hoc attribution of credit by an organized community—and opinions differ.[5]

These historical narratives, constructed by savants in the midst of their activities, are thus also a search for precursors. Since they aim less at recreating the consistency of a historical moment than at exposing how the "real world out there" was attained, a homogenous historic time is created—validating procedures and practices used in the past are spontaneously equated with those of today—and an everlasting and never-changing Science is made to emerge. These stories universalize today's practices (whether they be material, social, or literary); they act as if today's rules (moral and behavioral rules as well as proof criteria and procedures) have always been immutable. These stories are crucial to maintaining the values of the institution of science—the specificity and unique character of the knowledge it produces, for example, or the pivotal part played in its elaboration by the scientific method. They are essential to the smooth functioning and perpetuation of scientific communities.

## II. PHYSICISTS AT CERN AND THEIR OWN HISTORICAL NARRATIVES

Thus, history partakes of the structure of the scientific community; it is part and parcel of its daily practice, as necessary to its intellectual life as to its social one. Yet this statement does not imply that scientists' relation to history does not vary in intensity or nature over time or from one place to the next. The thesis I would like to develop here is that the interest in history has dramatically grown over the last decades—notably among particle physicists at CERN—and that a new need for history was felt in the 1980s in this community. (Let me remind you that CERN—Centre Européen de Recherche Nucléaire—is a very large European high-energy physics laboratory established in Geneva between 1949 and 1954. Financed collectively by numerous European states, CERN's first big accelerator was commissioned in 1959.) In the 1980s, CERN was becoming the most important international laboratory in the field.[6]

To try and illustrate this renewed interest in historical narratives among CERN physicists in the 1980s, let us begin by looking at the *Courrier CERN,* an in-house magazine published by the organization since 1960 with fifteen to twenty thousand copies distributed each month. A first conclusion I would draw relates to the celebration of physicists' birthdays.

[5] Augustine Brannigan, *The Social Basis of Scientific Discoveries* (Cambridge: Cambridge University Press, 1981). See also the very stimulating article by Mott T. Greene, "History of Geology," *Osiris,* 1985, *1* (2nd series):97–116.

[6] This article relies on documents covering roughly the years 1950–1990 and on my stay at CERN from 1982 to 1988 to write, with colleagues, a history of the organization. See Armin Hermann, John Krige, Ulrike Mersits, and Dominique Pestre, *History of CERN,* 3 vols. (Amsterdam: North Holland, 1987, 1990, and 1996).

Whereas the magazine only mentions one or two celebrations a year before 1975/80—celebrations that, in any case, did not take place at CERN—they have become more numerous since.[7] The first birthday celebrated at CERN was in 1978; the second was in 1980. Then we find one a year in 1983, 1986, and 1987; two in 1988, and two in 1989. Analysis of the homages paid by European physicists to the memory of outstanding members of the community corroborates this trend. The first commemoration took place at CERN in 1978, dedicated to Bernard Gregory, a former director-general who died in 1977. The second, held the following year, was for Lew Kowarski, one of CERN's pioneers. In 1980, a ceremony was held to honor the memory of Wolfgang Gentner; in 1984 John Adams and in 1985 Paul Musset was so honored. On these occasions, CERN published an historical brochure containing the texts of the ceremony's talks.[8]

What is more, the nature of these ceremonies evolved. Up to the end of the 1970s, their main objective was to offer technical assessments; to show how the scientific and technical questions asked by the honoree, and the solutions he or she proposed, had evolved since the honoree's first publications. In short, the talks remained scientific communications. When published, the proceedings were simply introduced by two or three pages recalling the career of that day's hero. In the 1980s, these jubilees began to undergo a change and became homages to a life, to history. The focus then tended to be on biography, and the celebration centered around the significant steps of a career. To assess the importance of this change, the reader could simply compare the records of a "classical" jubilee (for example, the twenty-three scientific communications published in 1970 on the occasion of Edoardo Amaldi's sixtieth birthday) with the speeches delivered to honor John Adams during the "memorial gathering" held at CERN in 1984. Here the celebration was focused on the man and his life as an engineer and scientist—and his commemoration was an opportunity to paint a historic panorama of the fields to which he had contributed.[9]

How birthday celebrations were staged is also worth analyzing in a somewhat more anthropological manner. Before becoming books, these ceremonies were highly codified public events.[10] At CERN, the meetings generally took place in the afternoon in the large amphitheater. Highly popular, they easily gathered two to three hundred people (as was the case for those I personally attended during my stay in Geneva from 1982 to 1988). The audience was partly elderly people and partly very young physicists. The ceremony was traditionally opened by CERN's director-general (which signals its importance), and it consisted of four or five talks, each centered on one aspect of the scientific or institutional life of the person being celebrated. The whole ceremony lasted about two hours. Carefully prepared (to meet the standards of a good lesson in physics, the use of an overhead projector and transparencies was obligatory), the narratives were often stimulating for those who tried to build an intellectual history combining the specific "ways of doing" of each individual with the overall development of a discipline. For example, the lectures deliv-

[7] Four are particularly mentioned in 1987: one at Columbia, one in Rome, one at Sussex, and one in Geneva.

[8] CERN did not publish proceedings on Musset.

[9] *Evolution of Particle Physics: A Volume Dedicated to Edoardo Amaldi,* ed. Marcello Conversi (New York: Academic Press, 1970) and *John Adams, 1920–1984* (Geneva: CERN, July 1984).

[10] Sharon Traweek, *Beamtimes and Lifetimes: The World of High Energy Physicists* (Cambridge: Cambridge University Press, 1988).

ered in 1989 for the sixty-fifth birthday of Georges Charpak—one of the best-known specialists in detectors in the 1960s and 1970s and since then a physics Nobel Prize winner—provided elements to reconstruct the logic of his inventions and to evaluate his originality in terms of the history of electronic detecting techniques. As the participants knew each other well—they had regularly collaborated—these meetings constituted rare and precious samples of interesting (self-) history.[11]

What triggered these lectures and made them so popular usually seemed to be an active and militant kind of humor. The ideal lecturer—who would be repeatedly invited—consistently knew how to make his audience laugh; he knew how to fascinate it thanks to an endless string of anecdotes, pseudoconfidences, and puns. For example, Georges Charpak's famed invention being the so-called wire chambers, everybody played (and had to play) on this theme. Carlo Rubbia, then director-general, reported in his introductory remarks that Charpak had confidentially told him before entering the room that his career had been determined by seeing the electric wires driving the tramways when he was a child in Poland—and the audience burst out laughing. Following parallel patterns, the papers were given elaborated titles, like "Fine Wines, Fine Wires: The Charpak Story," or "Of Mice, Wires and Men," with a photomontage on the cover of John Steinbeck's book. Each lecture was performed like a one-man show by one of the community's most famous scientists (the speakers were quite often Nobel Prize winners and outstanding scholars). The lecture exemplified why "doing physics was having fun" and why physics was mainly a family affair driven by the pleasure of knowing and of being together. It provided the community, through the relation of the glorious and fascinating story of one individual's life, the pleasure of attending a family event.[12]

This interest in historical narratives—which was quite widespread, as the high number of participants in the CERN meetings in the 1980s shows—did not call for any professional historians or specialists outside the community. The events we are considering were a private business; a way for the group to exist through the sharing of a memory. As it happened, when the historic 1984 exhibition at CERN was being prepared, or when the establishment of a CERN museum was being considered at the beginning of the 1990s, nobody asked us—the "foreigners" who were then writing the organization's history on the premises—for a contribution. Actually, these meetings' main function seemed less to record precise events than to collectively revive the greater moments of the recent history of physics. By adding new narratives to an already existing large corpus, these meetings helped strengthen the community's identity. They created opportunities for better enculturation of the younger generations, which in turn contributed to maintaining the group's dynamism and confidence. In this respect outsiders (who were not to be trusted anyway) were of very limited interest. The codified structure of the narratives, the exclusive focus on the relationships between individuals and on the struggle of the Scientist with Nature—as well as the favored form of humor—seemed to demonstrate how essential this function was.

The new relationship established between the circle of high-energy physicists and

[11] To me, George Charpak's celebration did not present unusual features compared with others.

[12] This analysis confirms the conclusions of Paul Forman, "Social Niche and Self-image of the American Physicist," in *The Restructuring of Physical Sciences in Europe and the United States, 1945–1960,* eds. Michelangelo De Maria, Mario Grilli, and Fabio Sebastiani (Singapore: World Scientific, 1989), pp. 96–104, quotation on p. 104.

their history is finally made more precise by analyzing the ceremonies held to honor the laboratory itself: for instance, the celebrations of CERN's anniversaries.

The 1964 tenth anniversary ceremony, held in CERN's council room, was quite a short one. Opened by the president of the CERN council, it consisted of just two communications dealing with the great ideological themes cherished by European scientists of the time: namely, "the role of pure science in civilization" (by Cecil F. Powell) and "the future of scientific collaboration in Europe" (by Edoardo Amaldi). Fifteen years later, in 1979 (the twentieth anniversary was not celebrated in 1974, which is significant), the ceremony was barely different, even if it was now held in CERN's great amphitheater for a larger audience. The two reports—whose topics were not innocent in the context of the financial difficulties then faced by high-energy physics—were devoted to what made CERN a prestigious place (by Victor Weisskopf, former director-general) and to the relations between science, technology, and development (by Hendrick Casimir, former managing director of the Netherlands' Philips Laboratories).

By contrast, the thirtieth anniversary made apparent both the importance of the political show and the renewed interest in history. First, the ceremony was staged in a huge laboratory space that could accommodate thousands of people, and the attendees included the king of Spain and several European ministers. Then, one of CERN's pioneers, the American Nobel Prize winner in physics Isidor I. Rabi, related how the organization began. In parallel, a historical exhibition was for the first time set up by CERN, and a one-day seminar on the prehistory of the center was organized by the group of historians. About forty people, including most of the still-living founding members of CERN, took part in this seminar.

In short, the 1980s brought the rise of history—which contrasted with its very discreet presence in the previous three decades—but also the need for glamorous celebrations, for spotlights, for wider social recognition. On the one hand, an ever-growing production of self-history boasting the deeds of the organization was manifest, and on the other—and their parallel occurrence is certainly not due to chance—the need to be recognized politically and by the mass media; the need to build a clear public image. It is interesting to note that this situation was not unique to CERN but, on the contrary, characterized other important European scientific organizations. If we were to study the European Space Agency (ESA), for example, history and political shows would appear simultaneously (and quite suddenly) in the 1980s too. The tenth-anniversary ceremonies of ELDO (European Launcher Development Organization—the ancestor of ESA) in 1974 were quite short, while the 1984 and 1989 celebrations were prestigious, presided over by Queen Beatrix of the Netherlands in 1984 and by Prime Minister Helmut Kohl and President François Mitterand in 1989. In the same way, historical articles about ELDO/ESA became more and more numerous in the agency's monthly journal as of 1984, whereas none had appeared before. If we add that since 1989 ESA has engaged a group of historians to write the history of Europeans in space, the picture seems quite complete.[13]

Let us stop briefly and summarize the first two sections. At CERN, opportunities

[13] This section relies on the study of *ESA Bulletin,* a periodical similar to *Courrier CERN.* Two historical articles were published in 1974 for the tenth anniversary, and the next was not published until 1984. Between 1984 and 1989, various articles were published on ESA programs (one in 1984, one in 1986, and one in 1989) and ESA stations (two in 1987 and two in 1989), etc.

to produce historical narratives multiplied in the 1980s, and self-history gained new ground. Obviously, this new interest among particle physicists—notably those working in Geneva—was manifest both in the organization of private meetings for the community and in the presentation of ceremonies to publicly and politically affirm CERN's excellence. Regarding the historical narratives, they seemed partly to originate from a need to cement the community, and partly to enable each of its members to participate in the life (and honors) of a prestigious, if increasingly anonymous, science.

As regards the highly publicized political shows built around historical narratives, it has to be remembered that CERN and ESA are large organizations that rely totally on public money for their funding, and whose financial needs keep soaring. A good public image and a positive appraisal of their activities are of the utmost importance, especially when social or economic tensions arise. Hence the importance of well-organized public relations (which have been in place at CERN since the beginning of the organization) and of history—since the function of history is to legitimate, and since the mere process of telling the history of something makes it exist and gives it weight, and since an historical narrative (whatever it says!) signals something worth keeping for posterity.

Finally, the physicists' demand for history—whether created by the community itself or by commissioned academics, as in the project set up to write the history of CERN—can be linked to the mutations this group underwent during the last fifty years. The relationship between physicists and the state dramatically changed with the Second World War, as we all know. Before 1939, what physicists did was of moderate interest to politicians and its political fallout was limited. With the development of the atomic bomb and radar during the Second World War, of electronics and large computing systems during the cold war, and then of rockets and other ballistic weapons, fundamental physics became deeply set at the heart of the political (and imaginary) vision of Western societies.

In the 1950s and 1960s, the number of physicists was booming, and the community enjoyed extraordinary prestige. It was the heyday of the nuclear sciences. Still in its prime, the group took an interest in its prestigious elders and in the prewar period—in Einstein, of course, in Niels Bohr, in quantum theory. Twenty years later its main spokesmen were between sixty and eighty years old; they had become the new tutelary heroes of the discipline. It was their turn to retire—but in a difficult context, from a discipline whose excessive ambitions were now regularly challenged. This is perhaps why they, and others, were led to think that the time had come for their own history to be recorded—a history so evidently crucial for any understanding of the contemporary world.[14]

An important change in physicists' self-image had developed in parallel with their rise in number, their increased specialization, and their closer relationship with the world of business and war. Unlike many of their predecessors, such as Henri Poincaré, Max Planck, or Einstein, renowned postwar physicists rarely became "savant-philosophers" writing on the nature of science or on its human values. Aiming at

[14] In his autobiography, Emilio Segrè says that he wrote his book "because I thought it might interest a public curious about the science-dominated period in which I lived." See Segrè, *A Mind Always in Motion: The Autobiography of Emilio Segrè* (Berkeley: University of California Press, 1993), p. xi.

efficiency and control over phenomena, pragmatic in their objectives and approaches, adopting any methodological solution provided it proved useful, they often became opportunistic technicians "who never worried about the logic of what they were doing." By way of testament, they no longer disclosed their philosophical reflections and they adopted an "attitude of detachment," which Paul Forman summarizes through one of their most common expressions: "physics is fun." They ended flattering autobiographies, tinted with hedonism, with historical narratives that never mentioned their connections with the military but told of the intellectual successes of the famous institutions they had contributed to establishing.[15]

## III. PHYSICISTS AT CERN AND HISTORIANS' NARRATIVES

Now, I would like to comment on how physicists reacted to the work and presence of the historians retained from 1983 to 1989 to write the history of CERN. First, it is worth noticing that, despite the regular release of intermediary studies and our daily presence on the site, few people showed an interest in us and in our work. We were particularly ignored by the younger physicists and engineers (say, those under forty)—and they mostly continue to do so. There is no denying that the first volume of the *History of CERN* sold rather well in Geneva but, unlike the previously mentioned commemorations, our seminars only attracted "historic" characters. And even among those people, we were to be dealt with warily. This could be readily understood. First, as we were not (or were no longer) professional physicists, we were strangers, people deprived of the necessary expertise and authority to handle the study correctly. For the people at CERN, as the essential task was to report on the part played by the laboratory in the development and achievements of high-energy physics and in acceleration and detection techniques, it was obvious that we could not do as good a job as any of them. In short, we were incompetent as far as technical matters were concerned.[16]

Most CERN physicists and engineers also thought that the prime role of the historian-stranger was to record the actors' memories—which alone could retain the truth of what had happened. Convinced that we would be led astray by the archives, CERN's scientists insisted on the importance of the interviews they were ready to grant us. Listening to what they said would constitute an essential means to prevent us from being deceived by documents that had often been "arranged" for political reasons. Finally, they feared that the written documentation would not convey the sense of adventure that had pervaded CERN; it would not allow us to find the pioneering spirit that had been at the root of the organization's strength and success.[17]

[15] Gerald Holton, "Les hommes de science ont-ils besoin d'une philosophie?" *Le Débat,* 1985, *35:*116–38, was the first to express this idea. It seems to be particularly true in the American case for reasons well explained by Sylvan S. Schweber, "The Empiricist Temper Regnant: Theoretical Physics in the United States, 1920–1950," *Historical Studies in the Physical and Biological Sciences,* 1986, *17:*55–98. The quotations in the text are from Holton and from Forman, "Social Niche," (cit. n. 12).

[16] Herman *et. al., History of CERN* (cit. n. 6).

[17] On oral sources and oral history, see Dominique Pestre, "En guise d'introduction: Quelques commentaires sur les 'temoignages oraux,'" *Cahiers Pour L'histoire du CNRS,* 1989, *2:*9–12, and Soraya de Chadarevian, "Using Interviews to Write the History of Science," in *The Historiography of Contemporary Science and Technology,* ed. Thomas Soderqvist (Amsterdam: Harwood, 1997), pp. 51–70.

Thus, as does any group sharing a set of strong values, CERN's physicists and engineers knew what mattered and what did not; they knew what was important to be recorded and written. They expected us to assimilate (and legitimate) the history they had always told themselves, and they questioned our ability to grasp the gist of the subject. Our first discussion meetings with them often lacked serenity and we had to wait until our first results were published for our relationship to become less unbalanced. Then, fruitful exchanges took place with the two or three dozen personalities who were interested in our work.[18]

As a whole, then, the European scientific community did not show a great interest in our work, even though some dozen people read and commented on it. How they reacted to it is telling. I will start with an anecdote. When we arrived at CERN, there was a standard narrative describing the genesis of the organization. It identified a string of rather individual actions and highlighted the part played by several great scientists, such as Louis de Broglie, Isidor I. Rabi, and the father of quantum physics, Niels Bohr. De Broglie initiated the center's founding in December 1949, when he sent a message to the European Movement Conference in Lausanne advocating the pooling of forces as a solution to the weaknesses of science in Europe. Rabi was celebrated for taking over the idea and including it in a UNESCO resolution that bore fruit in 1951. About Niels Bohr, people said (I quote Victor Weisskopf) that it was "his personality, his weight, his action which had rendered CERN possible," and that "if such a well-known man had not supported the project and . . . had not involved himself in each significant act of the foundation period, the enthusiasm and good ideas of the others would never have been sufficient."[19]

It was not long before we felt that we could not but drastically challenge these assertions. De Broglie's message had been sent, no doubt, but it had resulted from an explicit request made by Raoul Dautry, a former French minister, high commissioner of the Atomic Energy Commission (CEA), president of the European Movement in France, and that country's main force acting in favor of creating a European nuclear science center in the late 1940s. It was within the framework of his maneuvers within the CEA and the Quai d'Orsay that Dautry devised the idea that Louis de Broglie send a plea. As for Rabi, his initiative was revealed to be not only that of a disinterested scientist willing to help his fellow physicists in Europe, but it was also (and perhaps essentially) connected with a change in the United States' international scientific policy, motivated by the development of the cold war. Actually, the text Rabi proposed and got turned into a UNESCO resolution had been derived from an official American document released two months earlier. It was rather vague, and had UNESCO's Pierre Auger not seized upon it and acted at the very heart of UNESCO in the next eighteen months, nothing would have come of it. Lastly, in Bohr's case, his correspondence unambiguously showed that, far from being a supporter of the project, he was one of its main opponents. Deeply convinced that Europe's strength did not lie in technological prowess and that there was no point in trying to compete with the Americans in this field, he did all he could to stop a

[18] As an example, see Amaldi's reaction to the first volume of *History of CERN* in *CERN Courrier,* 1987, *4:*31.

[19] Victor Weisskopf's sentences were first written in the 1960s. Van Hove repeated them in 1985 for Bohr's one hundredth birthday party at CERN. They are reproduced in *Courrier CERN,* 1985, *10:*431.

project that he judged useless, ill conceived, and incompatible with Europe's financial capabilities.

Reactions to our new reading were varied. One of the first to respond, in a way that surprised us, was Edoardo Amaldi, the Italian cosmic ray specialist and one of this period's main activists in favor of CERN. As he could not believe what we had said of Bohr, he wanted to check our sources. He came to our offices, looked at our documents, and left puzzled at having forgotten such an opposition (or perhaps at having had just a partial knowledge of it).

Having expressed serious doubts and reservations about our findings, CERN's physicists then tended to minimize them, to play them down. Our writings overemphasized certain points, they argued—had not Auger himself claimed to agree with Bohr's support as early as 1950, for example? And we were sometimes biased or even malicious: de Broglie had effectively written the 1949 letter—that was undisputable, and Auger himself had recognized in a telegram to Rabi in 1952 that the latter had "fathered" the CERN project. There was then a reluctance to give up a version that had been true for so long, which made so much more sense since it had been given by the actors themselves and was in the "logic of things." As the capability to discover and to invent before anybody else is what characterizes great scientists—and since Bohr, Rabi, and de Broglie were great scientists—it was just plain "natural" that they should also invent CERN. And as for Bohr, the greatest of them all, it was quite impossible to figure out what could have led him to oppose the creation of something that was so obviously excellent.[20]

Our articles often triggered a cycle of such reactions. Suspicion dominated when the physicists were confronted with texts that challenged the current narratives. We were reproached for insisting on what separates people, stressing the points they disagreed about, and drawing attention to behaviors that, although sometimes deprived of nobility, were quite common and not worth reporting. We were said to have had an overly reductionist and narrow approach—which prevented us from seeing the positive trends along which knowledge expanded—and to have been more interested in sociology than in science. In the second phase of their reactions, habit (and perhaps the weight of certain historical proofs) being at work, our descriptions had lost some of their strangeness. More easily accepted, they tended to be rereported in a casual, banal way. Reduced to widely shared and received views, they were regarded as being obvious and not very surprising—and consequently dull and of little interest.

This analysis can be extended a little. As mentioned at the beginning of this article, the history generated by the scientific community is often one that tends to celebrate men and events and regard its object—science—as being independent of any local (cultural or social) determination. On the wings of the universal spirit, it tends to reject, as Joseph Priestley wrote in 1769, "what is the most tedious and disgusting in both [civil and natural history]." In this exercise of idealization, whose aim is transcendence, any narrative tends to be the illustration of what the commu-

[20] That version articulating the names of de Broglie, Rabi, and Bohr was definitely fixed as early as 1961 by Lew Kowarski, *An Account of the Origin and Beginnings of CERN* (Geneva: CERN 61-10, 1961). In the appendices, Kowarski only reproduced two texts: de Broglie's message and the UNESCO resolution proposed by Rabi.

nity considers the essence of things—what makes a true discovery, for example, or what makes the scientific method so efficient. This could probably help us understand why scientists' historiography often ignores what historians write, why it does not worry too much when it does not stick fully to historical "details," and why it is ready to accept what historians consider approximate narratives—provided they display a higher and more essential reality.[21]

An anecdote can help here. In 1985, CERN celebrated the one hundredth anniversary of Bohr's birthday. A renowned European theoretician, Léon Van Hove, was in charge of the talk entitled "Bohr and the Beginning of CERN." As he had read our texts attentively, he knew them perfectly well, and he began his lecture by mentioning our names to his audience. For once, he warned them, they would not be offered the usual, linear history of the beginnings of CERN that they had already heard several times, and he started telling them in detail a story directly based on our reports. He may have slightly changed some elements and disregarded our more political or sociological readings of the events, but he recognized the historical value of our work and accepted our "discoveries." But he did not quite draw the same conclusion as we had. Instead, he chose to end his report with the Weisskopf quote that I gave here earlier, which described Bohr as the man without whom nothing would have existed. For Van Hove, no doubt, we had become trustworthy people whose serious, documented work was to be respected and used. But this only implied one thing—that we could be trusted for chronological precision and factual details. But the heart of things, the true logic of events, a deep understanding of men and history—in short, what actually counts in life—were matters too complex and too serious to be left to historians. In short, history has a bedrock that could only be perceived from *the inside*—and historians' endless talk just remains at the surface of things.[22]

### IV. THE HISTORIANS' TRADE, LOCAL NARRATIVES, AND POWER RELATIONS

My account thus far is based on a real experience, that rather ordinary experience of a stranger who arrives in a place to study the natives and discovers people who do not exactly speak his or her language, who organize their narratives according to other rules, and who listen to the stranger only when it pleases and makes sense for them. (In the case of CERN, moreover, the natives were quite powerful and could bypass the stranger who pretended to speak about them.) The question then arises of what to do with the native narratives when they contradict those of the stranger/historian—which is what happened with Van Hove's and our texts. The first point, in my view, is to remember that all narratives are always spatially, socially, and temporally situated; that they are embedded in anthropological and intellectual ways of doing and thinking; and that they are elaborated in specific contexts and for given audiences (physicists in CERN's amphitheater, for example, or historians reading *Osiris* at home). They are geared to questions that matter today to the author (as professional, citizen, or intellectual); they assume beforehand what has to be

[21] This paragraph is directly inspired by Paul Forman, "Independence, Not Transcendence, for the Historian of Science," *Isis,* 1991, *82:*71–86. Priestley's expression is quoted by Forman on p. 84.

[22] That is also true for the exhibition set up at CERN in 1984, which presented documents inspired by the "standard history" but offered a text in the catalogue that was factually built on our narratives.

explained and what would count as an explanation; and they crucially depend on choices regarding what is pertinent in social actions.[23]

This statement might generate, if taken seriously, a certain uneasiness. If it is true that there are many possible narratives, and that each has its own legitimacy, then there is no longer a referent—Weisskopf is no more wrong or right than we are—and this is the end of any claim to demanding and serious thought. This disquiet is often grounded on the strongly affective conviction that a nonsituated Truth must exist and that it has to be the reference point to calibrate all claims. Without it, without a solid mooring post, mankind would soon fall back on savagery. Quite frankly, I do not share this exaggerated concern, nor the alternative that is proposed to us. My quite banal answer is that we have to accept, in the same move, two propositions. First, that any knowledge is produced in a given situation for a public able to understand it—this is the human condition. Second, that all knowledge claims are not equivalent—not everything goes, and differences matter enormously. What has to be abandoned in order to keep both statements is the conviction that rules exist that could *guarantee* us access to certainty or to what is right—rules that could make us *transcend* our human condition, become equal to God, and know the Truth (with a capital *T*).

If we admit, on the contrary, that looking for epistemological criteria of a universal kind is a dream (or an ideal), we must then learn to live with a principle of reality that is no more frightening than any other, once recognized: the fact that the social is forever-already in the realm of what we hold to be true or good. There is no use postulating, even in principle (but are we talking about anything else besides a matter of principle?) the existence of a "pure" knowledge (meaning unaffected by the social). This recognition, however, does not mean that life is without norms and rules, without ends and values. It has the ends, values, norms, and rules *we give and want for it.* Accepting the limits imposed on us by our anthropological and social situation does not lead, in logical as well as moral terms, to a relativistic stand. After all, many of us believe that moral and cognitive rules have always been human-made. Let us then be quiet and explicitly discuss them; let us openly defend and test them. The situation then appears less cut-and-dried than before. Knowledge and ethical claims are certainly local and determined—is not that our destiny?—but that does not mean they are devoid of pertinence or are simply relative. In fact, all relies on the norms and values we choose and fight for. If they are good or valuable—but there is, of course, no universal meaning to these words, and we have to defend our choices up to the end—the claims are likely to state something to be kept. Let us go back, after this theoretical detour, to the CERN physicists' narratives and ours and ask a simple question. What were their and our agendas concerning the history of CERN? What was essential to be remembered for them and us; what was at stake?

On the physicists' side, in the 1980s there was a sense of great progress, the certainty that fundamental knowledge had been fantastically advanced—a key value for mankind. And what was important to catch hold of when writing history was precisely this move towards an ideal, this fight towards transcendence, this achieve-

[23] Three exemplary books on these questions are Veyne, *Comment on écrit l'histoire* (cit. n. 4); Donna Haraway, *Primate Visions: Gender, Race, and Nature in the World of Modern Science* (New York: Routledge, 1989); and Jacques Rancière, *Les noms de l'histoire: Essai de poétique du savoir* (Paris: Seuil, 1992).

ment. On our side (that of young historians trained in the late 1960s/early 1970s), we were less interested in that story and wanted to know more about the way social and political questions were intertwined with scientific ones, or the role played by contingency in history. These attitudes refer to different situations. CERN physicists were caught up in a tension we evoked, and from which they needed to escape: they were now part of a technocratic world tied to high politics (the nature of their celebrations demonstrates that they were not unaware of this) and they enjoyed the prestige that flowed from it. At the same time, they still wanted to perceive themselves as the descendents of the moral savants of the first half of the century. Hence their stress on pure intellectual achievements, their preference for nobility, and their aversion to our institutional, political, and mundane stories—and their insistence on Bohr as tutelary hero of CERN.

For our part, we were not without political and professional agendas and our situation at CERN constrained us. Inheritors of the legacy of May 1968, we wanted to affirm—without being clearly aware of it—that the large scientific and technological systems established by Western societies during the cold war had a propensity to act for themselves, to view themselves as being above or apart from the rest of the social order, and to impose their own interests in the name of the public good or of science. We thought that these systems (of which CERN was only a very modest representative) should be subject to certain forms of control and be socially accountable. Against a dominant trend, we also wanted to show that science could be first studied as a social institution. From whence came our interest in processes of decision making, and our studies of the practical organization of experimental work and of how high-energy physics maintained itself as socially decisive and indispensable up to the 1980s. From whence, naturally, came questions different from those of our physicist friends, different intellectual interests and political preoccupations and, finally, very different narratives.

I would even go a little further and make the process of interaction with CERN physicists in Geneva from 1982 to 1988 an element of this story. Present at CERN for six years, we were caught in a logic of debate with our actors, a logic which was beyond us and which constantly remodeled us (as it remodeled them and the image they had of themselves). We had to respond to other narratives, as I have said, but our narratives provoked, in their turn, refusals or approvals. Our story, with its insistences and its blind spots, was constructed throughout this process, in a continuous reaction with a local audience that was reduced in number, to be sure, but active and interested all the same. Through that interaction, our historiographic and political demands were sharpened; our text took to emphasizing certain points and ignoring others, and our identity, questions, and certainties were regularly reshaped in that shared game between them and us.

But what about epistemological questions? Were there not also differences at that level between them and us, differences that would bring back some decisive asymmetry? Was there not, besides the choice of topics and frames of reference, some more systematic treatment on our side, some more serious attempts to consider every dimension, to confront interpretations and sources—in short, a better-grounded "scientific" approach? My answer would provisionally be yes—in the sense that we were ready to spend days and nights running after documents, which was not the case for CERN physicists; that we were ready to systematically reconsider interpretations; that we had a sense for the "situatedness" of every text or author, etc. In the sense, in short, that I firmly believe in some professional canons. Differences existed at the level of the *métier,* in other words, and I think they mattered. I said my answer is provisional, however, because that situation could be reversed and some CERN physicists, having enough free time and dedication, could answer the objection and become, technically speaking, professional historians.[24]

But even that is not enough, I fear. As outsider-historians, as people not sharing the culture of the group, I believe that "something" could not be reconciled with the memory of the actors, with their "intimate convictions and feelings." The way everybody makes sense of his or her own life cannot easily be recaptured, and it cannot just be dismissed as inadequate with historians' arguments. What is conveyed by Weisskopf's texts or Van Hove's remarks is more and other than factual information—it is an experience, the certainty of that which has been subjectively lived; it is a meaning given to life. The analogy with the narratives of those who went through the Holocaust is misplaced, but it indicates a direction of thought: in a sense, any historian's narrative cannot but be unbearable to actors, since it objectifies a past and puts it at too far a distance. It tries to impose a cold order on the hot memory of those who prefer to keep their shared souvenirs in all subjectivity, and it precisely adopts the (strange) posture of science to talk about things. The actors we talk about as historians are not all without defense (that is particularly true for high-energy physicists), but the relation between scholarly history and a group's narrative cannot but remain asymmetrical. Professional historians or anthropologists can irreparably undermine collective memory, and even if the consequences are not tragic, it is a responsibility they just cannot ignore.

[24] I am not unaware of the complexity of the question—even if I chose here to mark too clear a difference between "our" texts and those proposed by CERN physicists. Standard references on these questions include Michel de Certeau, *L'écriture de l'histoire* (Paris: Gallimard, 1975); Pierre Barberis, *Le prince et le marchand: Idéologiques: La litterature et l'histoire* (Paris: Fayard, 1980); Paul Ricoeur, *Temps et récit,* 3 vols. (Paris: Seuil, 1983–1985), and Rancière, *Les noms de l'histoire* (cit. n. 23).

# *Part III. Commemorating Scientific Disciplines: Memorializing Objectivity*

# Bourbaki's Art of Memory

*By Liliane Beaulieu**

## I. INTRODUCTION

SINCE THE MID-NINETEENTH CENTURY COMMEMORATIONS have proliferated, and lavish celebrations of the past mark the advent of the third millennium. Commemorations preoccupy contemporary researchers into memory, who have emphasized the past as recollected, represented, institutionalized. Yet societies do not commemorate if they no longer feel the presence of the past, or if they cannot define their future from the legacy of old lessons. How groups remember also depends on how they habitually record their present. Within the orchestral swell of our *fin-de-siècle* commemorations the Bourbaki group sings small, leaving its past as yet unexamined.[1]

"N(icolas) Bourbaki" is the pseudonym of a twentieth-century group of mathematicians, mainly French, who publish a work entitled *Eléments de mathématique* (*Elements of Mathematic*), an overview of several areas of mathematics that demonstrates the structures they have in common.[2] Although an expository work, the *Eléments de mathématique* nevertheless constitutes an original piece of craftsmanship. The writers innovated by choosing particular conceptual frames into which they cast whole mathematical theories; they also strove to give their presentation a clear but strict organic unity, and they achieved a particularly homogenous exposition by adopting the same approach throughout the treatise and by emphasizing the ties between different theories. In this way they present set theory, algebra, general topology, functions of a real variable, topological vector spaces, integration theory,

*Centre de Recherches Mathématiques, Université de Montréal, C.P. 6128 Succursale Centre-Ville, Montréal, Québec H3C 3J7 Canada. I wish to thank Jill Corner, Chandler Davis, Ivor Grattan-Guinness, Alexei Kojevnikov, Herbert Mehrtens, Lewis Pyenson, Erhard Scholz, Skuli Sigurdsson, and George Weisz for their invaluable comments. Pnina Abir-Am convinced me to explore more systematically the memory aspects of the Bourbaki group; her persistence must not go unacknowledged. I am also deeply indebted to the Bourbakis and their family or friends who provided access to their personal papers and allowed me to use the material liberally. It is a pleasure to thank Henri and Nicole Cartan, Pierre Cartier, Catherine Chevalley, Pierre Dugac, Micheline Guillemin-Delsarte, and Pierre Samuel.

[1] All of the Bourbaki manuscripts used for this article come from the personal papers of former members of the group whose names appear in each reference. Many of the quotations were as yet unpublished (in any language). All translations from the French are my own with the exception of the poems, which were translated by Jill Corner; kindly acknowledge when citing. The great number of quotations precludes proper referencing for each one of them; I trust the reader will nonetheless accept that they are genuine.

[2] Bourbaki deliberately uses the unusual singular rather than the common plural in the French word "*mathématique(s)*" in order to demonstrate a faith in the unity of mathematics. The French title *Eléments de mathématique* will be used here since only the volumes of the original French version will be discussed, rather than their later English translations.

 0369-7827/99/1401-0013$02.00

Lie groups and Lie algebras, commutative algebra, spectral theories, and differential and analytic manifolds. The work also contains a number of historical notes collected as a separate volume with the title *Eléments d'histoire des mathématiques.* For mathematicians, Bourbaki represents "pure" mathematics par excellence and its work is the paradigm of a rather dry mathematical "style," combining axiomatics with a meticulously abstract method of presentation. The Bourbaki group is a unique phenomenon in the recent history of mathematics: it has inspired legends and controversies, and it cultivated mystery even while achieving high visibility and its treatise was becoming a publishing success.

The team came together in the academic year 1934–1935 and continues to operate to this day (1999). The *Eléments de mathématique* were first published as a series of booklets, starting with a digest of results on set theory in 1939, reaching a publication peak in the 1950s, and persisting with a recently published chapter on commutative algebra. In 1948 the group established a seminar on cutting-edge mathematics, known as the Séminaire Bourbaki, which soon became the most notable mathematics seminar in the Western world, a position as yet unchallenged. Although the members of Bourbaki select the topics as well as the participants and edit the talks for publication, the activities of their public Paris seminar remain quite separate from the private writing of the *Eléments.* Indeed, this enterprise is a real and unusual collective effort: the team co-opts its members, keeps its membership secret, and does not acknowledge individual contributions.

The way Bourbaki has organized its behind-the-scene work sessions has varied over the years, but usually, especially since World War II, they had three to four general meetings a year at which members submitted contributions to the critical eye of their peers. Thus, each part of the *Eléments* went through many successive versions, being first written by individuals and then read, discussed, and reworked by the whole team before being handed over to a new writer who would more or less incorporate the group's recommendations into the revised text, and so on, with up to ten or more revisions occurring in some cases. These ongoing amendments have blurred the record, making it difficult to find traces of particular authors in the final palimpsests.

No application is required to become a member of Bourbaki. Recruits are coopted at the suggestion of a member and he (no women have been in the Bourbaki group) has to take an active part in the discussions; thus the recruits choose Bourbaki as much as the group chooses them. The draftees are either known mathematicians already advanced in their careers—some of whom are not French—or students of Bourbaki members (called "guinea pigs" in Bourbaki parlance), mostly young men from the Ecole Normale Supérieure on rue d'Ulm in Paris, one of the leading schools in France; the latter eventually constituted the majority.[3] At any point in time there are, on average, ten active members, and the membership forms a close-knit, inbred company with shared social and intellectual roots as well as mathematical tastes.[4]

[3] On internal cohesiveness in the elite of leading schools and its role in French intellectual life, see Pierre Bourdieu, *La Noblesse d'Etat: Grandes ecoles et esprit de corps* (Paris: Editions de Minuit, 1989).

[4] The story of the Bourbaki group, based on archival material, is told in Liliane Beaulieu, *Bourbaki: History and Legend* (Springer-Verlag, forthcoming). The complete list of Bourbaki members between 1934 and 1960 is featured as an appendix to this book.

Every group finds pretexts for being festive, and Bourbaki is no exception. Weddings, births, anniversaries, graduations, and the completion of theses or of particularly difficult chapters of the *Eléments* have been hailed in Bourbaki's intimate celebrations, which usually consisted of nothing more than partaking, while at a meeting, of a fine meal washed down with good wines. The arrival of new recruits and their eventual integration with the group were mentioned in Bourbaki's internal newsletter, but not celebrated, just as members' retirements and other departures passed by without pomp or ceremony. Anniversaries—of Bourbaki's so-called foundational meeting (1935), of its first published fascicle (1939), or of the creation of the high-profile Séminaire Bourbaki (1948)—spurred neither public nor private recollection rituals. Unlike most other groups, the Bourbaki team and their followers have not indulged in public celebrations. A vow of secrecy—not a serious ritual—allegedly forbids them to disclose the names of members, but this is not the only reason for their public shyness. For Bourbaki is not just a private coterie: the collective author of the monumental *Eléments* and organizer of a famed seminar has received peer recognition in reviews and even a prize.

Until recently, the only public celebration of Bourbaki was when the French Académie des Sciences awarded its Cognacq-Jay Prize to four of the nine founding members (Henri Cartan, Jean Delsarte, Jean Dieudonné and André Weil) for Bourbaki's work in general. For this occasion, a brief and modest ceremony was organized on 28 February 1967 in Nancy, where Bourbaki had its headquarters for many years. In their acceptance speeches, Delsarte and Dieudonné recalled Bourbaki's past and extolled the group's perennial youthfulness. They compared the impressive quantity of Bourbaki's publications (at the time, some thirty fascicles of the *Eléments de mathématique* totalling about five thousand pages) with the five to six times larger quantity of ephemeral texts put out by the members as trial drafts for the treatise, thus testifying to Bourbaki's hard work. They also spoke about members' anonymity, the ferocity of group discussions, and some of the personal quirks of the founders. Despite the mandatory comic elements, the ceremony was sober in overall tone, evoking the group's past and signal achievements.[5] In those days, Bourbaki's supremacy was unchallenged: the team, and its mathematical choices, dominated the French mathematical scene and were influential in many other parts of the world. Yet individual members were celebrated more than the Bourbaki institution: most members received several prizes for their personal contributions to mathematics, and some even earned the Fields Medal, the most prestigious international prize for mathematics. In contrast, even on the occasion of the Académie prize, the group as such did not revel in celebratory rituals.

Another public accolade came in the fall of 1994 when the Ecole Normale Supérieure marked the bicentennial of its foundation. Bourbaki decided to let its name be associated with the commemoration of the institution that is closest to its collective heart, and where it now has its headquarters and secretariat. Like the bicentennials of the French Revolution and of the Ecole Polytechnique that preceded it (in

[5] The award was sponsored by the Samaritaine Foundation. The announcement of the award was published in the *Comptes Rendus de l'Académie des Sciences de Paris,* vol. 263, "Table académique," p. 146. Details on the circumstances of this event appear in Beaulieu, *Bourbaki* (cit. n. 4), chap. 6, where I also explain why the prize was given to only four of the founders.

1989 and 1994 respectively), the Ecole Normale Supérieure's celebrations were enriched by historical studies.[6] In commemorating its inception, the Ecole Normale was particularly celebrating its second—and fundamentally republican, lay, and leftist—century, which consecrated its students (the *normaliens*) as "tribunes of new ideas" (a phrase borrowed from Sainte-Beuve and repeated in many historical studies) who dominated French intellectual and political life.[7] The ideas of the *normaliens,* it was held, represented the various currents of opinion in France and the sociological makeup of their cohorts reflected the composition of the whole French population.[8] Mathematics at the Ecole, and Bourbaki's role in particular, found a place in the context of these historical analyses.[9]

Yet on visitors' days when the august establishment threw open its doors to a curious public, Bourbaki's presentation consisted merely of a discreet showcase displaying a few old photographs from Bourbaki congresses and samples of unpublished texts. Bourbaki's devoted secretary had taken charge of the exhibition and asked some former members to lend their personal Bourbaki-related photographs for the occasion. This silent commemoration, however, did not galvanize latter-day members nor even attract older adherents. Nor did the group find many followers to ceremonially celebrate the name of Bourbaki or to obtain any worthwhile result from the occasion. Contrary to usual practice at a commemoration, the Bourbaki group did not try to attract attention in order to gain recognition or inspire eventual apprentices. Nor did anyone strive to demonstrate how topical the Bourbaki enterprise might still be.

As an aside to the celebrations, a colloquium on "The French School of Mathematics" between 1940 and 1970 set out to give a critical account of Bourbaki's role during that period.[10] The arguments raised by the talks and round-table discussions brought back famous battles of the past; there, of course, the memories of some former Bourbaki members as well as of outsiders, both defenders and detractors, were jogged into reminiscences. Indeed, Bourbaki is now a part of the common memory of mathematicians, especially in France where the group dominated the mathematical scene for several decades. This colloquium was not a commemoration *per se,* yet it may be considered as commemorative in nature—though not in pur-

[6] On the occasion of the bicentennial of the Ecole Polytechnique, historical studies giving the pride of place to mathematics were collected in *La Formation polytechnicienne 1794–1994,* eds. Bruno Belhoste, Amy Dahan, and Alain Picon (Paris: Dunod, 1994). On the centennial celebrations of the Ecole Polytechnique see the study by Anousheh Karvar, "Le Centenaire de l'Ecole Polytechnique: Rites d'une élite nationale sur fond de crise," *La Mise en mémoire de la science: Pour une ethnographie historique des rites commémoratifs,* ed. Pnina Abir-Am (Paris: Editions des Archives Contemporaines, 1998), pp. 191–206.

In addition to a reprint of the commemorative volume of the Ecole Normale Supérieure's first centennial, a series of articles collected under the title *Ecole Normale Supérieure: Le Livre du bicentenaire,* ed. Jean-François Sirinelli (Paris: Presses Universitaires de France, 1994) aims mainly to present a global history of the school's last one hundred years.

[7] "Republican," here, refers to the ideal of a republic in general and especially to the ideals of the French Third Republic.

[8] Jean-François Sirinelli challenges this standard image of the Ecole Normale as a mirror of French society and opinion in "Les Élites culturelles," *Pour une Histoire culturelle,* eds. Jean-Pierre Rioux and Jean-François Sirinelli (Paris: Seuil, 1997), pp. 275–96.

[9] In particular see the article by Martin Andler, "Les mathématiques à l'Ecole Normale Supérieure au XXe siècle: Une esquisse," in Sirinelli, *ENS: Le Livre du bicentenaire* (cit. n. 6), pp. 351–404.

[10] The colloquium was organized by Amy Dahan and Martin Andler, who are neither members nor followers of Bourbaki.

pose—inasmuch as it exhibited residual emotional ties between the Bourbaki group, which renewed French mathematics, and other mathematicians who defended various stakes in the mathematical science in France, one of the issues being the respective places of "applied" and "pure" mathematics. For many years after World War II, pure mathematics was triumphant and, in those days, pure mathematics in France meant Bourbakian mathematics; that is, the kind of mathematics that the *Eléments* exhibited and the Séminaire Bourbaki once fostered. In France, as elsewhere in the Western world, applied mathematics has for some time now won over pure mathematics to the point that the labels "pure" and "applied" with their opposition fading, have lost some of their purpose and meaning. Even the trend-setting Séminaire Bourbaki has followed suit by welcoming the presentation of algorithmic methods and other applications-related topics. At last, there is a relative truce between Bourbaki's defendants and opponents.[11]

Although Bourbaki has earned a place in the picture of French and *normalien* intellectual life, the group is presently fading into oblivion as the witnesses of its inception and heyday depart. For some time now the name of Bourbaki—in reference to a group of mathematicians—has not meant much to most educated Frenchmen. Even mathematicians are no longer informed; some wonder whether the group still exists while others are ignorant of the fact that no single author of that name ever lived.[12] In the pantheon of mathematics no lemma, theorem, method, or system of axioms bears the name of Bourbaki, and the monumental *Eléments de mathématique* are kept on library reference shelves and are sometimes useful, sometimes deemed rather out of date. The team may be considering an alternative vocation but, until that new order arrives, its task—the organization of the seminar notwithstanding—is basically to complete an old project. In such a context, had the Bourbaki group tried to stimulate public debate about its role; had it sought in any way to make a case for itself, it might have risked being considered out of place. A commemoration requires a particular blend of circumstances and stakes in order to be staged and prove beneficial to the congregation that carries it through, while another mix of conditions might have adverse effects.[13]

Any commemoration inevitably comprises a polemical, even combative, aspect,

[11] Such a truce has been in place for some time at France's famous Institut des Hautes Etudes Scientifiques, as David Aubin suggests in "Un Pacte singulier entre mathématiques et industrie: L'enfance chaotique de l'Institut des Hautes Etudes Scientifiques," *La Recherche,* October 1998, *313:*98–103. An indicator of how pure mathematics in France once corresponded to Bourbakian mathematics can be found in Jean Dieudonné, *Panorama des mathématiques pures: Le Choix bourbachique* (Paris: Gauthier-Villars, 1977). Martin Andler describes the supremacy of pure mathematics in France in "Les mathématiques à l'Ecole Normale Supérieure" (cit. n. 9). Elements for a history of the labels "pure" and "applied" mathematics in France and elsewhere are found in Amy Dahan, "Pur versus appliqué? Un point de vue d'historien sur une 'guerre d'images,'" *Gazette des Mathématiciens,* (Société Mathématique de France), 1999, *80:*31–46.

[12] However, the recent publication of articles on Bourbaki (see cit. n. 14) has rekindled the curiosity of mathematicians and historians of mathematics for whom Bourbaki is part of a collective past. It is also safe to conjecture that a new generation of mathematicians may have been enlightened about the singular group through the reminiscences of their teachers who, themselves, were students during the heyday of Bourbaki.

[13] In his case study of the great French Bellevue magnet and the singular subcommunity of scientists that clustered about the giant instrument, Terry Shinn has shown that, although demarcation and differentiation—usually gained from public commemorations—are often necessary to strengthen authority and recognition, they may interfere instead with the fulfillment of this goal, while discretion and omnipresence might succeed. See his "L'Effet pervers des commémorations en science," in Abir-Am, *La Mise en mémoire* (cit. n. 6), pp. 225–47.

since its main functions are to honor past or present protagonists, inveigh against opponents, and edify newcomers. What does it become, this act of commemoration, if there is no one to claim the status of heir and if contemporary issues leave little room for Bourbaki in either hagiography or condemnation? A commemoration would also be somewhat superfluous for a restricted group that, although self-aware in many ways, has not spent much time mulling over what its own history ought to be. So far, the group has neither written down its history nor chosen an official biographer from its ranks; only fragments of the past are featured in interviews, autobiographies, obituaries of former members, and public speeches given on the occasion of mathematicians' professional jubilees.[14] That the group still exists and attends to more pressing tasks is another explanation for the absence of standard celebrations. At this juncture, however, Bourbaki might have more to lose than to gain by drawing attention to itself and the singularity of its withered agenda.

Instead of having public celebrations and commemorations, the Bourbaki clan has deployed various devices for forging its own memory as well as the memories of several generations of mathematicians. Such memory-making procedures make up what I loosely call Bourbaki's "art of memory." Frances Yates defines art of memory as a procedure that "seeks to memorise through a technique of impressing 'places' and 'images' on memory."[15] Thus, an art of memory gives a set of tricks by which the mind can be trained to retain ideas or words that will henceforth produce a memory. Central to the technique is the art of identifying places—of making up a memory map or "memory palace"—and of creating strong visual images to attach to these places. In the Bourbaki context, the art-of-memory analogy has definite limits, however, since the devices they forged did not have the rote and rigor of mnemonics, nor were they intended for memorizing data with accuracy or for improving an orator's memory (the functions of the classical art of memory). Instead, they were mostly used to describe the group's activities and to constitute a group memory. As important ingredients in the group's identification process, these image-making devices emphasized continuity and repetition; some even clustered into genuine celebrations of the past, but none has yet served as a standard commemorative ritual.

## II. GROUP CHRONICLE AND GROUP MEMORY

> Group memory, after all, is no more than the transmittal to many people of the memory of one man or a few men, repeated many times over; and the act of transmittal, of communication and

[14] Speeches delivered at the jubilees of Henri Cartan or Laurent Schwartz, for instance, and the obituaries of Claude Chevalley, Jean Delsarte, Jean Dieudonné, and André Weil unveiled fragments of Bourbaki's history. Individual testimonies of former members are found in: Armand Borel, "Twenty-five Years with Nicolas Bourbaki," *Notices of the American Mathematical Society,* 1998, *45,* n. 3:373–80; Marjorie Senechal, "The Continuing Silence of Bourbaki—An Interview with Pierre Cartier, June 18, 1997," *The Mathematical Intelligencer,* 1998, *20,* n. 1:22–8; Laurent Schwartz, *Un Mathématicien aux prises avec le siècle* (Paris: Éditions Odile Jacob, 1997); André Weil, *The Apprenticeship of a Mathematician* (Boston-Basel: Birkhäuser, 1992), which is the English translation by Jennifer Gage of *Souvenirs d'apprentissage* (Boston-Basel: Birkhäuser, 1991), also Weil's extensive comments in his *Oeuvres scientifiques,* 3 vols. (New York-Heidelberg: Springer-Verlag, 1979/1980). The death of André Weil (1998) has provoked a great number of articles by colleagues and fellow Bourbakis; these appeared as a series in the *Notices* of the American Mathematical Society (April 1999, *46:*4, 6, 7) and in the *Gazette des Mathématiciens,* "André Weil (1906–1998): Adieu à un ami," Spring 1999, supplement to *80:*13–35. More are forthcoming.

[15] Frances A. Yates, *The Art of Memory* (London: Routledge and Kegan Paul, 1966), p. xi.

> therefore of preservation of the memory, is not spontaneous and unconscious but deliberate, intended to serve a purpose known to the man who performs it. —Moses Finley[16]

In so massive an undertaking as the writing of Bourbaki's *Eléments de mathématique,* with its incessant revisions and corrections of mathematical texts, details could be easily forgotten, lost in the intricacies of drafts and discussions. Memorable events nevertheless happened, and Bourbaki found chroniclers in its ranks to report on these, as well as on routine activities. Starting with accounts written in the plain style of minutes of meetings, in time the chronicle developed into a tapestry of fabulous stories that became more outlandish as the team's self-image became more clear.

While the classical principles of art of memory recommend that places of memory be organized, fixed, shorn of details, and made familiar in order to be easily grasped, they also stress how the images that stimulate memory must strike the imagination, arouse emotions, and excite the senses: they should be strong and sharp, strikingly beautiful or ugly, dramatic or even grotesque, comical or obscene; preferably, they should involve human figures in action.

Of all possible representations, the Bourbakian repertoire opted for the humorous and the ribald, on occasion ascending to the heroic contrasted with the loutish. Extraordinary characters, both real and invented, were skillfully put on stage on Bourbaki's private chronicle. Anecdotes merged strong visual images with playful stories, wordplay, and poems, which became set aids to memory as well as to group identification. Least of all was Bourbaki's "autobiographical" chronicle meant to inform readers about what had actually happened during meetings; this sterner duty was accomplished by the rosters of tasks, detailed corrections of texts, and lists of decisions that circulated internally along with the humorous narrative report. Although Bourbaki's chronicle is not the only way in which members recorded what happened, it nevertheless constituted the group's main testimonial, overriding and unifying other memory-making devices.

The team held an initial series of biweekly meetings in Paris during the academic year 1934–1935. Each record of these gatherings bears on its heading the date of the meeting along with the provisional designation "Analysis Treatise" or "Committee on the Analysis Treatise."[17] Beginning with its extended assembly at Besse-en-Chandesse in the summer of 1935, Bourbaki labelled each summer gathering—called "congresses" in the fashion of the numerous French political party meetings of the time—with the name of the place where it was held. In those days, shorter meetings took place throughout the year as well and their outcomes were briefly mentioned in the chronicle they called "Journal de Bourbaki." This newsletter, which circulated among members between 1935 and 1937, communicated organizational information as well as reports and drafts for the treatise.

The venues of the early congresses remained the landmarks of Bourbaki's initial work: in time, Besse (in Auvergne) and Chançay (in Touraine) became places of pilgrimage that Bourbaki members later visited whenever chance brought them in the vicinity, driven by their longing for the landscapes of Auvergne and their thirst

[16] Moses I. Finley, *The Use and Abuse of History* (London: The Hogarth Press, 1986, first published 1971), p. 27.

[17] The content of these initial meetings is described at length in Liliane Beaulieu, "A Parisian Café and Ten Proto-Bourbaki Meetings," *The Mathematical Intelligencer,* 1993, *15,* n. 1:27–35.

for the wines of Touraine. In 1940, when Jean Dieudonné took on the task of reviving Bourbaki's prewartime newsletter, the first issues of "La Tribu" (subtitled "an ecumenical, deadbeat and Bourbakic bulletin") identified congresses by the date and the place where they were held.[18] During those years, Clermont-Ferrand was the site of several meetings, as it was then in the unoccupied zone of France and some Bourbaki members lived or taught in the area. Congresses were rare and scarcely attended during the Occupation—most were labelled "rump" congresses—and they remained so until hostilities ceased throughout Europe. After 1945 Bourbaki once again met more often, usually three (sometimes four) times a year, in some secluded spot in rural France. The places chosen for meetings varied over time according to the group's whim or the availability of youth camps, monasteries, resorts, or hotels.[19]

Whereas in previous years the congresses were simply named after the places where they were held, in the 1950s and thereafter, as Bourbaki became more nomadic (within French continental territory and, later, elsewhere as well), "La Tribu" began sporting subtitles that illustrated a theme of the congress and were dictated at the whim of the reporter. Thus, we find the "Public Bench Congress," the "Extraordinary Congress of Old Fogies" (when anyone over thirty could be called a fogy), the "Congress of the Motorization of the Trotting Ass" (an expression used to describe the routine unfolding of a mathematical proof or process), the "Moon Congress" (in reference to Sputnik), the "Congress of Jointly Algebrised Universities" (in October 1968, following the student revolt of May and the proliferation of jointly managed academic institutions), and so on. When the number of meeting places had increased, these subtitles served to typify particular congresses in the minds of the attendees and readers of "La Tribu." They were complementary mnemonic signs and condensed interpretations of events that gave each congress an extra, memorable saliency.

After the title piece, "La Tribu" usually listed the members present (some branded "nearly absent" for their ineffectualness), the guinea pigs, the visitors, special or foreign guests, sometimes the "extras" (wives and other ladies, children, local characters, farm animals), and the props: cars, bicycles, prams, magazines, binoculars, cough drops, gargles, aspirins, poultices, a particular backdrop, along with sundry objects that complemented the standard Bourbakian paraphernalia—the blackboard and its duster (sometimes called "the flying saucer"), seat cushions (usually called "fanny pads"), the odd chaise longue, etc. Bourbaki even once "borrowed" French poet Jacques Prévert's racoon for its own imitation of his famous poem "The Inventory" as these lists, with their overabundance of ill-assorted objects, were highlighted by a hint of the absurd or surrealistic.

The wordy heading was then followed by a melodramatic narrative—usually one to three pages long—depicting or inventing the most ridiculous situations of the congress. As the writer of the bulletin gave free rein to his imagination, we find stories of members roaring like wild animals and prostrating themselves whenever a train entered the nearby station—especially when it was the South-Express, which they had allegedly greeted at their first meeting and thereafter. The Bourbakis are

[18] The subtitle reads "*Bulletin oecuménique apériodique et bourbachique*" in the original French. I translate the idiosyncratic "*bourbachique*" as "Bourbakic," a choice that gives the flavor of the jargon invented by the group.

[19] Bourbaki's meetings and work are analyzed thoroughly in Beaulieu, *Bourbaki* (cit. n. 4).

depicted playing chess, but more often boules, table soccer, volleyball, or Frisbee; embarking with gusto on mountain hikes, bicycle excursions, or swimming expeditions; having fun in bumper cars or setting out on butterfly hunts or mushroom picking; or just sunbathing, dozing off with text in hand, stuffing themselves, and getting royally drunk on local wines, Armagnac, champagne, or rum toddies, depending on the time of year and the means of the group. A congress, it was said, was content so long as "food was plentiful and the sun was shining." When resources were scarce, anything else might be sacrificed but not the *bon vivant* eating choice foods and drinking good wines, activities in line with French tradition and raised to a Bourbakian duty. Wine, it was proclaimed, was the much-needed fuel of Bourbaki's cogitation. Once under the influence—according to "La Tribu"—inebriated members sometimes worked themselves up to a virile French *cancan* or a lascivious belly dance.

From such descriptions one might think the Bourbaki group rarely worked and always played. The deliberately laid-back attitude—a typically *normalien* pose—gave the impression of insouciant genius at its most youthful, spontaneous, and elegant: powerful genius, beloved of the gods, that despises hard graft and operates on effortless, even drunken, inspiration. Indeed, when the bulletin mentions time spent during a congress, it is to claim that Bourbaki is wasting its time or is finding it a drag, rarely that it is spending it wisely. Hence, the congress itself is often called "nilpotent" (which also refers to a mathematical property), members' heads are in a "state of total vacuum," and dubious regrets are expressed when "A gust of panic had spread through the Congress before it broke up . . . there would be nothing to read!" Any excuse was good enough to pretend that work is something to be put off and avoided at all costs: adverse weather conditions and ill health are repeatedly blamed for Bourbaki's inactivity. One meeting reportedly found itself in dire straits due to the participants' ill health as "the Fate of old age, infirmity and Armagnac was in the ascendant. Soon we had a concert of coughs, sneezes and handkerchiefs, a procession of warmly wrapped persons twisted with lumbago and sprains. Finding a reader with a clear voice became a problem."

Here, humor disguised rather than emphasized the distinction between the serious and the nonserious: the genuine work that went into composing the *Eléments de mathématique* was usually passed over in silence or mentioned with disdain. In contrast, the large number of lengthy drafts and detailed critical comments from the group's readings are undeniable evidence of a lot of toil. Bourbaki was, in fact, a hard worker, and the pressure scarcely ever let up. Even mealtimes were occasions for long, drawn-out discussions over mathematical subjects, among others. The humorous writing of the bulletin, then, systematically reversed the work-play ratio and, since fun is more apt to be memorable than routine intellectual activities, the prevailing impression of the congresses becomes one of entertainment and leisure.

Like most closed societies, Bourbaki created a lingo for all its levels of production, even for the most mundane aspects of its collective life. The name "Bourbaki" may denote a member of the group as well as the whole group; thus one finds expressions like "the Bourbaki so-and-so" used within the group (whereas a "Bourbakist" is an enthusiast of Bourbaki's ideas). An original adjective and a verb were also derived: "Bourbakic" (*bourbachique*) and "to bourbakize" (*bourbachiser*). Bourbaki even had its *Khanonical Bible* (sometimes written with this peculiar, emphatic *normalien* spelling); that is, the collection of already-published volumes of the

*Eléments*. Notifications to attend the congresses were called *Diktat* ever since the 1930s, when the German word often appeared in French newspapers as Germany protested against the *Diktat* of the Treaty of Versailles. "Diplodoci" (there were even "menageries of diplodoci") were especially lengthy writing assignments; "sea monsters" were drafts that were long expected but had not materialized; and "carpets" were projects that their authors tenaciously defended. Many oft-used expressions, such as the untranslatable "*pôt*," "*bonvoust*," "*canuler*," and "*tapiriser*," come straight from the slang of their dear Ecole Normale Supérieure.[20] Even when the Bourbakis had not been students for a long time, they nevertheless remained *normaliens*, and references to the lingo of the Ecole Normale kindled within them a strong "old-boys" kinship while it immersed their ongoing work in the ever-juvenile world of overgrown *normaliens*. These verbal conventions, established through routine and time, acquire an evocative power and monitor set images within Bourbaki's group memory.

Bourbaki also invented its own mathematical terminology by which external realities gave a down-to-earth flavor to technical terms. For their expositions on topological vector spaces and integration theory, the group invented terms such as "barrel" and "barrelled" space (from wine barrels), as well as "bornographic" space. Right modules and left modules were once renamed "starboardules" and "portsidules" in quasi-naval terminology, and the resonant "QUASIMODOmorphisms" were once scrutinized during an Easter-break meeting. One finds "subversive mappings" in the theory of manifolds and "masterable spaces" (instead of measurable or metrizable spaces) in topology; a draft on sesquilinear forms proposes "sexylinear forms" and coins the novel "matrilinear" and "patrilinear" forms. These found their way into Bourbaki's "terminology casket," a list of imaginative terms suggested during a congress; some made it to final publication while most were kept strictly for ephemeral intramural use.

Bourbaki members often delighted in playing on the names of mathematicians, as in "hyperBORELic spaces" (after Emile Borel); elsewhere, some topologist suggested that congresses should henceforth be held in a luxurious "Hilton [referring to Peter Hilton] Hotel where one finds sphere bundles in every [Hermann] Weyl chamber." Imitating a popular parody of Marxist convictions, an inspired participant proclaimed, "I am a Marxist with a Groucho leaning; I am a LAXist with a Peter tendency." (Peter Lax is an American applied mathematician.) Judging the quality of its own witticisms (sometimes emphasized as Bourbaki's "WITT," after the name of yet another mathematician), one congress concluded that "Only the wordplay remained at the same level, but then it could hardly go any lower." The Bourbakis relished their own jokes nonetheless and obviously drew much pleasure from the mental and cultural agility that the wordplay required and nurtured.

Among this exclusively male coterie there is always a good deal of blue humor. Not only would they often use the word "member" to refer to the *membrum virile* as well as to any Bourbaki, but the group's comical memory arsenal is full of expressions in which mathematical terms are redeployed in suggestive ways. Different areas of mathematics and just about any situation could prompt the Bourbakis to

[20] Many of these, with some historical explanations, are found in Alain Peyrefitte, *Rue d'Ulm: Chronique de la vie normalienne* (Paris: Fayard, 1994 [revised and augmented edition of the ENS bicentennial]).

pass lewd comments. They complimented their colleague Henri Cartan, son of the famous mathematician Elie Cartan, on his own mathematical talent by saying that he "suckled differential geometry with his father's milk" and they claimed that, after reading a suggestive movie magazine, Cartan tried to show the formula Hom (B, Hom (B,B)) = Hom (B⊗B, B), in which "B. B." are the initials of famous French actress and 1950s sex symbol Brigitte Bardot and "Hom" (pronounced *'om* as in *homme,* the French word for man) designates, in mathematics, the homomorphisms—a special kind of mapping—of one set into another. In the same vein, one member once declared, "The spectral sequence is like the mini-skirt; it shows what is interesting while hiding the essential." The relatively taboo nature of sex and its incongruous application to mathematics create the expected humor in these more or less risqué expressions. Belted out in the heat of argument or gleaned from an unravelling of events, the saucy phrases are meant to strike the imagination, especially when they later appear out of context as gems of the congresses and are retold among Bourbaki members. Their omnipresence nurtures the impression that the Bourbakis did not meet to slave, but chiefly to have a good sporting time among men.

Mathematicians, it is known, are somewhat given to humor and wordplay related to their trade. Journals such as the *American Mathematical Monthly* (organ of the serious Mathematical Association of America) and *The Mathematical Intelligencer*—both regularly read and contributed to by practising mathematicians—occasionally intersperse their articles with mathematical jokes, poems, or anecdotes. The Bourbakis distinguished themselves from this general practice, however, in that they systematically integrated humor as part of their internal *modus operandi.* The habit of intense intellectual game-playing in youth, and especially in their Ecole Normale days, lead the Bourbakis to regard humor as a superior habit of mind that they used as the main ingredient of their memory-making devices.

Serious matters and mundane occurrences constitute markers in Bourbaki's time-keeping as well as favorite objects of derision. References to cultural or political news, in particular, provide additional, unusual temporal reference points in Bourbaki's reports, especially when woven into sections dealing with strictly mathematical output. For many years Joseph Stalin was the butt of Bourbaki stock jokes. In 1936 the "Journal de Bourbaki" threatened the group with a Stalin-like five-year plan, claiming that "[Stalin] is very interested in Bourbaki and envisages applying a systematic five-year rationalization of scientific productivity according to the principles of historic materialism." This was said in jest at a time when Stalin already ruled the USSR with an iron fist; the country was going through its second five-year plan and Stalin had ordered several purges. During World War II, Stalin took on the title of generalissimo of all the Soviet armies, and in 1945, after the victory of the Allies over the Axis, Bourbaki reported from its first "intercontinental meeting":

> Not to be outdone by other heads of state, the Congress decided at once, unanimously, to elect Bourbaki to the rank of generalissimo of the mathematical armies . . . and, in an order of the day addressed to all the faithful, solemnly announced that he will henceforth proceed only *from the generalissimo to the particular.*[21]

To "proceed only from the generalissimo to the particular" is a slight corruption of one of Bourbaki's own rules of expository writing, "to proceed from the general to

[21] "La Tribu," 1945, *8:*1. From the papers of Jean Delsarte. My emphasis.

the particular," as explained in the "Instructions to Readers" of the *Eléments* as of 1939.[22] As the whole group was meeting in 1945 for the first time in seven years, the phrase sounded like a battle cry.

Although Bourbaki is particularly interested in international news, French politics nevertheless finds many echoes in the group chronicle. When the French left won the elections that brought François Mitterand to the presidency in 1981, "La Tribu" hailed the event by pretending to rename left modules as "presidential-majority modules" and right modules as "departing-majority modules." Although Bourbaki's work, done behind closed doors, does not faithfully mirror current events in mathematics—a role that is better fulfilled by the public Séminaire Bourbaki—the outer mathematical world succeeds in obtruding into Bourbaki's reports in various guises. When the Bourbakis were not making fun of category theory, nonstandard analysis, or catastrophe theory, they were attacking Benoit Mandelbrot's fractals and, in line with the Bourbakian obsession with food, they considered, for instance, the "problem of confinement for fractal-like pancakes." Bourbaki's comments on current mathematical theories thus signal, indirectly, ongoing mathematical fashions.

Repeated allusions to the same subject over many years imprint the group's memory relative to changes in its own goals and achievements. The intimate relation between algebra and analysis, in Bourbaki's treatise, provide a reliable fall-back theme. Whereas at the first meeting in December 1934 Henri Cartan wanted "to eliminate algebra from the treatise," it was reported in 1938 that "[Claude] Chevalley promises to push algebra until it's flattened to the ground," while the first part of the treatise was earning the subtitle "Fundamental Structures of Analysis."[23] By 1940 it was said that "some fine minds are beginning to think that Bourbaki doesn't want any algebra in its analysis treatise, which, all things considered, might not be a bad idea." In the 1950s, while Bourbaki was working very hard on its algebra chapters, Dieudonné wondered "how can one say sensible things when one only does algebra" while another member queried "whether it is reasonable to go on much longer inserting analysis results in an algebra treatise" and someone declared "Analysis my arse, if we talk about it, it's just to piss our readers off!" Humor notwithstanding, these lines show a change in the nature of Bourbaki's opus: while the group had originally intended to write an analysis treatise, by 1940 the algebraic content of the work had expanded beyond the members' expectations and they knew, by then, what puzzles for their expository writing lurked within algebra. In the 1950s Bourbaki had already published many volumes of its *Eléments* and continued to work on a treatise that presupposed a lot of algebraic material; consequently, the volume on algebra was becoming so weighty that the question arose of whether they should set a limit on it or leave it open to further additions, according to the needs of the overall presentation. In the 1970s, when analysis had taken quite a turn within Bourbaki's enterprise, nonstandard analysis inspired the revealing notice that "timid attempts at

[22] N. Bourbaki, *Eléments de mathématique,* book I, *Théorie des ensembles: Fascicule de Resultats* (Paris: Hermann, 1939), p.v. ("N. Bourbaki" is the pseudonymic author [with no given first name] of the *Eléments de mathématique,* whereas articles, written by individual members and usually not discussed by the group, were published under the full name "Nicolas Bourbaki." The full name also designates the legal association: Association des collaborateurs de Nicolas Bourbaki.)

[23] I discussed the changes from an analysis treatise to an algebra-based treatise in the 1930s in "Dispelling a Myth: Questions and Answers about Bourbaki's Early Work," in *The Intersection between History and Mathematics,* eds. Joseph Dauben and Chikara Sasaki (Basel: Birkhäuser, 1994), pp. 241–52.

nonstandard analysis were cut short . . . and it was noted that for Bourbaki, even the other analysis is not standard." In the 1980s, it was suggested—in jest—that the subtitle of the first part of the *Eléments,* "Fundamental Structures of Analysis," be replaced by "Fundamental Analysis of Structures" which, in the minds of some, had become a more fitting description of the nature and content of Bourbaki's oft-revised books. Returning to the same theme, the role and importance of algebra and analysis in the *Eléments* not only marks distinct periods in the work of Bourbaki, but also produces a strong effect of continuity above and beyond ruptures and variations.[24]

These various devices are not essential to Bourbaki's timekeeping, since meeting reports usually bear dates and enough details to retrace a chain of events. They contribute to a group memory mostly because they produce the kind of humor with which the Bourbakis identify. Yet one must bear in mind that, at any point in time, the group was working on many different topics, some of which were reworked over and again for more than two decades, and that the different chapters of the *Eléments de mathématique* and their respective revisions and updates were not published in order of volumes, but rather in order of readiness. Robert Merton has shown how scientists systematically forget past solutions to a problem and will either reinvent them from scratch or run into the same pitfalls as their predecessors.[25] Bourbaki is no exception to this mode of functioning, and coordinating Bourbaki's activities with current events adds signs of the times to the chronicle scripts that can, at some later date, revive some of the details in individual members' reminiscences.

Although there are no rules for adopting the minutes of Bourbaki's meetings, the stories in the chronicle nonetheless acquire legitimacy within the group because of their circulation and their use of abundant humor, Bourbaki's second-favorite mind game after mathematics. The accounts of meetings imitate the verbal form; they were written to be read and reread, and to provoke readers to laughter. Yet humor also sets limits on the memories that it jogs, as some of the jokes are so pointed and so closely tied to the particular circumstances of a congress that they are not understandable to those who were not present, or even to those who have not looked at old issues of the newsletter in some time. Because the accounts of meetings—and especially their narratives—help them recall the ever-good old days, some members keep only the narrative part of "La Tribu" in their personal papers. This is partly due to a practical measure by which, in order to save on costs, Bourbaki once decided to stop sending the full reports to its retired or former members, who henceforth began receiving only the narrative section. Thus, the humorous accounts become mementos that the Bourbakis keep, and read on occasion. The chronicle earns the equivocal status of true caricature of the group's activities: no matter how grotesque or extravagant the descriptions are, their vivid and facetious tone creates an illusion of reality, at least in the eyes of the Bourbakis. Yet humor masks, as well as marks, the anecdotes it chronicles: jokes revise the events they encapsulate and hilarious stories overtake what was lived, as they reorganize and mitigate real-world

[24] The quotations in this paragraph are found in, respectively, "Meeting of the 10/xii/1934," p. 2, from the papers of René de Possel; "Dieulefit," 1938, "Note," from the papers of Henri Cartan, "La Tribu," 1940, *4:*2, from the papers of René de Possel; "La Tribu," 1955, *35:*1, from the papers of Henri Cartan; "La Tribu," 1959, *47:*2, from the papers of Claude Chevalley; "La Tribu," 1978, *102:*1, from the papers of Claude Chevalley; "La Tribu," 1984, *120:*1, from the papers of Samuel Eilenberg.

[25] Robert K. Merton, "Singletons and Multiples in Scientific Discovery: A Chapter in the Sociology of Science," *Proceedings of the American Philosophical Society, 105, 5:*471–87.

experiences. Bourbaki's group memory is not the collection of each and every member's own personal remembrances but, rather, a mediated set of recollections fixed by a chronicle that uses, over and over again, the same humor-based, memory-making devices.

### III. FOUNDERS AND FORERUNNERS

> The competitive society celebrates its heroes,
> the hierarchy celebrates its patriarchs,
> and the sect its martyrs. —Mary Douglas[26]

Literary allusions seldom appear in Bourbaki's mathematical publications, but the group has often used such references within its chronicle. Several Bourbaki members were keen on literature, and some even enjoyed making up verses inspired by their reading.[27] Among the poems written by Bourbakis, some follow the rules of classical French versification without any other literary allusion, while others are not only composed according to accepted rules but also imitate the sounds and even the content of known poems; such verses I call "pastiches." A pastiche thus contains two levels of literary allusion: a formal level at which set rules are followed, and another level at which meanings collide between the source poem and its imitation. For their pastiches, the Bourbakis are particularly fond of La Fontaine, Racine, the Symbolists, and Mallarmé among poets of the past; Valéry and Prévert among contemporary writers. They have been making up pastiches since their *lycée* years, and this sort of literary virtuosity is very common among students at the Ecole Normale Supérieure; it is not surprising, therefore, that they turn to versifying to hail Bourbaki's accomplishments and evoke pictures from the past. The rhythm of verses composed in classical forms—sonnets or alexandrines (the quintessential French meter, according to some)—make the pastiches easy to memorize and recite, the more so as the imitated poems are already part of the Bourbakis' shared culture and are known by them all.

In her study of how Victorian scientists used literary sources, Gillian Beer remarked that:

> Poetry offered particular formal resources to think with. Poetry works by cross-setting a considerable number of systems in simultaneity (natural speech word order, metric units, line units, grammatical units, cursive syntax—all play across each other). By means of metre in particular, and sometimes by rhyme, *the poet sets up multiple relations between ideas in a style closer to the form of theorems than that of prose.*[28]

In Bourbaki's circumstantial poetry, the grandiloquent tone and evocations of well-known poems create surprising, memorable simultaneities. Poetic forms sometimes lend their allusive power to self-critical expressions of difficulties in the collective work, but they also provide a formal idiom that suits Bourbaki's most extravagant hagiographic incantations.

[26] Mary Douglas, *How Institutions Think* (New York: Syracuse University Press, 1986), p. 80.

[27] Weil, for one, published two of his own poems in *Apprenticeship of a Mathematician* (cit. n. 14), p. 124. These had apparently not been circulated among Bourbaki members.

[28] Gillian Beer, *Open Fields: Science in Cultural Encounter* (Oxford: Clarendon Press/Oxford University Press, 1996), p. 210. My emphasis.

The first poem to be included in "La Tribu" evoked the founding and initial goals of the early group: the proto-Bourbakis, who had set out to revolutionize and recast the bases of university mathematics teaching in France by replacing Edouard Goursat's treatise on analysis (hitherto the standard text) with an up-to-date work, conceived and written collectively.[29] It also lauded one of Bourbaki's most noted successes, the first chapters on topology.

The Filter

O powerful, O formal, O Thou bright Bourbaki,
Wilt Thou not tear for us in an impulse of rage
The long-winded Goursat, mirror of Analysis,
Belated defender of a past that is long gone?

The sequence of former days thought itself infinity,
Useless, used without comprehension by
The clumsy freshman, impressed by Valiron
In his gloomy course, essence of tedium.

Ignorant of the secrets of Topology
To space inflicted, and of Thee Who studiest it,
He flounders in the error where his language is caught.

He views in stupefaction, as if drunk on a philtre,
Closure, a mantle he has never grasped,
Worn, in a compact space, immobile, by the FILTER.[30]

Pierre Samuel composed this pastiche of Stéphane Mallarmé's "The Swan" or "Sonnet in *i* Major" and gave it a title borrowed from the notion of filter, which Henri Cartan had invented during a Bourbaki congress in 1937 and which had been quickly used in the Bourbaki volume on general topology.[31] The words "space," "closure" (actually called *adhérence* in the original French, according to the name given to that notion by Bourbaki founder René de Possel), and "compact" all belong to the mathematical domain of topology. The unfortunate Valiron is a professor, first name Georges, who followed in the steps of Goursat at the Sorbonne and taught the

[29] The group's original goals, the standing of its members among other French mathematicians, and the content of the early meetings are studied in detail in Beaulieu, "A Parisian Café" (cit. n. 17) and in Beaulieu, *Bourbaki* (cit. n. 4), chap. 1.

[30] "La Tribu," 1945, *8:*3. From the papers of Jean Delsarte; translation by Jill Corner. The original French text is in Liliane Beaulieu, "Jeux d'esprit et jeux de mémoire chez N. Bourbaki," in Abir-Am, *La Mise en mémoire* (cit. n. 6), p. 110. The full text appears in English translation here for the first time. In his *Apprenticeship of a Mathematician* (cit. n. 14). Weil quotes the last tercet only and briefly describes the Paris congress of 1945, on p. 190.

[31] In English translation, Mallarmé's sonnet "The Swan" reads as follows: "The virginal, vigorous and beautiful today, / Will it tear for us with a blow of its drunken wing / This hard forgotten lake haunted under the frost / By the transparent glacier of flights that have not flown! / A swan of former days remembers it is he who / Magnificent but without hope frees himself / Because he did not sing of the country in which to live / When the tedium of sterile winter shone. / Phantom whose pure brightness assigns it this domain, / It stiffens in the cold dream of disdain / That clothes the useless exile of the Swan. / His whole neck will shake off the white agony / Inflicted by space on the bird that denies it, / But not the horror of the soil in which feathers are caught." In *French Poetry from Baudelaire to the Present,* ed. Elaine Marks (New York: Dell Publishing Co., 1962), p. 95.

Cartan published two articles on filters under his own name: "Théorie des filtres" and "Filtres et ultrafiltres," *Comptes Rendus de l'Académie des Sciences de Paris,* 1937, *205:*595–8 and 777–9, respectively. The concept of ultrafilter is a generalization of filter, suggested by Claude Chevalley, who renounced his intellectual rights over it.

students from the Ecole Normale Supérieure for many years; the basic course in analysis (which we would now call advanced calculus) that he gave in the 1940s was deemed traditional enough to deserve the same barbed criticisms that Bourbaki had once addressed to his master's work.

Composed by a newcomer to the group in 1945, a decade after the founding congress of Besse-en-Chandesse, this pastiche poem is commemorative in purpose: it linked the origins of Bourbaki to the present by conjuring up the spectre of enemies over which Bourbaki could triumph. It was also basically celebratory as it hailed the heroes of the past (namely, the founders of Bourbaki) and their battles, the expected success of their treatise, and the virtues of their new mathematics, which topology and the notion of filter represent. The poem's grandiloquent tone exhibits the customary Bourbaki playfulness, while its cleverness might have contributed, to a modest extent, to its author's social integration into the community: his mathematical talent and knowledge notwithstanding, Samuel was to become, for many years, Bourbaki's chronicler as the writer of "La Tribu's" narratives.

The summer of 1945 was a time for rejoicing since, at last, the end of the war enabled the team to start again. The group reunited for the first time since its interrupted meeting of September 1938, at the time of the Munich Pact. Weil had flown to Paris from Brazil, where he was teaching at the time, and when his friends Cartan and Delsarte learned of his arrival, they convened the rest of the group for an impromptu, though very well attended, meeting. Still mixing facts and fiction, the account of that meeting adds:

> Our readers know how, after lengthy negotiations, the United Nations finally recognized Bourbaki and placed at his disposal a four-engine bomber to allow him to hold his first intercontinental Congress. Loaded with Weil and *coffee,* the plane crossed the ocean without mishap and deposited its valuable cargo in Paris on June 20th; forewarned, members were arriving in the capital, and began to get together by the 22nd. The only absentee was Ehresmann, who claimed to have papers to mark; to show its disapprobation, the Congress *forbade him fire and water for the following winter.*[32]

A vague allusion to meager or nonexistent rations of fire and water in winter refer to past and even present miseries, while coffee—still a rare commodity in France in the immediate postwar period—became the centerpiece of the report, and several members later remembered that meeting as the "Coffee Congress."[33]

All told, the chronicler as well as the poet remained silent about other realities, such as war and occupation, the lives of the Bourbakis who had known internment or exile or who had joined the resistance, been "Germanized," or remained neutral. This apparent obliviousness was—unwittingly, perhaps—along the lines of the official political commemorations taking place in France in 1945, which systematically shut off memories of the grim and shameful daily struggles of a vanquished people whose main victory had been survival.[34] Omitted also are Bourbaki's earlier stumbling pronouncements, the projects abandoned, the opportunities missed, and

[32] "La Tribu," 1945, *38:*1. From the papers of Jean Delsarte. My emphasis.

[33] From my interviews with Henri Cartan (Paris, June 1986 and July 1992), Claude Chevalley (Paris, July 1983), Jean Dieudonné (Paris, July 1987), Pierre Samuel (Paris, July 1992) and André Weil (Princeton, N.J., May 1985).

[34] Gérard Namer analyzes the commemorations of 1945 in *Batailles pour la mémoire: La Commémoration en France de 1945 à nos jours* (Paris: Papyrus, 1983); see in particular p. 6 and pp. 13–141.

the departure of some disappointed or uninterested members. Overtaken by the prospect of a restored collective life, the poem and the narrative lines of the chronicle celebrate rather than retell Bourbaki's prewar years, expressing the hopes of the reunited team as it looked forward to a promising future as well as back on a glorified past. "The Filter" highlights the sweetest moments from Bourbaki's past while reinterpreting others; it depicts an heroic devotion to the Bourbaki cause, whereas in reality that supposed ardor had sometimes dwindled, and doubts had been openly expressed.[35] With some distance, the poem could thus evoke an embellished revolutionary past, still present in the guise of surviving heroes and enemies over which to triumph. A continuity of shared assumptions bridged any gap between former goals and the tasks at hand which the older members were then passing on to a new "generation". This was readily becoming the group's official past, the one that Bourbaki members could celebrate spontaneously and in unblemished unanimity; this chosen past, in which they all recognized themselves, was stamped on the memory they were building.

Another pastiche sonnet emphasizes, instead, a continuity between Bourbaki, its achievements, and a history of mathematics harkening back to ancient times. Here the poetic model itself moves along a generational line, as it borrows from one of Mallarmé's disciples, Paul Valéry.

Bourbaki
(by Anna of Noailles and Paul Valéry, discovered by L. Sartre)

On thy high forehead lit by a lightning flash
We read that Euclid was thy brother, Thales thy cousin.
Thy gods, in Crete and the Isle of Aegina,
Fail not to give thy vine abundant grape.

We see the backbone of the haughty unbeliever
Bleaching defeated on the reconquered fields
Of the divine Absolute whose Algebra Thou defendest,
And thy estate, Bourbaki, is a holding truly won.

Thou knowest all the riddles of Space and Time,
And yet I tremble on seeing that Thou canst,
Unaware of a winged presence hovering near,

Without a blink or a tremor in thy compass,
Glimpse in the same skies scanned by Zeno of Elea
An arrow that flies and yet that does not fly.[36]

The sonnet begins with a genealogy: Bourbaki descended from the ancient Greeks, the giants of pure mathematics on whose shoulders was erected the edifice of "THE"

[35] These are discussed in Beaulieu, *Bourbaki* (cit. n. 4), chaps. 1 and 2.

[36] "La Tribu," 1953, *30*:2. From the papers of Pierre Samuel; translation by Jill Corner. The original French poem features in Beaulieu, "Jeux d'esprit" (cit. n. 30), p. 112. The text appears in English here for the first time. The last strophe contains a pastiche of the following stanzas from Valéry's *Cimetière marin:* "Zeno! Cruel Zeno! Zeno of Elea! / Have you pierced me with the winged arrow / That vibrates, flies and does not fly! / Sound engenders me and the arrow kills me! / Ah! the sun . . . what a tortoise's shadow for the soul, / Achilles motionless with his giant stride!" In Marks, *French Poetry* (cit. n. 31), p. 159. The poem is said to have been written "by" (that is, in the style of) Paul Valéry and Anna, Countess of Noailles, who themselves often incorporated references to ancient Greece in their work; the countess claimed to have Cretan ancestors and the second quatrain adopts her style. Louis Sartre is a French mathematician who was not a member of Bourbaki.

mathematic, of which algebra constitutes one of the cornerstones. More specifically, the poem compares Bourbaki's expository work and Euclid's great compendium, itself possibly a collective endeavour. "Euclid", then, and "Thales" stand for Bourbaki's remotest historical antecedents and, for mathematicians, the nearest to gods are the great mathematicians of the past—the more remote they are, the greater. Bourbaki liked to believe that its own story stretched over centuries, going back to the beginnings of Western culture and to the initiators of mathematics, the Greeks, with whom—in the group's own mind—Bourbaki shared not only an ambitious vision but also the very essence of mathematics, pure and eternal.[37]

More subtle, and apparent only to initiates, is the allusion to the Cretan origins of one Nicolaïdes-Bourbaki, diplomat by profession and legitimate member of the family of General Charles Denis Sauter Bourbaki—whose family name the group had "borrowed"—who tracked down the team in 1948, thinking the author might be a distant cousin. On that occasion "La Tribu" had insisted that Bourbaki's Cretan background had nothing to do with any suggestion of paradoxes or untruthfulness—as in the case of Epimenides, the Cretan liar—it was made clear that "our Master's ancestors did not include an Epimenides or any other Cretan type-scrambler."[38] Imitating a passage from Valéry's *Cimetière marin,* the last two strophes of the poem proclaim Bourbaki's indifference to the paradoxes of set theory and, with irony, they celebrate Bourbaki's relative triumph over famous paradoxes: in choosing the concept of structure defined on a hierarchy of types of sets, Bourbaki had eluded set-theoretic paradoxes and the risks of contradiction they brought. Bourbaki's attitude towards the paradoxes also demonstrates confidence in the noncontradiction of set theory and of other mathematical theories that depend on it, as well as hope in the possibility of overcoming any eventual contradiction.[39]

When, in 1953, Bourbaki rallied the spirits of noncontroversial, long-dead mathematicians to furnish its genealogy, it was experiencing a serious break with its own immediate past. A controversy over category theory had begun within Bourbaki, and the unity of the *Eléments,* as well as the future of the project, were somewhat in jeopardy. It was not clear to the Bourbakis then how the concept of structure (based on the notion of set) and that of category could coexist in the treatise, which was supposed to lay down solid, coherent foundations and display a strong, unitary approach. Favoring structures over categories seemed to overlook much of recently developed mathematics, whereas choosing categories implied rethinking the whole edifice of the *Eléments;* mixing the two approaches challenged the unity of the math-

[37] A standard picture of the history of mathematics places its origins in Hellenistic Greece, a view which Bourbaki does not support blindly in the historical essays of the *Eléments de mathématique* but nevertheless endorses in the general introduction to the *Eléments.* The Hellenocentric view of the origins of Western science and mathematics has often been challenged; see *Isis,* 1992, *83:*547–607 (special section) for an update of the controversy and its historiography.

Leo Corry discusses Bourbaki's claim to eternal mathematical truth in "The Origins of Eternal Truth in Modern Mathematics: Hilbert to Bourbaki and Beyond," *Science in Context,* 1997, *10,* n. 2:253–96.

[38] "La Tribu," 1949, *17:*1, and "La Tribu," 1950, *22:*2–3; both from the papers of Henri Cartan. The group and its publisher, Enrique Freymann (director of Hermann in Paris) even entertained the diplomat and his family on Easter Monday while they were at congress at the Royaumont Abbey. Weil tells the story of the diplomat and gives a brief account of the hapless general's genealogy in *Apprenticeship of a Mathematician* (cit. n. 14), p. 107.

[39] See N. Bourbaki, *Eléments de mathématique* (cit. n. 22), book I, *Théorie des ensembles,* chaps. I–II (1954), introduction, pp. 7–9.

ematics thus presented. Bourbaki finally opted for structures, but was forced to abandon some of its ambitious comprehensiveness and even much of the unity it had once hoped for. While seeking an optimal solution in 1953 to this budding controversy—which continued for well over a decade—the group summoned the voices of its patriarchs as if to recast its present and recent past into the frame of a myth of origins, one in which time immemorial meets a Bourbaki-construed tradition.

Bourbaki's casting of its founders and forerunners thus had consequences for the present: it gave a particular portrait of the group, its standing in the field of mathematics, and its place in history. "The Filter" celebrated the heroes of Bourbaki's early years while the Greek poem hailed the patriarchs of a triumphant team. Initially the proto-Bourbakis had hoped their treatise on analysis would allow them to control "for twenty-five years" the teaching of differential and integral calculus at the undergraduate level; by 1945, Bourbaki already saw itself as a competitive society. Later, in the 1950s and 1960s, the group reached far beyond its initial goals: the *Eléments de mathématique* acquired an international reputation and Bourbakis stormed the French university system and the leading Parisian schools at which Henri Cartan, Laurent Schwartz, Roger Godement, and Jean-Pierre Serre, among others, held key positions. At the same time, the work of Bourbaki members Armand Borel, Alexandre Grothendieck, Pierre Samuel, and Jean-Pierre Serre was impressing mathematicians worldwide and receiving official recognition. In addition, the founding members, for their part, were still publishing notable work and spreading their ideas across the world. The Séminare Bourbaki had become the most famous in France and in the rest of the mathematical world, and it dictated mathematical fashions. In 1953 Bourbaki had started to control the field of mathematics and, like any hierarchy, it celebrated its patriarchs. Thus, the lofty images in the poems not only "propagated the illusion of a common memory" within the Bourbaki group but also symbolized the leading position that Bourbaki hoped to hold—and did hold for some time—in mathematics, as well as the high profile it wished to stamp on history.[40]

## IV. RAGS AND RELICS

> A jest's prosperity lies in the ear
> Of him that hears it, never in the tongue
> Of him that makes it. —William Shakespeare[41]

Not content with forming an original team, the Bourbakis also played at being naughty boys—even when well on in their careers—thumbing their noses at the mathematics establishment or indeed at everyone who was not Bourbakist. Their pranks took their cue from the folklore of the Ecole Normale, where ragging was an everyday event, even more a part of *normalien* customs than at other leading French schools.[42] A good rag (*canular* in French) is as public as possible; it's aim is to mock a particular authority figure, be that a politician, writer, scientist, the students of another elite school, or even the general population; the success of the

[40] The quote is an expression borrowed from James E. Young, "Ecrire le monument: Site, mémoire, critique," *Annales ESC,* May-June 1993, n. 3:736.

[41] William Shakespeare, *Love's Labor's Lost,* act V, scene 2.

[42] On the subject of rags, and especially political rags, in *écoles préparatoires* and at the Ecole Normale Supérieure between the two world wars, see Jean-François Sirinelli, *Génération intellectuelle: Khâgneux et normaliens dans l'entre-deux-guerres* (Paris: Fayard, 1988).

exercise is measured by the length of time the hoax goes undisclosed. More than any other form of joking, a rag expresses scorn towards power or authority that the perpetrators lack or do not yet possess, and it procures an acute sense of superiority over the victim. As Bourbaki's domination spread through the mathematical field, so the targets of Bourbaki's hoaxes proliferated.

Bourbaki's first public gesture was a rag, the long-term effects of which were perhaps even more significant than its initial impact. Once formed, the group had decided that the best way to announce its existence was to publish an article on mathematics. At the end of the summer of 1935, after their first general meeting, the members agreed to publish a text under a pen name, to appear as a note in the prestigious *Comptes Rendus des Séances Hebdomadaires de l'Académie des Sciences.*[43] André Weil wrote the article, and approached the academician Elie Cartan—whose son Henri was in the group—to ask him if he would present it signed, pseudonymously, "Nicolas Bourbaki." As the academy required each article to be accompanied by a biographical note on the author, Weil wrote the following lines to Elie Cartan:

> Dear Sir:
> I enclose for the C. R. a note that M. Bourbaki has asked me to send you. As you know, he is a former lecturer at the Royal University of Besse in Poldavia whom I met some years ago in a café in Clichy, where he spends most of the day and much of the evening. Having lost not only his job but almost all of his money during the troubles that have wiped the unfortunate country of Poldavia off the map of Europe, he now earns his living in this café by giving lessons in *belote,* a card game at which he is a master. Although he claims to be no longer concerned with mathematics, he has nevertheless been willing to discuss a few important questions with me, and even to let me look at some of his papers. I was able to persuade him, just as a start, to publish the enclosed note, which contains a very useful result for modern integration theory. . . .
> Please accept the thanks of both Mr. Bourbaki and myself. I remain, as always, yours truly with respect and affection. [signed] André Weil[44]

The "Royal University of Besse in Poldavia" was pure fantasy on Weil's part, but it was at Besse-en-Chandesse that the congress of summer 1935 was held. Poldavia was the imaginary martyred country invented in 1929 by Alain Mellet and his colleagues at *L'Action Française*—the organ of a right-wing movement—to puzzle radical deputies—leftist politicians whose political support was mainly rural at the time—and to denounce their gullibility and negligence by making fun of it. Clichy, where the old mathematician is said to have given lessons in playing-card games, is a proletarian suburb of Paris; it was common to find there Eastern European émigrés mixed with Armenians, Gypsies, and poor French workers in shanty *bistrots* where *belote* contests were the going fare.

The character of the foreign mathematician took its first inspiration from an old *normalien* hoax to which some of the future founders of Bourbaki (namely, Cartan, Dieudonné, and de Possel) had been subjected in 1923 as part of their freshmen initiation at the Ecole. Their class had been summoned to attend a talk delivered by

[43] At the time, one could get a note published in the *Comptes Rendus* quite easily and quickly by sending it to an academician. If the academician deemed it worthy of publication, he would read it to his peers and soon after, the note would be published. This procedure dated back to the founding of the *Comptes Rendus* by the physicist François Arago.

[44] The full original French text is in Beaulieu, "Jeux d'esprit" (cit. n. 30), p. 92. The quotation appears here in English for the first time. From the papers of Claude Chevalley.

one Professor Olmgren—really an older *normalien* in bearded disguise—who spoke on mathematics with a nondescript foreign accent, starting with real mathematics and ending in sheer nonsense in which theorems were named after different French generals (including a "theorem of Bourbaki"). The talk had been delivered in a tone that mocked the captain who, at the time, dispensed military training to unruly *normaliens,* who took pride in despising the military. While he was in India in the early thirties, before the Bourbaki venture was launched, Weil still relished the joke and suggested to his Indian colleague and friend, the mathematician Damodar Dharmanand Kosambi, that he refer to the then nonexistent work of a dead (and imaginary) Russian mathematician named "D. Bourbaki" in an otherwise serious mathematical article; the hoaxing reference was intended to confound one of Kosambi's pretentious rivals.[45] Thus, Bourbaki's first rag built upon a whole series of others, which it prolonged and recalled.

Weil's article—but not his letter—soon appeared in the august columns of the Académie, under the title "On a Theorem of Carathéodory and Measure in Topological Spaces" and signed Nicolas Bourbaki.[46] This text, which contained nothing but genuine mathematics, was the first hoax played by Bourbaki, as a group, at the expense of outsiders, who in this instance were certain members of the Paris Académie des Sciences and, indirectly, the readers of the *Comptes Rendus* at large. In the interwar period it was fashionable among rising stars in letters or sciences to decry the stifling authoritarianism of the academicians who reigned supreme over French literature and science. The Académie des Sciences of the Institut de France, then, symbolized a power that these "young Turks" did not yet hold in their hands. These academicians, however, no matter how smug and staid they might have appeared, were nonetheless mostly old *normaliens* themselves, and Weil's signature alone—Weil was already well known in Parisian mathematical circles—must have made them smell a rat, even without the ludicrous reference to *belote* lessons in Clichy. It seems unlikely, therefore, that men like Gaston Julia (then recently elected to the geometry section of the Académie des Sciences), Jacques Hadamard, Emile Borel, and Elie Cartan—all of whom supported the members of the nascent Bourbaki group and had doubtless already heard of the group's existence, even before they knew its name—were really fooled by Bourbaki-Weil, however sleepy they might have felt after the copious lunch during which this letter was read, according to Weil's own account.[47] Moreover, Elie Cartan probably told his fellow diners part of the situation. At least, they were tolerant enough of these goings-on to allow another article by "Bourbaki"—this time from the pen of Dieudonné—to appear in the *Comptes Rendus* in 1938, even before the Bourbaki team had published any part of its treatise.[48] Bourbaki staged, and later retold, the *bons coups* of its revolutionary days, when the highest-ranking French mathematicians were its main targets.

The creation of Bourbaki and the idea of an outlandish pseudonym induced a

[45] See Pierre Dugac, "Notes de la rédaction," *Cahiers du Séminaire D'histoire des Mathématiques de l'Institut Henri-Poincaré,* 1986, *7:*222, footnote 3; Weil alludes to this story in *Apprenticeship of a Mathematician* (cit. n. 14), p. 101. The full story of the pseudonym is told and analyzed in Beaulieu, *Bourbaki* (cit. n. 4), chap. 4.

[46] *Comptes Rendus, 201:*1309–11.

[47] Weil, *Apprenticeship of a Mathematician* (cit. n. 14), p. 106.

[48] "Sur les Espaces de Banach," *Comptes Rendus,* 1938, *206:*1701–04. The first fascicle of the *Eléments de mathématique* was published in 1939.

group of mathematicians at Princeton University to carry out their own hoax. Among them were postdoctoral fellows Ralph P. Boas and John Tukey, and the English mathematician Frank Smithies, who were visiting Princeton in 1937–1938. Together they revived an old Göttingen joke by developing a number of mathematical methods for lion hunting. They published their inventions, allegedly devised by one H(ector) Pétard—alias H(oist) W(ith) O(wn) Petard—who was writing under the subpseudonym of E. S. Pondiczery.[49] Pétard's existence was asserted by the publication of a spirited article on lion hunting ("A Contribution to the Mathematical Theory of Big-Game Hunting") in the *American Mathematical Monthly.*[50] That group wrote the notes of a course given at Princeton by one of the professors, but had no long-term mathematical project in common. They had learned about Bourbaki from Weil himself, who had visited Princeton the same year, and their play with pseudonymity evidently mimicked the original Bourbaki hoax. The lion-hunt theme later inspired a series of articles written by various authors between 1965 and 1985.[51]

In line with the Pétard joke, there followed yet another Bourbaki-related jape. When Boas visited Smithies in Cambridge during the Easter break of 1939, they met André Weil, his wife Eveline, Claude Chabauty (then a recent Bourbaki recruit), and Louis Bouckaert (from Louvain). In the course of conversation it was lightheartedly suggested that a marriage be arranged between Hector Pétard and Betti Bourbaki, daughter of Nicolas. The company wrote up a wedding announcement to be printed in the conventional French fashion. Weil even had calling cards printed in the name of Bourbaki.[52] The Princeton-Cambridge gang, together with other mathematicians who were in on the Bourbaki-Pétard joke, received the wedding invitation that read in part as follows:

> NICOLAS BOURBAKI, Canonical Member of the Royal Academy of Poldavia, Grand Master of the Order of Compacts, Conservator of Uniform Spaces, Lord Protector of Filters, and his spouse, née BIUNIVOQUE, have the honour to invite you to the wedding of their daughter BETTI to HECTOR PÉTARD, Associate Administrator of the Order of Induced Structures, Chartered Member of the Institute of Class-Field Archaeologists, Secretary of the Lion's Penny Charity Fund . . .[53]

The style is conventional of good society, and the invitation constitutes a parody as it mocks the customs of that society. Allusions are made to the Princetonian hoax, and the text presents many examples of the sort of mathematical wordplay in which Bourbaki—and especially Weil—delighted. Similar terms are found again in a 1948 issue of "La Tribu" in which all of Bourbaki's invented titles are enumerated.

[49] According to Smithies, these initials were made up from Shakespeare's line "For 'tis the sport to have the engineer, hoist with his own petard . . . ," *Hamlet,* act III, scene IV.

[50] *American Mathematical Monthly,* 1938, *45:*446–7.

[51] The articles are reproduced in *Lion-Hunting & Other Mathematical Pursuits: A Collection of Mathematics, Verse, and Stories by Ralph P. Boas,* eds. G. L. Alexanderson and D. H. Mugler, vol. 15 of Dolciani Mathematical Expositions (Washington: The Mathematical Association of America, 1992).

[52] According to Frank Smithies, "Reminiscences of Ralph Boas," and Ralph P. Boas, Jr., "Remarks: Memorial for Ralph P. Boas, 9 October 1992" in Alexanderson and Mugler, *Lion-Hunting* (cit. n. 51), pp. 25–31 and pp. 32–5 respectively. André Weil tells his version in *Apprenticeship of a Mathematician* (cit. n. 14), p. 139.

[53] Smithies, "Reminiscences of Ralph Boas," p. 28. The complete original French text of the wedding invitation is reproduced in Beaulieu, "Jeux d'esprit" (cit. n. 30), p. 95.

> We, Nicolas Bourbaki, Lord Protector of Filters, President of the Order of Induced Structures, honorary member of the Order of "Class-Filed Archaeology"—given the extreme pain caused us by seeing, among the European faithful, the most obvious signs of quarrels, schisms, discord, litigation and disagreement—enjoin our wise and beloved faithful in America to restore peace, order and good government among their more easterly brethren.[54]

Here, the Bourbaki group reveals its difficulties on a screen of humor. The "American faithful" were mainly André Weil and Claude Chevalley, who were at the time in the United States but were nevertheless asked for their opinions on some particular question.

Once more, in 1968, "La Tribu" recalled the lion-hunting technique and applied it to the group's current work. Repetition and overstatement are essential to memory-making. "The algebraists have invented a new method for lion-hunting: in the virgin forest there are free lions; we take their inductive limit and thus obtain a flat lion which can be used as a bedside rug."[55]

Within the group the wedding announcement, Bourbaki's calling card, and Weil's letter all ended up being souvenirs. In the 1950s some members who had joined after the war even asked for a second printing, so that they, too, could take part in the hoax of purported Bourbaki's rebellious days. Weil's letter was retyped and deposited with the archives of their publisher Hermann and with the Bourbaki secretary; copies of it were sent to members along with copies of letters that had been addressed by mathematicians—some of whom were in on the joke and others who weren't—to "Monsieur Bourbaki." In those days, Bourbaki's calling card was still sometimes sent along with complimentary copies of fascicles of the *Eléments* (see Figure 1). The joke was still extant and Weil's letter, the wedding invitation, and the business card became relics of times past, known to new Bourbakis only through the riotous stories and comic reminiscences of their elders.

Meanwhile, the hoax leaked outside the Bourbaki group and its tightly knit society of friends and official foes. An application for individual membership, twice sent by Bourbaki (Weil again) in 1947 to the secretary of the distinguished American Mathematical Society (AMS), is one of the better-known jokes that kept the Bourbaki myth alive among mathematicians in the Americas. The applications, made public within the council of the society, triggered indignant letters from American colleagues and Bourbaki finally met with a refusal: the AMS secretary suggested that an application for an institutional membership might meet with more success, but Bourbaki held on to its identity as an individual and never joined the society. In the United States, where in principle there are few hierarchies, some scholars were outraged that their high-minded national organizations were being derided by foreign mathematicians who seemed to delight in puerile behavior.[56]

Yet the hoax had not run its course; throughout the 1950s and 1960s—as Bourbaki's fame and production reached their acme—imitations, extensions, and recollections of the original rag abounded, some finding their way into mathematical organizations or even the newspapers. In 1968, French mathematician Jacques

[54] "La Tribu," 1948, *15:*19. From the papers of Henri Cartan.

[55] "La Tribu," 1968, *73:*1. From the papers of Claude Chevalley. By then, the *American Mathematical Monthly* had published yet another article on lion hunting.

[56] The incident is discussed at length in Beaulieu, *Bourbaki* (cit. n. 4), chap. 5.

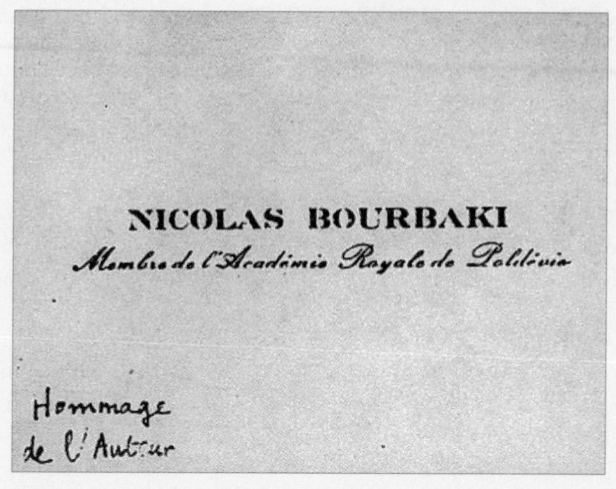


*Figure 1. One of Bourbaki's relics: Nicolas Bourbaki's calling card. The card was first printed in Cambridge in the summer of 1939 by André Weil. It was later used by the editor and by Bourbaki's secretary when sending complimentary copies of the volumes of the* Elements. *This sample was found among mementos that Marshall H. Stone brought back from his trip to France in 1951. Courtesy of Brown University Library.*

Roubaud—who was not a member of Bourbaki, but who belonged to the Bourbaki-inspired literary circle Oulipo—printed a formal death announcement of Bourbaki, which he distributed during a session of the Séminaire Bourbaki. This parody of the old wedding invitation reads in part as follows:

> The Cantor, Hilbert, Noether families;
> The Cartan, Chevalley, Dieudonné, Weil families;
> The Bruhat, Dixmier, Godement, Samuel, Schwartz families;
> The Cartier, Grothendieck, Malgrange, Serre families;
> The Demazure, Douady, Giraud, Verdier families;
> The Right Filtering families and the Strict Epimorphisms;
> Miss Adèle and Miss Idèle,
> regret to inform you of the death of Mr.
> NICOLAS BOURBAKI
> their father, brother, son, grandson, great-grandson, and second cousin respectively, [who] passed away in a state of grace on November 11, 1918 [sic 1968] (Armistice Day) at his home in Nancago. . . .[57]

To a recent mathematical genealogy—German mathematicians Georg Cantor, David Hilbert, and Emmy Noether—the author added a lineage containing representatives of each generation of Bourbakis, from the founding fathers to the latest newcomers.[58] Like the wedding announcement—which doubtlessly inspired it—the death announcement is an imitation of Bourbaki folklore. It even features Bourbaki's cherished "Nancago," a name invented by Weil and Delsarte from the combination of NANcy—where Delsarte had lived, taught, and administered Bourbaki's

[57] From the papers of Henri Cartan. A photograph of the full French version of this (false) death announcement is included in Beaulieu, "Jeux d'esprit" (cit. n. 30), p. 117.

[58] With one exception: J. Giraud had been a Bourbaki guinea pig and a guest but he was not, to my knowledge, a member of Bourbaki.

business—and ChiCAGO—where Weil was posted for a long time—for Bourbaki's imaginary, yet official university affiliation. In fact, the group had not stopped working at all: the false death announcement was a metaphor, just another hoax perpetrated by someone who had become critical of the group and its dominant role in French mathematics. Reading this anti-relic, as it were, some of the Bourbakis, it is said, tried to deny its murderous suggestion, claiming that the text of the announcement was neither witty nor well written. The counterhoax nevertheless shows that, in the 1960s, Bourbaki was already stepping on the other side of the looking glass.[59]

Still extant, the group continued to provoke interest and controversy until the late 1970s, as long as it functioned like a tyrannical religious leader, hurling anathemas and imposing *nihil obstats.* Among the mobs of non-Bourbakis the hoax of the pseudonym gave birth to others; countless anecdotes told how some people were taken in by the ambiguity between the famous and unique "N(icolas) Bourbaki" and the cohorts of his devoted, anonymous collaborators. For their part the presumed members, when questioned, claimed to be bound to silence by a heavy vow, as if they belonged to some secret society, or else they said "Bourbaki does not like people talking about him." In the 1970s stories about Bourbaki were still going the rounds, and they were highly successful, especially with junior mathematicians and students. If the *enfant terrible,* in its earliest years, did not entirely dupe the old academicians, it still mystified generations of apprentice mathematicians with a running gag. Like so many fabulous adventure stories, Bourbaki's rags were peddled like good news and exaggerated by the whole mathematical village. Even when Bourbaki was no longer at the height of its glory, the tales of its earliest battles, presumed victories (supposedly still fresh), and arcane activities found an audience that was eager to be initiated into the guru's teaching. While some did not appreciate this sport, others submitted to it as to a series of tests, the secret meaning of which they would gradually grasp. Collecting scraps of anecdotes, they could imagine themselves approaching the Holy of Holies, the mathematics star who, although he was not inviting them to the banquet, nevertheless allowed them a few leftovers.

The original rag had been successful, especially as an apparent triumph about which to crow, and its long-term effects have quite superseded the initial intention and reaction. Since the pen name cloaked the authors' identity, many mathematicians were puzzled by this equivocal being, whom rumor described as either an organization or "a many-headed mathematician," and they were often unaware of the identities of Bourbaki's collaborators. The Bourbaki hoaxes belittled those who were taken in by them, and equally subjugated those who asked for more. Each new joke reinforced, for some, a belief in the power of some elusive mathematician(s) and an acceptance of their views. For others, those in the know, it kindled a sense of belonging to the chosen few, the leaders of a pack of mathematicians who could abuse others. Even when intended as critical retaliation, Bourbaki-related jokes underscored the rigid hierarchy that prevails among mathematicians, where a deep

[59] In the third volume of his autobiographical trilogy, Jacques Roubaud only alludes to the circumstances of his hoax whereas he recalls, with detailed and perspicacious reflection, how he became a mathematician and a follower of Bourbaki in the heyday of the group; see Jacques Roubaud, *Mathématique: Récit* (Paris: Seuil, Fiction & Cie, 1997); the allusion to the hoax is on p. 148. David Aubin discusses aspects of the rise and fall of Bourbaki in "The Withering Immortality of Nicolas Bourbaki: A Cultural Connector at the Confluence of Mathematics, Structuralism, and the Oulipo in France," *Science in Context,* Summer 1997, *10,* n. 2:297–342.

division separates the best from the mediocre, those who know from those who do not. Indeed, the prosperity of Bourbakian jests lay in the ears of those who heard them, identified with them, and submitted to them. Thus was forged one of the grandiose figures in what may be called the "collective memory" of twentieth-century mathematicians.

### V. TERTIUM QUID

The staging of practical jokes also depended on the supposed agency of the elusive Nicolas Bourbaki. On the memory side of identification, the "core meaning of any individual or group identity, namely a sense of sameness over time and space, is sustained by remembering; and what is remembered is defined by the assumed identity."[60] A core sameness lies at the intersection of the various images that are deployed by the Bourbaki group to depict itself as a whole; the Bourbakian art of memory supported and strengthened this sameness.

From the start, the group or its self-appointed tricksters used the postulated existence of an eccentric mathematician. Maverick and antihero, the persona of the forlorn mathematician indicates how Bourbaki, in its earlier days, depicted itself as a pariah, both to its members and to outsiders. As a group, Bourbaki thus remained apparently—though not actually—rebellious and marginal, even though its members were themselves far from being so. The collective cultivated the posture of a sect, even as it achieved celebrity and widespread influence within the field of mathematics. The strategy was not exclusive to Bourbaki, however; other groups in France, such as the Durkheimians in sociology and later the *Annales* group of historians, among others, had depicted themselves as marginal or even as pariahs in the combative rhetoric of their early days and thereafter.[61]

Within the Bourbaki precinct the entity named "Bourbaki" took on other shapes and meanings as well. A striking representation of the group, its structure, and its functions emerged in the first issue of Bourbaki's internal newsletter, then called "Journal de Bourbaki." The sentence that frames the function of the "Journal" reads, "To establish by all effective means an intimate connection, a true, vital and concrete communion among the various members of the *body* of Bourbaki." The components of such a body are further detailed in rubrics: limbs, brain, nerves, and stomach all make up that body. This organicist metaphor borrows its imagery from an old judicial allegory which, since Roman times, has often been used to express the double nature of corporations: they are, at once, both one and many, a body politic. (The notion of body politic itself culls from theological thought about the "mystical body" of the church or of Christ.)[62] Bourbaki often expressed the double nature of the team when explaining how "A flash of inspiration, sprung simultaneously from the thought process, at once single and multiple, of the committee creates the Bourbakic functions." In the private Bourbakian lore, however, the organicist image, as well as

[60] John R. Gillis, "Memory and Identity: The History of a Relationship," introduction to *Commemorations: The Politics of National Identity,* ed. J. R. Gillis (Princeton, N.J.: Princeton University Press, 1994, reprint in 1996), p. 3.

[61] A more complete sociological comparison between different groups of French intellectuals is in Beaulieu, *Bourbaki* (cit. n. 4), chap. 5.

[62] Ernst H. Kantorowicz carefully traces the history of the notions of body politic and mystical body in *The King's Two Bodies: A Study in Mediaeval Political Theology* (Princeton, N.J.: Princeton University Press, 1957).

the character of the pariah mathematician, were eventually superseded by the evocation of a "Master" who ruled over his chosen disciples.

Sometimes a spirit, sometimes a hero or demigod, the Master had both human traits and divine attributes. Early accounts insisted on how this powerful being sometimes incarnated "Himself": to announce the time and place of a meeting, Delsarte wrote that "Bourbaki will manifest Himself in concrete space and time on Monday, 16 December, at noon at Capoulade's, the regular venue for His frolics." The formidable personage proved to be a harsh taskmaster who could as easily punish as reward: as of a good shadow one declared "the Spirit of our Master watches over us"; as of an evil force someone talked about "Bourbaki's demons." "He" interfered in the team's work; "His" wit or "His" whims provoked controversy within the group. Reports on meetings say things like, "The Master is very irritated by integration [theory] but satisfied with functional spaces." In contrast, "His" absence stymied the meetings: "Seriousness, discipline and argumentativeness dwindled dreadfully, and the spirit of the Master breathed only fitfully." The rebels of yesteryear accepted a staged tyranny when "Bourbaki, seeking in vain to keep His balance under a democratic regime, swooned with pleasure under the thrashing [administered by Dieudonné]." After having tortured his disciples, the figurehead triumphed over all obstacles. "Bourbaki appeasing (or opposing) the Elements" was a favorite phrase, and the wordplay, of course, quite intentional; someone once suggested (in jest) that a painting on that theme be commissioned from Picasso. Over time, the ineffaceable Bourbaki acquired the characteristics of a Zeus with a white beard and divine attributes, hurling thunderbolts and anathema.

In turn the Master was clad in military or evangelical garb. The first issue of "La Tribu" in 1940 readily hailed an almighty Bourbaki by mixing military and religious metaphors:

> To all our brothers in Bourbaki, greetings and blessings. It is six months now since Bourbaki unleashed His wrath, leaving His chosen people groaning in adversity, scattered to the four corners of the Universe. Some, with their noses in the dust yet unaware Whose hand has struck them down, are still bewildered, seeking in the rumblings of air and earth some dubious augury of a speedy end to all their woes and wasting their strength in menial tasks; others, isolated and deprived of all truly Bourbakic succor, have in their desperation gone so far as to take their offerings to false gods, in honor of whom they spit flame at the sky day and night. Despondency reigns everywhere: no more is hymned the glory of Bourbaki, and the enemies of true Mathematic rejoice in their hearts. . . . Amen.[63]

It was the time of the "Phoney War" in France (1940) and Dieudonné—who had decided to start the newsletter again—adopted the same evangelizing tone that Delsarte had used liberally when addressing his colleagues in prewartime years. While we sense an effort to present Bourbaki's goals as sacred missions, by the same token, terrible enemies are evoked for Bourbaki to vanquish, while good wishes are addressed to Bourbaki heroes and martyrs. The context of imminent war provided a real-life metaphor for more symbolic or alleged confrontations.

With the exception of a few narratives that used military metaphors instead, the

[63] "La Tribu," 15 March 1940, *1*:1. From the papers of Henri Cartan. The original French text appears in Beaulieu, "Jeux d'esprit" (cit. n. 30), pp. 98–9.

ecclesiastical tone prevails in "La Tribu" roughly between 1940 and 1960, when the Bourbaki chroniclers spoke of the Master as though he were a spiritual leader.[64] Such expressions came out of the French Catholic heritage (although several members were Protestants or Jews, many of them nonpracticing), and ecclesiastical notes laced with humor were common in postwar France where clerical as well as anticlerical feelings were strong. The religious overtones may also have indicated the convictions of the members, inasmuch as they felt driven by a sometimes blind faith in the Bourbaki enterprise and were convinced of its value and success. The religious imagery might have expressed, as well as mocked, their dedication to and belief in their goal-driven activity. An evangelical or military tone also connotes strictness or conformity to authority: the Bourbakis certainly wished others to accept Bourbakian ideas, as presented in the *Eléments,* like so many religious dogmas or military orders; even the work's introduction warned readers that the author had adopted a rather dogmatic attitude in mathematics.[65] Bourbaki's use of religious or military imagery, however, eventually disappeared from the chronicle, even before Bourbaki ceased to be influential in the field of mathematics.

No matter what tone the chronicle favored, it usually depicted Bourbaki as being dominated by a fictitious but strong figurehead. In contrast, the group perpetuated and defended a realm in which there was no sovereign distinct from the collective: the team, alone, had sovereignty. Unlike most intellectual clans in France, Bourbaki, the body corporate, had no head.[66] Superiority in mathematical knowledge or a strong personality might have given some members ascendancy over others, but it did not bestow the title of ruler on anyone. *De jure,* Bourbaki became a legal association in 1952 (hence a legal entity) in the sense defined by French law, yet even the statutes of the Association of the Collaborators of Nicolas Bourbaki, which detailed the mandates and responsibilities of the officers (president, secretary, et al.) did not impose an official leader on the group. These functions amounted to no more than having signature authority or the responsibility to report on financial activities; they did not confer on any individual the power to make decisions alone nor did they enhance anyone's personal dominance or charisma.

Within the group, creating a fictitious character had some psychological basis beyond the mere folklore of group kinship. When working for the collective, members often made mathematical choices that they would not have made for their own publications; they also wrote in a different style. What was done for Bourbaki, then, was rather different in nature from what one wrote on one's own behalf. In interviews, members readily expressed the feeling that, when working for Bourbaki, they were individually driven by the internal logic of the enterprise.[67] This is only one

[64] For military metaphors, see, for instance, the details on the meeting of 1945.

[65] In Bourbaki, *Eléments de mathématique* (cit. n. 22), book I, *Théorie des ensembles: Fasicule de Resultats,* "Mode d'emploi de ce traité," p. vii.

[66] One of the first to analyze this trait of the French university and of most fields of knowledge in France is Terry N. Clark, who summarized much of his work in "Le patron et son cercle: Clef de l'université française," *Revue française de sociologie,* 1971, *XII:*19–39. A counterfactual aside may be in order here: had André Weil remained in France and obtained the prominent position that he sought at the Collège de France, it is likely that Bourbaki would have had a designated leader, like any other group of French intellectuals.

[67] From my interviews with Henri Cartan (Paris, July 1986), Pierre Cartier (Montreal, October 1992), Claude Chevalley (Paris, July 1983), Roger Godement (Paris, November 1986), and Jean-Pierre Serre (Paris, July 1986).

step from saying, conveniently and truthfully, that some "spirit" guided them, inspired them, or dictated "His" will to them. The metaphor, then, worked hand-in-hand with a psychological, social, and political reality.

The Bourbaki egalitarian elite had a mock figurehead, which was nevertheless a *tertium quid.* For outsiders, the name and personage of Bourbaki created an aura of mystery around the group; it fascinated many and attracted their curiosity, yet it somewhat protected the team from intrusions. Although the Bourbakian social pact denied individuals any credit for their contributions, it also relieved them of any personal responsibility for the choices made by the collective.[68] Many a member did commit himself to a position in the eyes of his Bourbaki peers by defending one mathematical view over another, and by striving to convince others of his opinion. Yet, as seen from within the Bourbaki confines, the *tertium quid* stood as a scapegoat that carried the burden of mathematical mistakes or fruitless projects. As seen from without, it personified the collective that, as a corporate body, shouldered all the responsibilities of authorship for which individual constituents—the creators—refused to be accountable. *A contrario,* since Bourbaki had but a mock figurehead, the collective bore no responsibility outside mathematics and the strict writing of the *Eléments.* Members exercised individual moral judgement in the public arena, some making public statements on social or political issues (Henri Cartan and Laurent Schwartz on human rights; Claude Chevalley, Roger Godement, and Alexander Grothendieck against the military and nuclear armament), others taking part in mathematics curriculum reforms. The group itself, however, strove to remain amoral, apolitical, and disengaged, depicting its activities as being mostly leisure and fun.[69]

The mythic figure enabled the Bourbaki group to see itself as if through a third person's eyes, and this personage served as a group-memory screen and afforded protection against public involvement on the *agora.* For similar reasons, we may conjecture that disclosing the names of the members might have created a kind of "dismemberment" or dissolution of Bourbaki's corporate body: it might have been equivalent, in some way, to exposure of the members and death of the figurehead. No longer protected by anonymity, individual members would have become accountable for their Bourbakian activities on the same terms as they were for their personal research in mathematics. Bourbaki's masterpiece of mathematical exposition, signed with a pen name, thus concealed not only the identity of its creators but also the terms of Bourbaki's social pact and the integrity of its corporation.

According to Pierre Nora, "The less a memory is relived from within, the more it needs external props and tangible references."[70] Through the systematic use of a nominal figure, Bourbaki produced such artificial props and references. It created a memory screen that shielded individual members from private emotional reflections on the group, its activities, or its internal relationships. Although the group has now (1999) resolved to disclose its oldest documents to the eyes of serious historians,

[68] The conditions and forms of the Bourbaki social pact, an essential element of the group's social makeup, are given in Beaulieu, "Jeux d'esprit" (cit. n. 30), p. 98–109 and further analyzed in Beaulieu, *Bourbaki* (cit. n. 4), chaps. 1 and 4.

[69] In his studies of American scientists, Paul Forman analyzed the self-image, humor and amorality of physicists. See his "Independence, not Transcendence, for the Historian of Science," *Isis,* 1991, *82,* n. 311:75 and footnote 13.

[70] Pierre Nora, "Entre Mémoire et histoire," in *Les Lieux de mémoire,* vol. I, *La République* (Paris: Gallimard, 1984), p. xxvi.

some present-day members seem reluctant to associate themselves with the pronouncements or expressions of personal feelings of other (present or former) members. When former member Pierre Cartier published his reminiscences of André Weil—in which he mentions interpersonal relations within the Bourbaki alliance—he reportedly met the stern rebuttal of some colleagues who deemed the author's expressed feelings too personal and, thus, irrelevant for their profession.[71] To this day, it seems, the unshared or subjective past must be kept behind closed doors or otherwise dissociated from the Bourbaki enterprise, as well as from the French mathematical institutions to which it is, in some way, connected.

The pseudonym that protected anonymity also fostered interchangeability of contributors, a necessary condition for the succession in membership to evolve as naturally as in dynasties. The heirs had been tested on merit, and no outside influence could contest their having been thus designated by the group. Assuming the magic of a phoenix that rises from its own ashes, Bourbaki pretended to transcend mortality by renewal: the team perpetually rejuvenated itself by co-opting new, often younger members and by eventually retiring its older contributors.[72] The Bourbakis liked the idea of passing on the torch, and they chose their successors carefully. In this way they could keep their project going, and hope the team would not only reach its publication goals but continue to play an important role within mathematics. Certainly, no one seriously thought the venture or the group immortal, but the mechanism of succession by co-option preserved the group's identity despite changes in goals and personnel. Adverse circumstances delayed deadlines and writing efforts, or changes of objective postponed the completion of the enterprise to an undetermined date. All these factors nevertheless contributed to the perpetuity of Bourbaki, whose venture remains incomplete and open-ended to this day. Thus, the Bourbaki collective cast itself in a memorial limbo, lingering between an undefined past and a neverending future.

## VI. CONCLUSION

Placing memory in contemporary historiography, Patrick Hutton has shown how the recent literature on commemorations stresses the recollective aspect of memory and deals almost exclusively with memory *qua* representation while ignoring the presence, within memory, of repetition and "habits of mind."[73] In this paper, I have shown instead how Bourbaki's art of memory emphasizes repetition rather than recollection and commemoration.

Bourbaki's chronicle delivers the present-past in a whimsical fashion, and the discursive strategies it deploys create a vast fable with its own repetitions and predict-

[71] From a testimony of Pierre Cartier (May 1999). See his abridged "André Weil (1906–1998)" (cit. n. 14). The unabridged text is published as a preprint of the Philosophy and Mathematics Seminar of the Ecole Normale Supérieure, of which Cartier is an organizer; it is also available on the seminar's Web site. It is perhaps no mere coincidence that the group came to the decision to open its archive soon after the death of its chief trickster and leading founder, André Weil who, for a long time, had been the strictest about secrecy.

[72] It has often been said that Bourbaki members had to retire from the group at age fifty. This so-called rule had many exceptions; its initiation and enforcement are analyzed in Beaulieu, *Bourbaki* (cit. n. 4), chap. 5.

[73] Patrick H. Hutton, *History as an Art of Memory* (Hanover, N.H., and London: University Press of New England, 1993), p. 161.

able events. It records the present as a medley of melodramatic subplots and a profusion of wisecracks. No clear sequence of the events that had really been going on emerges from Bourbaki's humorous narratives. The narratives convey an intense impression, however, couched as they are in a style that verges on the grotesque, to the point that some of the Bourbaki congresses read like scenes from sixteenth-century French satirist François Rabelais's epics of the giants Gargantua and Pantagruel, in which food, drink, and all pleasures of the flesh are hailed whereas serious intellectual pursuits or moral commitments are scorned. Bourbaki's chronicle creates a lively interaction between past, present, and future: it is through humor and as humor that Bourbaki lays down the terms by which it wishes to be portrayed and remembered. Humor produces richly memorable images in Bourbaki's depiction of its own history—a story initially told and retold to the group through a mixture of comedy, memories, forgetfulness, and oblivion. Members share as well what they have forgotten of their common past; for, as there are always memories of pleasant or intense moments, straightening out the odd memories creates a past redolent of sweet evocations. Collective traits and feats are thus engraved the group's memory and stock the imagination of later recruits. If Bourbaki's record is a guardian of the social ties between the members, in turn, that party memory only works so long as it can count on a strong sense of belonging. Bourbakian humor both creates and presupposes a communality of thought among group members, while it serves as a ritual language for remembrance.

Bourbaki's art of memory served its identity: an integral part of the group's *normalien* heritage, it emphasized its ties to that institution. The egalitarianism of the Bourbakis, characteristic of old *normaliens,* in itself made the group rather different from the rest of the French scientific establishment where interpersonal relationships used to be rigid, formal, and highly hierarchical. The subversive power of humor played a considerable part in enabling the unfettered Bourbaki "spirit" to soar above the preestablished structures of the scientific field—and, even more, to represent itself as performing such an escape: Bourbaki's option in favor of secrecy, wit, and playfulness combined to identify a group that deliberately fought against the stranglehold of the scientific field to which, despite itself, it still belonged and which in many ways it reinforced.[74] Thus Bourbaki relieved itself of pressures from any authority—moral, political, or intellectual—and dictated the forms its own freedom would take.

This freedom has hitherto distanced the group from public commemorations, yet the relative segregation in which today's Bourbakis continue to immure themselves can only partly be justified by old habits, together with an elitist dislike of celebrations. At a time when public commemorations abound, we may wonder why a group so influential in twentieth-century mathematics has yet to have its own "party." Instead, the Bourbakis write down their own private, celebratory hymns in which a larger-than-life Bourbaki paints the picture of a shared fraternal past, reckoned in terms of brave founders to celebrate, imaginary enemies to vanquish, and ancient forerunners to worship. Bourbaki's pastiche poems have supported the collective appropriation of a group memory and identity; because of their formal structure and familiar sounds, famous poems have assisted Bourbaki's memory in much the same

[74] Pierre Bourdieu defines the notion of scientific field in "La Spécificité du champ scientifique et les conditions sociales du progrès de la raison," *Sociologie et Sociétés, 7,* 1:91–118.

way as do known places in the classical art of memory. They enabled the group to recall and celebrate its so-called revolutionary days; here, Bourbaki hailed living heroes. They also turned Bourbaki into a monument, conferring on him Greek descent and painting an unequivocal portrait of his vocation; there, Bourbaki ensured for itself a place of honor in the history of mathematics. At times of internal breakaway or of perceived discontinuity in its work, Bourbaki has presented itself as it wished to be remembered rather than as struggling. Founding moments and actors have taken on mythical proportions whereas a shaky, yet hegemonic team celebrated selected, noncontroversial patriarchs. Myths thus turned history into destiny and time into infinity.

Everyone's memory is complemented by the memories of others, to the extent that even a distant or unwitnessed past may become a part of one whole, communal memory. Bourbaki's hoaxes belonged first to a group memory, producing, in fact, the mementos of an elite team that handed down its visions of the past to the later generations of members it had co-opted. Through a persistent use of rumors, anecdotes, and pranks, the group also has contributed amply to the oral lore of the mathematical village. Repetitious rags, instigated by group members or copied by outsiders have, for many years, determined Bourbaki's self-built image and story, and constitute a kernel of collective memory in twentieth-century mathematics. For nearly four decades the proliferation of Bourbaki-related japes and anecdotes has kept the team in the limelight and emphasized the power that this mandarin group held over other mathematicians.

Through the rehearsals of fiction Bourbaki was rendered unchanged and immortal while the collective itself throve on plurality and succession. Frequent appearances of the Bourbaki personage, easily recognized by outsiders and willfully invoked in the Bourbaki internal chronicle, both indicated and fostered a form of unanimity within a tradition. In the accounts of their congresses as well as in their contacts with outsiders, the Bourbakis have regularly played on the multivalence between an elusive mathematician called Bourbaki, the team itself, author of the treatise, and the individual members, the creators whose identities were not revealed outside the Bourbaki precinct. The fictitious figure of Bourbaki has allowed the group to present itself, its discourses, and its practices as so many absolutes, while frequent appearances of the *tertium quid* in the accounts of meetings have constituted a memory screen on which fixed images were projected, over and over again. That personage also has erected an action screen, shielding the Bourbakian body politic from the many solicitations related to public moral, social, and political matters. In that context Bourbaki's art of memory has left as little room for group recollection as for individual reflection.

Present-day Bourbakis know little about the past of their group. With the exception of stock jokes and select anecdotes retold by their predecessors, members confess to being quite ignorant of what went on when former members were writing their parts of the *Eléments de mathématique*.[75] Gone are the discussions over mathematical issues—with the exception of those retold by the protagonists as epics; vanished are the questions and solutions, the crises and breakthroughs of the past. A

[75] This information I gleaned mainly from my interviews with Pierre Cartier, a recent former member (Paris, December 1986; Montreal, October 1992) and Jean-Christophe Yoccoz, a present member (Montreal, May 1997).

record is there, however, and, until recently, the Association of the Collaborators of Nicolas Bourbaki jealously guarded its collection of issues of "La Tribu," of draft texts, and of numerous lengthy, technical comments. While, until now, Bourbaki forbade nonmembers access to its archive, the members themselves felt little need to consult the accounts of the distant past, with the exception of former members Armand Borel and André Weil, who used the archive when they wrote down their recollections. What they identified with seems to be the past glory of the Bourbaki enterprise and, above all, the parallel success of the ongoing public Séminaire Bourbaki. What remains in this group's private memory, then, are fuzzy images, anecdotes with all their variants, worn-out jokes, scant relics, and vague grand narrations, all fading and shrinking into a small core: a tough kernel that the sharpest, even self-critical insights have not yet dented.

Their group memory has not yet faced up to the past and present-day members do not particularly seek the confirmation of their fleeing present, nor do they wish to associate themselves, as a team, to the work of historians. Moreover, although they may not yet perceive their task as being done, the Bourbakis are not setting a new writing agenda nor envisaging how an enterprise such as the old Bourbaki oeuvre could survive the plurality of concerns of the latter-day multivalent mathematical sciences. They do not peruse old lessons to map their future, thus they leave their past unexamined. Responding again to the demands of this timeline, the Bourbakis might attend to their group past once they see their group future as close-ended. For them, however, the time of group recollection and public commemoration has yet to come; perhaps it has even passed.

# Jocular Commemorations

## The Copenhagen Spirit

*Mara Beller**

> There are some things that are so serious that you can only joke about them.
>
> —*attributed* to Niels Bohr in Weisskopf, *Joy of Insight*

> Our laughter about Bohr was the escape route which enabled us to say that, although often we could not understand him, we admired him almost without reservation and loved him without limits.
>
> —Rudolf Peierls in French and Kennedy, *Niels Bohr: A Centenary Volume*

### I. INTRODUCTION: COMMEMORATIONS, CHARISMA, AND LAUGHTER

As do other public events, commemorations provide ample opportunities for reflecting on and understanding prevailing social structures. Commemorations direct the attention of their participants to objects and events of special significance and provide an occasion for a cohesive collective experience. Combining formal, ritualistic features with more spontaneous, emotive aspects, commemorations serve as a potent tool for defining individual and collective self-identity. Dense with symbols, associations, and meanings, commemorations—including scientific ones—provide a stage where people of a certain community can conceive and elaborate upon "a story they tell themselves about themselves."[1]

Public events address, and sometimes redress, the problems in the forms of life of a specific community. They point to, and sometimes even reproduce, the significant features of the world we live in. Yet as opposed to the multifaceted, incomplete flux of becoming, public events, including commemorations, are designed to convey

*Program in History and Philosophy of Science, Department of Philosophy, The Hebrew University, Mount Scopus 91905, Jerusalem, Israel.

I am grateful to Felicity Pors from the Niels Bohr Archives at the Institute for Theoretical Physics in Copenhagen, who kindly and promptly provided copies of the *Journal of Jocular Physics* (1935, 1945, and 1955), and to Ilana Silber for valuable discussions. I also wish to thank Eugene R. Speer for permission to quote from his poem, George Gamow's family for permission to reproduce his drawings, and Sam Schweber, Pnina Abir-Am, and Clark Elliott for their editorial remarks.

[1] Literature on the sociology and anthropology of public events is voluminous. Recent important works include Don Handelman, *Models and Mirrors: Towards an Anthropology of Public Events* (Cambridge: Cambridge University Press, 1994). Handelman's book contains an up-to-date, extensive bibliography on this and other relevant issues. The quotation is by Clifford Geertz in reference to the Balinese society in Clifford Geertz, *The Interpretation of Cultures* (New York: Basic Books, 1973).

messages with well-defined meaning. By freezing the ambivalent, open-ended experience into a narrative of rigid meaning, they impose the official version of events—the official "truth."

There is, however, a special kind of public event—jocular, or carnivalesque—that is uniquely suited to express the ambiguity and incompleteness of the live experience. In an atmosphere of gaiety and fun, it can question, in a seemingly nonoffensive way, the official authority and the prevailing order. In play and in laughter, jocular public events can offer a counter version to the official ideology, neutralize hierarchical distinctions, raise doubts about accepted conventions, and even concomitantly hint at alternatives. The jocular culture is, however, not merely subversive—it can also enhance the social order (Durkheimian view of comedy) and provide an outlet for the release of accumulated tensions and frustrations.[2] Both aspects were amply displayed in the jocular atmosphere during Niels Bohr's lifetime at the Bohr Institute for Theoretical Physics in Copenhagen.

In this paper I analyze the *Journal of Jocular Physics,* which was issued in Bohr's institute in 1935, 1945, and 1955 on the occasion of his fiftieth, sixtieth, and seventieth birthdays, respectively. (The three issues are available at the Niels Bohr Archives in Copenhagen.) I also analyze the parody on *Faust* written by Bohr's disciples and performed in Copenhagen to simultaneously commemorate the tenth anniversary of Bohr's institute and the hundredth anniversary of Goethe's death. These were not unusual events in the Copenhagen institute, which during the '20s and '30s was the most important center of theoretical physics worldwide. Each spring theoreticians from many countries assembled in Copenhagen to discuss recent developments in physics and it was customary, at the end of each conference, to produce a stunt performance—a jocular commentary on the field's advances and problems.[3]

I will also briefly comment on some expressions and arguments in two nonjocular commemorative volumes celebrating the hundredth anniversary of Niels Bohr's birth. One of the volumes contains proceedings of the American Academy of Arts and Sciences (AAAS) symposium held at the Massachusetts Institute of Technology in Cambridge. The other volume is an official publication of the International Commission of Physics Education of the International Union of Pure and Applied Physics.[4] The spirit of these volumes, which were published after Bohr's death, contrasts markedly with the jocular celebrations in Copenhagen.

Niels Bohr, the founder of the quantum theory of the atom, is one of the greatest scientists of the modern era, along with Isaac Newton, James Maxwell, and Albert Einstein. Bohr's model of a planetary atom serves as an icon of twentieth-century physics. As a founder of the Copenhagen Institute for Theoretical Physics and its director from 1921–1963, Bohr stood at the center of many pivotal advances in quantum physics. As the editors of the AAAS commemorative volume put it, "More than any other single individual, Bohr was responsible for the development of quantum

[2] Handelman, in *Models and Mirrors,* analyzes the Nuremberg Carnival and the ways in which social order was questioned by carnivalesque occurrences in keeping with the medieval European worldview. The classic study of the medieval folk culture of laughter and carnival is that of Mikhail Bakhtin, *Rabelais and His World* (Bloomington: Indiana University Press, 1984).

[3] Victor Weisskopf, *The Joy of Insight* (New York: Basic Books, 1991), p. 68; George Gamow, *Thirty Years that Shook Physics* (New York: Dover, 1985), p. 168.

[4] *Niels Bohr: Physics and the World,* eds. Herman Feshbach, T. Matsui, and A. Oleson (New York: Harwood Academic Publishers, 1988); *Niels Bohr: A Centenary Volume,* eds. A. P. French and P. G. Kennedy (Cambridge, Mass.: Harvard University Press, 1985).

mechanics and for many of its applications to the fundamental understanding of the physical world."[5]

The papers in these commemorative volumes describe Bohr's unique contributions to physics and his leadership of the community of theoretical physicists. They provide reminiscences of Bohr's life and science and simultaneously chronicle a great era in physics. No less important for the editors of these volumes was to give some sense of Bohr as a human being of unique personality, and as a role model to follow: Bohr set "an example of a scientist's concern with the human condition, one that every scientist should take to heart."[6]

Bohr is also widely known in educated circles as the main architect of the Copenhagen interpretation of quantum physics. Bohr's philosophy of physics implies a denial of the classical, observer-independent conception of reality and the overthrow of determinism. Connected with these issues is Bohr's assertion of complementarity between different atomic attributes. Complementarity implies that, in quantum physics, one can only have partial, equally correct, yet mutually incompatible perspectives, disclosed in mutually exclusive experimental arrangements (in some arrangements a microentity behaves as a wave, in others as a particle). It is not possible, according to the Copenhagen philosophy, to combine these partial pictures into a unified complete picture, and it is not meaningful to talk about physical reality as existing independently of the act of observation.

Bohr extended the philosophy of complementarity to serve as an overarching principle of knowledge, applicable not only to physics but to biology, psychology, and anthropology; he also expected complementarity to substitute for the loss of religion. Despite the heavy rhetoric of the "inevitability" of complementarity and its status until the early '60s as unquestionable dogma, the relevancy and correctness of Bohr's philosophy, in physics and elsewhere, is increasingly being questioned.[7]

As the founder of the philosophy of complementarity, Bohr was declared by his followers to be not merely a great philosopher, but a person of exceptional—perhaps superhuman—wisdom, both in science and in life. Some of Bohr's disciples did not hesitate to call him "the wisest of living men." Bohr's unprecedented authority resulted not only in an uncritical following of the Copenhagen philosophy, but sometimes even in a blind acceptance of Bohr's opinion on such matters as the physics of the atomic bomb. According to Richard Feynman's reminiscences about Bohr's visit to Los Alamos, "even to the big shots, Bohr was the great God."[8]

A touching and genuine affection, even love, shines in the words of many of Bohr's disciples. Yet these words also reveal an unbounded, intense admiration, often bordering on worship. Describing the atmosphere in Copenhagen, John Heilbron

[5] Feshbach, Matsui, and Oleson, *Bohr: Physics* (cit. n. 4), p. xii.

[6] *Ibid.*, p. xii.

[7] Today there are many alternatives to the Copenhagen interpretation that an increasing number of physicists find more promising. The Bohmian alternative and its numerous variants are presented in James T. Cushing, Arthur Fine, and Sheldon Goldstein, *Bohmian Mechanics and Quantum Theory: An Appraisal* (Dordrecht: Kluwer Academic Publishers, 1996). On the rhetoric of inevitability, see Mara Beller, "Bohm and the 'Inevitability' of Acausality" in the same volume. On Bohr's philosophy of complementarity, see Henry Folse, *The Philosophy of Niels Bohr* (Amsterdam: North Holland, 1985), and John Honner, *The Description of Nature* (Oxford: Clarendon Press, 1987).

[8] Richard Feynman, "Los Alamos from Below: Reminiscences of 1943–1945," *Engineering and Science,* 1976, *39:*11–30.

compared Bohr with a guru.[9] Abraham Pais, a prominent theoretical physicist and Bohr's faithful disciple, vigorously objected to such a characterization.[10] And while Bohr's image as a sage should not be taken too literally—there was more ambivalence and freedom in Copenhagen than the concept of the guru implies—it nevertheless aptly describes many aspects of Bohr's personality and the social atmosphere in Copenhagen.

Bohr perhaps comes closer than any other scientist to the Weberian notion of a charismatic leader. The concept of charisma refers to exceptional personal qualities of chosen individuals; simultaneously, it is a powerful tool for analyzing a typology of forms of authority. Weber defines charisma as "a certain quality of an individual personality by virtue of which he is set apart from ordinary men and treated as endowed with supernatural, superhuman, or at least specifically exceptional qualities."[11]

Charismatic religious leaders are regarded as being set apart from simple mortals; the devotee's enthusiasm is rooted in their belief that the leader is able to unlock the riddles of existence and of cosmic order, of what is fundamental and sacred. Charismatic leaders are not merely bearers of extraordinary traits, they are able to reshape cognitive, symbolic, and institutional orders. Bohr was surrounded by the mythology of his superhuman intuition, which enabled him to have direct access to the secrets of nature. Elevating Bohr's depreciation of mathematics into an admirable virtue, Bohr's devotees believed that, unlike ordinary mortals, Bohr did not need to calculate in order to obtain "the truth."

According to the Weberian concept, charismatic personalities do not have to establish a special hierarchy. Rather, the behavior of the disciples results in the establishment of order and protects the charismatic leader. According to Victor Weisskopf's reminiscences, "Without being assigned this task, the inner circle at the institute assumed the responsibility of protecting Bohr from visitors we considered unworthy of working with him." Among the members of the Copenhagen "clan" (Weisskopf's term), Bohr's assistants referred to themselves and were referred to by others—half tongue-in-cheek—as "slaves" or "victims."[12]

It is not merely a privilege but a duty of those who participate in the social order around a charismatic personality to recognize his unique powers and act accordingly.

[9] As John Heilbron noted, and as correspondence reminiscences of Bohr's disciples amply demonstrate, Bohr was considered by those close to him not merely as a physics teacher, but as a spiritual guide. The following quote, by the English chemist F. G. Donnan, is a striking yet characteristic example: "We all look up to you as the profoundest thinker in science . . . the 'Heaven-sent expounder' of the real meaning of these modern advances. . . . I can and will think of you walking in your beautiful gardens, and stealing some moments of peace whilst the leaves and the flowers and the birds whisper their secrets to you." Quoted in John Heilbron, "The Earliest Missionaries of the Copenhagen Spirit," in *Science in Reflection,* ed. Edna Ullman (Dordrecht: Kluwer Academic Publishers, 1988), pp. 201–33.

[10] Abraham Pais, *Niels Bohr's Times, in Physics, Philosophy and Polity* (Oxford: Clarendon Press, 1991), p. 435.

[11] Max Weber, *Theory of Social and Economic Organization* (New York: Free Press, 1969), p. 329. See also Max Weber, *Economy and Society: An Outline of Interpretive Sociology* (New York: Bedminster, 1988), pp. 241–51. See also a very useful selection of Weber's papers by Shmuel N. Eisenstadt, *Max Weber On Charisma and Institution Building* (Chicago: The University of Chicago Press, 1968).

[12] Weisskopf, *Joy of Insight* (cit. n. 3), p. 69 and p. 131. About assistants as "slaves," see further in this paper.

When faith in the wisdom of a leader is shaken, his followers experience intense guilt feelings. There is ample evidence of the guilt experienced by those followers who questioned Bohr's wisdom. When Carl Friedrich von Weizsäcker referred to the joking about Bohr's mathematical abilities, he called himself and his peers "bad boys."[13] When Bopp argued in a letter to Bohr against the Copenhagen interpretation, he prefaced his criticism with the following, guilt-ridden words: "A young Japanese colleague said once: 'We must do some things, that our parents do not understand, and it is very painful.' "[14]

Many physicists, including Wolfgang Pauli, Pascual Jordan, Abraham Pais, and Jørgen Kalckar, used the term "father figure" in reference to Bohr. As Otto Frisch recalled about his first meeting with Bohr, "To me it was a great experience to be suddenly confronted with Niels Bohr—an almost legendary name for me—and to see him smile at me like a kindly father."[15]

Einstein referred to Bohr as a prophet, and Bohr also playfully characterized himself as such. Prophets do not merely bear immense wisdom, but counsel individuals in their personal affairs, and they often make statements about public issues and ethical ways of life. All this applies in Bohr's case. As A. P. French summarizes in Bohr's centenary commemorative volume, "The physicists who as young men came to Copenhagen to work with Bohr speak of his having shaped their characters as well as their minds."[16]

Charismatic leaders, or prophets, might experience extreme emotional states in order to reach the mystical truth. They can enter into ecstatic trances, choose extreme forms of asceticism, or experience intense suffering and martyrdom. Bohr's devotees tell of Bohr's "suffering" in his search for truth. Thus, after his first encounter with Bohr, von Weizsäcker wrote in his diary, "I have seen a physicist for the first time—he suffers as he thinks." Similarly, suffering characterized Bohr's public talks: "At the 1932 conference Bohr gave a fundamental report on the current difficulties of atomic theory. . . . With an expression of suffering, his head held to one side, he stumbled over incomplete sentences."[17]

Bohr's devotees presented his difficulty expressing himself (which was naturally exacerbated when Bohr tried to talk about abstract mathematical theories in simple words) as an uncompromising search for the ultimate truth. Von Weizsäcker described Bohr's suffering when he tried to compare the process of thinking with the

[13] Carl Friedrich von Weiszäcker, "A Reminiscence from 1932," in French and Kennedy, *Bohr: Centenary* (cit. n. 4), p. 186.

[14] Bopp to Bohr, 4 February 1962, Archive for the History of Quantum Physics, American Philosophical Society, Philadelphia, assembled and edited by Thomas S. Kuhn, John Heilbron and Paul Forman and Lini Allen (hereafter cited as AHQP); *Sources for History of Quantum Physics: An Inventory and Report* (Philadelphia: American Philosophical Society, 1967): "Ein junger japanischer Kollege hat einmal gesagt: wir müssen etwas zu tun, was unsere Eltern nicht verstehen, und das schmerzt uns sehr."

[15] Otto Frisch, "The Interest in Focusing on the Atomic Nucleus," in *Niels Bohr, His Life and Work as Seen by His Friends and Colleagues,* ed. Stephen Rozental (New York: Interscience Publishers, 1967), p. 137. For Jordan's, Pais's, and Kalckar's comments, see Pascual Jordan, *Der Naturwissenschaftler vor der religiösen Frage* (Hamburg: Stalling, 1972), p. 2114; Pais, *Niels Bohr's Times* (cit. n. 10), Jørgen Kalckar, "Niels Bohr and His Youngest Disciples," in Rozental, *Bohr: Life* (cit. n. 15), p. 227.

[16] Heilbron, "Earliest Missionaries" (cit. n. 9), p. 221; Max Weber, "The Prophet," in Eisenstadt, *Max Weber* (cit. n. 11), pp. 253–67; A. P. French, "Some Closing Reflections," in French and Kennedy, *Bohr: Centenary* (cit. n. 4), p. 353.

[17] Weizsäcker, "A Reminiscence from 1932" (cit. n. 13), p. 183 and p. 186.

Riemann surface: "This was an abstract theory that he [Bohr] could not easily formulate in words; it was at the same time existential suffering. His stumbling way of talking . . . would become less and less intelligible the more important the subject became, and that came from this suffering." This suffering probably aroused empathy in Bohr's listeners, but also a great deal of frustration. But the more incomprehensible Bohr's words became, the more his devotees' conviction grew that a great truth, inaccessible to simple mortals, was hidden therein. They would spend hours, sometimes years, trying to recover the hidden meaning. Bohr could have said only one word—"thinking" or "harmony"—and it would set the ruminations of a disciple into motion.[18]

John Wheeler compared Bohr's wisdom with that of "Confucius and Buddha, Jesus and Pericles, Erasmus and Lincoln."[19] And yet it would be wrong to present Bohr as a guru or sage. While in matters of complementarity philosophy not directly relevant to research, physicists were willing to repeat "Bohr's Sunday word of worship," in physics proper they maintained a fruitful balance between humble reverence and free creativity.[20] Except for Bohr's temporary slaves or assistants, physicists kept a healthy distance from Bohr, which allowed them to pursue their work independently. It is this healthy distance that was sustained by, and revealed in, the jocular Copenhagen spirit. This jocularity permeates the lines of the students' parody on *Faust*.[21]

Healthy laughter is a sign of freedom of the spirit, freedom of expression. As opposed to the officialism of serious commemorations, the jocular ones revealed the living nerve, the deep-seated ambivalence, the genuine affection, and the self-mocking awareness that characterized the spirit around Bohr. In an atmosphere of gaiety and play, the young physicists made pointed commentary on the social hierarchy in Copenhagen and expressed their doubts about the official Copenhagen philosophy. Laughter was a respite from strenuous work, and at the same time an outlet for frustration. There is a delicious irony in that this laughter was often directed at exposing the pretense of truth of the complementarity principle—a principle that in an authoritative, dogmatic way denies the notion of a single truth.

Humor is, of course, multipurposeful and multifaceted. It can be recruited to support the existing order and ideology, and the *Jocular* journals contain examples of such humor. In fact, some of the contributions are not funny at all, such as the ode "The Atom that Bohr Built" in the 1955 issue, or the Japanese poem in the 1935 issue that celebrates Bohr's birthday in a worshipping manner: his students "follow the star [Bohr] who lights up their path."

The *Jocular* journals also contain contributions that reveal a touching, if somewhat obedient, affection for Bohr. One is the story titled "The New Elephant Child,"

[18] *Ibid.,* p. 185. As von Weizsäcker reported: "[H]e started a philosophical conversation with me by going to the blackboard and writing a single word 'Thinking.' Then he turned to me and said, 'I only wanted to say that I have written down something here which is quite different from writing down any other word.'" *Ibid.*

[19] John A. Wheeler, "Physics in Copenhagen in 1934 and 1935," in French and Kennedy, *Bohr: Centenary* (cit. n. 4), pp. 221–26, quotation on p. 226.

[20] "Bohr's . . . worship" is Alfred Lande's expression. Interview with Lande, AHQP (cit. n. 14).

[21] The parody on *Faust* was conceived and written by students of Bohr, notably by Max Delbrück and C. F. von Weizsäcker, and performed in Copenhagen in 1932. The original German version is available in AHQP (cit. n. 14), microfilm 66. The English translation is in Gamow, *Thirty Years* (cit. n. 3).

an improvisation on Rudyard Kipling's story in the 1955 issue. Instead of Kipling's bad elephant child, this story has a good elephant child—Bohr—who was never spanked (unlike the bad one) because he was so good and who asked ever so many questions in early childhood and new and unheard-of questions when he grew bigger. Eventually he grew into "a big and wise and peaceful uncle." He was surrounded by smaller animals; they all "picked up his habits, and they also grew big and wise. And the Elephant Child got a great big house, where he could dwell and make parties for all the greater and smaller animals. And he would play with the smaller animals whenever he had time."

Yet many other contributions send a message of disagreement. The criticism is not satiric; rather, the humor is good-spirited and good-hearted. It is a joyful, carnivalesque spirit in a Bakhtinian sense—the laughter simultaneously affirms and denies, celebrates and criticizes. It expresses at the same time genuine affection and meaningful reservation. The message of reservation and disagreement is emphasized in the epigraphs to *Faust* and to the 1955 *Journal of Jocular Physics.* In the first epigraph is a quote of Bohr's words "Not to criticize . . . ," and in the second epigraph there is delightful self-denial: "We agree much more than you think." Bohr used both expressions when he intended to voice total disagreement.[22]

## II. HERO WORSHIP: COMPLEMENTARITY BETWEEN SERIOUSNESS AND HUMOR

As is suitable for prophets, Bohr often used metaphors and allegories. He inspired by personal contact, endowing the existential message of complementarity philosophy with intense personal feeling. Bohr's authority was based on his outstanding past achievements, a formidable institutional power, and his unique personal charisma. Bohr could provide intellectual stimulation and help in advancing careers, spiritual fulfillment and down-to-earth fun, material benefits and psychological counsel. He became a father figure who many young scientists were eager to honor and whose authority not many dared to challenge.

Both the conversation and the environment in Bohr's home made a profound impression on young scientists who came to spend time in Copenhagen. Especially in his later years, Bohr's home was a center of social and intellectual life visited by famous politicians, artists, scientists, even Denmark's royal family. The discussion comprised philosophy, history, fine arts, religious history, ethical questions, politics, current events, and many other topics. Physicists found it a special, intoxicating privilege to be not simply in the Mecca of theoretical physics, but also at the intellectual Olympus. "One could correctly describe Bohr's house as resembling a Greek academy," wrote James Franck.[23] Sitting at Bohr's feet, his admiring disciples absorbed his wisdom. "Bohr had invited a number of us to Carlsberg where, sipping our coffee after dinner, we sat close to him—some of us literally at his feet, on the floor—so as not to miss a word. Here, I felt, was Socrates come to life again, tossing us challenges in his gentle way, lifting each argument onto a higher plane, drawing wisdom out of us which we didn't know was in us (and which of course, wasn't). Our conversations ranged from religion to genetics, from politics to art; and when I

[22] Pais, *Niels Bohr's Times* (cit. n. 10), p. 5; French and Kennedy, *Bohr: Centenary* (cit. n. 4), p. 188.

[23] James Franck, "A Personal Memoir," in French and Kennedy, *Bohr: Centenary* (cit. n. 4), pp. 16–18, quotation on p. 18.

***Figure 1.*** *Title Page of the Copenhagen* Faust.

cycled home through the streets of Copenhagen, fragrant with lilac or wet with rain, I felt intoxicated with the heady spirit of Platonic dialogue."[24]

Yet often Bohr's authority was anything but gentle. As von Weizsäcker put it, "He was one of those rare teachers who knew how to apply caution and, if necessary, force . . . . "[25] Bohr pursued his objectives forcefully, and, as many of those who came in contact with him remember, he rarely lost an argument, both when he was right and when he was wrong. We have the testimony of John Slater, who was "locked away in a back room" when he disagreed with Bohr on the contents of the Bohr-Kramers-Slater paper; and we have a reminiscence by Werner Heisenberg who described the "almost fanatical relentlessness" with which Bohr argued with Erwin Schrödinger. Heisenberg also described his own encounter with Bohr over the uncer-

[24] Otto Frish, quoted in French, "Some Closing Reflections," in French and Kennedy, *Bohr: Centenary* (cit. n. 4), pp. 351–53, quotation on p. 353.
[25] C. F. von Weizsäcker, quoted in French and Kennedy, *Bohr: Centenary* (cit. n. 4), p. 353.

***Figure 2.*** *Bohr's opponent, Landau, bound to a chair and gagged.*

tainty paper, when he "burst into tears, unable to sustain any more the pressure of Bohr."[26] By his inexhaustible determination, Bohr exhausted his opponents. Schrödinger, in Heisenberg's words, "became ill" from "over-exertion." Similarly, as Leon Rosenfeld described, Rudolf Peierls left Bohr's institute "in a state of complete exhaustion" after discussions with Bohr and Lev Landau over the content of a paper they had written (about uncertainty relations in relativistic quantum mechanics).

Friendly and mocking humor is a perfect outlet for releasing the tension between genuine affection and unbearable frustration. George Gamow, who was at the Copenhagen institute at the time of the Bohr-Landau-Peierls encounter, depicted the spirit of discussion in an eloquent cartoon (Figure 2).

The episode involving Bohr, Landau, and Peierls was commemorated in the Copenhagen parody on *Faust,* where Bohr, fittingly, was represented by God himself:

> The Lord: Keep quiet, Dau! . . . Now, in effect
> The only theory that's correct
> Or to whose lure I can succumb
> Landau: Um! Um-um! Um-um! Um-um!
> The Lord: Don't interrupt this colloquy!
> *I'll* do the talking. Dau, you see
> The only proper rule of thumb
> Is
> Landau: Um! Um-um! Um-um! Um-um!

[26] Interview with John Slater, AHQP (cit. n. 14). See also Werner Heisenberg, "Quantum Theory and Its Interpretation," in Rozental, *Bohr: Life* (cit. n. 15) and interview with Heisenberg, AHQP (cit. n. 14).

Especially complex was the predicament of Bohr's assistants. As Weisskopf depicts it, "The victim's life was not his own; he was at the mercy of Bohr's whims."[27] "When Bohr decided that he wanted to work on a paper the victim was called into his office. . . . There he stayed throughout the day while Bohr roamed around the room circling the victim's desk every few minutes. . . . By then I had already developed my lifelong habit of pacing while I worked. But Bohr would not have it. 'Only one of us is allowed to move' was his rule, so I sat there day after day in agony."[28] The victim's duty was to point to any lack of clarity in Bohr's formulation—a job that, in Weisskopf's words, demanded "a certain amount of courage." Sometimes Weisskopf would suggest ways to shorten Bohr's legendarily long sentences, yet Bohr refused. "Bohr would insist—and, naturally, he rarely lost." The impeccably honest Paul Dirac, not surprisingly, ended his role as a victim in half an hour. Those who stayed convinced themselves that their predicament was, in fact, a unique asset.[29]

In the Copenhagen *Faust,* Leon Rosenfeld, who played Mephisto (representing Pauli), sat—in true Copenhagen spirit—at the feet of the Lord and announced:

> since it seems you favor me a bit,
> Well, now you see me here [turns to the audience] among the slaves.

A peculiar dramatic tension must have been created by the juxtaposition between the slave Rosenfeld, who always fiercely and uncritically defended Bohr's stand, and Pauli (Mephisto), whose opinion Bohr respected more than anyone's. And a humorous inversion that contained a hint of truth was presented in *Faust:* Pauli dares to order, time after time, the Lord to keep quiet while he talks!

From the mid-'20s on, Bohr did not actively pursue research in quantum theory, devoting most of his energies to the creation of a unique form of life in the Copenhagen institute.[30] From 1922 to 1930 no paper by Bohr as a single author appeared, except his published talks. Bohr had many visitors and an extensive scientific correspondence (which he mostly dictated to his assistants), and he provided guidance to his younger colleagues. Many young physicists found contact with Bohr meaningful and most inspiring. Others were disappointed when, coming to Copenhagen, they learned that Bohr was working on complementarity—complementarity is not physics.

In the 1935 *Journal of Jocular Physics,* prepared for the occasion of Bohr's fiftieth birthday, Bohr's younger disciples mock him gently: While initially they had planned to present to Bohr a scientific Festschrift, they decided to relieve Bohr from such an

[27] Weisskopf, *Joy of Insight* (cit. n. 3), p. 70. Bohr's assistants, as their reminiscences indicate, had little time and space to pursue their own scientific work. Oskar Klein revealed that he accomplished his most original work when Bohr was away from Copenhagen. Similarly, Heisenberg was able to write his uncertainty paper when Bohr left Copenhagen for a skiing vacation.

[28] *Ibid.,* p. 71.

[29] "When the victim got over the feeling of being used, he realized that he had been given a glorious opportunity. He was witness to the mind of Niels Bohr in action." *Ibid.,* p. 71.

[30] "The time now approached [1923] when the main progress [was] comprised of scientists of a younger generation, but with Bohr still as a giver of ideas and increasingly as the philosophical integrator of the knowledge obtained." Oskar Klein, "Glimpses of Niels Bohr," in Rozental, *Bohr: Life* (cit. n. 15), p. 85.

"arduous task." They confined the Festschrift to subjects of a jocular character because of the danger that Bohr "would feel it as his duty to read the contents and even to try to learn something from them . . . . " The preface continues with a rationale for a jocular journal in a lovely mocking of Bohr's style: the celebration of Bohr's fiftieth birthday is "an eminently appropriate occasion for [such] a step, long since made an urgent need by the stormy development of our science"; the attempts to explore the jocular aspects of the physical world had been neglected until then "for reasons hitherto not quite clearly understood"; the authors strove "to preserve the very attitude of hopeful pessimism and serene preparedness" that characterized Bohr's approach to physics.

Commemorative talks and papers about Bohr often point out his "supernatural" intuition. It was this intuition that allowed Bohr, in 1913, to perform one of the most striking intellectual feats in the history of science: the construction of a theory of the atom in the image of and in contradiction to the dynamics of the solar system. Legend has it that Bohr's superhuman intuition allowed him to grasp the secrets of nature in some direct, immediate manner in cases where simple mortals, no matter how gifted, needed to calculate. And while on serious commemorative occasions Bohr's intuition is hailed without reservation, the jocular ones reveal the times when Bohr's intuition led him fundamentally astray. Bohr's denial of the existence of the neutron, positron, and neutrino are striking cases in point, as brilliantly depicted in the Copenhagen *Faust.*

A number of crucial problems characterized quantum field theories in the '30s. There were the problems of calculated infinite self-energies of an electron in the interaction of the electron with its own field, and of the negative energy solutions in relativistic quantum mechanics. As far as nuclear physics was concerned, the main issues revolved around the question of whether electrons, emitted in β-decay, are ejected from the nucleus. The prevalent assumption that the nucleus was composed of protons and electrons contradicted both the theoretical and experimental state of affairs. The problems were eventually solved by the incorporation of new particles into physics, to which Bohr vigorously and adamantly objected. Bohr attempted to solve all these puzzles in the same idiosyncratic way—simply by applying epistemological analysis to fundamental concepts and demonstrating the limits of their applicability. The idea was to dispense with the concept of a particle below the "classical radius" of the electron. This led to the resuscitation of Bohr's previously failed idea of nonconservation of energy.

Bohr's approach was a veritable failure. The discovery of the new particles was a triumph for those theoretical physicists who seemed to follow the motto that "mathematics knows better than intuition," even the intuition of the great Bohr himself. Many physicists (Dirac, Rutherford, Peierls, and Pauli) tried to warn Bohr that with his suggestion of nonconservation, he was, as Pauli put it, on a "completely wrong track."[31] It is in such cases that the physicists, as our epigraph by Peierls tells, could not understand Bohr and turned to humor as an "escape route."

If the official, serious commemorations celebrated Bohr for his unbounded curios-

[31] When Bohr sent his ideas on nonconservation in β-decay to Pauli, Pauli's reaction was curt. "The note about the β-rays . . . gave me very *little* satisfaction . . . *let this note rest for a good long time*" [Pauli's italics]. Pauli to Bohr, 17 July 1929, in *Bohr's Collected Works* (Amsterdam: North Holland), vol. 6, p. 443.

ity and desire to leave his mark on every problem in physics, the jocular were more reserved. In the Copenhagen *Faust* Mephisto—the ever-irreverent Pauli—scolds the Lord (Bohr) for ecstatically "sticking his nose" into every matter:

> Yet here you are in ecstasy again,
> Approving views that shatter like a bubble,
> Sticking your nose in every kind of trouble."[32]

A delightful exchange follows in which the Lord is bewildered by the possibility of a chargeless, massless particle and insists on the need to "remember the essential failure of classical concepts . . . . " Similarly revealing is the exchange between Dirac, who defends his idea of electrons occupying the negative energy states ("that donkey-electrons should wander quite aimlessly through space, is a slander") and Bohr, whose arguments against antimatter seem irresistible:

> But the point of the fact is remaining
> That we cannot refrain from complaining,
> That such a caprice
> Will reveal the malice
> Of devouring the world it is sustaining.[33]

This quote captures a characteristic feature of Bohrian arguments. Complementarity philosophy denies the very possibility of new, nonclassical concepts, just as this humorous, fictitious argument denies the possibility of positrons. Like many of Bohr's complementarity arguments, the reasoning seems to be highly persuasive, yet proves to be wrong. This presentation of Bohr's stand on the existence of new particles in the '30s creates a strong ironical effect, for these dialogues follow the opening words of the Copenhagen *Faust:* no matter how incomprehensible the Lord's word is, it is grand and inspiring, simply by virtue of being the Lord's word.

## III. JOCULAR CRITIQUE OF COMPLEMENTARITY

The jocular critique of Bohr's complementarity contained in George Gamow's *Mr. Tompkins in Wonderland* undermined Bohr's insistence on an "essential," "principal" cut between the classical and the quantum domain, on which Bohr built his arguments of "uncontrollability" of measurement interaction (from which other aspects of complementarity philosophy "inevitably" followed).[34] Changing at will the order of magnitude of Mr. Tompkins's surroundings, Gamow created hilarious effects. Yet this interpenetration of the micro and the macro, forbidden by complementarity, did not strike Bohr as funny—as Hendrik Casimir reported, "Bohr was irritated rather than amused."[35]

Another subtly humorous critique of the complementarity doctrine is contained

[32] The German reads, "Du bist doch immer wieder in Extase, / Beschwigtigst, wo nichts mehr zu retten ist; / In jedem Quark begrübst du deine Nase." (See cit. n. 21).

[33] AHQP (cit. n. 14), Microfilm number 62.

[34] George Gamow, *Mr. Tompkins in Wonderland* (1940; reprint, Cambridge: Cambridge University Press, 1965).

[35] Hendrik Casimir, "Recollections From the Years 1929–1931," in Rozental, *Bohr: Life* (cit. n. 15), p. 110.

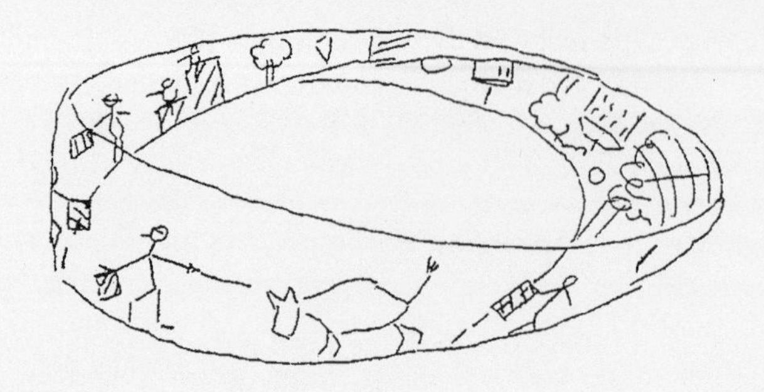

***Figure 3.*** *Pictures of a Möbius strip from the* Journal of Jocular Physics.

in Gamow's story "The Heart on the Other Side," printed in the 1955 *Journal of Jocular Physics.* In this love story, the intricacies of higher mathematics (and love, of course) reconcile what seem to be mutually exclusive propositions.

Stan, a poor mathematician specializing in topology, is madly in love with Vera Sapognikoff, daughter of a shoemaker. Vera's father prefers a son-in-law who will help him in the shoemaking business, so Vera sadly informs Stan that her father will never consent to their marriage. Yet Stan, who initially seems to agree that his mathematical research has no relevancy to the production and selling of shoes, is not inclined to lose his beloved Vera just because "topology and shoemaking seem to be two mutually exclusive propositions."

And Stan indeed proves that they are not by employing the "magical" properties of the Möbius strip. (Any two-dimensional object, sliding along the Möbius strip's surface until it reaches its initial point of departure, will turn into its own mirror image.) Stan tries to amuse Vera with upside-down pictures of matadors and bulls sliding on the Möbius strip, yet Vera remains unimpressed. "You can't give a Möbius twist to a shoe and make my father agree to our marriage, she says." It is at this moment that Stan is struck with a brilliant idea: one can save money on machinery and production by making only right shoes and transforming them into left ones by a trip on a three-dimensional Möbius surface.

As the history of science amply demonstrates, what seems at first to be an esoteric, abstract mathematical formalism often turns out to have direct relevance to the laws of the physical world (Riemann's geometry; algebra of matrices). Similarly, Stan calculates that three-dimensional Möbius twists of space must exist in unexplored regions along the Amazon River. Stan is unhampered by uncertainty; full of hopes, he starts his adventure. After a dangerous trip replete with alligators, mosquitoes, a life-threatening allergy, high fever, and delirium, Stan achieves the impossible: the right shoe does indeed, turn into its mirror image, the left shoe. Stan turns into his own mirror image as well—his heart is now on the other side. After some other dramatic twists of the plot, Vera and Stan are united and live happily ever after.

This story indicates, in a humorous way, the limits of reasoning based on everyday, commonsensical concepts. Intuition unaided by mathematics proves wrong again. What seems impossible with common-sense analysis is, in fact, achieved by mathematical ingenuity. Similarly, Bohr's philosophy of complementarity is permeated by "impossibility" proofs, which are based on an analysis of common-sense ideas (and their "extension"—classical concepts). These "proofs" forbid the very possibility of objective, deterministic, observer-independent ontology. Yet mathematical elaborations of quantum mechanical formalism and interpretations alternative to the Copenhagen one demonstrate that such ontologies are possible, despite Bohr's prohibitions.[36]

As noted in the introduction, jocular public events can not only question, but also support the official ideology. Ardently defending Bohr's philosophy of complementarity and the final overthrow of causality, Bohr's faithful disciple, Leon Rosenfeld, displayed his unquestionable loyalty even on jocular occasions. Rosenfeld contributed a story to the 1955 the *Journal of Jocular Physics* titled "A Voyage to Laplacia" in which he ridicules the deterministic state of mind. Yet, lacking a feeling of some distance from Bohr (which both Gamow and Weisskopf must have had), Rosenfeld tells a tedious story that ends with predictable preaching: one has to leave the deterministic world forever.

Rosenfeld's story describes his and Bohr's attempt in the autumn of 1931 to find a quiet retreat in a small Belgian town in order to work on their paper about the measurability of quantized fields. Needless to say, they do not make much progress in the gloomy land of Laplacia. The town in which they stop, they eventually discover, turns into a mecca for gamblers after the tourist season is over. They first try to stay in a luxurious hotel; dissatisfied, they leave for a romantic country inn. Both places are irritating, bizarre, and gloomy, and Bohr and Rosenfeld are just happy to escape.

The story begins with a labored tirade on the issue of determinism: "How did we ever, of all places, stumble upon that one? What unfathomable part in the predetermined concatenation of events which animate the scene of the universe did this trip fulfill? What may its necessary link have been with the tumultuous collisions of galaxies in the distant realms of space or the discreet embraces of earthworms in the cold recesses of the soil?" The style does not improve—the story continues in the same monotonous way, with a strained attempt to be funny. When the owner of the hotel demands that Bohr and Rosenfeld not use ink because a former guest spilled ink on a Persian carpet, Rosenfeld comments, "We could not help feeling that the assumption of a non-Newtonian attraction of Persian carpets for ink, which seemed

[36] The number of variants of such ontologies and the literature on this topic are voluminous. For a bibliography see Cushing, Fine, and Goldstein, *Bohmian Mechanics* (cit. n. 7).

to underlie his argument, showed excessive disregard of the great truth that we are not only spectators, but actors in the drama of existence."[37]

The visit is permeated with "ominous" deterministic signs: when Bohr and Rosenfeld reach the town by train, the station's clock, at which Rosenfeld "casually" glances, shows the same number as was written in the predetermined timetable! Rosenfeld initially does not realize the symbolic profundity of this occurrence, which constitutes, in fact, a warning about the gloomy Laplacian spirit. When Bohr and Rosenfeld leave the first hotel, Rosenfeld mutters something about unforeseen circumstances, at which moment he receives an unforgettable look from the hotel personnel—"a true Laplacian does not know of unforeseen circumstances."

In Rosenfeld's description, a determinist is, "in principle," a suspicious and unattractive creature. "When you come to think of it, this suspicious state of mind is a natural consequence of a Laplacian philosophy of life. If you go seriously about it, you must be asking yourself all the time: 'Didn't I overlook some causal connection?'" The conclusion of Rosenfeld's story is predictable: "We waved a hearty farewell to Laplacia—*for ever*!" (italics in original).

### IV. COMPLEMENTARITY OF "TRUTH AND CLARITY"

The ambiguity, opaqueness, and obscurity of Bohr's writings are legendary. The incomprehensibility of Bohr's talks and papers is repeatedly mentioned in commemorations of his life and work, both jocular and serious. In the 1985 centenary volume, D. H. Frish describes Bohr's 1957 lectures at the Massachusetts Institute of Technology: "People came from all over the Boston area and with great anticipation to hear the wisest of living men." Yet Bohr's wisdom was not easy to extract—the microphone was erratic, Bohr's diction was incomprehensible, and his thoughts were too "intricate even for those who could hear."[38]

Bohr's ambiguity was enveloped, by himself and others, in a rhetoric of profundity. The rhetoric of obscurity/profundity is exemplified in the principle of complementarity of truth and clarity: "Die Wahrheit ist komplementär zu Klarheit." A variant of this principle contrasts clarity and precision: "Complementarity suggests that the greater the precision of a statement the less its clarity."[39]

Taken seriously, the complementarity between truth and clarity, or clarity and precision, undermines Bohr's own philosophy. Bohr's philosophy is built on clear, precise, unambiguous statements of measurement readings. The statements are not merely clear and precise, they are also true, according to Bohr's own criterion of objectivity as intersubjectivity. Thus clarity, precision, and truth are presumed to peacefully coexist, rather than to contradict one another.

Bohr's statements of complementarity between truth and clarity were perhaps directed against the truth of clear and precise mathematical structures: being, in Heisenberg's words, a "natural philosopher rather than mathematical physicist," Bohr's access to truth was through philosophical analysis rather than through mathematical creativity. Pauli preferred to qualify Bohr's understanding of complementarity between truth and clarity as it applied to mathematical physics. In the volume of

[37] The expression about spectators and actors is one of the better-known Bohrian aphorisms.
[38] D. H. Frish in French and Kennedy, *Bohr: Centenary* (cit. n. 4), p. 247.
[39] Herman Feshbach, in Feshbach, Matsui, and Oleson, *Bohr: Physics* (cit. n. 4), p. xvi.

scientific essays dedicated to Bohr on the occasion of his seventieth birthday, Pauli stated that "a rigorous mathematical formalism and epistemological analysis are both indispensable in physics in a complementary way in the sense of Niels Bohr." While Pauli was trying to increase the clarity of this theory through mathematics, this clarity made him "aware that the final 'truth' on the subject is still 'dwelling in the abyss.'"[40]

Bohr's principle of complementarity between truth and clarity, though appealing at first glance, is in fact self-refuting. If the statement of complementarity of truth and clarity is obscure (unclear), it might be true, but then it is not clear what it means. Yet if the statement is clear, as all those physicists who quoted it without tongue in cheek seemed to assume, then it must be untrue.

This realization might underlie Weisskopf's delightful piece "Komplementäre Philosophie des Witzes" ("Complementary Philosophy of Jokes"), printed in the 1935 *Journal of Jocular Physics.* This is a self-referential, self-denying and self-affirming piece; it is mocking of (Bohr) and self-mocking. It affirms the elusiveness and ambiguity of truth, yet denies the possibility of expressing this by any finalizing principle (such as the complementarity of truth and clarity). It ridicules philosophical pretense and celebrates lighthearted laughter.

It is difficult, Weisskopf begins, to serve the cause of truth. Every sentence, once it is finalized in writing, is misapplied, overstated, and misunderstood. Every truth is meaningful only in becoming, when it is experienced with all its contradictory aspects. When one writes it down, it becomes trivial and false. (Here Weisskopf ironically refers to his own written statements that become false, including the statement that the written statement becomes false.)

Weisskopf continues in the same playful, jocular spirit. Before it is written, the statement still lives and breathes; after it is printed, its death begins ("das Sterben der Wahrheit beginnt mit Drucke.") With every new revision, it loses more and more life, becoming at the end the stiff corpse found in books and magazines. The statement's unequivocal expression through typed letters kills its essential meaning completely; Weisskopf's self-mocking conclusion is that "truth is complementary to clarity." Is there, then, no possibility of knowing the truth? Are there no rays emanating from the light of truth that preserve its living nerve with all its endless sources? There is such a light, answers Weisskopf; it is the magical light of jokes. The jocular truth is temporal; it is endlessly self-contained and self-correcting. In jokes and in laughter, concludes Weisskopf, the truth is revealed; even if distorted as in a curved mirror, it shines and lives here and now.

If initially the principle of complementarity between truth and clarity was a lighthearted, ambivalent and self-mocking way to sustain the difficulties and contradictions in quantum physics and to advance despite a lack of clarity, in later years it developed into a humorless legitimization of the Copenhagen philosophy and its shortcomings. The lack of clarity and consistency was elevated into a virtue; obscurity became equated with profundity; the incomprehensibility of Bohr's writings and talks became a sign of their subtlety and truth. Apologetics, ideology, and hero wor-

[40] Wolfgang Pauli, "Exclusion Principle, Lorentz Group and Reflection of Space-Time and Charge," in *Niels Bohr and the Development of Physics,* ed. Wolfgang Pauli (London: Pergamon Press, 1955). Pauli refers here to Bohr's favorite verses of Schiller: "Nur die Fülle führt zur Klarheit / Und im Abgrund wohnt die Wahrheit."

ship merged into an overarching "principle." Thus, in these words of the aging Weisskopf, one can no longer detect the jocular spirit of the '30s: "There is a complementary relation between . . . clarity and truth. That's why Niels Bohr was a very bad speaker, because he was too much concerned with truth."[41] Pais similarly presented Bohr's inability to deliver a coherent lecture as an "unrelenting struggle for truth."[42] Bohr himself probably encouraged this interpretation of his difficulty with expression. As Peierls noted, "Bohr's spoken or written words were not always easy to follow. As he liked to say, truth was complementary to clarity—and in his papers he leaned far towards the truth."[43] This mythology is often repeated in official commemorations of Bohr's science and life.[44]

## V. APPRAISAL OF COMPLEMENTARITY, OR COMPLEMENTARITY BETWEEN LOVE AND JUSTICE

Bohr and his faithful disciples, such as Pais and Rosenfeld, complained that philosophers and other nonscientific intellectuals did not comprehend the subtlety, profundity, and objectivity of the principle of complementarity. Recently Pais, in his biography of Bohr, voiced bitter disappointment that physicists of the younger generation seem to find no use for Bohr's philosophy of complementarity.[45] Unburdened by Bohr's intimidating authority and irresistible charisma (both in the personal and the sociological sense of the word), today's physicists and philosophers freely question, and even deny, the validity of Bohr's philosophy of complementarity. They build objective, causal, quantum ontologies that Bohr declared impossible; contrary to the complementarity principle, they devise experimental arrangements in which wave and particle attributes are revealed simultaneously. In other experiments with different wave and particulate definitions, waves and particles do seem, as Bohr claimed, to exclude each other. Bohr's wisdom seems to be applicable in some situations, and irrelevant or simply wrong in others.[46]

And not even Pais could avoid sounding ambivalent about complementarity.

[41] Victor Weisskopf, "Overview," in Feshbach, Matsui, and Oleson, *Bohr: Physics* (cit. n. 4), p. 5.

[42] Pais, *Niels Bohr's Times* (cit. n. 10), p. 11.

[43] Rudolf Peierls, "Some Recollections of Bohr," in French and Kennedy, *Bohr: Centenary* (cit. n. 4), p. 229. There are numerous reminiscences of Bohr's assistants and colleagues not only about the incomprehensibility of his talks, but also about his legendary difficulty with writing, be it a scientific article or a simple letter. I discuss this issue in ch. 12 of *Quantum Dialogue: The Making of a Revolution* (Chicago: University of Chicago Press, 1999).

[44] Thus, in a preface to the proceedings of the symposium celebrating the hundredth anniversary of Bohr's birth, Herman Feshbach wrote, "Complementarity suggests that the greater the precision of a statement the less its clarity. Bohr in his writings strove for precision with the result that his sentences were long, involuted and opaque." Feshbach, Matsui, and Oleson, *Bohr: Physics* (cit. n. 4).

[45] Thus, Bohr said "no man that is called a philosopher understands complementarity." Interview with Bohr, AHQP (cit. n. 14). Pais complained that Bohr's definition of phenomena "unfortunately, have not yet sunk in sufficiently among professional philosophers." Pais, *Niels Bohr's Times* (cit. n. 10), p. 23.

[46] The models are those of de Broglie-Bohm and their recent variants. The foundational status of the wave-particle complementarity rests on Bohr's controversial (and considered by many physicists and philosophers unfounded) doctrine of the necessity of classical concepts. When one gives up this doctrine, the wave-particle dilemma becomes a problem for experimental investigation. Today single photon states of light is considered particle-like behavior, while the tunnel effect is considered a wave phenomenon. Both are exhibited simultaneously in certain kinds of experiments. See Partha Ghose, Dipankar Home, and G. S. Agarwal, "An Experiment to Throw More Light on Light: Implications," *Physics Letters,* 1992, *A158:*95–9.

While declaring that the complementarity principle "makes him [Bohr] one of the most important twentieth-century philosophers," Pais has to admit that it is superfluous for the advance of physics "because it will not help in quantum mechanical calculations or in setting up experiments." (It is relevant though to the physicist who occasionally chooses to "reflect on the meaning of what she or he is doing.")[47]

Even on commemorative occasions, the analysis of Bohr's philosophy eventually became more detached and less celebratory—John Bell's and Abner Shimony's penetrating criticism of the central tenets of complementarity at the Niels Bohr Centennial Symposium are good cases in point. Bell questioned the cogency of Bohr's antirealistic response to the infamous Einstein-Podolsky-Rosen (EPR) argument, while Shimony argued, contra Bohr, that the EPR correlations can be accommodated in the framework of a realistic approach. Both found serious lacunae in Bohr's philosophy.[48] While Bohr's contributions to physics continue to be celebrated without reservation, Bohr's status as a philosopher already shows signs of decline.

If the appraisal of Bohr's philosophy of complementarity in physics is controversial, the verdict on Bohr's complementarity principle in biology is unequivocal and harsh. As Phillip Morrison put it, "The New Biology . . . is definitely anti-Copenhagen."[49] Bohr's complementarity in biology was a target of pointed jocular criticism by Hans Bethe and Edward Teller in the 1935 *Journal of Jocular Physics.*

Similarly, Bohr's analogies between physics and psychology are partial and limited. His principle of complementarity between reason and emotion is contradicted by contemporary psychological research, which explores the interconnection between cognition and affect. Bohr's inspiration was rooted in the wisdom of contemporary culture, and sometimes relied on its stereotypes. Thus, the complementarity between reason and feeling associates reason with masculinity and emotion with femininity. Similarly, the complementarity between love and justice draws its appeal from contrasting the just and rational (male) with the emotionally and intellectually inferior (female). While Bohr recognized societal prejudices about gender roles, he still felt that an "especially drastic" case is the one in which "the role of men and women are reversed."[50]

[47] Pais, *Niels Bohr's Times* (cit. n. 10), p. 23.

[48] John S. Bell, "Bertlmann's Socks and the Nature of Reality," in Feshbach, Matsui, and Oleson, *Bohr: Physics* (cit. n. 4), pp. 245–66. Bell stated, "While imagining that I understand the position of Einstein as regards the EPR correlations, I have very little understanding of the position of his principal opponent, Bohr. Yet most contemporary theorists have the impression that Bohr got the better of Einstein in the argument." Shimony (carefully) questions the coherence of Bohr's philosophy: "Although Bohr's writing is obviously philosophical in the sense of exploring fundamental questions concerning nature and human knowledge, it is far from clear how close he came to formulating a systematic philosophy." Abner Shimony, "Physical and Philosophical Issues in the Bohr-Einstein Debate," in Feshbach, Matsui and Oleson, *Bohr: Physics* (cit. n. 4), pp. 285–303 and p. 296.

[49] Phillip Morrison, "The Glimpse of the Other Side," in French and Kennedy, *Bohr: Centenary* (cit. n. 4), pp. 345–50, quotation on p. 348.

[50] The contemporary research about the connection between the cognitive and the emotional is by Lazarus and his group in Berkeley (Richard S. Lazarus, Allen D. Kanner, and Susan Folkman, "Emotions: A Cognitive-Phenomenological Analysis," in *Emotion Theory, Research and Experience,* eds. Robert Plutchik and Henry Kellerman (New York: Academic Press, 1980), pp. 189–243). The eloquent argument for complementarity between love and justice and between thought and feeling was made by Herbert Spencer: "The love of the helpless inevitably affects her [woman's] thoughts and sentiments, and this being joined in her with the less developed sentiment of abstract justice." Quoted in Brian Easlea, *Science and Sexual Oppression* (London: Weidenfeld and Nicholson, 1981). See Bohr's lecture "Natural Philosophy and Human Cultures," in *Atomic Physics and Human Knowledge* (New York: John Wiley & Sons, 1958), p. 30.

Bohr's attempts to extend complementarity reasoning to psychology were a target of jocular criticisms in Copenhagen. In "Psychologists' Revenge" the authors argue, in Bohr's characteristically obscure style, for complementarity between the notion of temperament and the behavior of the souls: "Notwithstanding the essential new situation created by the discovery of the quantum of action, the characteristic feature with which we have here to do is not unfamiliar in psychology. A typical example is afforded by the rational theory of emotions, according to which the vague concept of temperament stands in an exclusive relation to a detailed description of the behavior of the souls in the bodies concerned." The authors continue in the same "profound" style: "Psychological insight, as exhibited in the leveling of temperaments of a married couple, does not mean that a prediction of the course of events is impossible, but that the reversal of such a prediction must be part of any description involving a knowledge of the temperaments of the various persons." A strong ironical effect is gained by this contribution's close imitation of Bohr's structure of argumentation—in fact, as the authors point out, their text is often almost identical to Bohr's Faraday lecture. The authors' conclusion is a "revenge": if in physics complementarity reasoning leads to "an essential confusion in the physical interpretation . . . in psychology, contrary to quantum mechanics, the description leads to an essential mastery of our control of the events which is connected with the possibility of speaking of well-defined phenomena in the ordinary sense."[51]

The possibility that Bohr's complementarity might be not an objective, long-lasting contribution to science, and that Bohr's philosophical impact might have been a contingent phenomenon due to special historical circumstances and his unique personality, is sometimes raised even by Bohr's disciples. It is voiced respectfully and carefully, but voiced it is. Otto Frisch wrote, "Physicists from all over the world came to Copenhagen to work with Niels Bohr and spread his way of thinking, and naturally I do the same, having worked there for five years. But I think it is quite wrong to regard the Copenhagen School as an establishment set on perpetuating the views of its founder. . . . Newton may be accused of having suppressed the wave theory of light, simply by his great prestige. . . . Will the future lead to a similar accusation against Bohr? I doubt it; but only the future can tell."[52]

The affective connotations are, in large part, responsible for the attractiveness and longevity of complementarity. The wide appeal of complementarity is based on its existential message: an affirmation of the irreducibility of life, an invitation to sustain and accept contradictions, and a quest for unity that transcends the fragmented and painful nature of human existence. The attraction of complementarity is in its proclaimed goal of harmonizing all situations. When von Weizsäcker found himself enchanted by complementarity, it was because it promised to bridge his scientific and personal life. As physicist Daniel Kleppner put it in his talk at the Niels Bohr Centennial Symposium, "In large part it was this ability of Bohr to provide the affective underpinnings for quantum mechanics . . . that has made his role in physics so enduring."[53]

[51] "Psychologists' Revenge," communicated by E. Sparvehader and F. Vild, transmitted by Arrow. *Journal of Jocular Physics,* 1955.

[52] Otto Frisch quoted in Feshbach, Matsui, and Oleson, *Bohr: Physics* (cit. n. 4), p. 352.

[53] Daniel Kleppner, "Niels Bohr and Atomic Physics Today," in Feshbach, Matsui, and Oleson, *Bohr: Physics* (cit. n. 4), pp. 45–63, quotation on p. 45. See also Heilborn, "Earliest Missionaries" (cit. n. 9), and Beller, *Quantum Dialogue* (cit. n. 43).

The affective, personal connotations of complementarity were no doubt amplified when raised in live encounters between Bohr and his disciples. As Bohr's disciple Kalckar wrote, "The best of all was when the conversation turned to the so-called 'eternal questions.' . . . Nowhere was the overpowering intellectual and emotional impact of his personality more irresistible." Many disciples felt deep gratitude and affection towards Bohr, and justly so. Bohr took pains to secure funds and jobs for his disciples. A few of the young visitors to Copenhagen, among them Weisskopf and Pais, met their lifelong companions while visiting Bohr's institute. They were grateful to Denmark and, of course, to Bohr. Pais stated simply, "I loved Bohr." For Bohr's disciples, his words were imbued with intense meaning beyond their objective content.[54]

Today, with the distance of time and space, we encounter a very different kind of humor directed at Bohr and his teachings. This jocular spirit is a far cry from the affectionate one that reigned in Copenhagen—today we find humor that is unsympathetic, impatient, even perhaps disrespectful. This poem points out that Bohr's complementarity, and his operationalistic definition of "phenomenon," is of little help in solving the central interpretive problem of quantum theory—the measurement problem:

> There's an expert here we may call upon,
> But what shall we make of his advice
> That correctly defining phenomenon
> Will resolve our dilemma in a trice
>
> He speaks of uncertainty, virtue and vice,
> And complementarity. Come then, Sir,
> May we not ask you to be precise?

This loveless humor mirrors some of the harshest verdicts on Bohr's interpretation of physics, such as Nobel Prize-winning physicist Murray Gell-Mann's: "Niels Bohr brainwashed the whole generation of physicists into believing that problems were solved fifty years ago."[55]

Are these detached, unaffectionate evaluations of Bohr's philosophy the just ones? Or is Bohr's philosophy better estimated by his sympathetic, devoted, and admiring disciples, thus contradicting Bohr's own principle of complementarity between love and justice?

[54] Jørgen Kalckar in Rozental, *Bohr: Life* (cit. n. 15), p. 235. According to Weisskopf, there was an unwritten rule in Copenhagen: "Any physicist working with Bohr was certain to be married after no more than two years." Bohr traveled every year to England and America to find jobs for his disciples, and he worked tirelessly to get funds from Danish, American, and British sources for "his refugees" who could no longer work in their own countries. Weisskopf, *Joy of Insight* (cit. n. 3), p. 69 and p. 99. Weisskopf also wrote, "I am deeply grateful to his [Bohr's] country because I met my wife when I came to Copenhagen fifty-two years ago." Weisskopf, "Overview" in Feshbach, Matsui, and Oleson, *Bohr: Physics* (cit. n. 4), p. 1. The quote "I loved Bohr" is from Pais, *Niels Bohr's Times* (cit. n. 10), p. 5.

[55] The poem is "Ballade of the Copenhagen Interpretation" by Eugene R. Speer. The poem was distributed at the conference "Quantum Theory Without Observers," held in Bielefeld, Germany, on 24–28 July 1995. Gell-Mann's quote is from Murray Gell-Mann, "The Nature of the Physical Universe," *1976 Nobel Conference* (New York: John Wiley & Sons, 1979), p. 29.

## VI. AFTERWORD, OR COMPLEMENTARITY BETWEEN SERIOUSNESS AND HUMOR

After this paper was completed, I encountered two studies that resonate with the subject of my work. They are Paula Findlen's paper on the comic in the Scientific Revolution, and the last chapter of G. Nigel Gilbert and Michael Mulkay's book on discourse in science, which is devoted to humor in science.[56] These authors argue that the jocular in science should become a serious domain of the cultural and social studies of science. My paper can be perceived as joining that call.

The uncovering of the humorous part of scientific life can deepen our grasp of the cultural setting of science. The analysis of humor in scientific discourse can, at times, serve as an enlightening epistemological tool. Jocularity in science, as I have argued, can serve as a pointed commentary on the forms of authority and social hierarchies of scientific institutions. Jocular commemorations and narratives of the comic in science are a potent counterpart to the humorless, official, monological, Whiggish stories built around hero worship and a rhetoric of certainty.

Findlen's paper deals with the "tensions and pluralities" of the early modern world, as revealed in the ludic images of nature. Findlen invites the reader to follow the emergence of modern science as revealed in the battle between Carnival and Lent, between pagan indulgences and pious seriousness, between appetite and intellect. The early development of modern science is thus organically linked with the exorcism of the jocular, which is permeated with ambiguities and pluralities. This process culminated in the triumph of the ascetic Cartesian stance.

The jocular aspect was, of course, exorcised not from scientific life, but merely from the official narratives. Gilbert and Mulkay's description of humor in the twentieth-century scientific laboratory, as well as my description of the jocular Copenhagen spirit, are cases in point. Gilbert and Mulkay perceive the very existence of such humor as an indication of the scientist's awareness—even if for a short time only—of the contingency of scientific results. "Scientists, like other social actors, regularly employ divergent repertoires to construct versions of their social world which often appear to be literally incompatible . . . Much humor seems to depend on precisely the intimate juxtaposition of, and sudden movement between, divergent interpretative frameworks."[57] The authors identify these incompatible frameworks with two discourses: the contingent discourse, which acknowledges the interpretative leeway and contingency of scientific practice, and an empiricist discourse, which denies such a contingency.

Findlen invites the reader "to consider science in a *comic mode,*" and quotes the medievalist C. W. Bynum: "A comic stance knows . . . that doing history is, for the historian, telling a story that could be told in another way."[58] Yet what Findlen offers is not the "comic narrative" of the Scientific Revolution, but an outline for a serious narrative in which the jocular aspect is adequately represented.

What would a historiographic comic stance be like? How serious or humorous can we afford to be in our studies of the jocular in science? Would taking a comic

[56] Paula Findlen, "Between Carnival and Lent: The Scientific Revolution at the Margins of Culture," *Configurations,* 1998, *6:*243–67; G. Nigel Gilbert and Michael Mulkay, *Opening Pandora's Box: A Sociological Analysis of Scientists' Discourse* (Cambridge: Cambridge University Press, 1984).

[57] Gilbert and Mulkay, *Pandora's Box* (cit. n. 56), pp. 173–4.

[58] Findlen, *Carnival and Lent* (cit. n. 56), p. 246.

stance mean that we question our own narratives and ridicule our own arguments, and yet expect to be taken seriously? Or should we be serious, resisting the temptation of the comic stance, and tell well-defined, unambiguous, objective stories about the jocular aspect of science? Being that serious, we might end up looking ridiculous: while studies of science are objective, science itself is merely contingent, as the very presence of the jocular in science reveals.[59]

Here, ironically, Bohr's principle of complementarity between seriousness and humor turns out to be revealing: if we are tediously serious, this appears ridiculous, yet if we try too hard to be funny, this begins to look desperately serious.[60]

[59] This, of course, is a variation on the familiar objection to the strong program in the sociology of scientific knowledge: if all knowledge is contingent (context dependent and interest laden), so is the sociological account of science which cannot be objective and subsequently should not be taken too seriously. See David Bloor, *Knowledge and Social Imagery* (1976; reprint, Chicago: University of Chicago Press, 1991), p. 18.

[60] "Indeed, if we always endeavor to speak quite seriously, we run the risk of very soon appearing ridiculously tedious to our listeners and ourselves, but if we try to joke all the time, we soon find ourselves, and our listeners too, in the desperate mood of the jesters in Shakespeare's dramas." Niels Bohr, "Unity of Knowledge," 1955, reprinted in *Atomic Physics and Human Knowledge* (cit. n. 50).

# The 1959 Darwin Centennial Celebration in America

*By Vassiliki Betty Smocovitis**

Immortal Darwin,
Beloved scholar,
Modest and patient and wise;
This ancient school has chosen you
To honor and to eulogize.

—Lyrics to "Immortal Darwin"
from the musical play *Time Will Tell*

The party was an elbow-bending evolution of international intellectuals.

—Lois Baur, *Chicago's American,* 23 November 1959

The precedent for the successful celebration in honor of Charles Darwin was established in 1909. That year marked two critical anniversary dates in Darwin's "life story," or the narrative of his life and work: the centennial of his birth in 1809, and the fiftieth anniversary of publication of his *On the Origin of Species* in 1859. Both were celebrated across the world, but especially in England, Darwin's homeland, where scientists, dignitaries, and Darwin's living friends and family gathered under the auspices of Cambridge University to commemorate these events and to assess progress in evolutionary science. A well-known photograph from the Cambridge celebration of the centennial captures the historical moment: the aging Sir Joseph Hooker and Lady Hooker are shown expectantly looking over the shoulder of

* Department of History, University of Florida, Gainesville, FL 32611

A preliminary draft of this paper was read at the Boston Colloquium for the History and Philosophy of Science, at the Department of History and Sociology of Science at the University of Pennsylvania, and at the Department of History of Science, University of Oklahoma. Jeff Adler, Clark Elliott, Pnina Abir-Am, Susan Lindee, Kim Kleinman, Chris Koehler, and Gary Kroll assisted the project in some way. George Stocking, Joy Harvey, John Greene, Egbert Leigh, Frank Bonnacorso, Brian McNab, Harry Paul, Ernst Mayr, Susan Tax Freeman, and especially James Moore provided helpful comments on drafts. I wish to thank the Department of Special Collections of the Joseph Regenstein Library at the University of Chicago for providing assistance in obtaining documents, the film of the celebration, and permission to reproduce their documents; Deborah Day and Carolyn Rainey at the Carl Hubbs Papers at the Scripps Institution of Oceanography; and Beth Carroll-Horrocks at the American Philosophical Society. Research for this project was made possible by grants on the evolutionary synthesis from the National Science Foundation, a Research and Development Award from the University of Florida, and by the American Philosophical Society.

Mrs. Thomas H. Huxley, who is holding and gazing at the infant Ursula Darwin, the most recent offspring of Francis Darwin's son, Bernard.[1]

As the one hundredth anniversary of the publication of Darwin's *Origin* approached in 1959, plans for similar commemorative functions sprang up all over the world. An international committee called the Darwin Anniversary Committee, Inc., which included descendants Gwendolen May Raverat, Josiah Wedgwood, Lady Nora Barlow, Frances Darwin Cornford, Sir Charles Darwin, and Julian Huxley as honorary officers, was formed in the mid-1950s to oversee anniversary activities for 1959.[2] Since 1959 was also the 125th anniversary of the voyage of the HMS *Beagle,* the committee planned a spectacular, much-publicized reenactment of Darwin's voyage of scientific discovery around the world.[3] A "corps of leading scientists" hoped to retrace the voyage of the *Beagle* while engaging in "planned research" either "related to or stimulated by the original research of Darwin." Their goals were to "add greatly in a relatively short time to our knowledge of natural history and evolution."[4] Along with discussion of establishing Darwin fellowships and a Darwin memorial park on the Galapagos (pending the cooperation of the Ecuadorian government), the committee planned special lectures, seminars, conferences, brochures, and commemorative volumes.[5]

Even the most unlikely groups planned to commemorate the publication of Darwin's *Origin.* F. Alton Everest, one of the early leaders of the American Scientific

[1] As reproduced on p. 82 in Julian Huxley and H. B. D. Kettlewell, *Charles Darwin and His World* (New York: Viking Press, 1965). A volume published for the occasion included papers by leading scientists like J. Arthur Thomson, August Weismann, Hugo de Vries, William Bateson, and Ernst Haeckel, among others, and included contributions by Darwin's living academic sons, Francis and George. His "intimate" friend Joseph Dalton Hooker wrote an introductory letter to the editor of the volume. See *Darwin and Modern Science: Essays in Commemoration of the Centenary of the Birth of Charles Darwin and of The Fiftieth Anniversary of the Publication of the Origin of Species,* ed. A. C. Seward (Cambridge: Cambridge University Press, 1909). See also the special compendium of the centenary, *Memorials of Charles Darwin: A Collection of Manuscripts, Portraits, Medals, Books, and Natural History Specimens to Commemorate the Centenary of His Birth and the Fiftieth Anniversary of the Publication of "The Origin of Species,"* ed. W. G. Ridewood (London: Trustees of the British Museum, 1909). Other notable international commemorative activities included the special celebratory addresses given in honor of Charles Darwin at the American Association for the Advancement of Science meetings in Baltimore, Maryland, on Friday, 1 January 1909. A commemorative volume with the original papers was published as *Fifty Years of Darwinism: Modern Aspects of Evolution* (New York: Henry Holt and Company, 1909).

[2] The committee was legally incorporated in New York as a nonprofit organization on 23 December 1955. Document titled "Certificate of Incorporation," Bert James Loewenberg Papers, Darwin Anniversary Committee, folder 4, Library of the American Philosophical Society, Philadelphia, Penn. (hereafter cited as Loewenberg Papers). It drew on an international scientific advisory council that included explorers like Vilhjalmur Stefansson and Roy Chapman Andrews. Membership shifted as many of the officers and scientists dropped out of plans.

[3] A formal news conference and reception, with Julian Huxley and his wife present, was organized by the Darwin Anniversary Committee, Inc., and held at the Overseas Press Club in New York City on 13 November 1956. A prominent article appeared in the *New York Times* describing the plans. "Charles Darwin Commemoration Planned Next Year," *New York Times,* 14 November 1956; and "Darwin's Voyage to be Retraced," *New York Times,* 14 November 1956; also see Alvin Steinkoff, "Key to the Beginning: Scientists to Visit Darwin's Origin Area," *Los Angeles Times,* 31 August 1958, Carl Hubbs Papers, 1927–1979, MC-S, box 17, folder 25, "Darwin Anniversary, 1956," Scripps Institution Library, La Jolla, Calif., (hereafter cited as Hubbs Papers). Estimates for overall cost for the voyage ran at $350,000. Document titled, "How the Voyage of the Beagle will be financed, Nov. 13, 1956," Loewenberg Papers, Darwin Anniversary Committee, folder 1.

[4] Prospectus, Darwin Anniversary Committee, Inc., 1 March 1956, Hubbs Papers, MC-S, box 17, folder 25, "Darwin Anniversary, 1956."

[5] *Ibid.* The prospectus was circulated widely to all interested parties.

Affiliation, an organization founded in the United States in 1941 to explore the relationship between science and religion, focused his energy on the anniversary. Beginning as early as 1948, he planned an enormous commemorative volume assessing Darwin's impact on Western religion.[6]

In Canada, Australia, and Brazil, as well as in the United States and the Soviet Union—wherever Darwin and Darwinism were held in good favor—other plans were drawn up. Local agencies, institutions, and important scientific societies such as the Society for the Study of Evolution (SSE)—the international society founded in 1946 in the United States—were actively involved in both planning and coordinating activities across the world.[7] Journals of natural history, biology, general science, and even the occasional historical journal took special notice of the anniversary with lead articles, cover stories, and editorials assessing the current state of evolutionary knowledge.[8] Planned activities took many forms, from special lectures to conferences and books to the issuance of commemorative postage stamps and medallions with the Darwin figure prominently displayed.[9]

The activities planned on an international scale were so numerous and broad in scope that, as the centennial date approached, evolutionary biologists reeled from the work the occasion generated and suffered from stress precipitated by inevitable conflicts and rivalries. The minutes of the council meetings of the SSE in the mid- to late 1950s reveal the extent to which negotiations and coordination began to occupy the business of the society.[10] Individual scientists, especially more prominent evolutionary biologists, were additionally burdened with the demands of organizers who required their expertise. In an uncharacteristic letter of complaint, for example, Ernst Mayr expressed frustration with the demands placed on him as early as 1956. He wrote:

[6] Everest's plans for the anniversary, and the history of the American Scientific Affiliation (ASA), are discussed in James Gilbert, *Redeeming Culture: American Religion in an Age of Science* (Chicago: University of Chicago Press, 1970), see especially pp. 158–61. Although he and coeditor and president of the ASA, Russell L. Mixter, sought a secular press to increase the readership, they eventually published the volume with a religious press. The completed volume was edited by Mixter. See *Evolution and Christian Thought Today,* ed. Russell L. Mixter (Grand Rapids, Mich.: William B. Eerdmans, 1959).

[7] The minutes of the SSE council meetings in the mid-1950s reveal the extent of coordination required. Society for the Study of Evolution Papers, Series II and Series III-C, Library of the American Philosophical Society, Philadelphia, Penn. (hereafter cited as SSE Papers).

[8] See the *Proceedings of the American Philosophical Society,* 1959, *23.* The issue included an emblematic profile of Charles Darwin on the cover and was titled "Commemoration of the Centennial of the Publication of The Origin of Species by Charles Darwin: Annual General Meeting, April, 1959." Contributors included Wilfrid E. Le Gros Clark, I. Michael Lerner, Curt Stern, Arne Müntzing, Ernst Mayr, G. Ledyard Stebbins, Theodosius Dobzhansky, Norman D. Newell, George Gaylord Simpson, and Philip J. Darlington, Jr. Lerner's paper titled "The Concept of Natural Selection: A Centennial View" was especially topical. For the historical response to the centennial, see the September 1959 issue of *Victorian Studies* (vol. 3; n. 1). Contributors included Morse Peckham, Edward Lurie, and Sidney Smith. And see edited volumes like *A Century of Darwin,* ed. S. A. Barnett (Cambridge: Harvard University Press, 1958). For a review of the literature generated by the centennial see Bert James Loewenberg, "Darwin and Darwin Studies, 1959–63," *History of Science,* 1965, *4:*15–54; and *idem.,* "Darwin Scholarship in the Darwin Year," *American Quarterly,* 1959, *11:*526–33.

[9] Soviet organizers seemed especially keen to generate innovative memorabilia. See, for example, the medals struck in Moscow in commemoration of the centennial. Reproduced on p. 81 of Huxley and Kettlewell, *Charles Darwin and His World* (cit. n. 1). According to Sol Tax, they also issued special postage stamps.

[10] See especially the minutes of meetings held in 1956 and 1957. SSE Papers, Series II and Series III-C.

> I am afraid there will be so many Darwin celebrations that it will be difficult to decide where to say yes and where to say no. I have already committed myself to participate in the volume prepared by the American Philosophical Society where I am to discuss the role of isolation in evolution ... I am trying very hard to work on my own new book but find that all these other plans interfere with what is really, my A-1 project. This may sound unfriendly, but is merely the anguished cry of a harassed soul.[11]

Nor did the stress abate with the arrival of the critical year; in fact it seemed to accelerate to a fevered pitch and to push some people off the edge. No less a figure than the president of the SSE became the most visible victim of the celebrations. Edgar Anderson, the eminent botanist and evolutionary biologist at the Missouri Botanical Garden, found himself so emotionally stirred and excited by the University of Chicago's Darwin Centennial Celebration that he was unable to deliver his presidential address to the SSE at the Thanksgiving banquet ceremony. According to one eyewitness account, Anderson stood to speak, became agitated, faltered, and then abruptly terminated his lecture without apparent reason.[12] His wife's diary entry following his return from Chicago vividly describes her concern for Edgar, and hints at his report of a "religious experience" at the Darwin Centennial Celebration in Chicago.[13] He later described the convocation ceremony preceding the Thanksgiving banquet as "the most moving ceremony I have ever participated in."[14] According to disclosures Anderson made later to Anne Roe, he reported that the "large numbers of people" at the Darwin Centennial Celebration got him so excited that they eventually "did him in." He was subsequently hospitalized for three weeks.[15]

By the end of the anniversary year of 1959, the human and financial resources spent, and the level of emotion generated by the occasion, had gone far beyond the simple commemoration of the publication of a great book, even an epoch-making one at that. Why and how could a mere scientific anniversary place such extraordinary demands on its participants? Why were the celebrations so important that they received so much attention from vastly different, international audiences? What did the Darwin figure represent? And what exactly was being celebrated? While some of the festivities planned for the one hundredth anniversary were rather unassuming

[11] Ernst Mayr to Karl P. Schmidt, 25 September 1956, Darwin Centennial Celebration Papers, box 4, folder 15, Regenstein Library, University of Chicago, Chicago, Ill. (hereafter cited as Darwin Centennial Papers). Mayr's publication output increased dramatically as a result of the demand generated by the centennial celebrations. For a quantitative analysis of Mayr's publication record see Thomas Junker, "Factors Shaping Ernst Mayr's Concepts in the History of Biology," *Journal of the History of Biology,* 1996, *29:*29–77. The graphs on pp. 32–3 demonstrate sharp spikes in publication output for the years around 1959. Mayr's publication of historical articles was especially dramatic around the Darwin centennial year.

[12] Ernst Mayr, personal communication. This is confirmed by Anderson in a disclosure to Ann Roe. Undated document identified as "Notes, p. 1–19," Ann Roe Papers, folder on Edgar Anderson, Library of the American Philosophical Society, Philadelphia, Penn. (hereafter cited as Roe Papers).

[13] As cited in John J. Finan's biographical essay commemorating Anderson, "Edgar Anderson, 1897–1969," *Annals of the Missouri Botanical Garden,* 1972, *59:*325–45, quotation on p. 343. Finan also discusses Anderson's religious views. Anderson was a Quaker. See also cit. n. 14.

[14] Edgar Anderson to Sol Tax, 3 December 1959, Darwin Centennial Papers, box 3, folder 2.

[15] Document identified as "Notes, p. 1–19," Roe Papers, folder on Edgar Anderson. Anderson had a similar experience giving a public lecture at the 1937 meetings of the American Association for the Advancement of Science, when he became overcome with "something like stage-fright." For an eyewitness account of this "curious" behavior see the description on p. 153 in Una F. Weatherby, *Charles Alfred Weatherby: A Man of Many Interests* (no publisher given, 1951). Like Julian Huxley, Anderson suffered intermittent bouts of depression. Most probably, he suffered from bipolar (manic-depressive) disorder.

attempts to honor Darwin's life and to recognize his momentous book, others were more consciously designed to capitalize on the occasion and were carefully orchestrated to serve a range of functions. The most effective of the celebrations was planned to do both, and more: honor the great man of science—and his work—while at the same time serve a variety of personal, disciplinary, institutional, regional, and even national functions. Organized by anthropologist Sol Tax at the University of Chicago, The Darwin Centennial Celebration—as it came to be known—was filled with five days of scientific discussions, pageantry, ritual, and theatrical spectacle. Drawing at least 2,500 registrants with nearly 250 delegates representing 189 colleges, universities, and learned societies from fourteen countries, and at a hefty final cost of $59,022.08, it was the largest celebration of all.[16] Staged to begin exactly one hundred years to the day of the publication of Darwin's *Origin* on 24 November, and taking advantage of the Thanksgiving holiday to end on 28 November, the event was so well timed, and so carefully orchestrated to tap into lateral and adjunct concerns that spoke to many wide audiences, that it outshone—and arguably may still outshine—all other scientific celebrations in the recent history of science.

This paper explores the historical context of the University of Chicago's Darwin Centennial Celebration, with a view towards understanding the broad range of functions served by such commemorative and celebratory acts in the history of science.[17] In particular, it explores the varied interests served by this celebration, ranging from the national and regional to the institutional, disciplinary, and the personal concerns of organizers and participants.

## I. DISCIPLINARY MEMORY, UNITY, AND IDENTITY: THE CELEBRATION AND THE EVOLUTIONARY SYNTHESIS

One critically important reason for the intensity and the number of 1959 Darwin celebrations has to do with the timing of the anniversary within the larger history of evolutionary biology. In the wake of the evolutionary synthesis of the 1930s and 1940s, the anniversary of 1959, coming twelve years after the 1947 Princeton meetings (during which evolutionists celebrated the reconfiguration of biological disciplines around the new science of evolutionary biology), was perfectly timed to reassess the state of the art by the community of individuals that had worked to create a synthetic, unified science of evolution.[18]

Along with the belief that a modern synthesis of evolution was underway, there had come a restoration of Darwinian selection theory within a genetical and popula-

[16] For final registration numbers and Tax's personal account, see Sol Tax, "The Celebration: A Personal View," in *Evolution after Darwin,* 3 vols., eds. Sol Tax and Charles Callender (Chicago: University of Chicago Press, 1960), vol. III, *Issues in Evolution;* see also Memorandum dated 23 November 1959, Darwin Centennial Papers, box 8, folder 2; and Final Accounting Documents, Darwin Centennial Papers, box 2, folder 18.

[17] For an excellent theoretical discussion and explanation using historical examples from national studies of commemorative events (festivals, exhibits, and monuments) and cultural memories, see *Commemorations: The Politics of National Identity,* ed. John R. Gillis (Princeton: Princeton University Press, 1994); also see *The Invention of Tradition,* eds. Eric Hobsbawm and Terence Ranger (Cambridge: Cambridge University Press, 1983) for rituals, pageants, and other celebrations as invented traditions.

[18] The historical background to the evolutionary synthesis and the emergence of the scientific discipline of evolutionary biology is discussed in Vassiliki Betty Smocovitis, *Unifying Biology: The Evolutionary Synthesis and Evolutionary Biology* (Princeton: Princeton University Press, 1996).

tional framework that gave plausible accounts for the origins of biological diversity. It was from just this perspective in 1942 that Julian Huxley had looked back on the presynthesis era and coined the phrase "The Eclipse of Darwin" to denote the period's demise in popularity for Darwinian natural selection and for evolutionary studies generally. The "modern" synthesis of evolution, a term Huxley coined and popularized, saw a convergence of biological disciplines bearing on evolutionary studies and the end of a period of dissonance among biologists concerning the mechanism of evolution.[19] From that point on, evolution by means of natural selection, for evolutionary biologists, had become a fact.

Concomitant with the intellectual synthesis between disparate disciplines of biological knowledge came an organizational synthesis leading to the first society in support of evolution, the Society for the Study of Evolution (SSE).[20] The end result of the synthesis in the late 1940s was a new discipline with a self-aware community of individuals who identified themselves as "evolutionary biologists." By the mid-1950s, a well-defined, structured, and organized community of evolutionists had emerged who could not only serve as participants and organizers, but also as overseers and promoters of their field. The anniversary of the publication of the landmark work ushering in their field of interest was a well-timed opportunity to retell the life story of Charles Darwin, who was now reinvented as the "founding father" of their discipline.[21] It also served as the perfect opportunity to establish once and for all—for wide audiences—the facticity of evolution by means of natural selection.

Had this been merely an occasion for celebration within the new discipline, it would have been sufficient for the local audience of evolutionists, and possibly garnered the attention of other biologists and societies like the SSE. But so much of what it meant to be a twentieth-century evolutionary biologist hinged on identification with the narrative of Darwin's life and work. This might help to explain some of the emotion generated by the celebration, and the degree of conflict and rivalry over what was to emerge from it. Additional factors were clearly at play, and they contributed greatly to the particular success of the University of Chicago Darwin Centennial Celebration. Among these was the chosen site—the American Midwest—for the leading celebration, and the university's eagerness to promote itself as an intellectual center of the world. It was also because of the celebratory vision of the head organizer, Sol Tax, whose own discipline of anthropology—the area of inquiry that delved into human and cultural evolution—needed its own "evolutionary synthesis."

## II. RIVALRIES, CONFLICT, AND COORDINATION

The importance of the Darwin centennial to its community is underscored by the controversies sparked by the planning for the celebrations. Rivalries emerged be-

[19] For the historical account, see Julian S. Huxley, *Evolution: The Modern Synthesis* (London: Allen and Unwin, 1942).

[20] Smocovitis, *Unifying Biology* (cit. n. 18); also see Vassiliki Betty Smocovitis, "Organizing Evolution: Founding the Society for the Study of Evolution, (1939–1950)," *J. Hist. Biol.*, 1994, *27:*241–309; and Joseph Allen Cain, "Common Problems and Cooperative Solutions: Organizational Activity in Evolutionary Studies, 1936–1947," *Isis,* 1993, *84:*1–25.

[21] This phrase was explicitly used to describe Charles Darwin. See "Tax Statement for University of Chicago Reports," Darwin Centennial Papers, box 8, folder 1.

tween those who wanted to serve as the key organizers and their organizations, and, most importantly, over the national context for major activities.[22] Tensions emerged between English and American planners, who realized that their major celebrations would not only create inevitable conflicts for evolutionists in terms of resources, but might upstage more modest, local celebrations, if not lead to general confusion.[23]

One exchange, in particular, exposes the complex tensions between the various celebration committees and the American-based SSE.[24] Although much of this exchange was over which committee would serve as the central committee for all celebrations, it also brought up one especially volatile subject—Julian Huxley. The secretary of the SSE complained sharply that "Some of us [referring to the council members] were also disturbed because Julian Huxley—who has not been exactly cooperative in regard to his one-man show in London next summer—recently indicated to one member of the council that he, Julian, was going to run the celebration in Chicago (presumably as honorary chairman of the Darwin Anniversary Committee?)"[25]

The fear was legitimate: as the self-appointed leader of evolution in England, Huxley had, in fact, demonstrated a tendency to commandeer social activities pertaining to evolution. Any major effort at celebration had to at least make a nod to Huxley's eminence, but this also ran into the difficulties of working with an overbearing personality. Furthermore, a celebration officially sponsored by the SSE, the international organization devoted to the study of evolution, had the potential to upstage, if not undermine, Huxley's own efforts on behalf of his favored British societies and committees. The fact that Huxley had lost the bid to found an international society for the study of evolution in England, Darwin's homeland, did not sit well with the man who identified himself as the grandson and therefore living representative of Thomas Henry Huxley, known as "Darwin's Bulldog."[26] It also did not help that Huxley seemed to have an almost neurotic need to be the center of attention.[27]

Potential conflicts between the SSE and the Darwin Anniversary Committee, Inc.,

[22] See the letter from the program director for systematic biology, Rogers McVaugh, to Sol Tax informing him that the International Botanical Congress meetings to be held in Montreal in August, 1959 had as a central theme the origin of species and its effect on the biological sciences in the last one hundred years. He wrote, "It would be unfortunate from your standpoint to have events at the Congress make your November celebration anticlimactic, but presumably this can be arranged before hand in order to avoid duplication and confusion." Rogers McVaugh to Sol Tax, 6 June 1956, Darwin Centennial Papers, box 2, folder 19.

[23] At one point Ilza Veith suggested to Sol Tax that there were so many Darwin events planned in England and elsewhere that they should rethink their own celebration at the University of Chicago. Ilza Veith to Sol Tax, 9 August (undated, most likely 1957), Darwin Centennial Papers, box 6, folder 2. Also see the letter to the secretary of the SSE, Harlan Lewis, from Colin Pittendrigh, 22 August 1957, SSE Papers, Series III-C.

[24] Colin Pittendrigh to Harlan Lewis, 22 August 1957, and the response from Lewis, undated (most likely 1957), SSE Papers, Series III-C.

[25] *Ibid.*

[26] For historical explication of the tension between English and American workers in organizing evolutionary studies in the 1940s, and for Huxley's behavior, see Smocovitis, "Organizing Evolution" (cit. n. 20).

[27] At one point, Huxley demanded to be included in the musical production staged for the celebration, *Time Will Tell,* in some significant manner. Needless to say, this did not sit well with the writers. They wrote, "I am somewhat embarrassed by a conversation I had with Sir Julian Huxley. As I told you, instead of being bashful about appearing in the show, he seems as anxious as anything . . . The situation is further complicated by the fact that he implied that he wasn't particularly anxious to have Sir Charles Darwin appear on stage at all, which handed me quite a jolt. On the other hand I understand by the grapevine that Sir Charles is quite a comedian and will probably want to get into the act

which featured Huxley prominently, became increasingly attenuated as the committee began to develop internal conflicts. Not only had plans for the retracing of the *Beagle* gone awry, but the scientific status and authority of the directors, the head of whom was Bert James Loewenberg, a professor of history at Sarah Lawrence College, had been called into question by scientists like Henry Fairfield Osborn.[28] Osborn summarized his objections in a heated letter of resignation from the committee:

> I think the Board of Directors which will be responsible for the development and plans for this anniversary should consist of individuals of standing in the scientific world (more specifically in the natural sciences); in other words, individuals who through their own work and reputation are recognized as leaders in the natural sciences. It is my opinion that only through such a group can there devolve a series of plans in celebration of Charles Darwin's anniversary which will adequately reflect the meaning of his work as a scientist.[29]

As the anniversary date approached, a flurry of resignations were handed in as scientists like George Gaylord Simpson and Edward Weyer became increasingly uncomfortable with organizers' efforts.[30] Although the board of directors was reconstituted with prominent biologists including Theodosius Dobzhansky, John Moore, Edwin Colbert, and Colin Pittendrigh, The Darwin Anniversary Committee, Inc., became more and more of a clearing house for activities, rather than the focus for organizational efforts, as the anniversary date approached.[31]

Coordinating activities between competing groups required the concerted effort of organizers at the University of Chicago, who were eventually successful in convincing their administration, prominent societies like the SSE, and key scientists that the site of the major celebratory activity should be the University of Chicago.[32] Sol Tax, who would lead the American planners' efforts, won over competition by arranging to secure the presence of Huxley, the organizational pivot for English efforts, in Chicago.[33] With the offer of an endowed position and a sabbatical leave, Tax lured Huxley to the University of Chicago to be the Alexander White Visiting Lecturer in Biology and Anthropology for the fall semester during which the cele-

---

someplace too." Robert Pollack to Robert Ashenhurst, 26 October 1959, Darwin Centennial Papers, box 2, folder 1. Other indicators of Huxley's egocentrism include the demands he placed on Tax concerning his visiting appointment, and his much publicized "secular sermon."

[28] On the *Beagle* plans, see Carl Hubbs to Fairfield Osborn, 15 January 1957, Hubbs Papers, MC-S, Box 17, folder 26, "Darwin Anniversary, 1957–1960."

[29] Henry Fairfield Osborn to Bert James Loewenberg, 10 January 1957, *ibid.*

[30] George Gaylord Simpson to Bert James Loewenberg, 22 January 1957, *ibid.*, also see the explanatory letter to Julian Huxley from Fairfield Osborn, 28 January 1957, *ibid.* This file chronicles the internal problems with the Darwin Anniversary Committee, Inc.

[31] A report of the committee activities dated 15 May 1958 reveals that the committee was serving as an "informal clearing house for various groups and organizations interested in commemorating Charles Darwin and the theory of evolution." Kenneth Cooper to Carl Hubbs, 15 May 1958, *ibid.*

[32] Karl P. Schmidt wrote to inform the SSE of Chicago's plans and to suggest that Chicago's activities be under the "wing" of the SSE. Karl P. Schmidt to Harlan Lewis, 17 September 1956, Darwin Centennial Papers, box 2, folder 17. The SSE accepted the invitation in 1956. Minutes of Council Meeting, 26 December 1956, SSE Papers, Series II. See also Tax's handling of the "mild form of competition" with the Anthropological Society of Washington's plans: Betty J. Meggers to Sol Tax, 25 May 1959, Darwin Centennial Papers, box 2, folder 6.

[33] The invitation to Huxley went out on 20 January 1956. Darwin Centennial Papers, box 2, folder 7. A similar invitation went out to Sir Charles Darwin on the same day. Lawrence A. Kimpton to Mr. Charles G. Darwin, 20 January 1956, *ibid.*

brations would be held.[34] Nothing short of an organizational coup, this was the smoothest way of avoiding conflict with English organizers while winning their support through the person of Huxley. Huxley himself was able to have it both ways: in 1958 he held his "one-man show" in London under the auspices of the International Zoological Congress, which celebrated the anniversary of the joint reading of the Darwin-Wallace paper to the Linnaean Society, and he accepted the generous offer from the University of Chicago for the 1959 anniversary.

The competition with British efforts was thus smoothed away. Tax, who was a professor of anthropology at the University of Chicago, was completely candid about this competitive aspect of the celebrations. He later wrote in his public recollections that he was mystified by the British failure to secure the major celebratory site. He wrote, "The real mystery is why others did not pre-empt the opportunity of celebrating the publication event of the century. In England celebrations were planned, appropriately, for the 1958 Centenary of the reading of the Darwin-Wallace papers before the Linnaean Society. The centennial of Darwinism was, in fact, celebrated in many quarters, but November 24 1959—the one-hundredth anniversary of publication of *Origin of Species*—was left to the University of Chicago."[35]

With arrangements made for Huxley *in situ,* Tax began to lay the plans for the celebration, obtaining the official support of the SSE and widening potential audiences to garner support.[36] Critical to the success of his celebration was the strongest possible support he received from the University of Chicago.

## III. "WORLD SERIES" AND "INTERNATIONAL LIVESTOCK SHOW": THE UNIVERSITY OF CHICAGO AS INSTITUTIONAL SITE

If the celebration was characterized by fanfare and pageantry that drew much attention, it was because it was backed by a midwestern university eager to place itself on the intellectual map of the world. With the building of the first atomic pile in the 1940s and subsequent pathbreaking work in astronomy, physics, and chemistry, the University of Chicago had been gaining an international reputation as a leader in the physical sciences. In the biological sciences, too, it had established itself in areas associated with ecology and evolution, and by the 1950s it possessed a renowned zoology department that had included evolutionists like Sewall Wright and Alfred E. Emerson.[37] Even in the social sciences, the University of Chicago had been leading the way with renowned departments of sociology and anthropology.[38]

[34] Document to the Committee of the Darwin Centennial Celebration dated 12 October 1956, Darwin Centennial Papers, box 5, folder 9. The document relayed a conversation between Huxley, Emerson, and Schmidt that took place in St. Louis on 27 September concerning Huxley's tentative acceptance of the appointment.

[35] Tax, "The Celebration" (cit. n. 16), quotation on p. 273.

[36] The first letter revealing Tax's plans was from Chicago-based Karl P. Schmidt to Harlan Lewis, dated 17 September 1956, Darwin Centennial Papers, box 2, folder 17. See also the minutes of the SSE meetings for additional details (cit. n. 7).

[37] For the early history of the biological sciences at the University of Chicago see H. H. Newman, "History of the Department of Zoology in the University of Chicago," *BIOS,* 1948, *19:*215–39; William B. Provine, *Sewall Wright and Evolutionary Biology* (Chicago: University of Chicago Press, 1986); Philip J. Pauly, *Controlling Life: Jacques Loeb and the Engineering Ideal in Biology* (New York: Oxford University Press, 1987); Gregg Mitman, *The State of Nature: Ecology, Community, and American Social Thought, 1900–1950* (Chicago: University of Chicago Press, 1992).

[38] For a history of anthropology at Chicago see George W. Stocking, Jr., *Anthropology at Chicago: Tradition, Discipline, Department.* Prepared for an exhibition marking the fiftieth anniversary of the

In the 1950s, furthermore, the university had seen the pivotal—and much publicized—experiments of Harold Urey and Stanley Miller dealing with the origins of life. Given what was clearly pathbreaking recent work in a range of sciences, the university was especially eager to host a conference that would bring international attention to its intellectual activities, especially in science, the most technically demanding and prestigious realm of intellectual inquiry. Bringing leading, distinguished scientists to the university campus for an important and very public celebration would be one way of enhancing Chicago's reputation as a major center of learning. The fact that this commemorative event honored a "great book," at the very university that had seen Robert Maynard Hutchins's educational reforms in the preceding decade leading to the "great books of the Western world" focus, certainly made it a celebratory event worthy of notice by the administration.[39]

Such activities were critically important to the university administration in the late 1950s. Not only had the university neighborhood of Hyde Park seriously deteriorated, but it was also generally felt that Chicago and the Midwest did not "constitute a lure for some people."[40] These fears were foremost in Chancellor Lawrence Alphaeus Kimpton's mind during much of his administration from 1951–1960, and may help to explain his heavy involvement with and support of the celebration, which could only help further his goals of renewal through rebuilding. Furthermore, a precedent for the commemoration of Darwin had been established at the university in 1909 when The Biological Club had sponsored a series of special lectures for the fiftieth anniversary celebration.[41] It had drawn a number of notable biologists for the occasion and had been a successful event.[42] As part of the support, Tax was given an administrative assistant for the conference, Charles Callender, a recent Ph.D. in anthropology from Chicago, who had some experience in public relations.[43]

In fulfillment of the larger goal of promoting the University of Chicago, Tax demonstrated his political acumen by soliciting the support of the office of Mayor Richard J. Daley, whose persona was heavily identified with Chicago.[44] The city council

---

Department of Anthropology, 1979. Joseph Regenstein Library, University of Chicago. For a history of sociology, see Martin Bulmer, *The Chicago School of Sociology: Institutionalization, Diversity, and the Rise of Sociological Research* (Chicago: University of Chicago Press, 1984).

[39] Hutchins introduced the "Great Books of the Western World" courses in 1947 with Mortimer Adler. See William H. McNeill, *Hutchins' University: A Memoir of the University of Chicago, 1929–1950* (Chicago: University of Chicago Press, 1991); Milton Mayer, *Robert Maynard Hutchins: A Memoir,* ed. John H. Hicks (Berkeley: University of California Press, 1993).

[40] Lawrence Alpheus Kimpton, "State of the University, 3 November 1959," in *The Idea of the University of Chicago: Selections from the Papers of The First Eight Chief Executives of the University of Chicago from 1891 to 1975,* eds. William Michael Murphy and D. J. R. Bruckner (Chicago: University of Chicago Press, 1976), p. 498. His "State of the University, 11 November 1958" had echoed these concerns and looked forward to the rebuilding of the university and the city. For a history of the University of Chicago immediately preceding the celebration, and for a summary of the Kimpton administration, see McNeill, *Hutchins' University* (cit. n. 39); and Daniel Meyer, *The University and the City: A Centennial View of the University of Chicago* (Chicago: University of Chicago Library, 1996).

[41] See the "Programme of Darwin Anniversary Addressess," Darwin Centennial Papers, box 7, folder 1.

[42] The four-page "programme" indicates that it was a major, university-wide event open to the public that included fifteen lecturers on evolutionary topics. The lectures began on 1 February and ran to 18 March 1909. The dates were designed to coincide with the birthdate of Darwin.

[43] In the final accounts, Callender's salary was listed as part of the overall expense of the conference; it was $7,851.58. Final Accounting, Darwin Centennial Papers, box 2, folder 18.

[44] The transactions between Tax and the Office of the Mayor comprise a six-page fascicle of documents included in Darwin Centennial Papers, box 8, folder 2. The Office of the Mayor actually

had already designated the centennial year of 1959 as a special "jubilee" year: as early as 1957 they were hoping to see the opening of the grand St. Lawrence Seaway, a World Trade Fair, the Pan-American Games, and Festivals of the Arts and Sciences.[45] For a city known in American popular culture for its "frontier spirit," granaries, gangsters, and notoriously bad weather, these major events were to help reconstruct the identity of the city, placing it centrally within the postwar American mainstream then undergoing tremendous growth and prosperity.[46] Hosting such a celebration, which would bring luminaries from all over the world, was one way to assist in the reconstruction of Chicago as an intellectual center for the American Midwest. Chicago was, after all, home to the 1893 Columbian Exposition and, more recently, to the Chicago World's Fair of 1933, one of the Century of Progress Expositions. Both had drawn positive attention to the city.[47]

The prototypical model for this intellectual reidentification would be unabashedly English, as Americans could share some of the glory of the "empire," even if it appeared to be diminishing at the time. The University of Chicago, by Old-World standards a newcomer, could emulate the older and prestigious former school of Darwin, Cambridge University, the site of the 1909 celebration.[48] The very design of the buildings at Chicago, "Secular Gothic," which saw the "conjunction of gargoyles and test-tubes," had historically emulated such British ideals.[49]

---

contacted Tax first to request information on "American Indians" (Tax was a cultural anthropologist who had worked on Native American cultures). Tax brought up his celebration in the context of the planned activities involving native people. He wrote, "Are you aware that the University is planning the important international celebration of the centennial of the publication of Darwin's *Origin of Species*? . . . Scholars from all over the world will come; the list is terrific; including among many Sir Charles Darwin and Sir Julian Huxley and even a biologist from Russia.

"Should we make this a major intellectual-scientific-educational event of the year? There is even a connection with the Americas, since Darwin developed his theory after his famous Voyage of the Beagle, where he was impressed by what he saw in South America, especially in the Gallapagos [sic] Islands." Sol Tax to Jack Reilly, 13 January 1958, Darwin Centennial Papers, box 8, folder 2.

[45] Document titled "Resolution" from the Office of the Mayor, dated 7 June 1957, *ibid.*

[46] The scholarly literature on Chicago's history is vast. For a natural history of Chicago in the nineteenth century see William Cronon, *Nature's Metropolis: Chicago and the Great West* (New York: W. W. Norton and Co., 1991). See also James Gilbert, *Perfect Cities: Chicago's Utopias of 1893* (Chicago: University of Chicago Press, 1991); and Donald Miller, *City of the Century: The Epic History of Chicago and the Making of America* (New York: Simon & Schuster, 1996).

[47] See R. Reid Badger, *The Great American Fair: The World's Columbian Exposition and American Culture* (Chicago: Nelson and Hall, 1979); Robert Muccigrosso, *Celebrating the New World: Chicago's Columbian Exposition of 1893* (Chicago: Ivan R. Dee, 1993); see also Gilbert, *Perfect Cities* (cit. n. 46); Robert W. Rydell, *All the World's a Fair: Visions of Empire at American International Expositions, 1876–1916* (Chicago: University of Chicago Press, 1984). The "White City," which denoted the world's Columbian Exposition, and the "Gray City," which denoted the University of Chicago, were the progeny of businessmen hoping to transform Chicago into a city of refinement in the late nineteenth century. For a discussion of twentieth-century attempts at such transformation at the national level and the 1933–4 Chicago Century of Progress Exposition (the "Chicago World Fair" of 1933), see Robert W. Rydell, *World of Fairs: The Century-of-Progress Expositions* (Chicago: University of Chicago Press, 1993). Rydell argues that the 1933 fair, coming in the midst of the Great Depression, helped in "rebuilding confidence in the nation's future" (p. 9). According to Rydell, the Century of Progress Expositions promoted progress through the twin themes of science and technology.

[48] In his preface to the 1909 edited volume of the Cambridge meeting, A. C. Seward indirectly claimed Darwin as a "son" of Cambridge. See cit. n. 1.

[49] Jean F. Block, *The Uses of Gothic: Planning and Building the Campus of the University of Chicago, 1892–1932* (Chicago: The University of Chicago Library, 1983), quotations from "Foreword," p. xiii, by Neil Harris. Many of the buildings at the University of Chicago were modeled after Oxonian architectural forms.

Discussions in the early planning stages kept a constant eye on British efforts both to copy and to preempt any attempt at a greater celebration.[50] One possibility to increase visibility and further legitimate the celebration was to bring well-known British luminaries to Chicago. Winston Churchill, who was scheduled to visit the White House in the critical year, was briefly entertained as a possible honorary participant, but was rejected for fear of his "overshadowing the rest of the Celebration."[51] At least one document points to a brief discussion about bringing no less than Her Majesty Queen Elizabeth II to the celebrations.[52] After due consideration, the lesser figure of His Royal Highness Prince Philip was thought the perfect honorary participant, as he was a celebrated patron of the sciences. He was formally invited to attend, but he declined as he and the royal family did not have plans to travel to the United States that fall.[53]

Emulation of English sensibility was a conscious aim expressed publicly—and historically—in a documentary film of the Darwin Centennial Celebration produced by the *Encyclopedia Britannica.*[54] Beginning with the opening shots, Chicago is represented as the new intellectual center of the New World, the twentieth-century analogue to Darwin's Victorian London. The narrator reads from the following script:

> The year is 1859. Chicago, at the south end of Lake Michigan, is a flourishing city of 110,000. Still marked by a residue of the frontier spirit, it is already an important commercial and industrial center, but has few intellectual pretensions. The museums and universities are thirty-five years away.
>
> Across the Atlantic, in an England ruled by Victoria, London is in 1859 the center of a revolution in human thought. On November 24th of that year Charles Darwin had published the *Origin of Species;* the first printing was entirely sold out by the end of the day.
>
> One hundred years later, here is Chicago, the grain elevators that overshadowed the city in 1859 now replaced by skyscrapers; and the University of Chicago, which for one week has become the center of the intellectual world. Thousands of persons, some of them drawn from the far corners of the earth, have come here to celebrate the hundredth anniversary of Darwin's great book. Universities from every continent have sent delegates to this event. Newspaper headlines, stories in magazines, and television interviews have drawn attention to the celebration; excitement is in the air this week in November.[55]

[50] As early as 1956 Tax and the administration approached English organizers like Huxley with an eye towards sharing the planning or asking them to participate directly. Draft of letter by Tax to Lawrence Kimpton to notify Huxley, undated; Lawrence A. Kimpton to Julian Huxley, 20 January 1956, Darwin Centennial Papers, box 2, folder 7. Sol Tax to Julian Huxley, 27 April 1956, Darwin Centennial Papers, box 4, folder 9. According to this letter, Huxley appeared to have tentatively accepted the invitation by 4 April 1956.

[51] Organizers also thought that conferring a degree on Churchill would make other honorary degrees "anticlimactic." Document titled "Reports on Talk with Cannon," 1 May 1959, Darwin Centennial Papers, box 1, folder 2.

[52] Memorandum, William B. Cannon to Sol Tax, 10 September 1958, Darwin Centennial Papers, box 2, folder 6.

[53] Lawrence A. Kimpton to H. R. H. The Prince Philip, Duke of Edinburgh, 9 May 1958, *ibid.* Prince Philip's response on learning of the request of his presence was relayed to Chancellor Kimpton via the acting consul general as being "most appreciative." W. Macon to Lawrence Kimpton, 28 August 1958, *ibid.*

[54] Film number F-134, Darwin Centennial Papers. See discussion in this article on the history of the film.

[55] Transcript of film by *Encyclopedia Britannica,* Darwin Centennial Papers, box 9, folder 9.

Chicago newspapers, television and radio stations, and other local media needed little prodding to promote the event, though Tax extended himself, his staff, and the resources of the University of Chicago. Hiring an aggressive businessman named Sheldon Garber, whose official title was media services director, Tax launched a media blitz closely rivaling that for the coronation of Queen Elizabeth II.[56] Among Garber's duties was firing off press releases and letters to nearly every major American magazine in existence using an American-style hard-sell tactic. His letters began with a sales pitch worthy of the occasion: "It's going to be a scientific and intellectual world series. The world's greatest authorities—50-count 'em-50—will be at one place for five days of open talks to determine the score on evolution."[57] The official press release to United Press International shifted metaphors from the quintessentially American sport to the quintessentially American festival of the agricultural fair as Garber stressed the occasion's importance to Chicago: "Here is the University of Chicago's version of the International Live Stock Show. This will be a tremendous scientific gathering. I call it to your attention early because it may require some staffing. The stories in this gathering could be a great feather in the Chicago bureau's cap."[58] And Tax's media blitz served his purposes well: news of the celebration planning alone was picked up by major newspapers like the *New York Times* and minor local newspapers in the Chicago area, most of whom eagerly bought into his sales pitch that this event was to be "a milestone in the history of evolution."[59]

Careful not to leave any interested parties out of the celebration festivities, Tax approached institutions in the Chicago area that might potentially benefit from cosponsorship and would contribute to the success of the event. Tax wrote directly to the prestigious Chicago Zoological Park (also known as the Brookfield Zoo) to inform them of the meetings, and also officially invited members of the Illinois Academy of Sciences.[60] He also encouraged the Chicago Natural History Museum to coordinate its own commemorative functions with the University of Chicago celebration.[61] Even major Chicago booksellers were alerted by the organizers in time to prepare for the celebration.[62] Books and displays on evolution filled the shelves of major stores eager to sell copies. The Chicago area was thus perfectly poised—and primed—for a major intellectual event.

Although the publicity that the celebration was to receive (before, during, and after) was lavish to the extreme, Tax was exceedingly careful not to make preparations for the celebration around similarly wide-ranging sociopolitical motives.

[56] Garber's suggestions for publicity are relayed in a document titled "Publicity-Press," 11 September 1959, Darwin Centennial Papers, box 8, folder 1.

[57] Sheldon Garber to Robert Ajemian, *Life* Bureau, 15 September 1959, *ibid.* Magazines targeted included *Harper's, Look,* and *The Saturday Review.*

[58] Sheldon Garber to Jesse Bogue, United Press International, 15 September 1959, *ibid.;* also see an equally promotional letter to the Associated Press: Sheldon Garber to Carroll Arimond, 15 September 1997, *ibid.*

[59] Memorandum, Sheldon Garber to Editors, 14 September 1959, Darwin Centennial Papers, box 8, folder 5; and folder titled "news," Darwin Centennial Papers.

[60] Sol Tax to R. Robert Bean, 11 July 1958, Darwin Centennial Papers, box 2, folder 14. Bean was the director of the Chicago Zoological Park in Brookfield, Ill. For invitations to academy members, see "Addenda to work in progress," 19 August 1959, Darwin Centennial Papers, box 1, folder 4.

[61] Clifford Gregg to Sol Tax, 20 February 1958, Darwin Centennial Papers, box 6, folder 15; Theodor Just to Sol Tax, 15 August 1958, *ibid.*

[62] Document titled "To all Chicago area booksellers concerning the Darwin Centennial" dated 14 November 1959, Darwin Centennial Papers, box 8, folder 1.

Reasons for the conference publicized most often, included in Tax's letters of invitation and in the successive progress reports he and other organizers issued, stressed its intellectual nature and its contributions to scientific knowledge and to understanding the future of "man."[63] Of especial value to scientists was the maintenance and extension of the synthesis project: the gathering of leading figures from disparate disciplines to assess the current evolutionary state of the art. The official notice of the celebration defined its purpose thus: "The Celebration will bring together leading figures in the social and biological sciences to discuss the evolution of life, of man, and of the mind. Centering on common problems that cut across disciplinary lines, the discussions will comprise a series of five three-hour public panels."[64] In one advertisement following the celebration, the following were listed as representing the disciplines that touched on or informed knowledge of evolution: physics, biology, psychiatry, medicine, botany, zoology, embryology, photometry, anthropology, ethnology, cosmogony, biochemistry, genetics, archaeology, physiology, astronomy, and entomology.[65]

For Tax, the inclusion of anthropology was an especially important component of the celebration. One reason for his zeal in taking the lead in organizing efforts was his perception of the need to explore the relationship between biological evolution, human evolution, and cultural evolution.

## IV. THE CELEBRATORY VISION: SOL TAX, "MAN," AND ANTHROPOLOGY

The number of individuals involved in planning and hosting the celebration was considerable. One source lists, in addition to the six members of the Darwin Centennial Celebration Committee, five additional assistants, including Charles Callender as the conference director.[66] But despite the numbers of aides, assistants, and administrators involved, it was clearly a one-man operation. From its start to the very finish, the University of Chicago Darwin Centennial Celebration was masterminded by Sol Tax.[67]

Tax would probably have met with limited success had his motives for taking the organizational lead been simple. Personal glory, fame, attention, and such ends

[63] See, for instance, the notice of the 1959 SSE meetings. The planned celebration was "Emphasizing common problems that cut across disciplinary lines . . . "Announcement, undated, Darwin Centennial Papers, box 2, folder 17; also see Tax's official summary, "Tax statement for University of Chicago Reports, Darwin Centennial Papers, box 8, folder 1. See also document dated 25 September 1959, Darwin Centennial Papers, box 1, folder 1.

[64] Document dated 25 September 1959, Darwin Centennial Papers, box 1, folder 1.

[65] Advertisement from the University of Chicago Press, Darwin Centennial Papers, box 1, Addendum.

[66] These included Marie-Anne Honeywell as conference secretary, Jean Dames and Rose Weiner as assistants, Rochelle Dubnow as the director of volunteer work by students, and Marianna Tax as assistant. There is no indication of how many students assisted as volunteers. See the program guide and convocation guide for the official listing, Darwin Centennial Papers, box 7, folder 1.

[67] For biographical information on Sol Tax see *Currents in Anthropology: Essays in Honor of Sol Tax,* ed. Robert Hinshaw (The Hague: Mouton Publishers, 1979). See especially David Blanchard's contribution to the volume. See also Sol Tax, "Pride and Puzzlement: A Retro-introspective Record of 60 Years of Anthropology," *Annual Review of Anthropology,* 1988, *17:*1–21; and Robert A. Rubinstein, "A Conversation with Sol Tax," *Current Anthropology,* 1991, *32:*175–83. The son of Russian Jewish immigrants, Tax was born in Chicago in 1907 but grew up in Milwaukee; he was a graduate of the University of Chicago specializing in both North American and Meso-American cultural anthropology.

might have resulted in modest success for the celebration and may, in part, have explained his unbounded energy and organizational zeal, but as sole reasons they would never have led to the spectacular result Tax sought.[68] For this to be possible, Tax had to tap into some wider need that would draw in large numbers of participants.

For Tax, an anthropologist then at the height of his career, the celebration was the perfect opportunity to draw attention to his own discipline as well as to his home university. He was strongly identified with both.[69] Tax recounted his historical recollection of his motives and of his inception of the original idea for the anniversary celebration in 1955 (which came to him while he was visiting the library of the Wenner-Gren Foundation in New York) as follows:

> I remember thinking. Somebody would be organizing a celebration for the occasion. Why should not anthropology be center stage? An encyclopedia nearby revealed that the exact date was November 24, a good season in the academic world, since the school year is well underway and the holidays have not yet begun.
>
> My main train of thought carried me home to Chicago. What institution in this country was better suited to celebrate the centenary than the University of Chicago, born ten years after Darwin's death, far away—a celebration on behalf of the whole world? No personal interest; purely intellectual and scientific.[70]

And Tax's intellectual and scientific interests were genuine: as a talented organizer, editor, and effective networker, he recognized that such a public celebration would be the perfect opportunity to reengage anthropologists in evolutionary study.[71] As he pointedly stated in his recollections, twentieth-century anthropology, for historical, sociopolitical, and intellectual reasons, had seen the "complete separation" of "man as an organism from man as a member of society and bearer of culture." "Culture" and "evolution," although united in the thoughts of Darwin and Huxley in the nineteenth century, had gone entirely separate ways in the twentieth century as cultural anthropologists argued against rigid, overly deterministic, "evolutionary" explanations for culture.[72] As a result, anthropology had become a science with a "split personality." Tax wrote:

> So we come to a science which proclaims itself 'the study of man,' yet views culture as though it were not part of man; which studies the evolutionary process and traces the origin of man through the fossil record, yet steadfastly separates man from all other animals; generally denies social and cultural *evolution,* yet uses the word 'primitive'—apologetically—for most of the living peoples and cultures it studies.[73]

[68] Over the course of his long career, Tax repeatedly took the lead in organizing major conferences. See biographical material in cit. n. 67.
[69] See cit. n. 67.
[70] Tax, "The Celebration" (cit. n. 16), quotation on p. 272.
[71] See cit. n. 67.
[72] The postwar period saw a rise of anticultural evolutionary approaches that echoed earlier such approaches that had stemmed from Boazian reforms of anthropology. Reasons for this are discussed in George W. Stocking, Jr., *Race, Culture, and Evolution: Essays in the History of Anthropology* (New York: The Free Press, 1968; reprint Chicago: University of Chicago Press, 1982); see also *idem., Anthropology at Chicago* (cit. n. 38).
[73] Tax, "The Celebration" (cit. n. 16), quotation on p. 272.

The possibility and precedent for such an integrated study of anthropology, furthermore, had already been established at Chicago. Within the anthropology department, a course titled "Human Origins," taught by Robert Braidwood, Wilton M. Krogman, and by Sol Tax beginning in 1945, had paved the way for an integrated approach. Robert Redfield, an early member of the Darwin Committee, had been a strong advocate of "neo-evolutionary civilizational" approaches to culture, and with the addition of biological anthropologist Sherwood Washburn to the department in 1947, an integrative approach to culture and evolution was already in place as a backdrop to the Darwin Centennial Celebration planning in the mid-1950s.[74]

The schism in anthropology was also recognized, albeit from another perspective entirely, by evolutionary biologists. The fact was that anthropology, the subject that dealt most immediately with human evolution, had been curiously removed from organizational and intellectual efforts to synthesize evolution in the 1930s and 1940s (as historians of biology have long noted).[75] In the 1950s, the absence of anthropology—and anthropologists—at SSE meetings and in issues of *Evolution* (the leading journal for evolutionary study) had been noted by key figures in organizing evolutionists, like Ernst Mayr.[76] The absence was especially noted as anthropology logically *had* to be brought into the larger evolutionary synthesis, if the synthesis was to be as complete as its architects had begun to envision it should be in the 1950s. The paleontological and cultural components of human evolution, the understanding of the evolution of mind, and the concerns traditionally associated with culture and the social sciences had to be incorporated within the modern synthesis of evolution, especially as Theodosius Dobzhansky, George Gaylord Simpson, and Julian Huxley had paved the way for such a synthesis by addressing the evolution of man, mind, and culture in their semipopular and popular works.[77]

With the ambitious goal of opening discussion into the varied meanings of "culture" and "evolution," Tax consciously targeted and invited distinguished anthropologists to Chicago: Clyde Kluckhohn from Harvard University, A. Irving Hallowell from the University of Pennsylvania, and Alfred Kroeber from the University of California, Berkeley, all of whom would join an already impressive list of anthropologists at Chicago, including F. Clark Howell. Although most of them required little encouragement, Tax did his utmost to support their active participation by approaching each cordially as well as, of course, financing their participation generously. Funds were obtained for anthropologists primarily from a modest grant from the Wenner-Gren Foundation for Anthropological Research.[78] Despite Tax's gently coercive tactics, at least one of his anthropological invitees, Clyde Kluckhohn, hesitated to attend, and indicated that he would only come if he were not required to present a paper. Although sponsored participants were expected to contribute original articles, Tax agreed to these conditions as Kluckhohn was a leading ethnologist

[74] In his history of anthropology at the University of Chicago, George W. Stocking, Jr., states that these were the contributing factors leading to the Darwin Centennial Celebration. See his discussion on p. 39 (case 18) in Stocking, *Anthropology at Chicago* (cit. n. 38).

[75] See *The Evolutionary Synthesis: Perspectives of the Unification of Biology,* eds. Ernst Mayr and William B. Provine (Harvard: Harvard University Press, 1980) for a complete discussion. See also Smocovitis, *Unifying Biology* (cit. n. 18).

[76] See Smocovitis, "Organizing Evolution" (cit. n. 20).

[77] For more discussion on this see Smocovitis, *Unifying Biology* (cit. n. 18).

[78] They provided $5,000. Final Accounting Documents, Darwin Centennial Papers, box 2, folder 18. Grant materials are in box 1, Addendum.

who would add to the anthropological slant of the celebration.[79] Tax's commitment to bringing in anthropology and important anthropologists was therefore central to his organization and vision of the celebration.

Nor was Tax's vision solely confined to the narrower domain of academic anthropology. To extend the anthropological sphere of influence, and to draw the attention of the wider public, Tax secured special funds from the Wenner-Gren Foundation to invite Louis B. Leakey, an anthropologist who had vast international popular audiences in the 1950s. Taking advantage of the opportunity to visit the United States for the first time and to further promote his research, Leakey brought the latest of his sensational fossil hominid discoveries, *Zinjanthropus boisei,* with him to the celebration.[80] Both Leakey and "*Zinj*" were to serve as major highlights of the conference, drawing attention to anthropology as a central study in evolution and fulfilling Darwin's own notoriously elusive statement in *Origin* that "Light will be thrown on the origin of man and his history."[81] If the conference was successful at any *new* synthesis, it was in bringing anthropologists and biologists together to explore the biological and cultural aspects of human evolution, yet recognize the "non-genetical" processes operating at the level of culture. Tax's closing personal thoughts on the celebration made this much explicitly clear: "So, for me, the Centennial brought Darwin and evolution back into anthropology, not by resurrecting analogies, but by distinguishing man as a still-evolving species, characterized by the possession of cultures which change and grow non-genetically."[82] Certainly the number of anthropologists far exceeded what one would have expected at an anniversary celebration of Darwin's *Origin.*

## V. MAKING IT ALL HAPPEN: SOL TAX AND THE DARWIN CENTENNIAL CELEBRATION COMMITTEE

One reason that Tax's efforts to lead the celebrations paid off so well is that he had begun to plan the event as early as 1955. At that time he solicited the support of the chancellor and other university administrators to form a crossdepartmental committee that he would chair. The committee was officially appointed by the chancellor in January 1956.[83] The final university crossdepartmental committee consisted of Alfred E. Emerson in zoology, Chauncy D. Harris in geography, Everett C. Olson from geology and paleontology, H. Burr Steinbach in zoology, and Ilza Veith from the departments of medicine and the history of medicine and science. Two of the

[79] Sol Tax to Clyde Kluckhohn, 20 August 1959, Darwin Centennial Papers, box 4, folder 11; Progress report to Julian Huxley dated 3 October 1959, Darwin Centennial Papers, box 1, folder 6.

[80] For press clippings see Lois Baur, "It's easy to converse with scientists, if you know what Zinjanthropos means," *Chicago's American,* 23 November 1959; "Fossils Support Darwin's Theory of Man's Origin," *St. Louis Post-Dispatch,* 25 November 1959; and "Man Fossil Backs Up Darwin Idea," *Los Angeles Mirror News,* 25 November 1959, Darwin Centennial Papers. Leakey's tour is recounted in Virginia Morrell, *Ancestral Passions: The Leakey Family and the Quest for Humankind's Beginnings* (New York: Touchstone, 1995).

[81] Charles Darwin, *On the Origin of Species,* Facsimile of the First Edition (Cambridge: Harvard University Press, 1964), quotation on p. 488.

[82] Tax, "The Celebration" (cit. n. 16), quotation on p. 282. He closed with this additional reminder: "Human evolution includes the addition of culture to man's biology; 'cultural evolution' at the human level is quite a different matter. Anthropologists accept the first without question; they are divided about the second."

[83] This was officially noted in the program to the celebration. Darwin Centennial Papers, box 7, folder 1.

***Figure 1.*** *Darwin Centennial Committee. The text on the table is a copy of Darwin's* Origin of Species. *Standing, left to right: Sol Tax, Everett C. Olson, Chauncy Harris, Alfred E. Emerson. Seated, left to right: Ilza Veith and H. Burr Steinbach. Darwin Centennial Papers, Dept. of Special Collections, Joseph Regenstein Library, University of Chicago. Reproduced with permission.*

original members of the committee were Karl P. Schmidt and Robert Redfield, but both died before the celebration took place in 1959 (see Figure 1).[84]

By mid-September 1956, the committee sent out the first of the letters inviting key evolutionists like E. B. Ford, Ernst Mayr, G. G. Simpson, G. Ledyard Stebbins, and Charles Elton to the festivities.[85] Nearly all declined to participate in Tax's proposed precelebration planning conference, though all were supportive of the event, if not completely enthusiastic, as they already had heavy commitments demanded by the anniversary. The critical development in catalyzing Tax's efforts to lead the centennial celebrations appears to have been Huxley's agreement to visit the Univer-

[84] Schmidt died on 27 September 1957 from a snake bite; Redfield died 16 October 1958 from leukemia.

[85] Karl P. Schmidt to E. B. Ford, 14 September 1956, Darwin Centennial Papers, box 3, folder 16; Karl P. Schmidt to Ernst Mayr, 17 September 1956, Darwin Centennial Papers, box 4, folder 15. Charles Elton declined because he disliked conferences and because he had "nothing to say on the subject of evolution," as he had not written anything on the subject for years and "not closely enough concerned with the subject now." Charles Elton to Karl P. Schmidt, 3 October 1956, Darwin Centennial Papers, box 6, folder 15; Sol Tax to G. G. Simpson, 18 September 1959, Darwin Centennial Papers, box 5, folder 12.

sity of Chicago for the fall semester of 1959.[86] With Huxley in *situ* and on the payroll of the University of Chicago, plans to hold the biggest celebration became a possibility. Not only were the English efforts led by Huxley shifted to the United States, but American efforts could now reconfigure around Huxley and the midwestern location of Chicago.

Tax's efforts to bring Huxley to Chicago paid off generously.[87] In return for the visiting professorship, which provided a generous stipend, accommodations, and first-class airfare for he and his wife, Huxley would be required to teach a special seminar in the fall semester on evolutionary biology.[88] Tax put even this small requirement to good use in his organizational efforts. In 1958 Tax organized a special seminar for both faculty and graduate students.[89] The goal of this seminar was specifically to aid in the celebration organization. Participants in the seminar were required to read the conference papers circulated early, sort the topics into relevant categories, and then organize the conference program around presentation of the papers. In return, students received standard academic credit for the course and had the advantage of reading the conference papers in advance and contributing significantly to the celebration in their own way. Tax had been careful to select the participants in his seminar, going so far as to solicit applications in 1958 from students and interested faculty.[90] In addition to solving the major intellectual organizational problem of the conference, the seminar also served to expose new graduate students to the field of evolutionary biology and to involve them first-hand in the celebratory functions of the discipline. (One graduate student at the University of Chicago who participated in the seminar, Matthew Nitecki, went on not only to lead in his chosen area of paleontology, but to also become a master of such gatherings, organizing the Annual Spring Systematics Symposium sponsored by the Chicago Field Museum of Natural History and by the University of Chicago in the 1980s.)[91]

Once Huxley agreed to the visiting appointment, plans fell quickly into place. Although Huxley was clearly the most visible public figure in evolutionary biology—as well as the direct descendant of Thomas Henry Huxley—others were also needed to complete the slate of celebrants to represent the full Darwinian heritage. For this reason, Tax simultaneously sought to secure a living representative of the Darwinian legacy for the celebration: Sir Charles Darwin, the grandson of Charles

[86] See cit. n. 50 for the sequence of events leading to Huxley's appointment.

[87] It should have—Huxley came at considerable cost. The initial offer stood at $5,000 with an extra $1,000 to pay for his wife's travel expenses. Julian Huxley to Sol Tax, 12 May 1959, Darwin Centennial Papers, box 1, Addendum. The final accounting for salaries listed Huxley's as $1,335. The two remaining visiting professors received $784 and $250 (François Bordes and Alfred Kroeber, respectively). Darwin Centennial Papers, box 2, folder 18.

[88] For the exact arrangements between Huxley and Tax see Julian Huxley to Sol Tax, 12 May 1959, Darwin Centennial Papers, box 1, Addendum.

[89] See the announcement for the seminar, "To colleagues on the faculty" from Sol Tax dated 16 April 1959, Darwin Centennial Papers, box 2, folder 3. One account states that "thirty members of the faculty and fifty selected graduate students from twenty departments" were part of this seminar. Abraham Raskin, "Special Report on the Darwin Centennial Celebration: One Hundred Years Later," *The Science Teacher,* March 1960, quotation on p. 3 of the fascicle. Darwin Centennial Papers, box 8, folder 9.

[90] Application Form for Seminar on Evolution (Anthropology 425-Zoology 425), Darwin Centennial Papers, box 2, folder 3.

[91] Another graduate student who went on to a successful career as a geneticist and a publicist of science, David Suzuki, was included in the original list of students, but did not complete the seminar. Graduate student applications, *ibid.*

Darwin.[92] Although trained in mathematics and not fully cognizant of the latest developments in evolutionary biology, Darwin was invited to speak. His public lecture, titled "Darwin the Traveller," would share his grandfather's experiences aboard the HMS *Beagle,* complete with maps, diagrams, and family reminiscences of the Darwin adventure.[93] His mathematical expertise would be put to use in a forward-looking paper of some interest to evolutionists on the human population problem at the "present-day."

## VI. GARNERING SUPPORT: FUNDS, ALLIES, AND NEUTRALIZING CONTROVERSY

With two of the key individuals in place, Tax devoted his energies to securing funds and support for the conference. Although the University of Chicago made a considerable initial commitment of $25,000 (contingent on financing of the whole) toward the proposed final budget of $57,000, it could not finance the entire operation.[94] Tax therefore aggressively campaigned for external support from federal agencies and private foundations.

The emphasis of Tax's 1956 pitch to the National Science Foundation (NSF) was exclusively scientific in its scope and stressed the conference's potential for the furtherance of evolutionary study. Tax explicitly noted that the occasion was to emphasize "(1) the scientific outlook that has characterized these hundred years; (2) the progress made in the study of evolution and genetics; and (3) the effect on human sciences of concepts of evolution and the treatment of man as a part of nature."[95] Tax successfully secured NSF funds to host the Darwin celebration, in addition to funds from the Wenner-Gren Foundation for Anthropological Research and from the Markle Foundation.[96] An argument for synthesis was also used to incorporate other components of interest from medicine and psychology, the evolution of culture, the evolution of disease, and finally, the evolution of mind. Tax successfully obtained a grant from the National Institutes of Health (NIH) so that participants could be invited from these disciplines that he thought should rightfully be included in modern evolutionary study.[97]

Since the conference was the first major public demonstration of support for evolution following the "Scopes Monkey Trial," Tax was careful to address contemporary concerns dealing with the teaching of evolution and forged additional alliances.[98] One

[92] Lawrence A. Kimpton to Charles Darwin, 20 January 1956, Darwin Centennial Papers, box 2, folder 7; also see the draft to Kimpton from Tax, Darwin Centennial Papers, box 3, folder 10.

[93] Program, The University of Chicago Darwin Centennial Celebration, Darwin Centennial Papers, box 7, folder 1.

[94] Proposal to the National Science Foundation, 27 December 1956, Darwin Centennial Papers, box 1, Addenda. The final accounting shows that the University of Chicago's Division of Biological Sciences provided $7,500 and the University of Chicago's Division of Social Sciences provided $2,166. Final Accounting, Darwin Centennial Papers, box 2, folder 18. The bulk of the final financing came from external sources.

[95] Sol Tax to Alan T. Waterman, 27 April 1956. Waterman was director of the National Science Foundation. Darwin Centennial Papers, box 2, folder 20.

[96] The National Science Foundation provided $13,000. Financial Accounting, Darwin Centennial Papers, box 2, folder 18. The Wenner-Gren Foundation provided $5,000. See the grant material in Box 1, Addenda, Darwin Centennial Papers. The Markle Foundation provided $1,869. Final Accounting, Darwin Centennial Papers, box 2, folder 18.

[97] NIH provided $8,709. Final Accounting Documents, Darwin Centennial Papers, box 2, folder 8.

[98] For a recent scholarly assessment of this trial, see Edward J. Larson, *Summer for the Gods: America's Continuing Debate Over Science and Religion* (New York: Basic Books, 1997).

such alliance connected more theoretical scientific concerns with the promotion of evolution in the public schools. In 1958 a grant from NSF was obtained for the Institute and National Conference for High School Biology Teachers.[99] (This had been a second application, the first having been declined for unknown reasons.) The grant provided sufficient funds to have one high-school teachers' delegate from each state of the union attend the conference and special workshops. Thus, the celebration would quickly communicate the new understanding and status of evolution to teachers in the American secondary-school system. The very conference participants could directly address concerns raised by high-school teachers in such a forum. At the request of Tax, the most able spokespersons, Edgar Anderson, Theodosius Dobzhansky, Hermann J. Muller, George Gaylord Simpson, and Julian Huxley were chosen to act as direct intermediaries between all conference participants and high-school teachers at a special institute that he organized. The grant also provided financial support for the dissemination of newsworthy knowledge gleaned from the conference to high-school teachers across the country. One tangible result of the Darwin Centennial Celebration was a substantive thirty-six page booklet that was published and distributed widely to American high schools and educators.[100]

American high-school teachers were not the only adjunct group invited to participate in the celebration. As a way to smooth relations between evolutionists and theologians (or more accurately, to explore the relations between science and religion raised by Darwin's evolutionary theory), Tax and the University of Chicago also organized an Institute on Science and Theology in cooperation with the Federated Theological Faculty. This institute included formal lectures by the Reverend J. Franklin Ewing on "Creation and Evolution in Present-Day Roman Catholic Thought," and by Jaroslav Pelikan on "Creation and Causality in the History of Christian Thought."[101] A panel discussion titled "Warfare of Science with Theology" opened discussion on the subject between panelists including Harlow Shapley, Conrad H. Waddington, Sir Charles Darwin, Leo Strauss, and others. If these workshops reached consensus, it was, for the most part—with the exception of Huxley—that science and religion were not incompatible domains of knowledge.[102] The planned workshops were one way of facing head on and neutralizing potentially volatile subjects associated with study of evolution in the United States.

Although the institutes for high school teachers and science and theology were substantive components of the Darwin Centennial Celebration, both functioned as sideshows to the critical component of the celebrations: the panel discussions that included original papers from the invited participants.

[99] Document titled "A Proposal for the Support of a Special Program of Participation by Selected High School Teachers of Science," from Francis S. Chase and Sol Tax to Dr. James Phelps (at NSF), Darwin Centennial Papers, box 1, folder 14; also see the bid sent to C. Russell Phelps, Special Projects in Science Education, NSF: Sol Tax to C. Russell Phelps, 31 October 1958, Darwin Centennial Papers, box 2, folder 20.

[100] The booklet is included in the papers of the Darwin Centennial Celebration. Published under the auspices of the Graduate School of Education in 1960, the full title is "Using Modern Knowledge to Teach Evolution in High School: As Seen by Participants in the High-School Conference of the Darwin Centennial Celebration at the University of Chicago." The booklet featured a section summarizing relevant ideas from the conference, Julian Huxley's lecture titled "Evolution in the High-School Curriculum," and a list of books on evolution useful to teachers.

[101] These were published in vol. III of Tax and Callender, *Evolution after Darwin* (cit. n. 16).

[102] See the discussion here on Huxley's convocation address and the controversy it generated.

### VII. CENTER STAGE OF THE CELEBRATION: THE OFFICIAL PROGRAM, THE PANELS, AND THE SCIENTIFIC DISCUSSION

There was no doubt about it: for most of the parties involved, the most important part of the celebration was the panel discussions, which featured visiting scholars recognized for their work in a stunning assortment of disciplines bearing on evolution. Simply by attending and by contributing knowledge from their respective fields, the panelists served to make the celebration a success. Tax explicitly stated this himself:

> The Celebration *was* the Panel Discussions; the panels were participants; the participants were great scientists who did what no great scientist should be expected to do. The Celebration was good because, from beginning to end, the Committee and those they had chosen did so well.[103]

Although documents do not reveal exactly how panelists and representative topics were chosen, it is likely that Tax relied heavily on the committee and other Chicago faculty for specific recommendations. In his published recollection he admitted to relying on their expertise, especially because the broad crossdisciplinary nature of the celebration made it difficult for any one person to have thorough knowledge of all areas.[104] Their choices and representative topics were not finalized without some controversy, however, as was demonstrated by a minor conflict over the inclusion or exclusion of some topics. This conflict concerned two aspects of the celebration that were particularly objectionable to evolutionary biologists.

Working from H. J. Muller's suggestions, Tax had sought to invite biochemists, astronomers, and others who could explore the cosmic aspects of evolution, the origins of life, and probe into the plausibility of extraterrestrial life.[105] The inclusion of these topics for the panel discussions would allow exploration of the borderlands between the physical and the biological sciences. But instead of engendering approval from evolutionary biologists, the choice of these topics brought pointed disapproval. Writing from the 1959 Cold Spring Harbor Symposium, and having conferred with Sewall Wright, Th. Dobzhansky, Bernhard Rensch, Ernst Mayr, and G. Ledyard Stebbins, Alfred E. Emerson wrote to the Darwin Centennial Committee to object to the inclusion of the subject of extraterrestrial life. "The question of the existence of life outside the world is very speculative," he wrote. "It sounds sensational in your outline. I would feel along with the others [see above evolutionary biologists] that it would be a mistake to put too much emphasis on this question."[106]

[103] Tax, "The Celebration" (cit. n. 16), quotation on p. 275.

[104] *Ibid.*

[105] Muller recommended the addition of the special topic of the origin of life; he also recommended as possible contributors Joshua Lederberg, James Watson, Harold Urey, and Gerard Kuiper, among others, as well as Kuiper's "bright young graduate student," Carl Sagan. This was clearly very late in the planning stages, possibly because the topics were so new as to be less obvious for inclusion in the Darwin centennial to organizers not familiar with the subject. H. J. Muller to Sol Tax, 5 March 1959, Darwin Centennial Papers, box 4, folder 16.

[106] Alfred E. Emerson to Darwin Centennial Committee, 7 June 1959, Darwin Centennial Papers, box 3, folder 12. Emerson indicated that G. G. Simpson, who was known to object to biochemistry and the existence of extraterrestrial life, had not yet arrived. Evolutionary biologists increasingly had "tense relations" with biochemists and astronomers who opened discussions into the existence of extraterrestrial life. Reasons for this are discussed in Smocovitis, *Unifying Biology* (cit. n. 18).

There is some indication that this did not greatly deter Tax and the remaining committee, however, as they still attempted to secure Harold C. Urey as a panelist.[107]

Emerson's other objections pertained to the balance of topics; organic evolution appeared not to be given enough space on the panels, with Emerson suggesting that it "should be the focus of about half the panels." Another comment was that some "phases of biology" that were "not associated with modern evolutionary thought (e.g. development of the living organism)" should also not be "emphasized in a Darwin Centennial." Instead, Emerson suggested that "we should build upon Darwin, discuss contemporary evolutionary investigations and point toward the future, toward the solution of problems of evolution not now solved but solvable by scientific study."[108] Emerson's suggestion to exclude those topics bearing on development (i.e., embryology), to reduce the emphasis on biochemistry and questions into the origin of life, and to focus instead on evolutionary biology was in fact consistent with the larger rift that was widening in the late 1950s between the newer reductionistic sciences of molecular biology and biochemistry and the new category of "organismic biology."[109]

By far the most difficult part of the planning involved organization of the panels, however. Coordinating fifty participants' timetables, sorting their papers, and assembling the panels required extensive, focused energy on the part of the organizers. Because of the number of participants, the range of disciplines represented, and the intellectual scope of the presentations, Tax and the committee requested that each participant submit completed papers one year in advance of the celebration. Miraculously, nearly all of the panelists' papers were completed far enough in advance for Tax's special seminar course to work through them with enough time to plan the panels.

Despite all these difficulties, the final program was organized—thanks to the efforts of groups of students and faculty—around five topics:

**Panel 1: The Origin of Life.** This panel included biochemists and astronomers who discussed cosmic evolutionary processes and the biochemical conditions for the origin of life on earth and on other suitable planets. It included Sir Charles Galton Darwin, Th. Dobzhansky, Earl A. Evans Jr., G. F. Gause, Ralph W. Gerard, H. J. Muller, and C. Ladd Prosser. The chairs were Harlow Shapley and Hans Gaffron.

**Panel 2: The Evolution of Life.** The primary discussions addressed current understanding of evolutionary processes with natural selection as the dominant process. This panel included many of the "architects" of the evolutionary synthesis and other leading evolutionary biologists: Daniel I. Axelrod, Th. Dobzhansky, E. B. Ford, Ernst Mayr, A. J. Nicholson, Everett C. Olson, C. Ladd Prosser, G. Ledyard

[107] See the late invitation extended to Harold Urey, who declined because of heavy travel commitments. Sol Tax to Harold C. Urey, 16 October 1959, Darwin Centennial Papers, box 6, folder 1; Harold Urey to Sol Tax, 20 October 1959, Darwin Centennial Papers, box 6, folder 15; and see Tax's response: Sol Tax to H. J. Muller, 22 October 1959, *ibid.*

[108] Alfred E. Emerson to Darwin Centennial Committee, 7 June 1959, Darwin Centennial Papers, box 3, folder 12.

[109] "Organismic biology" began to be used as a category of the biological sciences in the early 1960s. See Smocovitis, *Unifying Biology* (cit. n. 18) for discussion of the growing split between molecular and organismic biology in the late 1950s. Embryology had held a problematic relationship to the evolutionary synthesis. See Mayr and Provine, *The Evolutionary Synthesis* (cit. n. 75) for more discussion on this.

***Figure 2.*** *Panel 2, "The Evolution of Life." Left to right: Daniel Axelrod, Theodosius Døbzhansky, E. B. Ford, Ernst Mayr, A. E. Emerson (chair), Sir Julian Huxley (chair), A. J. Nicholson, Everett C. Olson, C. Ladd Prosser, G. Ledyard Stebbins, and Sewall Wright. Darwin Centennial Papers, Dept. of Special Collections, Joseph Regenstein Library, University of Chicago. Reproduced with permission.*

Stebbins, and Sewall Wright. The chairs were Sir Julian Huxley and Alfred E. Emerson (see Figure 2).

**Panel 3: Man as an Organism.** This served as the transitional panel, bringing anthropological concerns to evolutionary biology. It included evolutionary biologists with an interest in human evolution, and biologically trained anthropologists and paleontologists: Marston Bates, Cesare Emiliani, A. Irving Hallowell and Louis B. Leakey, Bernhard Rensch, and C. H. Waddington. The chairs were George Gaylord Simpson and F. Clark Howell.

**Panel 4: The Evolution of the Mind.** This panel brought together psychologists and physiologists to discuss currents of thought on the evolution of the mind. It included Henry W. Brosin, MacDonald Critchley, W. Horsley Gantt, A. Irving Hallowell, Ernest Hilgard, Sir Julian Huxley, H. W. Magoun, Alexander von Muralt, and N. Tinbergen. The chairs were Ralph W. Gerard and Ilza Veith.

**Panel 5: Social and Cultural Evolution.** This panel represented Tax's bridge between biological and cultural evolution and brought together anthropologists and behavioral ecologists. It included Robert M. Adams, Edgar A. Anderson, Sir Julian Huxley, H. J. Muller, Fred Polak, Julian H. Steward, Leslie A. White, and Gordon R. Willey. The chairs were Clyde Kluckhohn and Alfred L. Kroeber.

Interestingly, the design of the panels followed both a logical sequence of the history of life on earth, and the logical ordering of knowledge: the physical sciences were followed by the biological sciences and the social sciences. The two central panels contained those individuals most closely associated with evolutionary biology, some of whom became historically designated as the "architects" of the evolutionary synthesis.[110] The centrality of evolutionary biology, located between the

[110] Mayr and Provine, *The Evolutionary Synthesis* (cit. n. 75).

physical sciences and the social sciences, was thus implied by the very organization of the panels. Emerson and other evolutionary biologists appeared happy with these final arrangements, or at least they did not appear to express any criticism of their roles on the panels.

Although Tax had envisioned the scientific discussions as leading towards a new understanding of evolution—at the very least bringing new insights into the evolutionary picture—the discussions and even some of the contributed papers were surprisingly flat. From the transcripts recorded of the discussions, it appears few genuinely original insights emerged.[111] The sole panel, with its suite of contributed papers, that made original insights into evolutionary history was possibly the first one, which brought in the latest research from biochemists, cosmologists, and origins-of-life researchers who were making tremendous advances in the 1950s—this, despite the objections of some of the biologists, like A. E. Emerson, about their inclusion.[112]

Much of what was said in the core panel discussions dealing immediately with biological evolution had been reported previously, however. Notably, the supremacy of natural selection was a dominant theme in all panel discussions bearing directly on the subject of biological evolution, with panelists agreeing that genetical understanding of evolutionary mechanisms was leading to major advances. The very spirit of consensus that had prevailed in evolutionary biology in the wake of the evolutionary synthesis may possibly have dampened animated discussions or disagreements. By the late 1950s, the group including Wright, Dobzhansky, Mayr, Stebbins, Simpson, and, of course, Huxley had worked out many of their differences and had carved out well-defined locations for themselves in evolutionary studies; little was left to spontaneous disagreement. With the exception of Conrad H. Waddington, who raised a voice of dissent at the failure of the synthesis to incorporate embryology, and Everett C. Olson, who pointed out fundamental weaknesses in the synthetic theory's response to the evidence from paleontology, there was little dissent in the panel discussions on biological evolution.[113]

Some of the more contentious evolutionists, furthermore, had been left out of the celebration entirely.[114] As noted by Huxley early on in the planning stages in a preliminary letter, R. A. Fisher and J. B. S. Haldane, who ranked alongside Chicago's Sewall Wright, had not been invited.[115] Reasons for this are not clear, but there is some indication that Fisher was not included because of his celebrated "quarrel"

[111] Tax and Callender, *Evolution after Darwin* (cit. n. 16).

[112] See cit. n. 106.

[113] Conrad H. Waddington, "Evolutionary Adaptation," in Tax and Callender, *Evolution after Darwin* (cit. n. 16), vol. I, *The Evolution of Life,* pp. 381–402; Everett C. Olson, "Morphology, Paleontology and Evolution," *ibid.,* pp. 523–45. For a historical commentary on Olson's paper see Stephen J. Gould, "Irrelevance, Submission and Partnership: the Changing Role of Paleontology in Darwin's Three Centennials and a Modest Proposal for Macroevolution," in *Evolution from Molecules to Men,* ed. D. S. Bendall (Cambridge: Cambridge University Press, 1983), pp. 347–66.

[114] One reviewer singled out Huxley's lively performance, however: "From time to time he [Huxley] asked incisive questions, fixed with mordant critique, and then disposed of loose ideas or loose thinking." William B. Bean, "Review of *Issues in Evolution* by Sol Tax," *Archives of Internal Medicine,* 1965, *115:*358–9.

[115] Julian Huxley noticed the absence of Fisher's name on the list provided to him by Tax. Huxley wrote, "I was sorry that the name of R. A. Fisher did not appear in your list. I know that he and Sewall Wright quarrelled, but after all he *did* write The Genetical Theory of Natural Selection which definitely linked genetics with Darwinian Evolutionary theory." Julian Huxley to Sol Tax, 24 August 1956, Darwin Centennial Papers, box 4, folder 9.

with Wright.[116] Fisher's collaborator, E. B. Ford, was, however, featured prominently. Consensus within the core group of biologist panelists was necessary, furthermore, in order to integrate those panelists at the periphery: those whose areas touched on the physical sciences on one side and on the social sciences on the other. At least one reviewer of the panels, however, noted the apparent consensus with unease and pointed to the rise of an "orthodox doctrine" with respect to the "synthetic theory" and its adherents. He asked explicitly "if it might not have been salutary to admit a real biological 'maverick,' to the proceedings and allow him to utter heresies?"[117] Certainly, it would probably have made for more lively discussions.

One such maverick did make an appearance in panel five, "Social and Cultural Evolution." Facing existing divisions in anthropology, Leslie A. White publicly reviled his nonevolutionary colleagues and the tradition in anthropology that was "definitely and vigorously antievolutionist." He expressed astonishment at finding himself in a public forum like the Darwin Centennial Celebration, which indicated that "cultural evolution is becoming respectable and therefore popular," but he also added the rather cynical remark, "I don't think cultural evolutionists are going to be made overnight by the popularity of television and other public ceremonies and exhibitions."[118] Despite White's outburst, most of the formal discussions and the contributed papers appeared to stay well within the existing divisions in anthropology, undermining Tax's organizational ambitions and occasional papers by anthropologists, like Alfred Kroeber, who called for a more synthetic long-term view.[119]

Another explanation for the rather tepid panel discussions is that they—and, to some extent, the contributed papers—were aimed for popular audiences. Original scientific research into evolution by the late 1950s would have been much too technical for a mostly general audience, even though the audience for the panels consisted of delegates and representatives of disciplines bearing on evolutionary study. The papers were also designed from the start to reflect the state of the art in each contributor's field, rather than to introduce novel insights. The panel discussions were also limited to five minutes and took place in front of no less than one thousand delegates (who held pre-arranged tickets) under the uncomfortable glare of klieg lights.[120] With eight or nine participants per panel, moreover, and with two chairs to guide discussion, spontaneous exchanges would have been difficult to generate. Discussion, for the most part, thus tended to be on such popular themes as the future of man, with special reference to general problems of population control, nuclear war, and other concerns stemming from the cold war.[121]

The panel discussions were not the only vehicle for scientific conversation. The Darwin Centennial Celebration included exhibits of an impressive collection of scientific literature, as well as of historical reissues of books that honored the publication of Darwin's *Origin.* Newer books on evolution, like Theodosius Dobzhansky's

[116] The scientific dispute between Fisher and Wright is discussed in detail in Provine, *Sewall Wright and Evolutionary Biology* (cit. n. 37).

[117] Thomas Goudge, "Darwin's Heirs," *The University of Toronto Quarterly,* 1961, *30:*246–50, quotation on p. 248.

[118] Leslie White, Transcripts to panel five, "Social and Cultural Evolution," p. 234, in Tax and Callender, *Evolution after Darwin* (cit. n. 16), vol. III.

[119] Alfred L. Kroeber, "Evolution, History, and Culture," in Tax and Callender, *Evolution after Darwin,* vol. II, *The Evolution of Man.*

[120] See Tax, "The Celebration" (cit. n. 16).

[121] *Ibid.*

*Genetics and the Origin of Species* (in its third edition), were part of the displays.[122] Several halls were set up with long tables, posters outlining the state of evolution, and displays including memorabilia representing the Darwinian heritage.[123] Of special significance was a hallway exhibit some 126 feet in length that consisted of aluminum beams supporting double rows of posters. Titled "The Darwin Exhibit," it traced out the story of evolution through successive representational images. It had been borrowed and transported at great cost from the University of California at Los Angeles Biomedical Library, which had erected it for its own celebrations.[124]

The University of Chicago Library held yet another exhibit on Darwin titled "The Influence of Darwin as Seen Through the Publication and Reaction to His Works." Books of historical interest included first editions of seventeen of Darwin's works from the University of Chicago Library and, of course, a first-edition *On the Origin of Species.*[125] The university bookstore held an additional display and made available books of historical importance and books by celebration participants for purchase; this was based on a list provided by Tax.[126] Other exhibits and displays in honor of the celebration were held far and wide in the Chicago area and included displays at Chicago booksellers. Nearby, the Chicago Natural History Museum created its own six-panel display on Darwin and evolution for participants and the public, and published a commemorative issue of its natural history journal assessing the current state of evolution by means of natural selection (see Figure 3).[127]

Perhaps not as customary as literature exhibits at conferences until the 1950s were special screenings of films. Two films were shown for participants on the official program: an advance screening, or sneak preview, of a very recent film billed as "evolution in action" and titled *The Ladder of Life,* and a film and lecture by ethologist Niko Tinbergen titled *Evolutionary Aspects of Social Communication in Animals.*[128] The latter film was officially part of the Institute for High School Biology Teachers.[129]

[122] Participants' books were displayed at the University of Chicago Bookstore for purchase, along with a list of additional titles for even "greater scope." Sol Tax to a Mr. Passmore, 5 November 1959, Darwin Centennial Papers, box 1, folder 3.

[123] For the listing of memorabilia included in seventeen display cases see "Darwin Exhibit," Darwin Centennial Papers, box 1, folder 13.

[124] The documentation for this exhibit is in Darwin Centennial Papers, box 1, folder 13. The exhibit required a considerable coordination of efforts to assemble, especially from West Coast evolutionists. Some of this correspondence is included in the Hubbs Papers, 1927–1979, MC-S box 17, folder 25, "Darwin Anniversary, 1956."

[125] Presented to the university by Col. William M. Spencer of Chicago. Program, The University of Chicago Centennial Celebration, Darwin Centennial Papers, box 7, folder 1.

[126] Sol Tax to Mr. Passmore, 5 November 1959, Darwin Centennial Papers, box 1, folder 3; and see the Program, The University of Chicago Centennial Celebration, Darwin Centennial Papers, box 7, folder 1.

[127] The special issue of the *Chicago Natural History Museum Bulletin* included a lead historical article titled "Centennial of Darwin's 'Origin of Species' Hailed," by the curator of vertebrate anatomy, D. Dwight Davis. *Chicago Natural History Museum Bulletin,* 1959, *30:*3–4. Darwin Centennial Papers, box 2, folder 13. It also included a description of the special exhibit on p. 4: "Six panels tell the story of *Origin* by explaining the meaning of, and the evidence for, natural selection—the book's theory that in the 'struggle for existence' those characteristics will be retained that best enable an organism to cope with life and to survive."

[128] *The Ladder of Life* had been prepared by the Columbia Broadcasting System and was set to air on *Conquest* on 29 November. Joseph Krumgold to Sir Julian Huxley, 14 October 1959, Darwin Centennial Papers, box 8, folder 3; Michael Sklar to Sol Tax, 20 October 1959, *ibid.;* Barbara Emerson to Sol Tax, 19 November 1959, Darwin Centennial Papers, box 8, folder 8.

[129] Program, University of Chicago Darwin Centennial Celebration, Darwin Centennial Papers, box 7, folder 1.

***Figure 3.*** *Cover of the* Chicago Natural History Museum Bulletin, *November 1959. Darwin Centennial Papers, Dept. of Special Collections, Joseph Regenstein Library, University of Chicago. Reproduced with permission.*

Tax sought to "generously" invite "the whole world" to attend the celebration including nuns, ministers, and even a cardinal.[130] By the time the last invitation had gone out, some fifty panelists/contributors had finally agreed to attend the meetings, along with delegates from 189 institutions from around the world.[131] The delegates had been invited to attend through an aggressive postcard campaign launched by

[130] Tax, "The Celebration" (cit. n. 16), quotation on p. 275. See also Cardinal Tisserant to Sir, 18 April 1959, Darwin Centennial Papers, box 6, folder 15.
[131] Tax, "The Celebration" (cit. n. 16), quotation on p. 275.

Tax that reached over one thousand institutions of higher learning worldwide.[132] The final tally of nearly 250 delegates and 2,500 registrants was remarkably large, considering that the conference was taking place at the height of the cold war. Bringing delegates and participants from iron-curtain countries was especially onerous for organizers and sometimes involved negotiations and interventions with the State Department.[133]

## VIII. SACRAMENT AND SACRILEGE: THANKSGIVING AND THE CONVOCATION CEREMONY

Conveniently, the five-day celebration at the University of Chicago coincided with the most American of holidays, and a secular one at that: Thanksgiving.[134] Participants could therefore observe dual celebratory functions: giving thanks for their American heritage, and giving thanks for their Darwinian heritage. The two had been brought beautifully in line on that day. In keeping with American tradition, the celebrants attended a banquet with the traditional Thanksgiving menu of turkey, mashed potatoes, and the like.The official banquet/dinner of the SSE, it also included the ill-fated presidential address by Edgar Anderson, which was to be the official toast to the anniversary celebration.

But the secular tone of this component of the celebration—and of the Darwinian heritage—was undermined by the grand culmination of the five-day celebration, also on Thanksgiving Thursday: the Convocation ceremony, which emulated the most sacred of ritual practices in evolutionary guise. An organizational problem of its own, the ceremony was staged with artful precision and was an exercise in the grandest of academic pomp and circumstance.[135] Dressed in full academic regalia (organizers had made prior arrangements with delegates for the renting of gowns; delegates provided their own hoods), delegates queued in orderly fashion as part of a grand procession from Mandel Hall, the site of the conference, to Rockefeller Memorial Chapel, whose bells and carillons were played for all in a "carillon concert and swinging peal."[136] Even though it was overcast and chilly, with light snow on the ground, the celebrants cheerfully marched in procession (see Figure 4).

[132] The letters of response are included in Darwin Centennial Papers, box 2, folder 8. Also see Sol Tax to Robert P. Houston, Jr., 5 November 1959, at the Department of State, in which Tax states that over one thousand delegates and learned societies had been invited. Darwin Centennial Papers, box 4, folder 3.

[133] The Soviet ecologist-mathematician G. F. Gause required special intervention with the Department of State. See the exchanges back and forth between interested parties in Darwin Centennial Papers, box 4, folder 3, especially Sol Tax to Robert B. Houston, Jr., 5 November 1959. So too did François Bordes, a French archaeologist from Bordeaux and a former Communist Party member. Sol Tax to François Bordes, 8 September 1959, Darwin Centennial Papers, box 3, folder 6. Tax secured a visiting professorship for Bordes for the fall, 1959.

[134] Tax held the conference during the Thanksgiving holiday in order to include high-school and college science teachers. Sol Tax to Detlev Bronk, 11 February 1959, Darwin Centennial Papers, box 2, folder 5. See also the "Memorandum to Participants," which explained reasons for the choice of dates, dated 24 October 1958, Darwin Centennial Papers box 6, folder 16.

[135] The program instructions are included in Darwin Centennial Papers, box 2, folder 10.

[136] A letter from Charles Callender to Sister Cecelia B. V. M. from the Department of Biology at Mundelein College, Chicago, reveals the extent to which the organizational details had been worked out in advance. Callender writes, "In answer to your earlier letter about academic dress at the convocation during the Darwin Centennial Celebration, the religious habit will be completely satisfactory. It will not be necessary to carry the hood, although you may do so if you wish." Letter dated 10 September 1959, Darwin Centennial Papers, box 2, folder 8.

***Figure 4.*** *Convocation procession to Rockefeller Chapel. Darwin Centennial Papers, Dept. of Special Collections, Joseph Regenstein Library, University of Chicago. Reproduced with permission.*

They filed into the imposing midwestern Gothic edifice and took their seats (prearranged according to a seating plan) to the moving sounds of the chapel organ playing Bach's Passacaglia and Fugue in C Minor. Chancellor Lawrence Kimpton presided over the ceremony and led the singing of "America." In what may have appeared a contradiction in terms, the congregation of evolutionists also shared a prayer led by the Reverend W. Barnett Blakemore, Jr.[137]

The mood of reverence abruptly ended, however, as Julian Huxley began to deliver his Convocation address.[138] From the pulpit of the chapel, Huxley delivered his lecture "The Evolutionary Vision" to an audience stunned by his revelations. Huxley's lecture boldly decreed that religion was an "organ of evolving man," and that the time had come for a "new organization of thought" based on the new evolutionary vision (see Figure 5).[139] In one especially inflammatory passage he proclaimed:

> Evolutionary man can no longer take refuge from his loneliness in the arms of a divinized father-figure whom he has himself created, nor escape from the responsibility of making decisions by sheltering under the umbrella of Divine Authority, nor absolve

The quote is from a memorandum to Charles Callender from Marilyn Lickfield, who was with the chapel administration. The university organist, Mr. Heinrich Fleischner, played for the convocation ceremony. Darwin Centennial Papers, box 2, folder 10.

[137] Convocation Program, Darwin Centennial Papers, box 7, folder 1.

[138] The committee had tried to secure Jacques Barzun for the convocation address, but he declined because of previous commitments. Lawrence Kimpton to Jacques Barzun, 19 May 1959, Darwin Centennial Papers, box 2, folder 10; Jacques Barzun to Lawrence Kimpton, 25 May 1959, *ibid.*

[139] See Julian Huxley, "The Evolutionary Vision: The Convocation Address," in Tax and Callender, *Evolution after Darwin* (cit. n. 16), vol. III, quotations p. 257 and p. 256, respectively.

*Figure 5. Sir Julian Huxley on the pulpit of Rockefeller Chapel delivering his address, "The Evolutionary Vision." Darwin Centennial Papers, Dept. of Special Collections, Joseph Regenstein Library, University of Chicago. Reproduced with permission.*

> himself from the hard task of meeting his present problems and planning his future by relying on the will of an omniscient, but unfortunately inscrutable, Providence.[140]

The "evolutionary vision," as "opened up to us" by Charles Darwin, he closed, his speech soaring to crescendo, "exemplifies the truth that truth is great and will prevail, and the greater truth that truth will set us free." "Evolutionary truth," he added, freed humans from "subservient" fears, and showed the way to "our destiny and our duty."[141]

Needless to say, Huxley's "secular sermon" was not well received. Many of the audience members later reported that they had been stunned by the lack of discretion that Huxley displayed, though none had expressed themselves so publicly or vociferously as had the popular press, which feasted on the newsworthiness of Huxley's fire-and-brimstone secular sermon.[142] One local press described the reaction thus:

[140] *Ibid.*, p. 253.

[141] *Ibid.*, p. 260.

[142] Historian John Greene still vividly recalls the aftermath of Huxley's "secular sermon." He writes, "My friends told me that the Chancellor of the University felt embarrassed by Huxley's performance ('right up there in that pulpit'!)" John Greene to Vassiliki Betty Smocovitis, 13 March 1996.

"Hush Shrouds Huxley's Talk: Evolution Talk Here Stirs Hushed Echoes"; another Catholic press from Cincinnati reported the news as "Science Doesn't Defeat Religion (Cheers did not Greet Atheist)."[143]

But the mood shifted to solemnity again as the names and achievements of the honorary degree recipients were announced.[144] The anthropological cast to the celebration was apparent from the choice of awardees, who shared not so much an expertise in evolutionary biology as an interest in the evolution of man.[145] The awardees were Sir Charles Darwin, Theodosius Dobzhansky, Alfred Kroeber, Hermann Joseph Muller, George Gaylord Simpson, Sewall Wright, and Sir Julian Huxley. Both living descendants of Darwin and Huxley were so honored; curiously excluded from notice, however, were two of the leading architects of the evolutionary synthesis: G. Ledyard Stebbins and Ernst Mayr, both of whom were panelists-contributors to the celebration. From documents deposited as part of the celebration proceedings, the two men appear to have never even been entertained as possible recipients, although both by then had been prominent and visible leaders of evolutionary biology.[146] Possibly, both Mayr and Stebbins had not distinguished themselves in anthropological areas dealing with the evolution and future of man in the way that the recipients had already done, and had therefore not been that central to Tax or the committee's vision of the event. Nor did they have especially close ties to the University of Chicago.

From the program, it appears that the Convocation ceremony was considerably long (it began at three o'clock and preceded the banquet at six).[147] Closing highlights recognized the host institution, the University of Chicago, by playing the "Alma Mater." The sacral function of the ceremony was once again restored as celebrants received a final benediction before they exited to the music of Max Reger's Toccata in D Minor.[148]

With all the sanctification of Darwin and evolution through ritual acts and soaring music in a sacred setting, it was small wonder that Edgar Anderson returned from the convocation feeling that he had a "religious experience."[149] The strong emotions that it generated in him may, in part, explain his inability to give his presidential address at the Thanksgiving dinner, which followed immediately after the Convocation.[150]

### IX. DARWIN'S LIFE STORY STAGED AS MIDWESTERN MUSICAL: *TIME WILL TELL*

As if the Convocation ceremony did not have enough pomp, pageantry, and spectacle, the five-day celebration included additional light entertainment for the cele-

[143] *Chicago's American,* 30 November 1959; *Catholic Telegraph Register,* 11 December 1959, Darwin Centennial Papers.

[144] The convocation ceremony was filmed by *Encyclopedia Britannica.* A copy of the original film is housed in the media room of the special collections. See film F-134, Darwin Centennial Papers.

[145] Sewall Wright is the possible exception, but he was a professor emeritus, having been on the faculty of the University of Chicago. He was the sole nomination by H. Burr Steinbach, the chairman of zoology who stated Wright was "the outstanding candidate." H. Burr Steinbach to Dean Wendell R. Harrison, 3 February 1959, Darwin Centennial Papers, box 2, folder 11.

[146] Document titled "Report on Honorary Degrees," undated, *ibid.* Other names considered included Bernhard Rensch, G. F. Gause, and A. Irving Hallowell.

[147] Draft of Convocation Program, Darwin Centennial Papers, box 2, folder 10. Also see the final program, Darwin Centennial Papers, box 7, folder 1.

[148] Convocation Program, Darwin Centennial Papers, box 7, folder 1.

[149] See cit. n. 14.

[150] See cit. n. 15.

brants. In the best spirit of American popular culture in the 1950s, Sol Tax had actively supported the production of a theatrical play as a special evening entertainment for celebrants. His initial desire was to stage a production of Jerome Lawrence and Robert E. Lee's well-known Broadway play dramatizing the 1925 Scopes Monkey Trial titled *Inherit the Wind.*[151] Studs Terkel was to be the director and Melvyn Douglas was to be an actor, but plans for this fell through.[152] Preliminary notes made by Tax indicate that his early plans for the entertainment portion of the celebration included a midwestern musical production—a "Darwin Musical Extravaganza"—with the tentative titles *Centennial Centenary* or *The Time is Back in Joint.* He wrote of his plans, "It is designed as a light musical but strongly integrated unitary piece rather than a 'review.' (Thus like Oklahoma, etc.)" In keeping with the academic gravity of the occasion, it would, however, "have an intellectual impact."[153]

His original concept of the play drew on a visionary, futuristic theme moralizing evolution and human society as reflected in the 1959 celebrations through the perspective of international celebrants in the year 2059. The theme of progress was strongly embedded in this scenario, and was indicated clearly in the transcript: "PROGRESS is assumed, and apparently with reason." If there was a message in the story relayed by Tax's futuristic scenario, it was to further endorse the success of the 1959 celebrations. Tax's vision of his celebration and its role in history was nothing short of grandiose and offers considerable insight into his motivations. Describing his futuristic play, he wrote:

> It becomes clear now that this is a world-wide celebration of the 'Centennial Centennial' (indeed it will come out later that we catch a salute from some not too distant planet). Why? because the 1959 Darwin Centennial turned out to be an event that changed the course of history by helping to resolve the crisis of the era.

Closing with the following, Tax thought to humanize his plot, leaving enough room for the indomitable human spirit:

> An argument for a 'personal plot'—a love triangle or other problem—is that it ought to become clear that no problems of the world at this level have been solved. A major lesson is that mankind lives with his imperfections, but he is free to live with them in a variety of ways—some more destructive than others. 'Human nature' does not change; whatever it is, it is irrelevant."[154]

[151] Jerome Lawrence and Robert E. Lee, *Inherit the Wind* (New York: Dramatist's Play Service, 1958); also see Jerome Lawrence and Robert E. Lee, *Inherit the Wind* (New York: Bantam, 1960). The play opened on Broadway in 1955. The first film was made in 1960. Edward J. Larson briefly discusses the significance of this play in the context of cold-war politics in *Summer for the Gods* (cit. n. 98).

[152] Sol Tax to Studs Terkel, 6 October 1959, Darwin Centennial Papers, box 8, folder 3; Melvyn Douglas to Sol Tax, 15 March 1958, Darwin Centennial Papers, box 2, folder 1. The precise reason for cancelling *Inherit the Wind* is unclear. A letter from John Reich, head of the Goodman Theatre in Chicago, to Tax dated 10 April 1958 indicates that there was a conflict in scheduling, with Reich having already committed his theatre to other productions for the month of November in 1959. *Ibid.* Tax did not easily give up his wish of staging *Inherit the Wind.* Documents indicate that he was trying to stage *Inherit the Wind* in addition to his *Time Will Tell* until late into the planning stages of the celebration.

[153] Notes by Sol Tax for a musical play to be written for the Darwin Centennial Celebration, November 1959. Darwin Centennial Papers, box 1, folder 7.

[154] *Ibid.*

For reasons not entirely clear, however, and perhaps fortunately for the celebrations, Tax's strange play did not materialize.[155] Instead, the Darwin Centennial Committee (including Tax) sought to produce a more traditional, yet festive reenactment of Darwin's life story with the title *Time Will Tell* (picking up the theme of time from Tax's original scenario). (Earlier titles considered included "Where's Charlie," "The Monkey Century," and "A Century Note".)[156] What could be more fitting entertainment for celebrants than a play on Darwin's life in musical form? And the life of Darwin did lend itself readily to such a production. Rodgers and Hammerstein being beyond the committee's budget, a local duo of musician and lyricist—billed in one advertisement as "those gifted upper primates"—were hired instead.[157] Ashenhurst and Pollack were an unlikely combination who were to have only the briefest of musical careers beyond their staging of *Time Will Tell:* Robert Ashenhurst was an assistant professor in the Graduate School of Business Administration, and Robert Pollack was a local investment broker and former drama critic for the *Chicago Sun-Times.* With a cast of forty-five "talented homo sapiens," the lead of Charles Darwin fell to a Chicago restaurateur by the name of Rick Riccardo.[158] Tax had designated his play as "a serious, though light-hearted entertainment for the celebration participants," and that is certainly what he got.

The play was an elaborate production complete with original costumes and staging (see Figure 6).[159] It was widely publicized with posters, flyers, and even mobiles of replicas of the *Beagle* hanging in prominent locations.[160] The play was a tremendous success, playing every night for celebrants (see Figure 7). The musical scores and librettos preserved in the archival documents of the celebration indicate an imaginative and lively reenactment of Darwin's life story (see Figure 8).[161] Songs included titles from key events in Darwin's life, like "Homeward Plows the Beagle," "Trilobite," "Marry, Marry, Marry," and "The Meeting of the BAAS." The title song, "Time Will Tell," played on an historical inversion as the Darwin celebration was validated by the actors' reference to the future (twentieth-century) assessment of Darwin. Members of the audience thus heard what they already knew: that the century had made a difference:[162]

> A hundred years hence
> Will Darwin make sense
> To the likes of us
> Who are on the fence.

[155] The change in plays appears to have taken place after Ashenhurst and Pollack took over the musical production.

[156] Memo, Carl W. Larsen to "Belowlisted," 23 July 1959. Darwin Centennial Papers, box 2, folder 1.

[157] Illustrated advertisement for *Time Will Tell.* Darwin Centennial Papers, box 2, folder 2.

[158] *Ibid.* (quotation).

Rick Riccardo was a well-known Chicago resident (he was sometimes called "Ricky"; I could find no connection to the character of the same name in the popular television comedy *I Love Lucy*); his restaurant, Riccardo's, was very popular at the time. The part of Emma Wedgwood Darwin was played by Jo Anne Schlag, and Win Stracke, a TV and radio folk singer, was cast as the bosun on the *Beagle.*

[159] *Time Will Tell,* Program from the Play, Darwin Centennial Papers, box 1, folder 16.

[160] Memo, Carroll G. Bowen to Marion Carnovsky, 31 August 1959, Darwin Centennial Papers, box 2, folder 2.

[161] Darwin Centennial Papers, box 1, folder 15.

[162] *Time Will Tell,* a musical play by Robert Pollack and Robert Ashenhurst, *ibid.*

***Figure 6.*** *Program from the musical play* Time Will Tell. *Darwin Centennial Papers, Dept. of Special Collections, Joseph Regenstein Library, University of Chicago. Reproduced with permission.*

> Will a century make a difference?
> Only time will tell
> Only time will tell![163]

As a souvenir of the celebration, a long-playing vinyl record of the musical was made available for purchase.[164]

On the surface, celebrants were treated to a lively, imaginative, and clever form of entertainment. But the theatrical production also saw, briefly and metaphorically, the resurrection of the Darwin figure and his life story playing itself out—if in a somewhat modernized, sensationalized, and Americanized form.

[163] *Ibid.*

[164] At the time of writing, I have been unable to locate a copy of the record. Registrants could purchase the album of two records for $8. Registrants Order Form, Darwin Centennial Papers, box 2, folder 4. Another advertisement is for a twelve-inch, long-playing record that sold for $5. Advertisement, Darwin Centennial Papers, box 2, folder 2.

***Figure 7.*** *Scene from the musical. Darwin Centennial Papers, Dept. of Special Collections, Joseph Regenstein Library, University of Chicago. Reproduced with permission.*

## X. MAXIMIZING PROFIT: IMMORTALIZING THE CELEBRATION AND SPREADING THE NEWS FAR AND WIDE

Despite a final total accounting cost of $59,022.06 and a hefty $7,032.46 total net deficit, Tax—and apparently most of the university administration—felt that "nobody doubts that the Centennial was worth whatever it cost."[165] The celebration's "value" to the university, he wrote, "must heavily outweigh the University funds invested in it."[166]

The investment in material, personnel, and financial resources made by the University of Chicago thus required the maximization of profit, through extensive press coverage and some sort of permanent reminder of the celebration. In the spirit of maximizing profit, ensuring permanence, and transmitting news of the conference

[165] Final costs given in Final Accounting, Darwin Centennial Papers, box 2, folder 18.
[166] Memorandum, Sol Tax to R. Wendell Harrison, 5 May 1960, titled "Darwin Centennial Celebration accounting and business affairs," *Ibid.*

*Figure 8. Musical score and lyrics to the title song, "Only Time Will Tell." Darwin Centennial Papers, Dept. of Special Collections, Joseph Regenstein Library, University of Chicago. Reproduced with permission.*

far and wide, Tax and his committee planned to continue to cash in on the celebration after it ended.[167]

They approached the *Encyclopedia Britannica* (EB) to produce a twenty-five minute documentary film of the occasion. Its purpose, as restated by the president of EB, was "to further education and interest in the subject of evolution."[168] As a result, in Pathe news fashion, cameras were turned on the celebration, covering especially the panels and the Convocation ceremony. The completed 16-mm sound film ran for 28.5 minutes and, according to advertisements, attempted to "capture the spirit of the Celebration."[169] Later, when the deficit from the celebration had become apparent, organizers recalled that the film was a "magnificent continuing public relations asset,

[167] Whether the play operated at a profit or sustained a loss became a point of disagreement between Robert Pollack and Sol Tax. See the exchanges between them, Darwin Centennial Papers, box 2, Addendum.

[168] Maurice Mitchell to Sol Tax, 20 June 1960, Darwin Centennial Papers, box 9, folder 5; this echoed Tax's earlier letter: Sol Tax to David Ridgeway, 5 May 1950, *ibid.*

[169] Notice, Sol Tax to "Dear Registrant," *ibid.* The film could be purchased or rented for use.

that could in addition supply cash income to the University."[170] The University of Chicago did not hesitate to cash in on the occasion yet again (though it charged no admission) when, at the premiere of the film in 1960, it invited former geology student John Thomas Scopes as a special speaker to further legitimate the screening and the celebration.[171] Scopes probably helped to draw a larger audience to the screening than would otherwise have attended.

But perhaps the most effective way of transmitting and, at the same time, recording news of the celebration was through the media campaign, the structure of which Tax carefully used to his advantage. In addition to promotional news alerting local and national audiences of the event (prepared by Sheldon Garber), the day-to-day celebration happenings were reported widely in newspapers and on radio news.[172] The actual press corps at the celebration numbered some twenty-seven reporters, most of whom stayed for the full five days.[173] No less than three full archival folders hold newspaper clippings covering the occasion. Items featured photographs of participants, news of discoveries of importance to evolution, and pronouncements on the status and future of modern man. Suitably sensationalistic titles like "Society to Toast Darwin's Ghost" and "It's easy to converse with scientists, if you know what Zinjanthropos means" drew special attention to celebratory events.[174] Some also transmitted the sober message that evolution by means of natural selection was a fact. The *New York Times* column reporting the event ran the title "Evolution a Fact, Darwin Fete Told."[175]

By far the most noteworthy item of the celebration was Huxley's pronouncement of the end of conventional religion and its substitution with evolutionary humanism, the coverage of which was unfavorable to Huxley (and frequently to Tax).[176] These news items generated further interest and response from the public, which was especially vocal about its displeasure with Huxley's negative attitude towards conventional religious belief. But perhaps the most interesting—and revelatory—of the clippings, from the pages of the *Milwaukee Journal,* demonstrated the themes of generational continuity in the celebration. The nineteenth-century Darwin and Huxley were shown alongside their living representatives, bringing renewed life to the Darwinian heritage for American popular audiences. The caption read, "Gentle Darwins and Fierce Huxleys: Old Tom, the Tiger, and His Intense Grandson Are a

[170] Memorandum, Sol Tax to R. Wendell Harrison, 5 May 1960, titled "Darwin Centennial Celebration accounting and business affairs." Darwin Centennial Papers, box 2, Addendum.

[171] The advertisement flyer includes the following: "The Darwin Centennial Committee presents Encyclopaedia Britannica's Documentary Film of the Darwin Centennial Celebration. Guest: John Thomas Scopes (Dayton, Tenn. 1925 'Monkey Trail')." Flyer, Darwin Centennial Papers, box 9, folder 5. A panel discussion on "The Scopes Trial of 1924 as Seen Today" was scheduled following the film. Letter dated 18 October 1960, *ibid.* Also see the press release titled "Immediate," dated 19 October 1960, Darwin Centennial Papers, box 8, folder 7.

[172] Many of the press releases are in Darwin Centennial Papers, box 8, folder 6.

[173] Document titled "Press and Public Response to the Darwin Centennial," Darwin Centennial Papers, box 8, folder 2.

[174] *Chicago's American,* 23 September 1959, Darwin Centennial Papers; Lois Baur, *Chicago's American,* 23 November 1959, Darwin Centennial Papers.

[175] *New York Times,* 26 November 1959. The *Milwaukee Journal* picked up the same theme: "Aims Stated in Evolution: 'Teach as Fact'," *Milwaukee Journal,* 2 December 1959, Darwin Centennial Papers.

[176] Document titled "Press and Public Response to the Darwin Centennial," Darwin Centennial Papers, box 8, folder 2.

THE MILWAUKEE JOURNAL Wednesday, February 10,

Four top scientists (from left), evolutionist Charles Darwin; his grandson, Sir Charles Galton Darwin; Sir Julian Huxley, and his grandfather, biologist Thomas Huxley.

Gentle Darwins and Fierce Huxleys

Old Tom, the Tiger, and His Intense Grandson Are a Striking Contrast to the Quiet Evolutionist and His Soft Spoken Relative

***Figure 9.*** The Milwaukee Journal, *10 February 1960. Darwin Centennial Papers, Dept. of Special Collections, Joseph Regenstein Library, University of Chicago. Reproduced with permission.*

Striking Contrast to the Quiet Evolutionist and His Soft Spoken Relative" (see Figure 9).[177] All told, the clippings compiled by the University of Chicago public relations department numbered some 196 items, from fifty-seven publications.[178] There were over 4,147 inches of standard newspaper type; as the official public relations analysis put it, "more than enough to fill a 23-page paper without ads."[179]

Of interest to more intellectual American audiences was the production at WBBM-TV (a CBS affiliate) of a talk show featuring the most "distinguished" (and photogenic) of the participants. At the formal invitation of the producer Jerry Levin, who viewed it as an "intellectual calamity" to "confine" the "distinguished men" only to campus meetings, Tax was encouraged to bring his celebration "to an extremely wide audience, which has proven by its unprecedented size that there are literally hundreds of thousands of people in this city who would be quite disappointed were they not to be given the chance to hear and see the men involved in this

[177] *Milwaukee Journal,* 10 February 1960, Darwin Centennial Papers.

[178] Document titled "Press and Public Response to the Darwin Centennial," Darwin Centennial Papers, box 8, folder 2.

[179] *Ibid.*

event."[180] Noted Chicago host and *Chicago Sun-Times* columnist Irving Kupcinet brought Sol Tax, Sir Charles Darwin, Sir Julian Huxley, astronomer Harlow Shapley, and no less an American luminary than Adlai Stevenson (who as former governor and presidential candidate in 1952 and 1956 also attended the festivities) to Kupcinet's late Saturday night "conversation program," *At Random.*[181] Later aired repeatedly to American audiences, it was reviewed as being one of the most intellectually stimulating and engaging conversation programs aired that year. Recalling the occasion on the death of astronomer Harlow Shapley in his *Sun-Times* column some thirteen years later, Kupcinet wrote, "We often cite that panel as most [sic] articulate and intelligent we've ever assembled . . . "[182]

Radio programs were also kept abreast of the celebration and transmitted not only news, but also talk-show programs with the visiting scientists. The local university-based (WTTW) radio series *All Things Considered* scheduled several programs around the panel discussions. Visiting scientists like Huxley and Alfred Kroeber joined Sol Tax and Ilza Veith for a round-table discussion transmitted to Chicago audiences.[183]

The most permanent of recordings, however, and possibly the one having the most intellectual influence, was the publication of the conference papers in a three-volume, comprehensive work. Titled *Evolution After Darwin,* 1,250 copies of the gray-and-blue volumes were published quickly by the University of Chicago Press and transmitted to libraries and individuals far and wide. In preparing the volumes, Tax once again put his organizational talents to work: not only were the scientific details of the conference included with all of the articles, but also transcripts of the panel discussions. Tax's historical reflections on the celebration, event memorabilia, photographs, and transcripts from radio and television talk shows were all included in the last of the three volumes with the assistance of Charles Callender. The profits generated by the volumes would help to defray celebration expenses and reduce the deficit.[184]

## XI. AFTER THE DARWIN CENTENNIAL: "REFLECTED GLORY" AND THE "WORDS OF THE DEVIL"

In addition to the daily commentary that filled the newspapers and the letters of thanks that poured in, response to the celebration was significant. Articles reporting

[180] Jerry Levin to Sol Tax, 2 October 1959, Darwin Centennial Papers, box 9, folder 6; Jerry Levin to Sol Tax, 7 October 1959, *ibid.;* also see the letter of invitation to Julian Huxley: Jerry Levin to Julian Huxley, 24 September 1959, *ibid.*

[181] The transcripts of *At Random* were published in Tax and Callender, *Evolution after Darwin* (cit. n. 16), vol. III, pp. 41–65.

[182] "Kup's Column," *Chicago Sun-Times,* 7 December 1972, Darwin Centennial Papers, box 3, Addendum.

[183] Lee Wilcox to Sol Tax, 26 May 1959, Darwin Centennial Papers, box 8, folder 3; document from Lee Wilcox, Office of Radio-TV, undated, *ibid.* The transcripts of the talk show were published in Tax and Callender, *Evolution after Darwin* (cit. n. 16), vol. III, pp. 263–70.

[184] Memorandum, Sol Tax to R. Wendell Harrison, 5 May 1960, titled "Darwin Centennial Celebration accounting and business affairs," Darwin Centennial Papers, box 2, Addendum. According to this document, the long-playing vinyl record of *Time Will Tell* had "broken even." As of 1968 the total royalties earned from *Evolution after Darwin* were $11,117.41. Of the three volumes, volume I, *The Evolution of Life,* sold the greater number. This is not surprising as this volume dealt with more conventional topics in evolutionary biology. Memorandum, J. Ratuszny to Maurice English, 21 October 1968, Darwin Centennial Papers, box 3, Addendum. Tax signed his royalties over to the centennial celebration to decrease its deficit.

on the festivities took many forms for different audiences, from high-school teachers to alumni of the University of Chicago to scientists reading technical journals the world over.

The response was overwhelmingly positive in nature. Many local dignitaries wrote to Tax to state that the celebration had worked to establish the University of Chicago as the primary site of learning in America's Midwest. One memorandum to Tax from Marie-Anne Honeywell stated this explicitly. She wrote, reporting on other conversations (including among Citizens Board Members at the University of Chicago's Quadrangle Club), that "The concensus [sic] is that this event has reaffirmed the position of the University as the center of learning in the Middle West."[185] Others wrote to say how much they enjoyed not only the scientific sessions, but also the musical productions, which had packed audiences.

Negative responses were also notable, but were mostly directed to Huxley's "pulpit" speech on evolution and not to the celebration as a whole. Huxley's oration could not have found a more unsympathetic audience. A "regional evangelical stronghold," Chicago had been proclaimed "the evangelical capital of the U.S.A." by *Christian Life* magazine only in 1952.[186] Even the University of Chicago had a strongly defined religious history.[187] It is not surprising, then, that large numbers of those who learned of the occasion reacted negatively: the archival file folders bulge with responses, many from outraged individuals who had read of Huxley's controversial "vision" and what it meant for conventional religion in newspaper reports. The official public relations analysis determined that this, and the Institute on Science and Theology, were the "outstanding spot news events of the conference."[188] A copy of a note to the British Embassy sent to Tax urged the immediate deportation of Huxley because of his Convocation address. It stated Huxley "should be reprimanded and sent home under censure for daring enter and enjoy the hospitality of the people of the United States and at the same time endeavoring to undermine the moral foundation of our being, daring express his theory of Godlessness, all has been evolved, rather than of God." The letter closed with an official challenge to a public debate.[189] Another letter addressed to the president of the University of Chicago, from a self-described Cuban writer, poet, and newspaperman, referred to Huxley as a "Super Imbecile."[190] Yet another letter condemned the celebration generally:

> God is speaking to you now thru me—because for you to deny 'The Truth' As I am telling it—you are condemning yourself & others to utter darkness for ever . . .
>
> All the money that is being spent on this useless 5 day convention should go to an orphanage of some church. because know [sic] good will come of it as it is only the

[185] Memorandum, Marie-Anne Honeywell to Sol Tax, 3 December, 1959, Darwin Centennial Papers, box 9, folder 1.

[186] Joel A. Carpenter, "Fundamentalist Institutions and the Rise of Evangelical Protestantism, 1929–1942," *Church History,* 1980, *49:*62–75. Chicago was also home to that "Vatican of Fundamentalism," the Moody Bible Institute.

[187] James P. Wind, *The Bible and the University: The Messianic Vision of William Rainey Harper* (Atlanta, Ga.: Scholar's Press, 1987).

[188] Document titled "Press and Public Response to the Darwin Centennial," Darwin Centennial Papers, box 8, folder 2, quotation on p. 3.

[189] William J. Taylor, Sr., to the British Embassy, 2 December 1959, Darwin Centennial Papers, box 9, folder 2.

[190] J. Campillo to the President of the University of Chicago, 28 November 1959, Darwin Centennial Papers, box 9, folder 3.

> work & words of the [sic] Devil—who is behind it all to destroy the Holy Bible & cause dis belief [sic].[191]

Tax's response to the letters of outrage and complaint from theologians and the lay public is not revealed through the documents deposited with the Darwin Centennial Papers, but there is some indication that Tax was upset by Huxley's inappropriate speech. Possibly, he chose not to respond to the most outrageous of these letters; he did, however, respond thoughtfully to one request for further conversation on the issue of science and religion precipitated by the Institute on Science and Theology. Writing firmly, yet gently, Tax defended the "fact" of evolution as it had been emerging from the panels. He wrote, "The Darwin Centennial showed that by now the basic scientific issue is settled as much as scientific issues are ever settled."[192]

The "fact" was far from "settled," however. Huxley's controversial lecture, the Convocation ceremony, and the excessive attention given to the Darwin centennial was noted by a growing and vociferous minority of evangelicals, who were outraged by the apparent and public triumph of the Darwinian world view. In response, a stream of antievolution literature began to appear shortly after the celebration, the most influential of which was John C. Whitcomb and Henry Morris's *The Genesis Flood.*[193] More than any other, this book ushered in the next wave of "scientific creationists," who launched a very public critique of the legitimacy of evolutionary science. The galvanizing effect of the Darwin centennial was explicitly noted by Henry Morris in later reflections, who stated that "The Great Darwinian Centennial" was only matched by the Scopes Monkey Trial in bringing to light issues on the fundamental differences between creation and evolutionary science. But while the Scopes Monkey Trial had brought embarrassment to creationists and had had a dampening effect on their organizations, he noted, the Darwin centennial had a catalyzing and unifying effect, bringing coherence to the group threatened directly by extremes of evolutionary belief. Targeting Huxley's "blatantly atheistic pronouncements" in his Convocation address, Morris described the history of scientific creationism thus:

> Evolutionism had, indeed, apparently become triumphant by this time. Creationism, except for isolated pockets of fundamentalists, seemed dead—most certainly among scientists! The immense and favorable publicity accorded to the Darwin Centennial year, especially the great Darwin 'worship service' at Chicago, where speaker after speaker rhapsodized about Darwin's contributions to the life of mankind, and exhorted each other and all their disciples to go on to further glories of evolutionary achievement, seemed to be the final nail in the coffin of creationism and even of meaningful Christian theism.[194]

[191] This letter was addressed "To Whom It May Concern: 1. Dr. Leakey, 2. Mr. Huxley, 3. Sol Tax, 4. Everett Olson, and many more." It was signed "From one who knows the 'Truth' that our Lord Lives & Know it all. You don't that is for sure." Dated 26 November 1959, Darwin Centennial Papers, box 9, folder 2.

[192] Sol Tax to Monsignor John M. Kelly, 14 December 1959, Darwin Centennial Papers, box 9, folder 1.

[193] John C. Whitcomb and Henry M. Morris, *The Genesis Flood* (Philadelphia: Presbyterian and Reformed Publishing Co., 1961).

[194] Henry M. Morris, *A History of Modern Creationism* (San Diego: Master Book Publishers, 1984), quotations on pp. 74–5. Morris also felt that the Darwin centennial fed into the Biological Sciences and Curriculum Study (BSCS) that was launched in 1959. The importance of the Darwin centennial to creationist causes in the later half of the twentieth century is noted in James Moore, "The Creationist Cosmos of Protestant Fundamentalism," in *Fundamentalisms and Society,* eds. Mar-

Huxley's own response to the controversy his lecture had generated was consistent with his egocentric personality: while recognizing the bad feelings his speech had provoked he still refused to acknowledge that it had marred the celebration. Reflecting on the event years later in his *Memoirs II,* he admitted that the speech was "perhaps not appropriate in a chapel." Referring to his "evolutionary vision" for human affairs, he stated:

> I had already affirmed this view in various books and essays, but here in Chicago, from a chapel pulpit, it shocked many orthodox Middle-Westerners—and much upset Professor Tax. He feared that the whole celebration would be damaged by my utterance, in spite of the fact that I insisted, here as elsewhere, that some form of religion, some self-transcending experience, was universal and apparently inevitable—and psychologically rewarding—in all human societies.

In an all too characteristically dismissive manner, he added, "Actually, I do not think that my preachment damaged the celebrations. Anyhow, a book, edited by Sol Tax and containing a record of all the discussions and speeches, including my 'sermon,' was published early in the following year and had a wide circulation."[195]

As the 1,384-page, three-volume set of the proceedings, *Evolution after Darwin,* made its way to readers, both the news and scientific content of the celebration reached an even wider audience. Book reviews were, for the most part, favorable and praised the breadth of the articles and the range of expertise held by the contributors.[196] A 1960 review in *Science* by anthropologist W. W. Howells, for example, singled out the contributions concerned with the evolution of human culture as the most "interesting" because they drew attention to the diversity of anthropologists' perspectives on the parallels between biological and cultural evolution.[197] Yet another reviewer for the *University of Toronto Quarterly* stressed the importance of the synthetic theory of evolution and singled out biologists like Muller, Dobzhansky, Simpson, Mayr, Huxley, Stebbins, and Wright, whose "outlook dominated the proceedings."[198] The celebration was praised as a whole and described as an organizational *"tour-de-force."*[199] According to Tax, the sales of the book were "good," and it continued to sell into the next decade.[200]

Of considerable interest is Tax's most historical move of all, the orderly preservation of relevant documents from the celebration. Catalogued and housed in the Uni-

---

tin E. Marty and R. Scott Appleby (Chicago: University of Chicago Press, 1993), pp. 42–72; Willard Gatewood, Jr., "From Scopes to Creation Science," *The Proceedings and Papers of the Georgia Association of Historians,* 1983, pp. 1–18; and Gilbert, *Redeeming Culture* (cit. n. 6). For a history of creationism in the United States see Ronald L. Numbers, *The Creationists* (New York: Alfred A. Knopf, 1992).

[195] Julian Sorell Huxley, *Memories II* (New York: Harper & Row, 1973), quotation on p. 192.

[196] George Basalla, "Review of *Evolution After Darwin,*" *Library Journal,* 1961, *86:*588; Bean, "Review of *Issues in Evolution*" (cit. n. 114); W. W. Howells, "Review of *Evolution After Darwin,* volume 2, *The Evolution of Man: Mind, Culture and Society,*" *Science,* 1960, *131:*1601–02; Goudge, "Darwin's Heirs" (cit. n. 117).

[197] Howells, "Review of *Evolution After Darwin,*" cit. n. 196.

[198] Goudge, "Darwin's Heirs" (cit. n. 117).

[199] *Ibid.,* p. 246.

[200] Sol Tax to Kate Brown, 6 May 1970, Darwin Centennial Papers, box 3, Addendum; also see Memorandum, Sol Tax to R. Wendell Harrison, 5 May 1960, Darwin Centennial Papers, box 2, Addendum. For sales figures on the volume see cit. n. 184.

versity of Chicago, the documents retell a history of their own, the most interesting of which relates, once again, to the tense—and complex—relations between international organizers.

A curious thing happened: in 1961 Tax sent three cartons of celebration documents back to England to be housed permanently at the Darwin Museum in Down House.[201] According to Tax's historical recollection, this was at Huxley's request.[202] If this were the case, then one may view this act as a return of the new century's Darwinian heritage to its original home in England. But another source indicates that this may have happened after Tax learned that a delegation of twenty Soviet scientists had personally deposited considerable memorabilia of their own Soviet celebration, including medallions, posters, and stamps, at Down House in a "special pilgrimage."[203] Tax may have responded to what he perceived as a competitive gesture that undermined the permanence of the American celebration by depositing the cartons with the American memorabilia to Down House.[204] Possibly some combination of both may explain Tax's action, but his magnanimous efforts came to little, as the documents were returned to the University of Chicago for permanent safekeeping for unknown reasons. Tax did, however, secure the participation of Americans in the historic memory of Darwin by sending copies of *Evolution after Darwin* to be permanently housed at Down House. Even this small gesture was perceived as a newsworthy event.[205]

## XII. A SURPRISINGLY "AHISTORICAL" EVENT

If the Darwin Centennial Celebration was any indication, just about everyone appeared to be involved in the one hundredth anniversary of Darwin's epoch-making book by the closing of 1959—all, that is, but the most obvious group: historians of science. Considering that the celebration was to serve an historical function, the paucity of historians of science as formal celebrants and contributors is notable. With the exception of Robert Stauffer, who served as the official delegate to the celebration from the History of Science Society, few historians of science participated formally or contributed scholarship.[206]

In the early stages of planning Tax had envisioned that the conference would have

[201] Sol Tax to the American Express Company, 20 October 1961, Darwin Centennial Papers, box 2, Addendum. The letter authorized and gave instructions to ship the three cartons to the curator of the Darwin Museum. The items were insured for $500.

[202] Sol Tax to unknown recipient, August 1985, Darwin Centennial Papers, box 3, Addendum; Tax informed Huxley that he was returning the material. Sol Tax to Julian Huxley, 20 September 1961, *ibid.*

[203] Newspaper clipping titled "Books Mark Darwin Centennial Fete" by Ruth Moore, undated and unidentified. *Ibid.*

[204] The competition with the Soviet Union is clearly stated in Ruth Moore's newspaper account. See cit. n. 203.

[205] See cit. n. 203.

[206] Robert Stauffer was listed as a delegate from HSS in the convocation program. Darwin Centennial Papers, box 7, folder 1. Veith, the historian of medicine on the Darwin Centennial Committee, wrote to Dorothy Stimson early in 1956 to informally request cooperation from the History of Science Society. Ilza Veith to Dorothy Stimson, 27 February 1956, Darwin Centennial Papers, box 2, folder 6. In a letter dated 18 July 1956, Veith reports that "The History of Science Society has, through Dr. Stimson, indicated its willingness to cooperate in our celebration." Ilza Veith to I. Bernard Cohen, 18 July 1956, Darwin Centennial Papers, box 6, folder 2.

three topics that would be published in advance of the meeting in three volumes.[207] The third of these he planned as an historical assessment.[208] Yet the completed conference three-volume set, *Evolution after Darwin,* had no formal historical contribution at all, let alone an entire third volume devoted to historical aspects of evolution, a fact that drew some negative criticism from Chicago humanities faculty who charged Tax and the committee with "scholarly parochialism."[209] Ilza Veith, one of the committee members, had an interest in the history of medicine, yet her own contribution "Creation and Evolution in the Far East" seemed only tangentially relevant to the central function of the occasion.[210]

Historical documents reveal that Tax and Veith had tried valiantly to secure noted historians of science for the celebration and support from the History of Science Society for the historical portion of the conference and volume, but ultimately failed. Beginning with I. Bernard Cohen in the summer of 1956, Tax and Veith sought historians of science to play an active role.[211] According to Veith's account, Cohen seemed very interested initially, but after consideration declined, because he could not "take on another job."[212] According to Veith, both Cohen and Charles C. Gillispie, whom Veith next approached, were equally interested, but like Cohen, Gillispie seemed "baffled by the task of getting ten authors to write something new in this field." She concluded, "[N]obody seems to be able to think of any other American historian of science who has worked on evolution. Therefore he [Gillispie] seems our only bet at the moment."[213] Tax formally invited Gillispie to Chicago in the fall of 1956 to discuss the historical portion of the celebration. Gillispie traveled to Chicago, but ultimately expressed reservations about taking on the editorial work until funds were available to pay potential contributors and until a firm commitment to proceed was made by organizers. He also added that "the possibility must be faced that I shall prove unable to get some or all of the contributors whom we want—the number of qualified historians of science is strictly finite. In fact, it is about the same as the number of essays that we want."[214] Tax made one more formal approach when he invited John C. Greene to serve as editor and organizer, but

[207] The first topic was biological evolution, the second was the impact of evolution on the study of "man" and his culture, and the third was the history of evolution and its consequences. Tax hoped that Karl P. Schmidt would edit the first portion and Robert Redfield would edit the second topic as well as serve as general editor. Ilza Veith to I. B. Cohen, 18 July 1956. Darwin Centennial Papers, box 6, folder 2. No doubt, the fact that both died during the early planning stages altered these plans.

[208] *Ibid.*

[209] Leonard B. Meyer to Sol Tax, 6 May 1959, Darwin Centennial Papers, box 2, folder 6; Memorandum, Sol Tax to Leonard Meyer, 8 May 1959, *ibid.,* Memorandum, Sol Tax to Marvin Mikesell, 30 December 1958, *ibid.;* Memorandum, Sol Tax to Walter Johnson, 30 December 1958, *ibid.* The absence of philosophy was noted by Goudge in his review in the *University of Toronto Quarterly.*

[210] Ilza Veith, "Creation and Evolution in the Far East," in Tax and Callender, *Evolution after Darwin* (cit. n. 16), vol. III, pp. 1–17.

[211] Ilza Veith to I. B. Cohen, 18 July 1956, Darwin Centennial Papers, box 6, folder 2. In one memorandum to Tax, Veith referred to Bentley Glass as an "ideal person," with the names of Harcourt Brown, Henry Guerlac, and Conway Zirkle as possible historians of science to contact. Memorandum, Ilza Veith to Sol Tax, 6 June 1956, *ibid.*

[212] Ilza Veith to Sol Tax, 8 September (undated, most likely 1956), *ibid.*

[213] *Ibid.*

[214] Charles C. Gillispie to Sol Tax, 15 November 1956, Darwin Centennial Papers, box 2, folder 5. Tax reported on Gillispie's letter to the committee soon after. Memorandum, Sol Tax to members of the Darwin Committee, 23 November 1956, *ibid.* 5.

Greene declined because he had "too much unfinished business with respect to publication." In fact, Greene was in the throes of completing his own pathbreaking study of the history of evolutionary thought in *The Death of Adam: Evolution and Its Impact on Western Thought.*[215] All hope of obtaining an external historian of science gone, Tax finally approached Veith in the hopes that the committee might be "one happy little family," but she too declined because of other commitments.[216]

By the end of 1958 Tax had abandoned all hope of producing an historical volume.[217] Possibly the deaths of his fellow committee members and potential editors Karl P. Schmidt and Robert Redfield contributed to the decision not to publish a centennial volume in advance of the conference. The decision was probably also motivated by the fact that there were simply too few historians of science available for the project, and few of these had a real specialty in Darwin or in the history of evolution. As John C. Greene noted in a letter to Tax, "Most American historians of science are not working on 19th century biology and certainly not on 20th century biology."[218] Another strong possibility is that the Darwin centennial was duplicating historical efforts elsewhere. Johns Hopkins University, for instance, was planning an historical celebration.[219] The most likely explanation for the absence of history of science (or "intellectual history," as the documents state) at the Chicago celebration was a combination of the difficulty of securing a historian along with the plans at Johns Hopkins. The absence of historical treatment may have had, however, a positive long-term consequence in that it drew attention to the paucity of scholarship on the important figure of Charles Darwin and on the history of evolution in the history of science. One outcome of the anniversary date as a whole, in fact, was to generate enthusiasm for the subject, and for the body of historical literature on Darwin and on evolution that grew throughout the 1960s.[220] By 1982, the one hundredth anniversary of Darwin's death, a veritable "Darwin Industry" of historians turned out to properly commemorate him and to examine the "Darwinian heritage."[221]

[215] John Greene's letter is reproduced by Tax in a letter to Veith. Sol Tax to Ilza Veith, 25 July 1957, Darwin Centennial Papers, box 6, folder 2. Greene did, however, drive from Ames, Iowa to attend some of the sessions. John Greene to Vassiliki Betty Smocovitis, 13 March 1996. See John C. Greene, *The Death of Adam: Evolution and Its Impact on Western Thought* (Ames: Iowa State University Press, 1959).

[216] Sol Tax to Ilza Veith, 25 July 1957; Ilza Veith to Sol Tax, 9 August (most likely 1957). Darwin Centennial Papers, box 6, folder 2.

[217] Sol Tax to John C. Greene, 10 November 1958, Darwin Centennial Papers, box 2, folder 5. An earlier letter to Gillispie following his visit to Chicago indicates that Tax was having a difficult time raising funds for the historical volume early on in the celebration planning. Sol Tax to Charles C. Gillispie, 23 November 1956, *ibid.*

[218] John C. Greene to Sol Tax, 5 November 1958, *ibid.*

[219] Memorandum, Sol Tax to Marvin Mikesell, 30 December 1958, Darwin Centennial Papers, box 2, folder 6.

[220] For a historiographic exploration of evolutionary biology see Smocovitis, *Unifying Biology* (cit. n. 18). Some of the books published around the anniversary date or shortly after include Loren Eiseley, *Darwin's Century: Evolution and the Men Who Discovered It* (New York: Doubleday, 1958); C. D. Darlington, *Darwin's Place in History* (Oxford: Blackwell, 1959); Gertrude Himmelfarb, *Darwin and the Darwinian Revolution* (New York: Doubleday, 1959); Gavin de Beer, *Charles Darwin: Evolution by Natural Selection* (Edinburgh: Nelson, 1963). See also the contemporary reviews of this literature by Bert James Loewenberg, included in cit. n. 8.

[221] *The Darwinian Heritage,* ed. David Kohn (Princeton: Princeton University Press, 1985); also see Bendall, *Evolution from Molecules to Men* (cit. n. 113); and Timothy Lenoir, "The Darwin Industry," *Hist. Biol.,* 1987, *20:*115–30.

## XIII. CLOSING THOUGHTS

What general conclusions can we draw from this examination of the largest of the celebrations for the one hundredth anniversary of the publication of Darwin's *On the Origin of Species?* From the historical record, it appears that much was being celebrated, though clearly celebrants—or participants—may not have been rejoicing in the same thing. From the scientists' perspective, the celebration honored Darwin and his great work and was an opportune time to reflect on the status of their evolutionary art.[222] Central to their new synthetic evolutionary theory was Darwin's evolution by means of natural selection, in its synthetic and very modern guise. Their own identities were secured through the biological and scientific legitimacy conferred by the presence of Sir Charles Darwin and Sir Julian Huxley, who were living heirs of the original story and whose presence preserved historical continuity. Within this new synthesis of evolution, room was finally made for human evolution, which had been left out of the synthesis—and out of Darwin's account in *Origin.*[223] Tax's disciplinary goal of bringing anthropology and evolution together was thus fulfilled, if only partially, by bringing anthropologists to the celebration and by showcasing their field. Cultural evolution—what it was, how it operated, and the extent to which it hinged on biological evolution—remained a problematic and divisive subject.

In keeping with the evolutionary synthesis as the convergence of biological disciplines, the conference also served to reinforce not only the unity between biological disciplines of knowledge, but also the continuum between the physical sciences and the social sciences, especially with the inclusion of anthropology, cultural evolution, and the evolution of mind. Evolution could claim to be the "central unifying principle of biology" at the same time that it shed light on the future of "modern man," even though there were bumpy borderland regions between the physical, biological, and social sciences.[224] A. E. Emerson's post-celebration musings to Tax were optimistic on this note, however. "Out of it all, I think there is a chance for a new emergent understanding through a synthesis from different fields, particularly between the biological and the social sciences," he wrote.[225]

From the perspective of the discipline as a collective, functioning community, the Darwinian life story could play itself out at the celebration, reifying the field's founding father. Along with this narration, moreover, a canon of relevant scientific and historical literature was established to lend a feeling of coherence and unity to the community. From Darwin's *Origin* to Dobzhansky's *Genetics and the Origin of Spe-*

[222] A memorandum from the University of Chicago Press to Sol Tax concerning the planned volume explicitly began with an attempt to be clear and consistent about what was being celebrated. The memorandum was referring directly to the publication of Darwin's *Origin* and requested consistency in titles of the book. Cathy Nissen to Sol Tax, 30 October 1959, Darwin Centennial Papers, box 9, folder 7.

[223] But discussed, of course, in Darwin's *Descent of Man* (1871).

[224] Smocovitis, *Unifying Biology* (cit. n. 18).

[225] A. E. Emerson to Sol Tax, 21 December 1959, Darwin Centennial Papers, box 3, folder 12. Emerson continued, "Ten years from now I shall tell you whether I think we achieved this 'breakthrough.'" His biographical "memoir" for the National Academy of Sciences was cowritten by Edward O. Wilson (and Charles D. Michener) who argued for just such a synthesis between the biological sciences and the social sciences in 1975. See Edward O. Wilson, *Sociobiology: The New Synthesis* (Cambridge: Belknap, 1975); also see Edward O. Wilson, *Consilience: The Unity of Knowledge* (New York: Alfred A. Knopf, 1998).

*cies,* presented in the hallway exhibits and book displays, celebrants could see the historical continuity that led to the construction of their modern identities. Critical participants—in the symbolic image of Darwin—were marked as heirs by receiving honorary degrees, and new members, including graduate students, were enrolled in the community in the celebratory process. New members were also enrolled into the formal societies that supported evolution: the Society for the Study of Evolution, for instance, saw a significant increase of new subscriptions right around the anniversary date of 1959.[226] News of the celebration would also help enroll a wider public community to belief in evolution by means of natural selection; they would validate the process of legitimation at the same time as they "witnessed" the event.

All of the celebratory activities were part of an historical process of constructing disciplinary identities for evolutionary biologists and building a coherent identity for the collective community of scientists. If the celebrations, as a whole, were a success of great magnitude, it was because the evolutionary synthesis, which was well underway, had been a success in unifying evolution and biology. Ultimately, at a deep level, celebrants rejoiced and at the same time reified their own new identities as sons—and, literally, as grandsons of the protagonists of the original story. Hence, Darwin and his life and work held powerful symbolic meaning for postwar evolutionary biologists, who were eager to unify, strengthen, and promote their new-found community. Although the agreement over the centrality of natural selection as *the* mechanism of evolution certainly did much to enhance this sense of unity and consensus, it also may have given rise to a constricting new orthodoxy—as the panel discussions and somewhat exclusive invitation list indicate. This "hardening" of the synthesis around a selectionist orthodoxy has been long noted by historians of evolutionary biology like Stephen J. Gould.[227]

Finally, the celebration was not the exclusive domain of scientists and academics assembled at the University of Chicago. The rather tepid, nontechnical scientific discussions were intended for mostly general audiences. The numerous press releases, reports, and commentaries on the occasion made evolution and evolutionary culture the stuff of public consumption. As the University of Chicago Press public relations department summarized, "One hundred years ago the meaning of evolution for the future of the human race was seen and considered by only a few intellectuals. Today it is meat for the popular press."[228] Although the same popular press capitalized on the occasion with trivial and sometimes inflated reports, it also frequently transmitted the scientific consensus that had emerged. Evolution, it reported, despite the continuing controversy over science and religion, was a recognizable fact. With proper knowledge of this evolutionary past, humans would be able to control their evolutionary future. Evolutionary progress and social progress were thus inextri-

[226] See Smocovitis, "Organizing Evolution" (cit. n. 20); see the data on membership growth in the SSE in Figure 2, "Growth of Membership in the Society for the Study of Evolution, 1946–1969," on p. 297. The sharp increase in membership around 1959–1960 probably resulted from the interest generated by all the celebrations and commemorative activities combined.

[227] See Stephen J. Gould, "G. G. Simpson, Paleontology, and the Modern Synthesis," in Mayr and Provine, *The Evolutionary Synthesis* (cit. n. 75). See also Stephen J. Gould, "Irrelevance, Submission, and Partnership: The Changing Role of Paleontology in Darwin's Three Centennials and a Modest Proposal for Macroevolution," in Bendall, *Evolution from Molecules to Men* (cit. n. 113), pp. 347–66. Gould explicitly notes that the synthesis had "hardened" at the time of the Darwin Centennial.

[228] Document titled "Press and public response to the Darwin Centennial," Darwin Centennial Papers, box 8, folder 2, quotation on p. 30.

cably linked for American popular audiences; their own increasingly technological culture had, after all, progressively evolved.

But not all celebrated, however, and the Darwin centennial had at least one noteworthy but unexpected consequence. While it promoted the respectability and legitimacy of Darwin's evolution by means of natural selection to a wide audience, it also drew attention to a theory of longstanding unpopularity and controversiality. The excessive attention and promotion given to evolution by natural selection at the Darwin centennial galvanized into action the very group that most opposed it on religious, philosophical, and moral grounds. One outcome of the Darwin centennial was a regrouping of Christian evangelicals, led by individuals like Henry Morris, into the movement that came to be known as scientific creationism. No longer would they accept the rhetoric of scientific legitimacy given to evolution by means of natural selection; evolution was to remain a problematic concept subject to scientific, as well as religious and philosophical, criticism. For them, the Darwin Centennial Celebration had a unifying effect, lending coherence and purpose to the group, and it would be matched *only* by the Scopes Monkey Trial in highlighting the tensions between science and religion for vast American audiences.

Much was therefore at stake in the celebrations commemorating the publication of Darwin's *Origin,* especially the one at the University of Chicago. On the surface appearing to be a mere scientific anniversary, the celebration served to highlight a range of concerns in addition to the formal science. For American organizers, it was an attempt to share some of the glory associated with Darwin and the English empire: at the very least, some of the patina of the authority and antiquity associated with established English universities like Cambridge University, the site of the 1909 centennial, rubbed off onto the midwestern school. Organizers were conscious of this end, and worked towards the appropriation of cultural artifacts—and English bodies—for their success. So, too, the University of Chicago benefited from serving as host to such an international and newsworthy event. As one memorandum to the British Broadcasting Corporation inviting them to film the occasion put it, "the BBC might well be interested in an English occasion as celebrated at the University of Chicago."[229]

The Darwin Centennial Celebration—and other similar celebrations across the world—thus brought out a complex interplay of interests at varied levels, from the personal to the institutional to the disciplinary and even to the national level of hosting, so great was the figure of Darwin and his theory of evolution. The fact remained, however, that the grandest of all such celebrations did not directly address the development of Darwin's ideas or his work. What exactly, then, did the Darwin figure represent to general audiences to have drawn such attention and generated so much emotion? This is a question best left answered by one of the most memorable images produced in honor of the centennial: the cover of the *Bulletin* of the Chicago Natural History Museum (Figure 3). The enormous and somewhat grotesque alabaster bust of Darwin figures centrally against the backdrop of organismal representatives from his major works: insectivorous plants, pigeons, chimpanzees, and other "lower" forms of life. Dominating this diversity of life, the Darwin figure is grossly out of scale. "Man's place in nature," the message tells us, is as the dominant life form.

[229] Lee Wilcox to Aubrey Singer, 4 September 1959, Darwin Centennial Papers, box 8, folder 8.

And what does the Darwin Centennial Celebration tell us about commemorative events in the history of science? For one thing, they may serve a range of functions having little to do with the historical event defining the occasion. This celebration, for example, had little to do with the historical Darwin or the development of his work; instead it revealed much about postwar American culture and its embrace of a new synthetic science of evolutionary biology, a science that could potentially redirect the future of "modern man." The Darwin Centennial Celebration also demonstrates once again that science is not only a part of wider culture, but is itself a culture, worthy of focused anthropological study.

# The First American and French Commemorations in Molecular Biology

## From Collective Memory to Comparative History

*By Pnina G. Abir-Am**

### I. INTRODUCTION: SCIENTIFIC COMMEMORATIONS AND THE CONSTRUCTION OF COLLECTIVE MEMORY: FROM COMMEMORATIVE TEXTS TO SOCIAL PRETEXTS AND POLITICAL CONTEXTS OR VICE-VERSA?

THE INCREASE IN OVERALL COMMEMORATIVE activity since the 1980s due to a convergence of historical trends—most notably the post-cold war globalization—with major anniversaries, such as the bicentennial of the French Revolution and the half-centennial of World War II, was reflected in an upsurge of commemorative activity within the scientific community.[1] This upsurge was further sustained by a historical shift in social utility, and hence in cultural hegemony, from the physical to the biological sciences, in particular from nuclear physics to molecular biology.

The declining social utility of physics in the post-cold war era, best symbolized by the refusal of the United States Congress to allocate the megabudget for the superconductor supercollider, and the parallel, spectacular rise in the social, economic, and cultural utility of molecular biology, especially as reflected in its biotech applications, manifested in heightened commemorative activities within these two disciplines.[2]

* Office for History of Science and Technology, 470 Stephens Hall, University of California, Berkeley, CA 94720; e-mail pgabiram@socrates.berkeley.edu

[1] For examples of the rise in commemorative activity in both society and science since the 1980s, see Pnina G. Abir-Am, "Introduction," this volume of *Osiris.* For examples of studies published in conjunction with the bicentennial of the French revolution and the half-centennial of World War II, see notes 1 and 2 of this "Introduction."

[2] On the decline in the social utility of nuclear physics, see chapters by Robert W. Seidel and by Dominique Pestre in this volume. On the decline of nuclear physics and the rise of molecular biology see Daniel J. Kevles, "Big Science and Big Politics in the United States: Reflections on the Death of the SSC and the Life of the Human Genome Project," *Historical Studies in the Physical and Biological Sciences,* 1997, *27:*269–97. On the rise in the utility of molecular biology, see a review of six books from the early and mid-1990s by Pnina G. Abir-Am, "'New' trends in the history of molecular biology," *Hist. Stud. Phys. Biol. Sci.,* 1995, *26:*167–96. See also Dorothy Nelkin and M. Susan Lindee, *The DNA Mystique: The Gene as a Cultural Icon* (San Francisco: Freeman, 1995); Robert Bud, *The Uses of Life: A History of Biotechnology* (New York: Cambridge University Press, 1993); Robert Teitelman, *Gene Dreams: Wall Street, Academia, and the Rise of Biotechnology* (New York: Simon & Schuster, 1989).

 0369-7827/99/1401-0016$02.00

In particular, molecular biology became the arena for numerous commemorative events, culminating with multiple celebrations of the fortieth anniversary of the discovery of the double helix in 1993. One of these events held at the UNESCO headquarters in Paris, signalled with unmistakable clarity the *new* status of molecular biology as a nexus of the multinational biotech industry and governmental science policy.[3]

Yet molecular biology has actually had a longer and rather impressive tradition of commemorative activity. Since the mid-1960s, anniversaries and commemorations have emerged as a favorite forum for groups seeking public recognition as founders of this multidisciplinary, agonistic field. Eventually, the quest for founder status through the process of staging a commemoration and publishing a collective memory of the participants emerged as a historiographical rite of passage; the success of the first group in so validating its founder status compelled contender groups to follow suit. Within two decades of the institutionalization of molecular biology in the mid-1960s, over half a dozen commemorative projects took place, preceding a wider movement of return to collective memory in the late 1980s in society at large.

Ironically, commemorative activism in molecular biology appears to stand in inverse relationship to international recognition. For example, American efforts took the lead in the mid-1960s, with the French following in the 1970s and the British joining the commemorative bandwagon only in the 1980s.[4] This order contrasts with the predominance of the British in 1962, the French in 1965, and the Americans in 1968, among Nobel Prize winners for research in molecular biology.

This essay will compare the very first collective memories in molecular biology in the United States and France, in order to shed light on their relationship to pervasive commemorative practices and their implications for the writing of a comparative history of a paradigmatic field that featured claimants to founder-group status in the above-mentioned three countries. Other collective memories were published in these three countries from the late 1960s to the late 1980s that enriched, modified, or subverted the import of the very first collective memories.[5]

For reasons of methodological clarity, this study will proceed in three stages. First, it will compare the commemorative *texts* published as the outcome of prolonged commemorative efforts. Second, it will compare the social *pretexts* used to assemble the memories of many scientists for publication. Third, it will compare the political

[3] For overviews of molecular biology see Michel Morange, *A History of Molecular Biology* (Cambridge, Mass.: Harvard University Press, 1998; French edition published by La Découverte in 1994); Hans-Jorg Rheinberger, *Toward a History of Epistemic Things: Synthesizing Proteins in the Test Tube* (Stanford: Stanford University Press, 1997), and the ten books and articles reviewed and references included in Abir-Am, "'New' trends" (cit. n. 2); *idem,* "Themes, Genres and Orders of Legitimation in the Historiography of Molecular Biology," *History of Science,* 1985, *23:*73–117. For a survey of commemorations in molecular biology see Pnina G. Abir-Am, "Entre mémoire et histoire en biologie moleculaire: Les premiers rites commemoratifs pour les groupes fondateurs" in *La mise en mémoire de la science,* ed. Pnina G. Abir-Am (Paris: Editions des Archives Contemporaines, 1998), pp. 25–79.

[4] See "Introduction" in Abir-Am, *La mise en mémoire,* (cit. n. 3), pp. 1–14; *idem,* "A Historical Ethnography of a Scientific Anniversary in Molecular Biology: The First Protein X-ray Photograph (1984, 1934)," *Social Epistemology,* 1992, *7:*321–54; see also the three commentaries by Barbara Fraenkel, Kenneth Gergen, and Hugh Gusterson and Abir-Am's response to each, pp. 355–82.

[5] For the ensemble of commemorations in molecular biology see Pnina G. Abir-Am, *Research Schools of Molecular Biology in the United States, United Kingdom, and France: National Traditions or Transnational Strategies of Innovation?* (Berkeley: University of California Press, forthcoming). A detailed version of the American case study and a listing of the other relevant case studies is also available in Abir-Am, "Entre mémoire et histoire" (cit. n. 3).

*contexts* enabling and constraining the pretexts and texts at particular points in space and time, namely the mid-1960s in the United States and the late 1960s, especially the so-called post-1968 era, in France.[6] In order to engage this problem of the relationship between text, pretext, and context, it is useful to elaborate a bit on the three key categories as they will be used to compare the first American and French collective memories:[7]

**The symbolic texts** pertain chiefly to the published range of collective memory including the semantic, the organizational-rhetorical, and the visual (or photographic) dimensions of the commemorative texts. (The crucial dimension of orality, which is examined elsewhere, was omitted from this essay since speeches were delivered orally in the French case only.) The texts of the American and French collective memories are further compared in terms of both their message and their medium or narrative form. First, I will compare the texts' contrasting historiographical messages regarding the rise of molecular biology as a revolutionary modernist or a continuously decadent, scientific event. Second, I will compare their narrative forms as structured or unstructured, coherent or disjointed, monolingual or bilingual.

**The social pretexts** pertain to the discrepancy between the public rationale given by the organizers for promulgating a collective memory that usually involves mobilizing three or more dozen scientists—namely, to honor a senior figure at a turning point in his or her career—and the private reasoning that surfaces in unpublished correspondence, according to which the orchestration of such commemorations was driven by other social pretexts. These pretexts include, in each case, a disciplinary component (i.e., defining the public imagery of molecular biology in the scientific community and in society at large), an institutional component (i.e., securing a power base in the discipline in the form of the directorship of a major laboratory, and a genealogical component (i.e., using the honoree as a link between the heir presumptive and an ancestral figure who was associated with a scientific revolution that legitimizes the heir's quest of leadership).

**The political contexts** are traced to historical forces affecting the organizers of the American and the French collective memories when they began their efforts. For example, major scientific change in the early 1960s was a crucial determinant of the

[6] Though the heuristic value of the analytical categories of text, pretext, and context initially emerged in a noncomparative, detailed narrative of the construction of the first American collective memory in molecular biology (which was the first worldwide), the three categories have been retained as an organizational device in this American-French comparative study. This procedure may offer an additional benefit since the three categories also resonate with the three basic aspects of social action as described by the theory of structuration, namely the production and reproduction (hereafter the re/production) of meaning, social order, and power relations. That theory was previously applied to explain an earlier, primarily British phase in the history of molecular biology. (Pnina G. Abir-Am, "The Biotheoretical Gathering: Transdisciplinary Authority, and the Incipient Legitimation of Molecular Biology in the 1930s," *Hist. of Sci.,* 1977, *25:*1–70.) Therefore, the new empirical evidence from American and French commemorative practices, compared along the key indices of text, pretext, and context, can offer further support to a theory of scientific change, one which explaines a broader range of phenomena since it insists on the interdependence and simultaneity of the (triple) re/production of meaning, order, and power relations in science.

[7] On the relationship between text, pretext, and context see Jacques Derrida, "Signature, Event, Context" in his *Limited Inc.* (Evanston: Northwestern University Press, 1990), pp. 1–23 (French edition: Paris: Galilée, 1990); *idem, Writing and Difference* (Chicago: The University of Chicago Press, 1978); French edition: *L'écriture et la difference* (Paris: Seuil, 1979). For a different use of Derridean ideas in the history of science see *Inscribing Science, Writing Texts and the Materiality of Communication,* Timothy Lenoir, ed. (Stanford: Stanford University Press, 1998) and the forthcoming review of this book by Pnina Abir-Am in *Journal of Interdisciplinary History.*

American effort's precocious timing, but played a lesser role in the French effort since by the late 1960s, the discipline was already consolidated institutionally and otherwise. At the same time, in both the American and the French cases, international science policy and domestic cultural politics converged to play a key role.[8]

The following narrative entails an inverted chronological order, since it reflects the choice to first compare the published texts, then to examine the pretexts operating behind the public scenery, and lastly to recover the contexts that enabled or constrained the texts and pretexts. Though this choice made the writing much more challenging, as the historian Carlo Ginzburg observed, this literary device, called "estrangement," may have certain merits. At least, it was used by some of the greatest creative minds in writings dealing with the very subject of memory:

> [Proust] was suggesting that historians and novelists (or painters) share a cognitive purpose . . . In order to describe my own project, as a historian, I would not hesitate to use, with a small change, a few lines from the passage by Proust I quoted earlier: "And even supposing *history* to be scientific, one would still have to depict it as Elstir did the sea, in inverse order [italics in original]."[9]

## II. CONTRASTING COMMEMORATIVE TEXTS: FROM AMERICAN AND FRENCH NARRATIVES OF R/EVOLUTIONARY SCIENCE TO A SHARED TYPOLOGY OF MEMORIES

The two texts to be compared in this section, *Phage and the Origins of Molecular Biology* and *Les microbes et la vie/ Of Microbes and Life,* were published in 1966 and 1971, respectively.[10] This relatively small difference in timing had major implications for shaping the main message that each collective memory sought to convey and its eventual reception. The priority—even precocity—of the American text, as well as its resonance with the counterculture of the 1960s, gave it a great deal of latitude in defining its historiographical message as a revolutionary one while enjoying an untapped and as yet uncontested market. The implications of this crucial difference in time and place are systematically explored in section IV, where the contrast between the early-to mid-1960s in the United States and the late 1960s in France is examined as an input into the analytical category of "political context."

[8] Since the 1960s are within the range of memory of many living scientists, this section also benefited from oral history (the author and collaborators conducted an oral history with almost every scientist who wrote for the American or the French collective memory between 1986 and 1995). This method poses the problem of conflicting accounts between different interviewees, as well as of potential contradictions between the oral and scriptural record, among other problems. See the recent discussion of such issues in *The Historiography of Contemporary Science and Technology,* ed. Thomas Soderquist (London: Harwood, 1997) including papers by Larry Holmes, Soraya de Chadarevian, and Jean-Paul Gaudiliere that are particularly relevant for issues arising in this study. See also Abir-Am, "A Historical Ethnography" (cit. n. 4), for such a methodological discussion.

[9] See Carlo Ginzburg, "Making Things Strange: The Prehistory of a Literary Device," *Representations,* 1996, *56:*8–28, quotation on pp. 22–23. Ginzburg substituted "history" for "war" in Proust's narration in the final section of *Remembrance of Things Past:* "And even supposing war to be scientific, one would still have to depict it as Elstir did the sea, in inverse order, starting with illusions and beliefs that one corrects bit by bit, as Dostoyevsky would relate the story of a life." Ginzburg gives the original text in French in his note 51.

[10] *Phage and the Origins of Molecular Biology,* eds. John Cairns, Gunther S. Stent, and James D. Watson (Cold Spring Harbor: Cold Spring Harbor Press, 1966; second revised edition in 1992); *Les microbes et la vie/Of Microbes and Life,* eds. Jacques Monod and Ernest Borek (New York: Columbia University Press, 1971).

### *Contrasting Narratives: From an American Modernist Plot to a French Postmodern Ploy*

The contrasting forms of the American and French texts of collective memory had major implications for conveying rather different historiographical messages. On the one hand, the American text conveyed a modernist aura of revolutionary progress as it was structured into six parts that advanced abruptly from "Origins" and "Renaissance" in the 1940s to "Ramifications" in the 1960s (as the titles of parts 1, 2, and 6 clearly indicate). Moreover, the core message sandwiched between "Origins" and "Ramifications" (presumably referring to developments in the 1950s) had an unmistakeable quality of clarity, simplicity, even single-mindedness. As the titles of parts 3, 4, and 5 (namely, "Phage Genetics," "Bacterial Genetics," and molecular genetics, shortened to "DNA") suggest, various phases of a single discipline—genetics—were sufficient to account for the revolutionary rise of molecular biology.

This rise was further designated by the editors as a revelation, or *denouement,* to suggest its comparability to a mystical experience and/or hint at the cultivated status of those who first publicly proclaimed to have had such an experience. The editors' clear and simple message was conveyed in a uniform style, using colloquial American English as an unifying idiom. Since a third of the thirty-three memories came from foreign scientists who had spent a sabbatical or shorter research stay in the United States—while the editors referred to most of the American scientists as being able to write "disconnected anecdotes" only—a serious investment in a prolonged process of translation and stylistic homogenization was undertaken. Indeed, the tight editorial control over both theme and style that the three closely cooperating editors (and especially the executive editor, Gunther S. Stent) maintained is evident in their correspondence with prospective authors. Perhaps as a way of getting some recognition for the editors' daunting task, one author was allowed to reveal on the last page of the volume the common knowledge that Stent rewrote many of the essays, regardless of their substantive merit or stylistic flaws.[11]

By contrast, the text of the French collective memory was not structured by any guiding table of contents or by divisions into thematic qua chronological parts with informative subtitles. (An initial blueprint with five topical decades was dropped upon publication; see discussion in section III.) Furthermore, the titles of the various memories were often enigmatic, reflecting on the wider context rather than being merely descriptive (e.g., "The Unmasking of the Unseen," "When a Thing's Really Good it Cannot Die," "La Vie est dur" ("Life is Tough"), "Le model" ("The model," i.e., both theory and role model), or the ambitiously apocalypotic "A Latter-Day Rationalist Lament" and the vaguely encompassing "Du microbe à l'homme"

[11] See the personal papers of Gunther S. Stent, the Bancroft Library, University of California (UC) at Berkeley; the personal papers of Salvador Luria and Thomas Anderson, American Philosophical Society Library, Philadelphia, Penn. (hereafter APSL); and the personal papers of Max Delbrück, George W. Beadle, Jean Weigle, and Robert Sinsheimer at Caltech Archives, The Beckman Institute, California Institute of Technology, Pasadena, Calif. See also the following (in the files of Pnina G. Abir-Am): oral history with John Cairns in January 1991, School of Public Health, Harvard University, Boston, Mass.; with Gunther Stent, UC-Berkeley, in January 1988, December 1997, and April 1998; with Aaron Novick, University of Oregon in January 1988 and May 1998; with Thomas Anderson in Fox Chase, Penn., in December 1989; among others. See also section III which quotes some letters. Additional relevant correspondence is quoted or cited in Abir-Am, "Entre mémoire et histoire" (cit. n. 3).

("From Microbe to Man"). Where disciplinary themes were mentioned, they ranged widely from parasitology to microbiology, physiological evolution, biochemistry, genetics, and virology.

Last but not least, the text of the French collective memory emerged as a bewildering bilingual mosaic. Memoirs written in French by scientists working in France alternated in no evident order (other than the one in which they were probably submitted) with English-language memoirs by foreign scientists. Hence, the first French collective memory consisted of a fragmented, complex, opaque, perhaps inadvertedly postmodern text; neither themata nor chronology nor even a single language ensured minimal coherence in the underlying message.

As a result of these organizational features, decoding the message of the French collective memory would be a very tough call for the casual reader. Even a determined, bilingual reader and veteran of comparative crossnational research in the history of molecular biology, such as this author, had to resort to prolonged oral history and extensive archival research on the personal papers of the two editors and other members of the international organizing committee in order to decode the opaque and convoluted meaning of the French collective memory.[12]

### *Contrasting Historiographies: A Scientific Revolution or a Gradual, Multidisciplinary Synthesis?*

Perhaps the most important difference between the American and French texts pertained to their respective message on the revolutionary or gradual nature of the scientific change underlying the rise of molecular biology. The American collective memory sought to convey a message of revolutionary or discontinuous scientific change, while projecting the rise of molecular biology as the outcome of the nonconventional or radical scientific pursuits of a new social formation, the so-called Phage Group. Established through recruitment at annual summer courses held since 1945 at the Cold Spring Harbor Laboratory (hereafter CSHL) on Long Island in New York, the group focused on the phenomenon of phage multiplication (namely, how viruses that multiply within bacteria manage to make 200 or so copies within 20 minutes) and further coordinated its findings by restricting the work to certain series of phages. This was a relatively new pursuit at the time since the study of replication patterns, known as classical genetics, had by then been restricted to select model organisms in plants and animals.

Furthermore, the group's rationale for focusing on phage emphasized phage's capacity to serve as a model for genetic replication, one of the key unsolved problems of classical genetics that had attracted the attention of many theoretically minded

[12] Oral histories were conducted in France with Andre Lwoff, Germaine Stanier, François Gros, François Jacob, Georges Cohen, and Elie Wollman, among others, in June 1986, October 1987, June 1988, and May to December 1993; in the United States with Seymour Cohen and Annamaria Torriani on several occasions between 1986 and 1998, with Neil Groman in 1988, and with Robert Austrian, Sylvia Kerr, and Hilary and Irena Koprowsky in 1995; in Canada with Louis Siminovich in 1989 and 1992; in Holland with Andreas Querido in 1995. All are filed with the author. Archival research was conducted in France between 1993 and 1995 on the personal papers of Jacques Monod and Andre Lwoff, and the directorial records of l'Institut Pasteur, all at the Service des Archives, l'Institut Pasteur, Paris; and in the United States on the personal papers of Ernest Borek (courtesy of Sylvia Kerr), and of Seymour Cohen at the APSL (cit. n. 11), in addition to the personal papers cited in note 11.

geneticists and physical scientists in the interwar period, when physico-chemical approaches to biology became a major research agenda.[13] However, that problem was not valued in its own right but rather because it seemed to hold the promise of exposing paradoxes, *à la* quantum physics, that might require for their resolution the uncovering of new laws of nature. Biology was an underdeveloped domain, Warren Weaver of the Rockefeller Foundation, a major sponsor of physico-chemical research in biology, described it as "backward and lacking laws."[14] Hence, the still unexplored biological territory allowed for a redeeming hope on the part of those willing to seek highly valued and epistemologically revelatory paradoxes.

In contrast to a transformation from classical into molecular genetics, which was the disciplinary core of the revolutionary message in the American collective memory, the French collective memory conveyed a message of continuity and convergence. There, molecular biology was described as a multidisciplinary synthesis of microbiology, comparative biochemistry, and genetics. This convergence presumably occurred in response to many comparative studies of physiological and biochemical functions in various monocellular organisms that aimed to explain the devolution of parasitic forms relative to the nonparasitic ones. The frame of reference for the French collective memory was not the potentially illuminating paradoxes of quantum physics, but rather the paradoxes or anomalies of regressive evolution, both organic and molecular, that challenged the theory of evolution's predicate of progress.[15]

[13] There is a large literature on the rise of a research agenda in applying physico-chemical approaches to biology. For the 1930s see the books reviewed in Abir-Am, "'New' trends" (cit. n. 2); *idem,* "Deconstructing the Rhetoric on Biology and Physics at Two International Congresses in 1931," *Humanity and Society,* 1985, *9:*341–82. For the post-World War II period, especially the 1950s, see Nicolas Rasmussen, "The Mid-century Biophysics Bubble: Hiroshima and the Biological Revolution in America, Revisited," *Hist. Sci.* 1997, *35:*247–93; Lily E. Kay, "Cybernetics, Information, Life: The Emergence of Scriptural Representations of Heredity," *Configurations,* 1997, *5:*23–91. For molecular overviews of biology in the twentieth century see *Molecularizing Biology and Medicine: New Practices and Alliances, 1910s–1970s,* eds. Soraya de Chadarevian and Harmke Kamminga (London: Harwood Publishers, 1998); Evelyn Fox Keller, *Refiguring Life: Metaphors of 20th Century Biology* (New York: Columbia University Press, 1995); Robert Olby, *The Path to the Double Helix* (London: Macmillan, 1974; reprint, New York: Norton, 1994); *idem,* "The molecular revolution in biology," in *Companion to the History of Modern Science,* eds. Robert Olby and others (London: Cambridge University Press, 1990), pp. 503–20; see also Abir-Am, "From Multidisciplinary Collaboration to Transnational Objectivity: International Space as Constitutive of Molecular Biology, 1930–1970" in *Denationalizing Science,* eds. Elisabeth Crawford, Terry Shinn, and Sverker Sorlin (Dordrecht: Kluwer Academic Publishers, 1993), pp. 153–86 (also available in Spanish in *Arbor,* February 1997, 111–50), *idem,* "The Molecularization of Biology in the 20th Century," in *Science in the 20th Century,* eds. John Krige and Dominique Pestre (London: Harwood Publishers, 1997), pp. 495–520.

[14] For further details on Weaver's role in the rise of molecular biology see Abir-Am, "From Physical Power to Biological Knowledge: Reassessing the Rockefeller Foundation's 'Policy' in molecular biology," *Social Studies of Science,* 1982, *12:*341–82; special section of five authors commenting on the previous paper, "Responses and Replies," *Soc. Stud. Sci.* May 1984, *12; idem,* "The Assessment of Interdisciplinary Research in the 1930s: The Rockefeller Foundation and Physico-chemical Morphology," *Minerva,* 1988, *26:*153–76; Robert E. Kohler, *Partners in Science: Foundations and Natural Scientists* (Chicago: Chicago University Press, 1991); and Lily E. Kay, *The Molecular Vision of Life: Caltech, The Rockefeller Foundation and the New Biology* (New York: Oxford University Press, 1993), both reviewed and compared in Abir-Am, "'New' trends" (cit. n. 2).

[15] For details on this framework see Andre Lwoff, "From Protozoa to Bacteria to Viruses: Fifty Years with Microbes," *Annual Review of Microbiology,* 1971, *24:*1–26; *idem, Jeux et combats* (*Games and Struggles*) (Paris: Fayard, 1981); Roger Stanier, "L'evolution Physiologique: A Retrospective Appreciation" in Monod and Borek, *Les microbes et la vie* (cit. n. 10), pp. 70–76; Martin R. Pollock,

The contrast in the texts between revolutionary and gradual scientific change was also reflected in the historicity of the origins of molecular biology. While the American collective memory traced the rise of molecular biology to the annual phage summer course given since 1945 at the Cold Spring Harbor Laboratory, thus spanning a coming-of-age period of twenty-one years (1945–1966), the French collective memory traced the discipline's rise to a period spanning half a century (1921–1971), beginning with the arrival of Andre Lwoff (1902–1994) at the Institut Pasteur in Paris. Lwoff, the founder of the French school of molecular biology, together with his younger research associates Jacques Monod (1910–1976) and Francois Jacob (1920–), became in 1965 the first French Nobelists in thirty years.[16]

Indeed, the careers of the two honorees of the American and the French collective memories resonated quite well with the revolutionary versus evolutionary pattern of scientific change ascribed to the rise of molecular biology. While Andre Lwoff's career, which stretched over half a century at the Institut Pasteur (the only change being his move from the department of tropical parasitology to a new department of microbial physiology established for him in 1938) can easily convey continuity; the career of the American honoree, Max Delbrück (1906–1981), the "prime mover" of the Phage Group, can easily convey discontinuity. Delbrück migrated from quantum physics to phage genetics in the late 1930s, and from his native Germany to the United States following marriage to a United States citizen in 1940.

However, neither the arrival of a 19-year-old student in a laboratory (i.e., Lwoff at the Institut Pasteur in 1921) nor the delivery of annual summer courses (i.e., Delbrück at the CSHL since 1945) are conventional commemorative objects in science. Usually, commemorations in science focus on discoveries, theories, or instruments. How did Delbrück and Lwoff come to be selected as honorees or icons of posterity for the American and French collective memories, in 1964 and 1968 respectively, at a time when both were still active, youthful, and openly critical of the pervasive molecularism that seemed to be engulfing biology?

In order to answer this question, one needs to examine the individual memories constituting the collective memory, especially the memories of those who orchestrated the assembly of a collective memory and selected the honoree. Following a comparison of such key, master, defining, or "totemic" memories in the American and French efforts, this essay will further compare the visual memories of the American and the French commemorative efforts, since they condense (and occasionally subvert) the totemic memories in more memorable images. Lastly, the textual comparison will focus on a key category of missing memories: those of women scientists.

### *A Genealogical Imperative: Comparing the American and French Totemic Memories*

In the American collective memory, the key or totemic memory was provided by coeditor James D. Watson, (1928–), who in the early 1960s possessed the greatest

---

"Back to Pangenesis?," ibid., pp. 77–88; Seymour S. Cohen, "Are/Were Mitochondria and Chloroplasts Microorganisms?," *ibid.*, pp. 129–49; Jacques Monod, "Du microbe à l'homme," ibid., pp. 3–9.

[16] On the French Nobelist trio of 1965 see Michel Rouzé, *Les Nobel scientifiques Français* (Paris: La Découverte, 1988), pp. 178–97; Oral history with Andre Lwoff, Paris and Banyuls-sur-mer, 1986

amount of scientific capital, having become the first (and for a while, sole) American Nobelist in molecular biology. (Watson shared the Nobel Prize in 1962 with Francis Crick and Maurice Wilkins, both of the United Kingdom, for elucidating DNA structure). In his memory titled "Growing up in the Phage Group," Watson alone, among the almost three dozen contributors, traced his scientific legacy to his "descent" from the Phage Group.

According to Watson, the feat of "growing up" (a reference to his unusual youth, since he began doctoral research at the age of nineteen) was accomplished by his exposure to the Phage Group's spirit—a general sense of intellectual adventure mixed with a specific quest for theoretical paradoxes that was conveyed by the group's prime mover-turned-honoree, Delbrück. As Watson recalled, Delbrück never failed to inform his associates about his own inspiration while in Europe before World War II: Niels Bohr's complementarity principle in quantum physics, the core of the famous "Copenhagen spirit."

> Some weeks later, in Luria's flat, I first saw Max Delbrück . . . Then, as on many subsequent occasions, Delbrück talked about Bohr and his belief that a complementarity principle, perhaps like that needed for understanding quantum mechanics, would be the key to the real understanding of biology. Luria's views were less firm, but there was no doubt that on most days he too felt that the gene would not be simple and that high powered brains, like Delbrück's or that of the even more legendary Szilard, might be needed to formulate the new laws of physics (chemistry?) upon which the self-replication of the gene was based. So, sometimes I worried that my inability to think mathematically might mean I could never do anything important. But in the presence of Delbrück I hoped I might someday participate just a little in some great revelation.[17]

The totemic memory thus intimates that Watson experienced an imported version of the Copenhagen spirit via Delbrück's insistence that theoretical paradoxes similar to those associated with the complementarity principle in quantum physics might be found in biology in general, and in the problem of genetic replication in particular. Watson further understood that in order to be accepted as a leader with the reputation of a revolutionary and a theoretician, he needed a genealogy that legitimized his increasingly marginal status while neutralizing the contingencies of his rise to celebrity—contingencies that he planned to reveal in his soon-to-be-published, sensational autobiography. This ingenious quest for a genealogy that might confer upon him the status of a theoretical revolutionary, at least by metaphorical association, occurred at a time when Watson's theoretical contributions to molecular biology did

---

to 1993 (cit. n. 12); Pnina G. Abir-Am, "Nobelesse Oblige: Lives of Molecular Biologists" (review of seven autobiographies including Monod's and Jacob's), *Isis,* 1991, *86:*326–43.

[17] James D. Watson, "Growing up in the Phage Group," in Cairns, Stent, and Watson, *Phage and Molecular Biology,* (cit. n. 10), pp. 239–45; quote on p. 240. For biographical inroads into Delbrück's career see his correspondence with the oral historian and archivist Carolyn Kopp Harding in Delbrück's personal papers, Caltech Archives (cit. n. 11). See also Peter Fischer and Carol Lipson, *Light and Life: Max Delbrück and the Origins of Molecular Biology* (Ithaca: Cornell University Press, 1988; German edition by Fischer-Verlag in 1985); Lily E. Kay, "The Secret of Life: Niels Bohr's Influence on the Biology Program of Max Delbrück," *Rivista di Storia della Scienza,* 1985, *2:*487–510; *idem,* "Conceptual Models and Analytical Tools: The Biology of Physicist Max Delbrück," *J. Hist. Biol.,* 1985, *18:*207–47; Finn Aaserud, *Redirected Science: Niels Bohr's Institute in the 1930s* (New York: Cambridge University Press, 1990); Abir-Am, "Themes, genres and orders" (cit. n. 3), 73–117. On the Copenhagen spirit see Mara Beller, "Jocular Commemorations: The Copenhagen Spirit," in this volume of *Osiris.*

not keep up with that of other leading figures, most notably Crick, his coauthor of the double helix papers, who went on to propose the "central dogma" as well as theorize on the features of the genetic code; and Monod and Jacob, who proposed the operon model and theorized on structural versus regulatory genes and the repressor, among other concepts.[18]

Such a genealogical quest may also explain why Watson alone elevated the Phage Group, an informal gathering in the 1940s, to the status of his scientific origins. No one else other than the organizers in their joint preface (a text that was finalized by Watson) mentioned the Phage Group or its "spirit and folklore" as something worthy of preservation for posterity. In addition, the scientists contributing to the collective memory, like the organizers, would have known that a persisting tradition in science centered on the commemoration of concrete and conventional objects of scientific accomplishment, such as discoveries, laws, techniques, or instruments. Nevertheless, the spirit of the Phage Group was singled out for commemoration due to its capacity to evoke, via metaphorical and professional kinship, the better-known Copenhagen spirit, which by the mid-1960s was already entrenched as a symbol of revolution in science and an emblem of wider spiritual grandeur.

Associating oneself and one's heritage with such an image was a clever way to avoid the harsh scientific realities of the early 1960s. In addition to Watson's gap in theoretical achievement, the biochemical decoding of the genetic code, especially by macromolecular biochemists whose practicality, "messiness" and lack of abstraction were derided in the Phage Group, meant that those who invested their effort in seeking paradoxes via a purely abstract genetic approach, like Watson, would have to do a lot of catching up or retooling in order to remain in business.[19]

Though Watson's attempts to emulate Delbrück were quite moving at both the historical time (the late 1940s and early 1950s) and the memorial time (mid-1960s)—colleagues recalled him trying to play tennis, cut his hair, even speak with a German accent so as to look and sound like Delbrück—such efforts, however genuine, still provided for a very selective memory. Particularly incomprehensible is Watson's silence or absence of memory about his formative background, the places where he spent his daily professional life: namely, the University of Chicago and Indiana University in Bloomington. The Midwest not only hosted the greatest contingent of phage workers in the country when Watson "grew up" in the late 1940s, but also featured a great concentration of leading geneticists.

In addition to two founders of the Phage Group—Salvador Luria, Watson's Ph.D. adviser at Indiana University and Alfred Hershey, then at Washington University in St. Louis, Missouri (who both shared the 1969 Nobel Prize with Delbrück)—the

[18] On molecular biology in the early 1960s see Horace F. Judson, *The Eighth Day of Creation: The Makers of the Revolution in Biology* (New York: Simon & Schuster, 1979, 1994) part II, "RNA"; see also Rheinberger, *Toward a History of Epistemic Things* (cit. n. 3). See also Francis Crick's autobiographical *What Mad Pursuit?* (New York: Basic Books, 1988); and François Jacob's autobiographical *The Inner Statue* (New York: Basic Books, 1988; French edition as *La statue intérieure,* Paris: Gallimard, 1987, 1990), both reviewed in Abir-Am, "Nobelesse Oblige" (cit. n. 16). See also François Gros, *Les secrets du gene* (Paris: Seuil, 1986, 1991).

[19] For further details on the tension between molecular biologists and biochemists see Arthur Kornberg, *For the Love of Enzymes: Adventures of a Biochemist* (Cambridge, Mass.: Harvard University Press, 1989); see also its review in Abir-Am, "Noblesse Oblige" (cit. n. 16); *idem,* "The Politics of Macromolecules: Biochemists, Molecular Biologists and Rhetoric," *Osiris,* 1992, *7:*167–91; "*The Tools of the Discipline: Biochemists and Molecular Biologists,*" eds. Soraya de Chadarevian and Jean-Paul Gaudelliere *J. Hist. Biol.,* Fall 1996, *17.*

most legendary of the Phage Group associates, Leo Szilard (1898–1964), was also in the region. (Szilard switched to biophysics at the University of Chicago once he completed his historical role in the Manhattan project on the disappointing note that, despite his advice, the atomic bomb he had helped to conceive and build was actually going to be used.)[20] The Bloomington campus featured the greatest living American classical geneticist, and after Thomas Morgan's death in 1945, the sole American Nobelist in genetics, Herman J. Muller (1890–1967). As the discoverer of the mutagenicity of X-rays in the late 1920s, Muller was an early and persistent advocate of physical and chemical approaches to genetics, as well as a renowned exponent of gene theory.[21]

In the same vein, at the nearby University of Wisconsin at Madison was Joshua Lederberg, a pioneer of microbial genetics who in 1958, at age 32, became the youngest American Nobelist. Lederberg's wife and collaborator Esther Zimmer Lederberg, who discovered the phage lambda, a major experimental system in molecular biology, had been a fellow student of Watson's. Though the Lederbergs were frequent visitors at the midwestern meetings of the Phage Group, neither they nor Muller have any place in the first American collective memory in molecular biology, despite their absolute centrality in theoretical and microbial genetics, respectively, and their geographical proximity to the midwestern contingent of the Phage Group.[22]

Since Watson obviously had plenty of local exposure to both legendary physicists-turned-biologists and great geneticists, why did he choose to recall Delbrück so selectively as his sole ancestor? Unless one concludes that, for reasons of aesthetics or social prejudice, Watson's memory had no place for short figures, especially roly-poly "ethnics" such as a Luria, a Lederberg, or a Szilard, the choice of slim, tall, Nordic Delbrück can only be explained in two ways. On the one hand, this choice serves to obscure more serious symbolic debts that Watson incurred vis-à-vis other

[20] See William Lanoette, *In the Shadow of Genuis: A Biography of Leo Szilard* (New York: Scribner's, 1992). By 1945, when the first annual summer course began to recruit the post-World War II generation, the Phage Group already included almost half a dozen partners. The earliest collaboration between any two of them is usually traced to joint experiments by Delbrück and Salvador Luria in 1941. On Luria see his personal papers at the APSL (cit. n. 11) and his autobiography, Salvador Luria, *A Slot Machine, a Broken Test Tube* (New York: Basic Books, 1984). See also the eulogies at the memorial service for Luria, including one by Watson, held at Kresge Auditorium, M.I.T., 1 April 1992 (M.I.T. Archives, Cambridge, Mass.).

[21] On Muller see Elof Axel Carlson, *Genes, Radiation, and Society* (Bloomington: Indiana University Press, 1987); for arguments on his role in molecular biology see Elof Axel Carlson, "H. J. Muller, Gene Theory and the Origins of Molecular Biology," *J. Hist. Biol.,* 1984, *17:*410–30; Evelyn Fox Keller, "Physics and the Emergence of Molecular Biology: A History of Cognitive and Political Synergy," *J. Hist. Biol.* 1990, *23:*389–409.

[22] See Joshua Lederberg, "Genetic Recombination in Bacteria: A Discovery Account," *Annual Review of Genetics,* 1987, *21:*23–46; Esther Lederberg, "The Mutability of Several Lac-mutants on E. coli," *Genetics,* 1948, *33:*617. For joint work see Esther Lederberg and Joshua Lederberg, "Genetic studies of Lysogenicity in E. coli," *Genetics,* 1953, *38:*51–64; *idem,* "Replica Plating and Indirect Selection of Bacterial Mutants," *Journal of Bacteriology,* 1952, *63:*399–406. The Lederbergs won the Pasteur Prize jointly in 1958 from the Society of Illinois bacteriologists; later in 1958, Joshua shared the Nobel Prize with George W. Beadle and Edward L. Tatum. See also oral history with Joshua Lederberg at the Rockefeller University in 1989, 1991; with Esther Lederberg at Stanford University in 1988, 1997, 1998; and with their collaborator Norton Zinder at UC-Berkeley, 1997 (on file with the author). For an overview of microbial genetics and the transition from classical genetics see Thomas Brock, *The Emergence of Bacterial Genetics* (New York: Cold Spring Harbor Laboratory, 1990); see also the personal papers of Milislav Demerec, director of CSHL in the 1940s and 1950s, at the APSL (cit. n. 11).

geneticists who, had they been acknowledged, might have overshadowed Watson's claim to monopolize the genetic origins of molecular biology by having shared in the molecular solution of genetic replication. Though Watson credited Delbrück with making the problem of genetic replication look important, the fact remains that it was sufficiently stressed by interwar geneticists, especially Muller and the Klampenborg assembly, so that it could be considered as being in the public domain, or not requiring attribution. Invoking such an influence on Delbrück's part could be said to contain an element of gratuity, especially since Delbrück's ideas turned out to be scientifically wrong or nonconvertible into lasting scientific capital; hence, one need not pay "interest" in the form of acknowledging actual scientific results. Instead, soft, symbolic currency such as "spirit" and "folklore" was pedalled as the basis for a genealogy that seemed to be more imaginary than real.

While an appeal to Delbrück did not entail any concrete scientific debts, it still conferred upon the organizers a respectable geanealogy, both scientific and social. In addition to his carefully cultivated scientific aura as a disciple of Bohr, Delbrück also had social aura since his family was part of the German intellectual aristocracy, which included renowned scholars and scientists. His background contrasted with the more modest, usually middle-class background of other scientists. A social preference for Delbrück was congruent with the derogatory references that abound in the correspondence of Delbrück's disciples to "New York type chemists," who were associated with low prestige but high-paying industrial positions.[23]

Watson's choice of a former quantum physicist and Bohr disciple as the first public hero of molecular biology may have also been a clever effort to create for himself and his new discipline a respectable genealogy that appealed to physicists. By positioning themselves as followers of the physicists' own great saga, the Phage Group not only basked in the reflected aura of revolutionary science, but indicated its acceptance of the reductionist hierarchy or social order of science in which biology presumably progressed by imitating physics.[24] Such a rhetorical maneuver further helped the leadership quest of the Phage Group spokesmen since it demonstrated that they knew how to use a language that the physicists, who invariably controlled the allocation purse for new disciplines, both understood and greatly valued as a *mythology.*[25] Veiled allusions to the Copenhagen spirit in biology were not only a way to get accepted by those who mattered most, but served as a timely reminder of the influential physicists' own historical role at a time when Bohr's recent death (in 1963) would have stirred great nostalgia.

Conceivably, Watson and Stent possessed greater awareness of the Copenhagen spirit than other members of the Phage Group (who did not mention it in their recollections), since both did their postdoctoral studies in the department of biochemistry at the University of Copenhagen. Though this institution had no direct link to the

[23] See the personal papers of Stent, the Bancroft Library, (cit. n. 11).

[24] See Donald Fleming, "Emigré Physicists and the Biological Revolution," in *The Intellectual Migration: Europe and America, 1930–1960,* eds. Bernard Bailyn and Donald Fleming (Cambridge, Mass., 1969), pp. 51–82. See also Pnina G. Abir-Am, "An Intellectual History of the Discovery of the Double Helix" (M.Sc. thesis, The Hebrew University of Jerusalem, 1976); *idem,* "From Physical Power to Biological Knowledge" (cit. n. 14); Robert Olby, "The Molecular Revolution in Biology" (cit. n. 13); Fox Keller, "Physics and Molecular Biology" (cit. n. 21).

[25] For the physicists' own mythologies see the papers by Mara Beller, Dominique Pestre, and Robert Seidel in this volume of *Osiris.*

Copenhagen spirit, by the time Watson and Stent arrived there in 1950, Bohr's influence had extended well beyond physics.

Moreover, as Watson recalled in his totemic memory, the Phage Group scions in Copenhagen actually had a direct link to Bohr's Institute for Theoretical Physics, the cradle of the Copenhagen spirit, as the brother of their post-doctoral adviser, the biochemist Herman Kalckar, was a research associate in Bohr's institute.[26] (This invocation of "kinship" in science recalls a Sinhalese myth in which the prosperity and status of a certain village is explained by claims of descent from the sister of the wife of the myth's mighty hero, who had offered gracious hospitality to the hero during a trying time in his rise to kingship.)[27]

Indeed, this quest for a genealogical filiation with the Copenhagen spirit may explain Kalckar's subversive presence in the collective memory (he offered a counter-memory that placed the origins of molecular biology in bioenergetics [part of biochemistry, the very field declared to be the *bête noire* of the Phage Group]). Kalckar's role in providing a direct genealogical link to Bohr via his physicist brother, as well as in providing a testimonial to the honoree's and the organizers' links, however tenuous, to the Copenhagen spirit, was apparently deemed more crucial for the collective memory than the specific details of his memory, which challenged the organizers' view of the genetic origins of molecular biology and placed them in the very field of biochemistry that they tried so hard to avoid.[28]

By selecting Delbrück—who had presumably sought to transfer Bohr's philosophical teachings from physics to biology—as the first public incarnation of molecular biology, Watson intimates that he too, like Bohr, should be regarded as an icon of twentieth-century science. Indeed, Watson's autobiography, written in the mid-1960s after Bohr's death, mentioned a letter from Delbrück to Bohr in which he referred to the double helix as comparable to Rutherford's model of the atom.[29]

A similar quest for genealogy could also be detected in the French collective memory. There, however, the most visible and vocal within the French Nobelist trio, Jacques Monod (1910–1976), then recently appointed director of the Institut Pasteur, did not need a genealogy leading to Niels Bohr but rather to Louis Pasteur (1822–1895), the legendary founder of the Institut Pasteur and the ultimate source of legitimation within "la Maison" or the institute's family. Much as Watson had transformed the honoree Max Delbrück into a genealogical link to Bohr, Monod transformed the French honoree Andre Lwoff into a genealogical link to Pasteur.[30]

[26] See Watson, "Growing up in the Phage Group" (cit. n. 17); see also Herman M. Kalckar, "High Energy Phosphate Bonds: Optional or Obligatory?" in Cairns, Stent, and Watson, *Phage and Molecular Biology* (cit. n. 10), pp. 43–52.

[27] See Margaret Robinson, "The Myth of the Mighty Hero," in *Structuralism,* ed. Martin Lane (London: Hutchinson, 1970), pp. 253–84.

[28] See Kalckar, "High Energy Phosphate Bonds" (cit. n. 26). Kalckar's contribution is examined in more detail in Abir-Am, "Entre mémoire et histoire" (cit. n. 3).

[29] Conceivably, Watson came to believe in the singular importance of the double helix, and hence that his own place in twentieth century science was on par with Bohr's, following this letter from Delbrück to Bohr in which Delbrück described the double helix as resembling Rutherford's model of the atom, which was a prelude to Bohr's. Delbrück to Bohr, 13 April 1953, in Delbrück's personal papers, Caltech Archives (cit. n. 11). It is difficult to reconcile this letter with Kendrew's recollections that in 1953 Delbrück wondered whether the double helix had anything to do with biology. See John Kendrew, "How Molecular Biology Started?" (a review of *Phage and the Origins of Molecular Biology*), *Scientific American,* 1967, *216:*421–43.

[30] On the affinity between Monod and Pasteur see Patrice Debre's preface to his *Jacques Monod* (Paris: Flammarion, 1996); *idem, Louis Pasteur* (Paris: Flammarion, 1995). On Monod see *Origins*

This effort took place at the expense of using his introductory essay to confer coherence upon the linguistically fragmented and historically dispersed French collective memory, as some far-sighted contributors had initially hoped.[31]

Monod's task was accomplished in two stages. First, he repeatedly emphasized his close rapport with the French honoree, Andre Lwoff through a frequent use of "me too" (moi-aussi) markers, while conveying the impression that their careers were always close and for several decades parallel and hence, that he was well informed about Lwoff's career and life. Indeed, the two met as early as 1931 when they collaborated on a paper with Edouard Chatton, a professor of protozoology then at the University of Strasbourg and after 1937 at the Sorbonne. Moreover, Monod's career was profoundly influenced by Lwoff's suggestion in 1937 that he shift his growth experiments from protozoa to bacteria. In the same vein, Lwoff's discovery of the induction of lysogeny in 1950 benefited from conversations with Monod.

Nevertheless, the two had only sporadic contact during the first two crucially formative decades of the honoree's career (1921–1941), since it was not until 1945 that Monod relocated from the Sorbonne to the Institut Pasteur to become Lwoff's *chef de laboratoire* (head of laboratory) or second-in-command in the department of microbial physiology, which Lwoff had headed since 1938. Similarly, the relationship became again more detached after 1954 when Monod became head of his own department of cellular biochemistry and relocated his group to another laboratory. Although they retained the custom of having joint daily lunches for their groups, in 1954 Lwoff changed his research effort from bacterial viruses to animal viruses, especially polio, which further distanced him from the topics in Monod's and even from those in his own laboratory.

In the late 1950s both became increasingly busy, adding to their responsibilities as department heads at the Institut Pasteur professorships at the Sorbonne, membership in science policy committees and participation as speakers in demand on the international circuit, especially in America. Though their public rapport culminated in 1965 when Lwoff, Monod, and Lwoff's student Francois Jacob shared the Nobel Prize, they drifted apart again in the late 1960s, first in connection with the strategy for reforming the Institut Pasteur and later in connection with their diverging responses to the events of May 1968.

Therefore, it is plausible to conclude that these two enjoyed a close scientific and personal rapport only in the decade between the mid-1940s and mid-1950s. This period, when administrative demands were limited since their groups were relatively small—as were the financial resources for research—produced the only collabora-

*of Molecular Biology: A Tribute to Jacques Monod,* eds. Andre Lwoff and Agnes Ullmann (New York: Academic Press, 1979; French version Paris: La Découverte, 1980); see also the review of this book by Pnina G. Abir-Am, "How Scientists View their Heroes:", *J. Hist. Biol.,* 1982, *15:*285–315. See also Jean-Pierre Soulier, *Jacques Monod: Le choix de l'objectivité* (Paris: Frison-Roche, 1996). See also oral histories conducted by Pnina G. Abir-Am with Monod's associates, most notably Andre Lwoff, Annamaria Torriani, Melvin Cohn, Germaine Stanier, Georges Cohen, François Jacob, François Gros, Agnes Ullmann, David Perrin, Jean-Pierre Changeaux, and Antoine Danchin, among others, mostly at the Institut Pasteur in Paris, June 1986, October 1987, June 1988, May to December 1993, July and December 1994, July and December 1995 (on file with the author). See also Monod's and Lwoff's personal papers and the directorial papers of Institut Pasteur (cit. n. 12).

[31] The idea that Monod's essay should tie together the other essays was expressed in a letter from Luria to Borek, 20 November 1969, in which Luria had requested an extension in view of his need to prepare his Nobel lecture (delivered in December 1969) as well as in view of campus disturbances at M.I.T. that required that he sit on many committees. See Luria's personal papers, APSL (cit. n. 11).

tive papers the two ever did (in 1946 and 1947). A joint trip to the United States in 1946 was a peak in their relationship. This limited rapport explains why Monod's discourse on the honoree often sounds as if Monod was talking about himself.

On the other hand, Monod emphasized the early arrival and long stay at the Institut Pasteur of the French honoree, and his status as the protégé of Felix Mesnil (1868–1938), a relatively obscure but well-connected biologist who was a former personal secretary of Louis Pasteur and a collaborator with Pasteur's Nobelist associates Alphonse Laveran and Elie Metchnikoff. As Lwoff always acknowledged, Mesnil's patronage was indispensable for his career at Institut Pasteur. As head of a department of microbial physiology, established for him through Mesnil's intervention in 1938, Lwoff was able to give a home to many associates, including the future stars of molecular biology, Monod and Jacob.[32]

Having established a social-professional genealogy, Monod went on to establish a scientific one. In a bold rhetorical move he redefined Pasteur's legacy as being in basic biology rather than in applied bacteriology. This topic had great relevance for Monod, then a new, controversial Institut Pasteur director who was soon to embark on a policy of prioritizing basic research while relegating applied research and products to a separate industrial unit. Anticipating resistance to his reorganization plans from the institute's veterans, Monod tried to undermine his critics' legitimacy by suggesting that his associate, Andre Lwoff, had all along pursued Pasteur's authentic legacy in basic biology, while the others had pursued a less authentic and more applied line in bacteriology.

Furthermore, according to Monod, Lwoff revived and long maintained Pasteur's legacy in the aftermath of the directorship of Emile Roux (1906–1933), a physician who had assisted Pasteur in the dramatic vaccination against rabies and prioritized the production of vaccines and serums as the institute's major activities. As Monod explained, Pasteur's authentic legacy was initially pursued by the director who came after Pasteur's death, the eminent biochemist Emile Duclaux, but that legacy was eventually eclipsed under the subsequent long directorships of physicians, most notably Roux.[33]

Monod thus intimated that the time had come to return, once again, to Pasteur's authentic legacy of studying microbes as model organisms for the fundamental questions of biology, rather than regarding them as resources in the production of vaccines and serums. As the honoree's close associate, he, Monod, was eminently suitable to head such a task. As he put it:

> In the world of microbes that he had discovered, Pasteur did not see the agents of infectious diseases only, but the most elementary forms of life, which consequently could reveal the most fundamental secrets of life. The tremendous success of the bacteriology of infectious diseases, of epidemiology, and of immunology led in part to forgetting this veritable Pastorian tradition . . . Pasteur was not a "bacteriologist." He was a biologist, in all the meanings of the term, fascinated by the strange logic of the living world and inhabited by the need to perceive its mysteries. Andre [Lwoff] too, and this is why it

[32] See the obituary of Felix Mesnil (1868–1938) by Andre Lwoff, *Annales de l'Institut Pasteur,* 1938. See also correspondence between Mesnil and the director of l'Institut Pasteur in the directorial papers of l'Institut Pasteur (cit. n. 12). For an overview of l'Institut Pasteur see *L'Institut Pasteur: Contributions à son histoire,* ed. Michel Morange (Paris: La Découverte, 1991).

[33] Monod, "Du microbe à l'homme" (cit. n. 15), pp. 3–9.

> became his task to recover this tradition and soon illustrate it with illuminating discoveries.[34]

Pursuing the same commemorative strategy as Watson of recasting the honoree's scientific legacy from the viewpoint of his own career—after all, the research program that he claimed was most authentic in the Pasteurian tradition was the one which he himself pursued as the honoree's associate—Monod's quasi-biographical essay zeroed in soon on agendas other than his primary goal of reforming the Institut Pasteur. For example, he went on to castigate the Sorbonne, then also seen as being in great need of reform, as an obstacle in the rise of molecular biology. As Monod explained, Lwoff's "destiny, or genius, compelled him to circulate across several disciplines"; thus the honoree succeeded in synthesizing the separate disciplines of biology, which was required for the rise of molecular biology, just by avoiding the Sorbonne, where presumably each discipline continued to remain isolated from its neighbors.[35]

Rather than developing this intriguing theme of how Lwoff's unusual career shifts across various fields may have contributed to the unification of biology, Monod used the honoree's pioneering research in comparative microbiology as a launching pad for pursuing yet another agenda of his own: the reform of the French university system. This was a highly debated topic in the aftermath of the student rebellion of 1968, a period that coincided with the prolonged assembly of the French collective memory. Hence, Monod castigated the Sorbonne for having failed to teach either microbiology or genetics in the 1920s, when he and the honoree were students there. (These subjects were introduced at the Sorbonne only after World War II with Lwoff becoming the first professor of microbiology in 1959.) Monod asserted that had Lwoff remained at the Sorbonne, he could only have failed, since that "complacent" university could not anticipate the later union of microbiology and genetics into modern biology, as accomplished by the honoree and his associates at the Institut Pasteur.

Since the last phrase describes rather well Monod's own experience at the Sorbonne (he was told at his Ph.D. thesis defense in 1941 that the Sorbonne had no interest in his research on microbial growth, and he eventually left for the Institut Pasteur in 1945 after a decade of modest progress), it appears that Monod's totemic memory, much as Watson's beforehand, used the honoree's career to score rhetorical points relevant to his own present agendas. These included proposed reforms of the Institut Pasteur and of the university system, and characterization of the origins of molecular biology as a Pasteurien unification of several disciplines—which would have been inconceivable in another, rival institution such as the Sorbonne).

In comparing the totemic memories offered by Watson and Monod, the

[34] Ibid., p. 5. The original French text is: "Dans le monde des microbes qu'il decouvrait, Pasteur ne voyait pas seulement les agents des maladies infectieuses, mais les formes les plus élémentaires de la vie, celles par consequent qui pouvaient en reveler les secrets les plus fundamentaux.

Les prodigieux succès de la bacteriologie des maladies infectieuses, de l'épidémiologie, et de l'immunologie avaient, en partie, fait oublier cet aspect si important de la véritable tradition pastorienne . . . Pasteur n'était pas un "bacteriologiste." C'était un biologiste, dans tout l'acception du terme, fasciné par l'étrange logique du monde vivant et habité par le besoin d'en percer le mystère. Andre [Lwoff] aussi, et c'est pourquoi il lui fut donne de retrouver cette tradition et de l'illustrer bientôt des lumineuses découvertes."

[35] Ibid., p. 3.

organizers-turned-heirs apparent who served as the key editors for the collective memories, it is striking to note their similar concerns with defining the honoree's overall legacy, the origins of molecular biology, and the role played by a key institution. Totemic memories thus contrast with ordinary memories by rank-and-file scientists, which usually combine recollections of the honoree during a limited period of collaboration with discussion of the author's most important piece of research, for which the author sacrifices taking credit in order to acknowledge the honoree's influence. Totemic memories transform the honoree into a totem by declaring him to be a founder, or a totalizing symbol, of the new discipline.

Once so annointed, the honoree's totemic value for the organizers is extracted in the form of a genealogy reaching uniquely from the honoree to an ancestral figure, usually a legendary scientist associated not only with great discoveries but also with heroic social and cultural status. While the French collective memory transformed its honoree—Andre Lwoff—into a genealogical link between the key French organizer, Monod, and Louis Pasteur who was known worldwide as a savior of humanity from various infectious diseases in humans, animals, and plants, the American collective memory transformed its honoree—Max Delbrück—into a genealogical link between the key American organizer, Watson, and Niels Bohr, the creator of the Copenhagen spirit and an icon of a twentieth-century theoretical revolution in science.

The authors of totemic memories further refer to themselves as disciples of their honoree-turned-totem, thus demonstrating that the organizer's quest for leadership in the new discipline and/or one of its key institutes is legitimate, since it flows from a proper scientific genealogy. This strategy of elevating the honoree to the status of a totem or symbol is a necessary prelude to displacing him as leader of a group via a smooth rite of succession. Therefore, both the American and the French totemic memories were profoundly ahistorical, since their goal of extracting genealogical "surplus value" from the honorees' legacies drove them to reify those legacies to fit the heirs' agendas in the present. As the following section suggests, this effort is perhaps even more evident in the visual memories included in each collective memory, since visual memories condense the heirs' present agendas into easier-to-grasp but influential images.

### *Visual Memories in the American and the French Texts: Encoding or Subverting?*

The American collective memory contains seven pictures (see Appendix 1) that reflect, as an ensemble, a certain thesis about people, places, and discoveries in the Phage Group and its projection as the locus for the origins of molecular biology. This is the thesis of coeditor Watson, who had the last word in the matter of "picture control." A full-page, half-bust picture of the honoree is placed, as is customary, opposite the frontispiece. Taken in the early 1960s, it conveys Delbrück's informality (open shirt, sweater, no tie) and relative youthfulness, as if to signal the irony in the collective memory's aim of sending him to posterity while still active.

Two additional pictures placed at the beginning of parts I ("Origins of Molecular Biology") and II ("Phage Renaissance"), respectively, show Delbrück and his key collaborator, Salvador Luria, first doing experiments in 1941 and second, sunbathing in shorts in 1953. Both photographs were taken at the Cold Spring Harbor Laboratory. The picture dating from 1941 shows a seated Luria examining colonies of lysed bacteria on a petri dish, the standard way to uncover the presence of phage, while a standing Delbrück looks at the dish from above. Since their best-known and perhaps most important paper from that time carried the qualifying statement "experiment by Luria, theory by Delbrück," this picture appears to have been selected as a mocking reminder of Delbrück's tendency to oversee the work of others, presumably in the name of his superior theoretical or critical faculties.

The other picture showing Delbrück and Luria sunbathing suggests that after the summer of 1953 (when the Watson-Crick papers on the double helix were distributed to the phage summer course attendees, thus introducing the double helix to American scientists), the only things left to be done (other than sunbathing) were to pass the torch from the phage genetics they had created since 1941 to the new molecular genetics initiated by Watson and Crick, and to solve the key problem of replication at the molecular level.

Another picture of a duo, taken in 1951, shows Leo Szilard and Alfred Hershey, other prominent members of the Phage Group's founder generation. While Szilard was the "man behind the bomb" whose roving activities as a science policy adviser did not fail to impress Watson, Hershey designed the experiment that proved phage DNA to be the hereditary component, a finding that was of major importance in bolstering Watson's confidence to focus on DNA structure alone. Thus, the gallery of portraits selected by Watson to represent the Phage Group is composed of two "pictorial pairs" of scientists from the previous generation (Szilard, Delbrück, Luria, and Hershey were born in the years 1898–1912, while Watson was born in 1928).

Since there are no pictorial representations of Watson's generation other than a full-length picture of Watson with his back to the viewer, wearing shorts and a loose shirt and pointing to a double-helix model on a screen in a lecture room at Cold Spring Harbor, Watson appears figuratively as the sole heir of the generation of four founders. Furthermore, the image of his solitary interaction with the famous model is suggestive of both his singular place as the only American coauthor of the double-helix structure of DNA, as well as of the American cultural bias—namely, that introducing an item to America is a more significant event than the item's original discovery elsewhere by a majority of non-American scientists.

The last two pictures are an aerial view of the Cold Spring Harbor Laboratory, of which Watson became director in 1968, and a frontal view of the biology department at Caltech as it looked in the 1930s, when Delbrück arrived there as a visiting fellow in genetics. The two concluding pictures thus symbolize the transition from the honoree's institution, Caltech, a site of pilgrimage by the Phage Group in the 1940s and 1950s, to the heir apparent's institution, Cold Spring Harbor Laboratory, which became a major center of research and instruction in molecular biology in the late 1960s when its activities were put on a solid basis of endowments and long-term grants.

The pictorial representation of the American collective memory can be said to express in condensed images its key agenda of legitimizing a multiple transition rite: from the honoree (Delbrück) to the main organizer (Watson) as the publicly acknowledged leader; from phage genetics to molecular biology; from Caltech to Cold Spring Harbor Laboratory; from crucial experiments on organisms of choice, such as phage, to molecular models of nucleic acids and genetic codes.

In contrast, the pictorial representations of the French collective memory are more numerous (there are twelve) and diffuse (they range from group pictures of many scientists, both in the 1920s and in the 1950s, to several portraits of the honoree). Thus, they lack the punchy conciseness of the American pictorial message. This occurred in part because the production process was in the hands of a foreign coeditor who had a rivalrous relationship with the French coeditor (Monod), to the effect that no picture of the latter would appear alone, as did Watson's to signal the succession rite behind the collective memory.

Five photos include individual images of the honoree, one roughly from each of the five decades covered by the commemorative occasion. The inner front-cover photo shows him near the micromanipulator he used in discovering the induction of lysogeny in 1950, which indicates the key organizers' agenda of portraying the honoree as primarily an experimentalist so as to leave the theoretical spots to themselves. The inner back-cover photo shows an aged and sober Lwoff at his desk as director of the Cancer Research Institute in Villejuif (a suburb of Paris) since February 1968, an anticlimactic phase of his career as it meant that a directorial position, presumably a career peak, could only be achieved outside the Institut Pasteur.

The other three photos of the honoree alone capture his appearance at different historical moments: a reserved outlook during his army service in the late 1920s; an expression of euphoric irony in the laboratory in the 1930s; and a bewildered youthful outlook as the gowned recipient of an honorary doctorate from Oxford University in 1959. These photos convey some of the physical qualities and attributes of the French honoree: his tall, slim figure; handsome, youthful face; angelic, curly blond hair; ironic, aloof attitude; reputation for being a superb experimentalist; and his receipt of honors, mostly outside of France.

On the other hand, the four group photos and two that depict Lwoff and one other colleague suggest the atmosphere of *gai savoir* that reigned in the honoree's scientific environment both in the 1920s (one photo shows Lwoff taking a coffee break with eight colleagues during the summer program at the Marine Biological Station Roscoff in Normandy), and in the 1950s (another shows the honoree in a comparable company of relaxed colleagues during the much-remembered group lunches held in his department at the Institut Pasteur, when stimulating discussions ranged from science to cultural affairs and politics).

Other group photos reinforce the image of informal camaraderie prevailing in the honoree's company. But they also suggest that key individual colleagues, most notably Boris Ephrussi (1901–1979), who appears in three "sociability" pictures, and Odette Touzet, a collaborator and later professor of microbiology at the University of Montpellier who is shown with Lwoff in the laboratory, were missing from the collective memory assembled by the honoree's presumptive heirs. The next section focuses on the excluded memories of women as a case study in categorical exclusion, or exclusion as members of a social group rather than as individuals.

### *From Totemic to Taboo Memories: The Missing Memories of Women Scientists*

The pursuit of multiple agendas by presumptive heirs, such as Watson and Monod, under the guise of a commemorative effort includes not only the production of totemic memories and the alignment of various ordinary memories, but also strategies to exclude historical witnesses who might offer countermemories. Excluding strategies may focus on individuals—often personal rivals of the organizers—or on groups deemed capable of counteracting the "official" or consensual collective memory assembled for publication.

For reasons of space, the process of excluding various categories of memories cannot be elaborated on here, beyond noting that the names and topics that failed to materialize were often more significant scientifically than many that were included. Still, it is of special interest to consider the "taboo memories," or the memories of those excluded *en bloc* by virtue of a common attribute, e.g., gender.

Although women were not the only potential authors to be excluded from the final text of the American and French collective memories, their total absence is an amazing phenomenon. First, the complete invisibility of women scientists at the level of public memory raises the question of whether they were also absent from history. In view of the fact that the number of women mentioned by various authors in each collective memory is comparable to the total number of authors, or almost three dozen in each case, women scientists appear to have been present in history but absent from memory. The question then persists whether their exclusion from public memory occurred because their work was not so important. After all, many men were excluded for this very reason, since the orchestrators of the collective memories selected authors who were both associated with "crucial experiments" and who maintained a close, disciple-like rapport with the honoree.[36]

Still, some women scientists did fit this dual, stringent criterion. Indeed, bibliographic references to various crucial experiments done in the American and the French groups invariably include women coauthors. For example, the discovery of the induction of lysogeny by the French honoree was coauthored with Antoinette Guttman; the discovery that phage DNA was responsible for phage multiplication in 1952 was coauthored by Alfred Hershey (a founder and would-be Nobelist of the Phage Group) and Martha Chase; the plaque technique in animal virology was perfected in the mid-1950s by Renato Dulbecco (a future Nobelist who contributed to both the American and the French collective memories) and his long-term collaborator, Marguerite Vogt; Greta Kellenberger collaborated with Jean Weigle on the genetics of the phage lambda, among other examples of crossgender collaboration.[37]

On the whole, the visibility of women in the collective memories is low. For example, only Esther Lederberg and Martha Chase were mentioned more than twice

[36] The final number of authors was 32, (out of an initial list of 60 in the French case and 41 in the American case). Though a token woman scientist was included in each initial list of invited contributors, those women declined (in the French case, she cited her lack of a distinguished career; in the American case she give no reason, although the fact that an apparent breakthrough turned out to be a fiasco played a role, although similar circumstances in the case of the American [male] organizers did not deter them). Ironically, the two token women who were invited were not the most accomplished among those available, but were known to the organizers and apparently fit their image of women scientists as helpmeets or followers of men. On the link between gender and collective memory see Joy Harvey, "A Focal Point for Feminism, Politics, and Science: The Centennial of Clemence Royer in 1931," in this volume of *Osiris;* see also Abir-Am, "Introduction," this volume of *Osiris.*

[37] For further information on collaborative couples in science see *Creative Couples in the Sciences,* eds. Helena M. Pycior, Nancy G. Slack, and Pnina G. Abir-Am (New Brunswick, N.J.: Rutgers University Press, 1996), especially the appendix which lists over fifty such contemporary and historical couples. See also oral history conducted by the author with Dr. Elizabeth Bertani, March 1998 at Caltech. (She was the first woman to obtain a Caltech Ph.D. in biology in 1957 and collaborated with her husband Joe Bertani, a University of Illinois Ph.D. who did postdoctoral studies at Caltech with Delbrück. The first female Caltech Ph.D. was Dorothy Semonov in chemistry.) See Margaret W. Rossiter, *Before Affirmative Action: Women Scientists in America, 1940–1972* (Baltimore: Johns Hopkins University Press, 1995), p. 82.

in the American collective memory; while only Annamaria Torriani and Germaine Stanier, who worked in Lwoff's department at the Institut Pasteur for several years in the early 1950s and later continued their careers in the United States (the former at the Massachusetts Institute of Technology [M.I.T.] and the latter at the University of California [UC]-Berkeley); as well as Marguerite Lwoff, the honoree's long-term collaborator and spouse, were mentioned more than twice in the French collective memory. Another point of interest is that the memories of native male scientists, whether in the United States or France, were more exclusive of laboratory women than the memories of foreign scientists. For example, Niels Ole Kjelgaard, a Danish visiting microbiologist who collaborated with the French honoree on the crucial papers on the induction of lysogeny, noticed seven women working in the laboratory, both scientists and technicians, while the majority of authors noticed none. The issue of gender invisibility was not only a product of crosscultural sexism but was further compounded by nationality and marital status, with foreign nationality and married status likely to increase visibility. For example, about one quarter of the women mentioned in the American collective memory were French, while about a third of those mentioned in the French collective memory were foreign, mostly American. A third of the thirty or so women mentioned in each group were mentioned by both the American and French groups, indicating a certain level of overlapping memories as well as a role for women in brokering crosscultural relationships.[38]

The hegemonic position of married status and the prevalence among women scientists of marriage to a collaborator, especially in the French group, in the period covered by the collective memories (pre-1960s) meant that women scientists were seen first and foremost as spouses of male scientists who typically did either collaborative or derivative work, and hence who could be subsumed largely, if not entirely, in their husbands' reputations. This also meant that the mostly married, male scientists were likely to ignore single women scientists to avoid the liability of entanglement in sex-related scandals, given the pervasive cultural association of unattached women with temptation, diversion from serious work, and even sin.

The overall invisibility of the women scientists as authors of memories, despite their historical presence in comparable numbers, raises the question of the organizers' outlook on gender issues. For example, Watson's exclusion of Esther Lederberg, who had been his fellow graduate student, is incomprehensible in view of her subsequent association with important research in phage genetics. Such an exclusion can be attributed not only to the prevailing sexism of the mid-1960s but also to Watson's agenda of acquiring sole custody of the genetic origins of molecular biology. This agenda could not have been implemented if other major spokespeople for the genetics of microorganisms, such as the pioneering Lederbergs, had had the opportunity to share their memories in the same collection.[39]

In contrast, the exclusion of women from the French collective memory can be explained not only by the level of sexism prevailing before the 1970s, but also by the institutional agenda of the heir presumptive. While Monod's agenda of tracing

[38] Some of the women excluded from the volume honoring Lwoff were included by Lwoff in a volume honoring Monod eight years later. See especially the essays by Annamaria Torriani and Germaine Stanier, who worked in Lwoff's department, in Lwoff and Ullman, *Origins of Molecular Biology* (cit. n. 30). See also Abir-Am, "How Scientists View their Heroes" (cit. n. 30), pp. 285–315.

[39] Stent explained the exclusion of the Lederbergs in terms of their supposedly tense relationship with Delbrück; this issue needs further clarification.

his lineage directly from Louis Pasteur had little to gain by including women, it had a lot to lose by including some women but not others, especially in view of Monod's reputation as a womanizer. Nevertheless, his exclusion of women recalled multiple times by others, most notably Annamaria Torriani and Germaine Stanier who collaborated with Monod himself, while working in Lwoff's laboratory, remains incomprehensible, especially since neither presented a gender-related "liability" (both were happily married to other important scientists—Luigi Gorini of the Sorbonne and Harvard Medical School and Roger Stanier of University of California-Berkeley, respectively).

Furthermore, each woman had a pioneering career both in France and in the United States, the former becoming a professor at M.I.T. and the latter head of a department at the Institut Pasteur. Hence, if these two remained outside the realm of memory, then it is obvious that women were excluded from collective memory because of their gender, or regardless of their national affiliation(s), scientific seniority, or marital status. Gender as a category obviously upset the nexus of power and memory in science, because totalizing symbols in science were assumed to be male.

Interestingly enough, the second French collective memory in molecular biology, published as two monolingual collections (one in English in 1979 and the other in French in 1980) in honor of the prematurely dead Jacques Monod, featured seven women out of almost three dozen contributors. This difference from the collective memories of mid- and late-1960s is accounted for by the rise of gender consciousness throughout the 1970s with the women's liberation movement, which made the silencing of women's voices no longer acceptable.[40]

Having seen three out of several categories of memories (totemic, ordinary, and taboo) which together produce the objectifying effect of collective memory, the next section explores how the concerted action of the organizers managed (or failed) to assemble and align such diverse memories into a coherent whole.

## III. DIVERGING SOCIAL PRETEXTS: FROM AN AMERICAN BUSINESS ALLIANCE TO A FRENCH IDEOLOGICAL BETRAYAL

In the American collective memory, the highly structured and stylistically coherent text gradually unfolded as a "rags to riches plot" from modest origins in the 1930s to great ramifications in the 1960s, as told by almost three dozen contributors who each concocted his own blend of personal-best contributions to scientific progress and confession of admiration for the honoree. This seamless narrative was the product of an effective business alliance among the three organizers turned coeditors, John Cairns, Gunther Stent, and James Watson. This triumvirate jointly presented a rather persuasive front and very few would dare refuse its call, contest its final product, or intervene with the monitoring of the product's distribution.

The success of the precocious American commemorative effort was predicated upon the capacity of these three (all of whom were part of the second generation of Phage Group members and were born in the 1920s) to sustain a strategic alliance that benefited all but especially Watson, who initiated the project and had the highest stakes in it.

By the editors' own account, the DNA faction or section V in the American collec-

[40] See especially Rossiter, *Before Affirmative Action* (cit. n. 37).

tive memory titled "DNA" amounted to only one-sixth of either topicality or membership (i.e., one part out of six, or five chapters out of thirty-two). Yet its authors were a strategically situated faction *vis-à-vis* the overall claims to have originated molecular biology, since they were the only ones to work directly on DNA, the "mastermolecule" that emerged by the mid-1960s as a symbol of the new molecular biology.

Though this DNA faction included, besides Watson and Cairns, other renowned DNA scientists who would later become known as public figures—most notably Mathew Meselson of "yellow rain fame" (who recalled jointly with Frank Stahl their crucial experiment demonstrating the semiconservative mode of DNA duplication) and Robert Sinsheimer, later known as the first organizer of the human genome project (who recalled his discovery of single-stranded DNA rings in small viruses)—the composition of the organizational and editorial triumvirate was determined by specific institutional, genealogical, and personal considerations on the part of Watson, who proposed the project to the other two.[41]

As the member who had the idea of transforming the Phage Group lore into a scientifico-historical commodity, a collective memory to be marketed for fund-raising purposes, Watson called the shots, especially since he enjoyed both prestige and power as the only Nobelist member of the Phage Group at that time, as a nominator for future Nobel Prizes, and as head of a laboratory at Harvard University. Watson first selected Cairns in view of the latter's strategic position as the new director of the Cold Spring Harbor Laboratory, targeted by Watson as the host institution for the projected commemoration.

However, Watson had a larger design for this laboratory, to which he had been attached ever since his student days there at summer courses in the late 1940s. Watson's interest intensified after 1963 when the pull-out of CSHL's long-term patron, the Carnegie Institution of Washington, forced CSHL's longtime director, the classical-turned microbial-geneticist Milislav Demerec, to retire angrily, leaving the laboratory financially and adminstratively troubled.

Moreover, the Nobel experience in 1962, in which Watson was the only non-British scientist among five laureates in molecular biology, convinced him that he should emulate the British laboratory that had produced the majority of the Nobelists, the Medical Research Council (hereafter MRC) Laboratory of Molecular Biology in Cambridge. It was in that laboratory that Watson shared in the discovery of the double helix in 1953; in 1962 the MRC Laboratory moved to new, lavish quarters on the outskirts of Cambridge and continued turning out Nobel laureates, producing seven by its fortieth anniversary in 1987.

Watson's concern with transforming CSHL into a comparably prestigious institution led him to try and help first on the administrative front, by bringing in Cairns as a new director. But Cairns turned out to be an interim solution only, since the laboratory's difficult condition required a more well-connected director than the twice-expatriate Cairns, whose previous career in Australia and Britain did not lead

[41] On Matthew Meselson and Franklin Stahl's experiment demonstrating the semiconservative replication of DNA see the forthcoming book by Larry Holmes of Yale University; see also the joint essay, "Demonstration of the Semiconservative Mode of DNA Duplication" in Cairns, Stent, and Watson, *Phage and Molecular Biology* (cit. n. 10). On Robert Sinsheimer see his autobiographical *The Strands of a Life: The Science of DNA* (Cambridge, Mass.: Harvard University Press, 1994).

to a strong power base in the United States. Indeed, Watson's larger design for CSHL became obvious in the mid-1960s when he first volunteered as part-time director, presumably to assist Cairns with grant writing. In February 1968, less than two years after the publication of the American collective memory—whose good sales were regarded as fund-raising for CSHL, Watson left Harvard to become the laboratory's full-time director.[42]

While Cairns's presence among the organizers enabled them to use CSHL as a base, thus conferring a more formal allure upon an essentially private initiative on the part of Watson, Stent's presence enabled them to do, as Stent put it, the "hell of a job" that the actual editing of so many memories turned out to be. Stent had facility with writing and editing, as well as the prior experience of packaging Phage Group meeting reports with pictures and philosophical allusions into a successful textbook, published in 1963 with a dedication to Delbrück.[43] Stent had also edited a collection of key papers in microbiology that put him in touch with many scientists in and out of the Phage Group, a form of networking that he was particularly adept at cultivating and which would prove crucial for the delicate effort of assembling the private memories of many scientists for publication.[44]

Several other elements colluded to make Stent the quintessential executive editor of their venture. These included Stent's long-standing friendship with Watson, which dated from their time together as postdoctoral fellows in Copenhagen (Watson wrote part of his autobiographical *The Double Helix* in Stent's house in Berkeley, California); his reputation as a devoted disciple who kept in close touch with Delbrück in whom he had found a father figure; and his relatively light standing in science (unlike Cairns, who did a classical experiment in autoradiography of DNA in the bacterial chromosome, and Watson, Stent had not been involved in a major discovery or a crucial experiment). Furthermore, the trio worked efficiently since there was no room for conflict. A clear sense of hierarchy prevailed with Watson as the remote control or *eminence grise* who did not like to handle the daily organizational details or the interim drafts, but who was always there to put his final touches on the big picture and final manuscript, or even to be known as the excluding "stinker," if needed.[45]

The alliance's capacity to accommodate the organizers' respective priorities was demonstrated when they agreed on a dual pretext for mobilizing the memories of Phage Group members: namely, to mark both the twentieth/twenty-first anniversary of the first phage course at CSHL—a priority for both Watson and Cairns, who

[42] See Watson's contribution, as the only non-British scientist, to the fortieth anniversary of the MRC Laboratory of Molecular Biology in the special issue of *New Scientist,* 1987, *47:*3–40. See also Cairns, "Memorandum on CSHL," [1967], copy with the author; the correspondence of Cairns, Stent, and Watson with Delbrück in Delbrück's personal papers, Caltech Archives (cit. n. 11) and the correspondence among Cairns, Stent, and Watson in Stent's personal papers, The Bancroft Library (cit. n. 11).

[43] See Gunther S. Stent, *The Molecular Biology of Bacterial Viruses* (San Francisco: Freeman, 1963).

[44] Stent's large correspondence with colleagues invited to submit papers to an edited volume on phage research testifies to his ability to maintain a wide network of colleagues turned friends. See Stent's personal papers, The Bancroft Library (cit. n. 11).

[45] See acknowledgment in James D. Watson, *The Double Helix* (New York: Atheneum, 1968). For further details on the correspondence among the editors see Abir-Am, "Entre mémoire et histoire" (cit. n. 3).

wished to draw attention to CSHL as a cradle of molecular biology—and Delbrück's sixtieth birthday, a priority for Stent, who was a particularly loyal disciple, de facto impresario, and keeper of Delbrück's aura.[46]

The dual pretext was useful not only as an objectifying reinforcement that expanded the justificatory range of the effort to combine genealogical, institutional, and disciplinary elements, but also as a face-saving device. When Luria questioned the wisdom of sending Delbrück to posterity at a time when he was still active, even youthful, and when the honoree refused to show up for a ceremony in which he was to receive a collective tribute, the trio had a fall-back position: marking the anniversary of a summer course, which required permission from the CSHL director only, who was one of them.

As the coeditors admitted, the overt pretext of honoring their one-time inspiring colleague was just a pretext. The more covert desire was the then-timely need to publicly project a powerful consensus around a particular definition of molecular biology. The strategic alliance within the trio, as well as its timeliness, were evident in the trios' correspondence, as the following letter from Stent to Cairns indicates:

> Nothing, but nothing, would give me greater pleasure than to have you organize the Max Memorial Festival that Jim has proposed. I tried to do something like this already for Max's 50th birthday back in 1956, but my suggestion for a Festschrift went over with the boys like a lead balloon. Actually, I think the 21st anniversary of the phage course (the real coming of age) might be a more diplomatically chosen occasion . . . if the Laboratory of Quantitative Biology could get a financial boost through this venture, all the better! (I am sure that the authors would be glad to forego royalties for The Cause.) . . . I also think that the editor(s) should suggest to each contributor more or less what he is to write about so the book could be more useful than a disconnected series of anecdotes. I would envisage it as a brief history of molecular biology, as told by the Folks Who Made IT What IT Is Today. I had made up already such an assignment of topics, in fact.[47]

In contrast to the compact alliance among the organizers of the American collective memory, who shared a wide consensus on both goals and means while further complementing each other's primary agendas, the French collective memory was orchestrated by an eight-member international organizing committee and was headed by two coeditors who not only pursued diverging, even conflicting agendas, but who barely communicated with each other.

This unusual situation for a commemorative effort occurred because the initiative for assembling a collective memory in honor of Lwoff did not come from his French colleagues but from overseas veterans, the so-called *combattants outre-mer.* The initiative came from Ernest Borek (1911–1986), a visiting professor in the department of biochemistry at Columbia University and a veteran science popularizer who would eventually edit the English-language essays in the bilingual French collective memory. He also oversaw the collective tribute's publication process with Columbia

[46] Watson, *The Double Helix* (cit. n. 45). Stent wrote several obituaries of Delbrück and edited the manuscript left by Delbrück that became *Mind from Matter?* (Palo Alto: Stanford University Press, 1986). See also their correspondence which testifies to a close relationship, especially on Stent's part: Stent to Cairns, 2 October 1964; Cairns to Stent, 21 May 1965; as well as Stent to Luria, 28 May 1965. Stent's personal papers, The Bancroft Library (cit. n. 11).

[47] Stent to Cairns, 2 October 1964, Stent's personal papers, The Bancroft Library (cit. n. 11).

University Press, for which he had long been an advisory editor of its Science Series.[48]

Borek's initiative was triggered by his opportunity to review *Phage and the Origins of Molecular Biology.* Borek criticized POMB for its claim that the origins of molecular biology were to be found in genetics, whether in the theoretical speculations of Schrödinger on a code or in the experimental research of the Phage Group. He suggested instead that molecular biology merely continued the spirit of systematic and imaginative biochemistry, a view prevailing among macromolecular biochemists yet one which failed to explain the great excitement generated by molecular biology in the 1960s:

> I must take issue, however, with an assertion in Stent's essay: he overstates the influence of Schrödinger's book, *What is Life,* on the growth of biological science . . . This is an egregious exaggeration. Molecular biology is the fruit of the union of imaginative biology and systematic biochemistry of the past century . . . Actually, the need for what today is called molecular biology was part of the Zeitgeist of imaginative biochemists in the 1940s.[49]

However, unlike other distinguished reviewers, most notably the Nobelists John Kendrew and Arthur Kornberg (who also criticized POMB's narrow view of molecular biology's origins, its total lack of reference to structural contributions, and its derogatory reference to biochemical contributions, Borek, being an experienced science popularizer, understood right away that the only effective way to contest the partisan claims of the Phage Group (which had derided his own field of biochemistry as "messy" and uninspired by theory) was to produce an equally impressive collection of memories.[50] Borek recognized the need to assemble a de facto countermemory by another group of distinguished scientists who would chart their own alternative or biochemical path to the promised land of molecular biology.

Just as Watson and Cairns had understood that such a collection could not be floated under the flimsy pretext of the anniversary of a summer course, but required the anniversary of a major, or at least popular, figure in science to galvanize prospective authors, Borek too realized that he needed, first and foremost, a suitable honoree.

In April 1968, Borek encountered in *Science* magazine a stronger and longer version of POMB's claims on the exclusively genetic origins of molecular biology, written by Stent and provocatively titled "That Was the Molecular Biology That Was."[51] This essay elaborated on the thematic chronology embedded in the table of contents of *Phage and the Origins of Molecular Biology* (see section II), which was briefly

[48] On Borek see Sylvia J. Kerr and Opendra K. Sharma, "Ernest Borek, 1911–1986," *Trends in Biochemical Sciences,* 1986, *11:*397–8; oral history with Sylvia Kerr (Borek's widow), June 1995, on file with author. See also the considerable correspondence of Borek with Seymour Cohen in Cohen's personal papers, APSL (cit. n. 11). See also Ernest Borek, *The Code of Life* (New York: Columbia University Press, 1965); *idem,* "The transfer of Biological Stress in Microorganisms," in Monod and Borek, *Les microbes et la vie* (cit. n. 10), pp. 157–63.

[49] Ernest Borek, "Review of *Phage and the Origins of Molecular Biology,*" *Bioscience,* April 1967, *17:*273.

[50] See Kendrew, "How Molecular Biology Started?" (cit. n. 29); Arthur Kornberg, "Review of *Phage and the Origins of Molecular Biology,*" *Nature,* 1966, *260:*206.

[51] See Gunther S. Stent, "That Was the Molecular Biology That Was," *Science,* 1968, *160:*390–5; see also idem, "The Rise and Fall of Molecular Genetics" in *The Coming of the Golden Age* (New York: Westview Press, 1969), pp. 1–74.

discussed in Stent's "Introduction" to POMB in terms borrowed from art history, such as the classical and the romantic periods (presumably corresponding to the 1930s and the 1940s, respectively). More recent periods received scientific or institutional designation, e.g., the "dogmatic" (a reference to the central dogma of unidirectional and irreversible transfer of biological information that was advanced by Crick in the 1950s, which guided some of the efforts to decode the genetic code); and the "academic" (a reference to the academic institutionalization of molecular biology throughout the 1960s).

Since this essay mentioned its origins in a lecture given in June 1967 at the Collège de France in Paris, Borek realized that the previously esoteric or in-group manifesto of the Phage Group and its claim that phage geneticists had originated molecular biology were now being brought not only to worldwide attention (*Science* magazine enjoyed a wide, international distribution, perhaps only second to *Nature*), but specifically to the center of the French scene in Paris. Borek, whose own commemorative essay in the French collective memory opened with an epigram from Ernest Hemingway ("If you are lucky enough to have lived in Paris as a young man, then wherever you go for the rest of your life it stays with you"), turned his attention to Paris and learned that Andre Lwoff had just retired from the Institut Pasteur, where he had spent a remarkable career of almost fifty years, in order to become director of the Cancer Research Institute in Villejuif.[52]

Borek understood that such a momentous retirement of a recent Nobelist provided an excellent opportunity for launching a countermemory of the origins of molecular biology. Furthermore, since Borek had been an early (1951–1952) overseas visitor in Lwoff's department of microbial physiology at the Institut Pasteur, he could assume the mantle of spokesman for Lwoff's overseas' disciples. Borek's previous experience with several volumes of science popularizations meant that, like Stent, he was both able and willing to do the considerable editorial work entailed in publishing a multiauthor volume of memories. And using the same noble justification of assembling a collective tribute to a scientist on the verge of retirement, the contributions of comparative biochemists and microbiologists could be publicly displayed to establish them as founders of molecular biology on the same par with the phage geneticists.

However, the logistical problems associated with assembling the memories of Lwoff's disciples were much greater than those encountered by the three organizers of the American collective memory, who were bound by long-term friendship, compatibility of scientific interests, institutional interdependencies, and a clear sense of hierarchy that prevented conflicts of authority from torpedoing the project. In contrast, even if Borek proved capable of handling the overseas disciples, he still had to secure the cooperation of the French disciples, who were much closer to the honoree and whose knowledge about a career stretching over half a century was indispensable. (The overseas disciples, Borek included, did not spend more than one or two years at a time at l'Institut Pasteur.)

Moreover, authority conflicts were bound to develop between the foreign and the French organizers because the French had a different agenda than the one that animated Borek (namely, to establish comparative biochemistry as the origin of molec-

[52] Borek, "The Transfer of Biological Stress" (cit. n. 48); quotation on p. 157.

ular biology and thus provide a countermemory to the one published by the phage geneticists). Since the French group in microbial physiology at the Institut Pasteur was multidisciplinary and had no serious contenders for the status of founder of molecular biology in France, it had no need for a partisan disciplinary agenda. Instead, it focused its efforts on conquering the institutional arena, first and foremost the Institut Pasteur itself, where they and Lwoff had spent a long time, albeit under conditions of utmost marginality.

Due to geographical distance, differences in cultural mentality, and the national hostility instilled by De Gaulle's regime in the 1960s toward Americans, British, and other "Anglos," the rapport between Borek and potential French co-organizers was much more remote than the close "buddy-ness" shared by the three coordinators from the Phage Group, with the result that their joint product could hardly convey a coherent message.

Furthermore, unbeknownst to Borek, the leading members of the French group were highly divided, especially in the aftermath of the 1965 Nobel Prize that not only propelled Lwoff, Monod, and Jacob into the national limelight, but formally minimized the traditional seniority enjoyed by Lwoff as the only one among the three who had accomplished a great deal in the interwar period and who had a considerable individual record. (Monod's and Jacob's overall output was collaborative, whether with each other or with other authors.)

Possibly under the stimulus of the Nobel citation, which repartitioned their joint interdisciplinary accomplishment in creating regulatory molecular biology into traditional disciplines ("assigning" Lwoff to microbiology, Monod to biochemistry, and Jacob to genetics), the two younger co-laureates, Monod and Jacob, began to prioritize a public carving-out of their own distinct legacies. They had to distinguish themselves not only from the honoree but also from each other, since their names had been closely linked in the early 1960s as authors of the highly acclaimed model of the operon, which had arisen from an intense five-year collaboration that ended in 1963.

In addition to these disciplinary forces pulling apart the three French stars, especially the more ambitious Monod and Jacob who felt compelled to produce some individually stamped legacies in congruence with their new public positions (both received chairs at the prestigious Collège de France, the former in molecular biology in 1967 and the latter in genetics in 1964), there was further divisiveness among them in connection with their efforts to reform l'Institut Pasteur.[53]

Much like the Cold Spring Harbor Laboratory, the Institut Pasteur faced major financial and administrative difficulties in the mid-1960s. Initially, the three Nobelists collaborated with the union of scientific workers to draft a resolution demanding the resignation of the institute's administrative council. However, the resolution was narrowly defeated late in 1965 in the institute's general assembly, as was the candidacy of Lwoff for the directorship of the institute in mid-1966. His departure from the Institut Pasteur early in 1968 reflected his bitterness about his younger associates' lack of effective support, which became tantamount to a betrayal when it

[53] See *Le Monde,* 18–23 December 1965, p. 14; Debre, *Jacques Monod* (cit. n. 30), pp. 316–30; see also Wollman to Stent, 20 December 1965; 10, 14, and 28 January 1966; and 1 and 12 July 1966; in Stent's personal papers, the Bancroft Library (cit. n. 11).

became obvious that two of them, Monod and Elie Wollman, were pursuing directorial positions for themselves.[54]

Last, Borek's relationship with Monod, who had emerged in the aftermath of the Nobel Prize as the most visible of the French laureates and who was to emerge as the French co-organizer, was one of tension and rivalry, since both were domineering types whose large egos had clashed on previous occasions.[55] This difficult relationship provided for minimal or superficial cooperation only. Essentially, it meant a division of labor across linguistic lines with little opportunity for crosscultural integration: Borek was in charge of the English-language contributors, and Monod was in charge of the French-language contributors.

As a matter of fact, Borek had initially planned to avoid Monod altogether and approached instead Elie Wollman, (1917–) who, like Monod, had joined Lwoff's laboratory shortly after World War II. Wollman had considerably more experience in the United States (two years at the California Institute of Technology [Caltech], 1948–1950, and one year at UC-Berkeley, 1958–1959) than anyone else in the Institut Pasteur, and he had better command of colloquial American English than any other French disciple, including the laureates. Moreover, as the honoree's first doctoral student and a war orphan strongly attached to l'Institut Pasteur as a symbolic home, Wollman was uniquely suited to act as the French partner in Borek's envisaged foreign/French coproduction of collective memory in comparative biochemistry and microbiology.[56]

However, at the time Wollman was totally consumed by a desire to participate in reforming the Institut Pasteur, of which he had become vice-president for academic affairs. Thus he preferred to pass the responsibility on to Monod, who as the most outspoken Nobelist commanded a greater public persona anyway. But Monod, too, became involved in reforming the Institut Pasteur, and in pursuing its directorship, once it became clear that that was the only way to accomplish the reforms he had in mind.

Therefore, after mid-1966 Monod and Wollman's focused their efforts on that goal, and eventually achieved it in April 1971 when Monod was appointed director and Wollman became associate director of the Institut Pasteur.[57] This goal, however, was complex and time-consuming to achieve since, unlike the Cold Spring Harbor Laboratory, which had primarily a regional importance in the mid-1960s, l'Institut Pasteur was a national monument (on a par with the Eiffel Tower, as De Gaulle himself put it) so even the prime minister (Georges Pompidou) would get involved in the decision on its directorship.[58]

[54] See note 53. See also Debre, *Jacques Monod* (cit. n. 30); oral history with Elie Wollman at l'Institut Pasteur, June 1986 and June 1988 (cit. n. 12).

[55] Oral history with Sylvia Kerr (cit. n. 48); oral history with Seymour Cohen, June 1995, October 1998 (cit. n. 12).

[56] See notes 53 and 54. Elie Wollman's parents, Eugene and Elizabeth, worked at l'Institut Pasteur (the former as head of a department) as a collaborative couple throughout the interwar years, coauthoring several key papers on phage. In 1943 they were denounced and deported to Auschwitz from where they did not return. Elie Wollman continued his medical studies in the unoccupied zone of France, eventually joining l'Institut Pasteur after the war to earn a doctorate in microbial physiology and genetics with Lwoff. See Andre Lwoff, "In Memory of Eugene Wollman," eulogy delivered in 1958; manuscript with the author.

[57] Ibid. See also the correspondence of Wollman and Stent in the period 1965–1971, Stent's personal papers, The Bancroft Library (cit. n. 11).

[58] See further details in Debre, *Jacques Monod,* (cit. n. 30).

Without detailing here the various difficulties that Borek experienced in seeking to meet the deadlines set by the international committee of four French, one British, and three American scientists, it became clear that Monod was a major source of delays since he was busy with other, more pressing agendas, most notably pursuing the directorship of l'Institut Pasteur and preparing his popular book *Chance and Necessity.*[59] Hence, Monod had little time or interest to pursue slow contributors or integrate the submitted manuscripts.

Indeed, the attrition rate among the invited authors in the French case proved double that in the American case. Moreover, no effort was made either to integrate the French- and the English-language contributions into a meaningful sequence, or to secure both a French and an English version of the collective memory. The resulting volume was not only delayed for three and one-half years (as opposed to one and three-fourths years in the American case for an equal number of essays), it was published as a bilingual mosaic with very limited market appeal in either the United States and France. As Borek put it in September 1970, while defaulting on the deadline of December 1969 for a year:

> Andre's book is indeed a nightmare. Three-quarters of the French have defaulted. Jacques Monod's essay is still not here. I now double sympathize with Winston Churchill. But we will have a fairly decent book by brute force on my part.[60]

To what extent was Monod's conduct (delaying the collective tribute, vetoing a title that explicitly associated the honoree with leadership of an international school of molecular biology [even though by Monod's own account his view was in minority], and using the tribute to legitimize a succession rite grounded in subversion) an opportunistic response to the May 1968 events, which took place a month after Borek approached the French?[61] Section IV will examine the impact of the post-1968 political context on the French collective memory, and compare it with the cold war political context in the early-and mid-1960s in the United States and its impact on the American collective memory.

## IV. CONVERGING POLITICAL CONTEXTS OR "THE SIXTIES": FROM THE COLD WAR AND THE FREE SPEECH MOVEMENT IN THE UNITED STATES TO THE FIFTH REPUBLIC AND THE MAY 1968 EVENTS IN FRANCE

In assessing the political context that shaped the social pretexts and commemorative, symbolic texts of the first American and French collective memories in molecular biology, it is useful to distinguish the three major categories of historical forces that fed into the act of creating these collective memories:

[59] Jacques Monod, *Le hasard et la Nécessité* (Paris: Seuil, 1970, 1973; New York: Atheneum, 1973 as *Chance and Necessity*). See also Louis Althusser, *Philosophie et philosophie spontanée des savants* (Paris: La Découverte, 1974); Gunther S. Stent, "An Ode to Objectivity" in Lwoff and Ullmann, *Origins of Molecular Biology* (cit. n. 30), pp. 231–39; *From Enzyme Adaptation to Natural Philosophy: Heritage From Jacques Monod,* eds. Ernesto Quagliariello, Giorgio Barnardi, and Agnes Ullman (Naples: Einaudi, 1987); Stephen Toulmin, "Review of *Chance and Necessity,*" *New York Review of Books 17,* 6 December 1971, p. 17; Debre, *Jacques Monod* (cit. n. 30). See also Soulier, *Jacques Monod: Le choix* (cit. n. 30); Edward Yoxen, "The Social Impact of Molecular Biology" (Ph.D. diss., Cambridge University, 1978), chapter 9.

[60] Borek to Cohen, 2 September 1970, Cohen's personal papers, APSL (cit. n. 12).

[61] Ibid., 7 May 1968, 4 December 1968. See also Debre, *Jacques Monod* (cit. n. 30). There is no clear answer to this question, as of yet.

1. Scientific breakthroughs in the early- and mid-1960s in the form of a series of far-reaching discoveries that boosted the disciplinary identity and autonomy of molecular biology, further propelling it to the center of public scientifico-philosophical debates;

2. International science policy initiatives that suggested concrete channels for institutionalizing molecular biology as a new, integrative science that required new policies and political agendas, especially in the European arena; and

3. Domestic cultural politics that mobilized students and intellectuals around issues such as the Free Speech movement in the United States (1964) and the May 1968 events in France, and manifested as a general climate of counterculture that rejected tradition and authority while embracing novelty and "rootless hybrids," such as molecular biology.

These historical forces shaping the political context of "the Sixties" are indispensible for understanding the timing of the drives to publish and to capitalize on the first collective memories in molecular biology. They also help elucidate the symbolic outcomes produced by the interaction of the heterogeneous political context with the social position of the main orchestrators, who translated the political opportunity into collective memories, histories, and destinies for the founders of molecular biology, one of the quintessential scientific fields of the twentieth century.

### *Scientific Breakthroughs and their Implications for the Public Imagery of Molecular Biology*

The early 1960s were a period of major scientific change, while the late 1960s, if not quite the "desert" described by Morange, were of much less significance in terms of input into the scientists' efforts to define the new field of molecular biology.[62] Essentially, the three major directions of spectacular progress in molecular biology during the decade were:

1. The solution of the structure of the first two proteins, myoglobin and hemoglobin, which consolidated the structural perspective in molecular biology. This was largely a British effort; four of the first five Nobelists in molecular biology, in 1962, were British (John Kendrew [1917–1997] and Max Perutz [1914–] shared the chemistry prize for their work on the above-mentioned two proteins, and Francis Crick, James Watson, and Maurice Wilkins shared the physiology prize for their work on DNA structure);[63]

2. The discovery of the operon (or coordinated cellular regulation, including the role of messenger RNA) in 1961, which established the regulatory perspective within molecular biology. This largely French effort was rewarded

[62] Morange, *A History of Molecular Biology* (cit. n. 3); see review in Abir-Am, "'New' trends" (cit. n. 3).

[63] See *Nobel Lectures in Physiology or Medicine, 1942–1962* (Amsterdam: Elsevier, 1962) for 1962.

by the Nobel Prize of 1965. The recipients, Andre Lwoff, Jacques Monod, and Francois Jacob, were all of the Institut Pasteur in Paris;[64] and

3. The biochemical decoding of the genetic code between 1961 and 1965, which established a molecular-genetic perspective in which macromolecular biochemistry and enzymology were necessary tools. Marshall Nirenberg of the National Institutes of Health, H. Gobind Khorana of M.I.T. and Robert Holley of Cornell University shared the 1968 Nobel Prize for this line of work.[65]

These three major lines of research determined a new scientific frontier in molecular biology, and left behind the paradigm of transmission genetics and the attendant expectations for paradoxes or profound revelations that had animated the Phage Group. Furthermore, the biochemical decoding of the genetic code, first announced at the International Congress of Biochemistry in Moscow in the summer of 1961, shocked Crick and presumably other theoretically ambitious molecular biologists, Watson included, who championed an abstract, genetic approach to this key problem. Also in 1961, a team based in Watson's laboratory at Harvard lost priority for the description of messenger RNA, a line of work that appeared to be Watson's last hope to participate in another major discovery.[66] (Unlike Crick who made subsequent contributions to the work on the genetic code, Watson has not done any significant work since the double helix.)

The converging losses around RNA work, whether structural or regulatory, raised serious questions about Watson's future prospects. At the same time, his minority status at the 1962 Nobel ceremony as the only recipient without a solid legacy in structural studies, coupled with the inability of the genetically oriented molecular biologists, such as himself, to take part in the biochemically oriented race to decipher the genetic code, imprinted upon Watson the serious ramifications of the new scientific constellation.[67] Unlike those who were too busy exploiting their unexpectedly great fortunes in the present, Watson and his partners in the Phage Group saga had very good reasons to seek solace in the past—to compensate for a far-from-glowing present. As Stent put it so unprophetically in a letter to Delbrück while reflecting on the mood of the Phage Group faithful, "Now that the code is broken what else remains to be done?!"[68]

[64] See *Nobel Lectures in Physiology or Medicine, 1963–1970* (Amsterdam: Elsevier, 1970) for 1965.

[65] Ibid., for 1968.

[66] See the sections on m-RNA in Judson, *The Eighth Day of Creation* (cit. n. 18) and in Rheinberger, *Toward a History of Epistemic Things* (cit. n. 3). The complex story of m-RNA remains to be told; for a preliminary account see Pnina G. Abir-Am, "The Discovery of m-RNA as a Multinational Collaboration of Four Research Schools" (paper read at the First International Conference on Integrative Approaches to Molecular Biology, Cuernavaca, Mexico, 20–27 February 1994); see oral histories by the author with Howard Hiatt, Woods Hole and Boston, Mass.; Walter Gilbert and Mathew Meselson, both of Harvard University's Fairchild Biochemical Laboratories; and François Gros, Paris (cit. n. 12).

[67] See Cairns, Stent, and Watson, *Phage and Molecular Biology* (cit. n. 10), especially the essays by Delbrück, Stent, and Benzer; see also oral history with Brenner in Judson, *The Eighth Day of Creation* (cit. n. 18); Crick, *What Mad Pursuit?* (cit. n. 18); see also review of the latter in Abir-Am, "Nobelesse oblige" (cit. n. 16).

[68] Stent to Delbrück, 17 January 1962 in Delbrück's personal papers, Caltech Archives (cit. n. 11).

While Delbrück, Stent, Seymour Benzer, Sidney Brenner, Crick, and others who regarded genetics as the thoroughfare into molecular biology—while detesting the biochemical route—had no choice but to relocate to another biological frontier (neuroscience), Watson turned to a consolidation of his one and only gain, DNA, via a variety of pedagogical, institutional, financial, and literary-autobiographical strategies.[69] Assembling and disseminating a partisan collective memory, ostensibly as a fund-raising effort for the then-troubled CSHL, was part and parcel of an overall salvaging strategy in which the past had emerged as the one asset that might be capable of arresting the rapid, disturbing scientific shifts in the present. As the editors' preface revealed, such a rapid change might soon place the past beyond recall.

Watson's first step in preventing a fall into oblivion was to write a textbook, an undertaking that busy research scientists never want to do unless they have absolutely nothing else to do. In this textbook, *The Molecular Biology of the Gene* (1965), Watson used graphical and other attractive presentation techniques to depict the rise of molecular biology from a genetic perspective, so as to ensure that future students of molecular biology would perceive that competing and presently more compelling perspectives were derivative of the primordial contribution of genetics.

Second, Watson began maneuvering to obtain the directorship of the Cold Spring Harbor Laboratory, succeeding within half a decade and retaining the position for an unusually long period of time (over 30 years). Moreover, under Watson's directorship, CSHL extended into the wider community via an outreach program called the DNA University. Since the director and his family proceeded to collaborate on various publishing ventures, it is perhaps not out of place to suggest that CSHL might have become not only a pillar of the local community, but also a family business.[70]

Third, Watson began writing his tell-all autobiography, *The Double Helix* (1968), a winner-take-all saga. Though the book angered many leading scientists and triggered more than a few letters to the editors of scientific journals, it eventually became a bestseller, largely because its candor and single-minded, Cinderella-like hero appealed greatly to the countercultural demobilized masses of the late 1960s with their consumerist zeal for scandals in the highest echelons, science included. The *New York Times,* while seeking to explain why President Nathan Marsh Pusey of Harvard had ordered Harvard University Press not to publish *The Double Helix* (calling it "the book that couldn't go to Harvard"), may have precipitated a financial windfall.[71]

It was precisely this sort of fund-raising that Watson and his two co-organizers had initially hoped for through their carefully monitored sales of the collective memory of the Phage Group. (Though the sales were good, CSHL's future was secured

[69] See note 67; also Delbrück, *Mind from Matter?* (cit. n. 46).

[70] See, for example, the photographic album of Cold Spring Harbor produced by Mrs. Elizabeth Watson with text by James Watson and various outreach program literature in CSHL's publications and public relations offices.

[71] Watson, *The Double Helix* (cit. n. 45). See also the large numbers of reviews and commentaries, some assembled in a "critical" edition of *The Double Helix* edited by Stent (New York: Norton, 1980). Many others are available with the author thanks to the generosity of the late D. J. de Solla Price.

by more conventional financial means, such as long-term grants and endowments).[72] At the same time, the collective memory served to back up Watson's later-published, scandalous autobiography by grounding it in a wide memorial substratum qua social consensus on the past. It might have never worked the other way around; i.e., the collective memory had to be a prelude to the autobiography. This further corroborates the necessity of its precocious timing.

Precocious timing was thus tied to two intertwined aspects of the collective memory's singular emphasis on genetics. On the one hand, if it had to back up an autobiography in which the author, Watson, highlighted his ignorance in structural matters as part of his Cinderella character (not to mention his ethically still-unresolved use of structural results by unknowing colleagues), then the collective memory had to supply a credible but different scientific background for a hero who had come from a structural nowhere. This background emerged as a fascination—even obsession—with genetics, a field ignored by most of the British structuralists, yet one that could be portrayed as central to studies of DNA as well as to Watson's contribution to discovering DNA structure—a contribution that, by his own account, was hard to pinpoint in serious structural terms.[73]

On the other hand, the singular emphasis on genetics may have also stemmed from the realization that, regardless of the real or apparent role of genetics (whether classical or microbial) in paving the road to molecular biology in earlier decades, it held the key to capturing leadership within biology in the 1960s. That leadership was held in the early 1960s by organismic biologists, especially evolutionary theorists, who had just demonstrated their accomplishments by creating the evolutionary synthesis and by linking biological evolution with both cultural evolution and evolution of the mind at the worldwide centennial celebration of Darwin's *On the Origin of Species* in 1959.[74] In contrast, Watson's colleague at Harvard, Ernst Mayr, argued at the time that molecular biology was just physics and chemistry. Underscoring the genetic dimension of molecular biology was thus a strategy that was likely to command attention from evolutionary theorists, whose grand synthesis depended on population genetics.[75]

The argument of whether molecular biology belonged to physics-chemistry or to biology was not purely theoretical but had implications for the make-up of biology departments, as well as for the prospects of molecular biologists who sought academic jobs. While molecular biologists with a structural or biochemical background often sought and obtained their own niches in chemical or biochemical establishments, the biological establishment could only be conquered via a route that

[72] See the correspondence between Cairns, Watson, Stent, and Delbrück in 1966 and 1967, and the correspondence between Stent and reviewers of *Phage and the Origins of Molecular Biology,* Stent's personal papers, the Bancroft Library (cit. n. 11).

[73] See Olby, *The Path to the Double Helix* (cit. n. 13); Watson, *The Double Helix* (cit. n. 45); *The Race for the Double Helix* (film), British Broadcasting Corporation.

[74] See V. Betty Smocovitis, "The Darwin Centennial in America," in this volume of *Osiris; idem,* "Unifying Biology: The Evolutionary Synthesis," *J. Hist. Biol.,* 1992, *25:*1–65.

[75] For major statements on the relevance of molecular biology to central issues in organismic biology see Theodosius Dobzhansky, "Biology: Molecular and Organismic," *American Zoologist,* 1964 4:443–52; *idem,* "Genetics: The Core Science of Biology" (a review of five books), *Science,* 1961, *134:*2091–92; Ernst Mayr, "The Recent Historiography of Genetics," *J. Hist. Biol.,* 1974, *6:*125–54; *idem, Toward a New Philosophy of Biology* (Cambridge, Mass.: Harvard University Press, 1988).

challenged the grip of evolutionary theory, biology's main theoretical framework. Molecular genetics promised to be such a route, especially in view of the implications of the genetic code for taxonomy and evolution, not just for protein synthesis and the structural aspects of molecular biology.[76]

This complex and pressing scientific dimension of the wider political context explains both the sense of urgency in the orchestrators' hastened publication of the collective memory of the Phage Group, as well as the group's priority in capitalizing on the past over more accomplished groups, which were much too busy pursuing scientific opportunities in the present to excavate their own past. The orchestration of a collective memory that not only anticipated or "scooped" better-situated contenders for the status of founders, but made money doing so, was the ultimate irony, but it was also an excellent way to compensate for strategic losses in the present.

The orchestrators' gamble that "the past does matter" paid off since the past, as they represented it, put all contender groups on the defensive. Essentially, the contenders had to go public with their own past memories under conditions of disadvantage, while their very joining of the memorialization bandwagon served to validate the first such effort. Moreover, since latecomer efforts rarely display historiographical expertise in deconstructing earlier collective memories, they became de facto countermemories that are suspected of having political agendas, whereas their predecessors' agendas remain hidden by the very precocity of their commemorative pretexts.

The invocation of the past by circulating the first collective memory in molecular biology, however premature or otherwise imperfect it might have been, still had conclusive social, political, and ethical implications, as collective memories of any sort invariably do.[77] If, as Watson stated in his autobiography, DNA was "up for grabs" in the early 1950s, then the public image and institutionalization of molecular biology were also up for grabs in the mid-1960s, when ambitious leaders could transform the era's dramatic scientific changes into new institutions while capitalizing on parallel, dramatic events in both global and domestic politics.

The sense of urgency that enabled the Phage Group spokesmen to orchestrate the first collective memory in molecular biology was also influenced by decisive events outside of science. As detailed in the next section, developments in both international and domestic politics in the 1960s helped shape the production of the first American and French collective memories in molecular biology.

### *International Science Policy for Molecular Biology*

Awareness of the international prospects of molecular biology was greatly amplified, especially for the discipline's first Nobelists in 1962, by events related to the Cuban missile crisis, which unfolded at that time as a climax of the Cold War that threat-

[76] See note 75; also Thomas S. Jukes, *Molecules and Evolution* (New York: Rockefeller University Press, 1966); Ernst Mayr, "From Molecules to Organic Diversity," *Federation Proceedings of the American Society of Experimental Biology,* 1964, *23:*1231–35.

[77] See in particular Steven Knapp, "Collective Memory and the Actual Past," in *Memory and Counter-Memory,* eds. Natalie Zemon Davis and Randolph Starn, special issue of *Representations,* 1989, *26:*123–46; Amos Funkenstein, "Collective Memory and Historical Consciousness," *History and Memory,* 1989, *1:*5–26.

ened to trigger another world war. Though eventually the crisis was averted, it precipitated the efforts to establish an independent European molecular biology facility, later known as the European Molecular Biology Laboratory (EMBL).[78]

Shortly after the first Nobel Prizes in molecular biology were awarded in December 1962, two of the five recipients Kendrew and Watson, flew from Stockholm to Geneva to meet with Leo Szilard and Victor Weisskopf, who were among the most influential figures in science policy. Their purpose was to discuss Szilard's idea for a supranational institute to be modelled on CERN (the European Nuclear Research Center near Geneva), which at the time was under the directorship of Weisskopf. Convinced that the Cuban missile crisis would escalate into World War III, Szilard (then a roving science policy adviser and founder of the Salk Institute in La Jolla) left for Western Europe, which he apparently considered safer in the event of an American-Soviet escalation.[79]

While Szilard had conceived of the idea of a European center for molecular biology, Kendrew, then founding editor of the *Journal of Molecular Biology,* was instrumental in its implementation, and he became EMBL's first director in 1974. Watson knew Szilard from his days as a graduate student at Indiana University in the late 1940s, since Szilard had occasionally hosted gatherings of the Phage Group at his biophysics laboratory at the University of Chicago. Watson also knew Kendrew from his postdoctoral time at the Cambridge Laboratory in Molecular Biology in the early 1950s, where his initial assignment was to assist Kendrew with his studies on the structure of myoglobin.

As a result of this meeting in Geneva, held under the dramatic circumstances of a possible world war, Watson—whose mentors at Harvard were part of the "shuttle" to Washington, D.C. during the Kennedy administration in the early 1960s—had an early exposure to international science policy in molecular biology. He could thus notice that although molecular biology was not yet established in American or other academic institutions (for example, Watson's home institution, Harvard, established a department of biochemistry and molecular biology only in 1967), the discipline could be boosted via science policy means.[80]

Indeed, Kendrew and Weisskopf later organized more formal meetings related to Szilard's initiative, inviting (among others) Delbrück, who was then on leave from Caltech, to establish a new institute of genetics and molecular biology in Cologne, Germany. Delbrück, who considered his role in establishing this institute as the pinnacle of his career, wrote about that meeting to Stent, who later joined Watson in co-organizing the first American collective memory in molecular biology. Both were thus informed of the European initiatives in science policy for molecular biology. At a time when the Europeans were thinking about the future while other Americans were busy capitalizing on opportunities in the present, Watson and Stent came up

[78] See further details in the correspondence of Victor Weisskopf with Kendrew and Delbrück, e.g., Weisskopf to Delbrück, 4 March 1963; Kendrew to Delbrück, 25 November and 15 December 1965; Delbrück to Kendrew, 10 December 1965; 13 March 1967, and 30 October 1967; all in Delbrück's personal papers, Caltech Archives (cit. n. 11). See also Pnina G. Abir-Am, "From Multidisciplinary Collaboration" (cit. n. 13).

[79] See notes 78 and 18.

[80] Oral history with Paul Doty and Helga Doty, Fairchild Laboratory of Biochemistry, Harvard University, 1987 (on file with the author).

with the original idea of pulling ahead in some unexpected direction, such as excavating the past, packaging it, and circulating it as memories for sale.[81]

Despite a difference of only half a decade in the timing of the first French and American collective memories (1971 versus 1966 in terms of publication or 1968 versus 1964 in terms of initiation), the historicity of each was grounded in a different political and scientific configuration. While science policy impinged indirectly on the American political context, the French political context included an alliance of longer standing between the leaders of molecular biology and the state. The inauguration of the Fifth Republic with General de Gaulle as president late in 1958 marked the arrival of a regime that prioritized turning France into a superpower, which soon led to concerted action in science policy headed by one of de Gaulle's companions in the World War II resistance, Pierre Pigagnol.[82]

Most of the ideas underlying French science policy during the early days of the Fifth Republic originated in initiatives of the former government of Pierre Mendes-France. These initiatives were known as the Colloque de Caen, after a science policy gathering of leading scientists and political figures held in 1956. However, their implementation began only after 1959, when a stable government with a specific mission to restore France's international grandeur came to stay in power for two decades.[83]

As one of several disciplines targeted for special support under the auspices of a newly established governmental agency attached to the Office of the Prime Minister, the Délégation Générale pour la Recherche Scientifique et Technique (DGRST), molecular biology received a great boost. Many of the leading French molecular biologists, including Lwoff, Monod, and Jacob, as well as other Pasteuriens, most notably Elie Wollman and François Gros, were members of the DGRST's science policy committee in molecular biology from its inception in 1959. Monod and Wollman were particularly active, serving as vice-chairs of the committee. When after six years of rubbing shoulders with political power in the name of science policy, three of the committee's members returned from Sweden as the Nobel laureates, their views commanded considerable public attention. Indeed, after 1965, Monod displaced Rene Wurmser, a veteran biophysicist from the l'Institut de Biologie Physico-Chimique, as chair of the DGRST committee on molecular biology.[84]

Science policy channels such as the two Colloques de Caen (the second took place in March 1968) and the DGRST committee on molecular biology gave Monod and Wollman a systematic exposure to science policy as well as to science politics, and this experience eventually brought them together in a crucial alliance for reforming the Institut Pasteur. The road proved slow, in part because Monod's critique of French governmental policy toward science alienated Prime Minister Georges Pom-

[81] See the correspondence of Stent, Cairns, and Watson with Delbrück in the 1960s in Delbrück's personal papers, Caltech Archives (cit. n. 11), and between the first three in Stent's personal papers, the Bancroft Library (cit. n. 11).

[82] See oral histories by the author with Pierre Pigagnol and Philip Ziegle, Institut Pasteur, Paris, June 1986; see also Jean-François Picard, *La Republique des savants* (Paris: Flammarion, 1991); *Cahiers pour l'histoire du CNRS* (Paris: Editions du CNRS, 1990), especially the essays by Xavier Polanco and Jean-Paul Gaudilliere. See also Morange, *A History of Molecular Biology* (cit. n. 3).

[83] See *The Fifth Republic at Twenty,* eds. William G. Andrews and Stanley Hoffman (Albany, N.Y.: New York University Press, 1981).

[84] See note 82; see also the archives of the Délégation Générale pour la Recherche Scientifique et Technique, the Research Ministry, Paris.

pidou, but eventually, in 1971, Monod and Wollman became director and associate director, respectively, of the Institut Pasteur.

Still, their activism in science policy was not limited to the national scene as, among other efforts, they made an overture to EMBL to accept a site in Nice. This role in science policy had an impact on their position on the international organizing committee for the first French collective memory, for example, in defining its commemorative pretext in relation to the Institut Pasteur, the site and source of their own power. However, the French collective memory was also influenced by domestic cultural politics emanating from the May 1968 events, especially since they formalized the gradual drift that was developing between Monod, the co-organizer, and the honoree.

### *Issues of Domestic Cultural Politics: The Free Speech Movement at UC-Berkeley (1964) versus the May 1968 Events at the University of Paris*

Issues of domestic cultural politics, especially campus revolts, loomed large in the background of both the American and the French efforts to assemble collective memories in molecular biology. On the American side, the Free Speech movement that erupted under the eloquent leadership of Mario Savio (1942–1996) in 1964 on the Berkeley campus of the University of California greatly affected Gunther Stent, who had emerged by then as a popular teacher of microbiology whose methodical course syllabi were sought by colleagues in both the United States and France.[85]

Stent, who had been at UC-Berkeley since 1952, had just completed one of the very first textbooks in molecular biology, *The Molecular Biology of Bacterial Viruses* (1963).[86] His considerable effort as coeditor of the first collective memory of molecular biology can be seen as part of an effort to ensure the survival of a cultural heritage that appeared to be threatened by the drastic, violent, and uncompromising, nature of the student antiestablishment movement that was to expand from Berkeley to all American campuses throughout the 1960s. As he stated in one of his essays on the cultural implications of molecular biology, the impact of the Free Speech movement led him to reconsider the foundations of civilization:

> The trauma inflicted on the University of California professors by the 1964 Free Speech Movement of the Berkeley students forced upon me and many of my colleagues an agonizing reappraisal of earlier and now obviously outdated attitudes regarding our life's work. At the outset, the prevalent faculty view of the Movement had been that it was just one more tempest in a teapot. But by the time its revolutionary drama—with our then Chancellor in the role of Louis XVI, our Academic Senate as the Estates-General, our Administration Building as the Bastille, and our charismatic student leader Mario Savio as Danton—had run its course, most of us had finally appreciated the cosmic significance of those events. For it appeared to us that Berkeley had become the stage on which the global future of higher education was being acted out.[87]

Stent took Monod (who conducted numerous research trips to the United States and to California in particular, having been a founding member of the Salk Institute in

[85] See William J. Rorabaugh, *Berkeley at War: The 1960s* (Berkeley: University of California Press, 1989); 1968: *A Student Generation in Revolt,* ed. Ron Fraser (New York: Columbia University Press, 1988); Terry Anderson, *The Movement of the Sixties* (New York: Norton, 1978); *The '60s Without Apology,* eds. Sohnya Sayres et al. (New York: Basic Books, 1978).
[86] Stent, *Molecular Biology of Bacterial Viruses* (cit. n. 43).
[87] Stent, *Golden Age* (cit. n. 51), p. ix.

La Jolla) on a guided tour of the beatnik-turned-hippie center of counterculture in the Haight-Ashbury enclave in San Francisco. Though Monod was greatly interested in the Free Speech movement at Berkeley, understandably he was even more affected by the May 1968 events in Paris.[88] Not only had he been a student at la Sorbonne, but since 1959 he had also taught there. Moreover, as someone who had been long active in science policy, he had already been involved in various efforts to reform the French universities.

Once the students' rebellion of 1968 had begun, Monod crossed the barricades to help wounded female students reach the hospitals. He also met with members of the National Assembly (who did not seem to appreciate the gravity of the situation), sent messages and eventually a collective demand to resign to the minister of education (Alain Peyerfitte), and a cable to the president of the Republic (de Gaulle) to release the revolting students. He also met with the rector of the Université de Paris and took part in joint student and faculty committees. The May 1968 events resonated with Monod's desire to get involved in politics and they came at a time when he could afford to do so, having reached the peak of the Nobel Prize three years earlier:

> In France, at the beginning of May 1968, the risks of an upcoming student uprising were considerable . . . Daniel Cohn-Bendit, at the head of students from Nanterre [a suburb of Paris] lands in the yard of the Sorbonne, but is dislodged by the police, against the rule that the police cannot operate in a university. Hence the situation could not but deteriorate with street demonstrations and arrests multiplying . . . Jacques Monod arrives at the National Assembly at the head of a delegation of professors from the faculty of sciences. He insists on bringing an urgent message to Alain Peyerfitte demanding to make public the government's decision to give amnesty to the rebelling students . . . Monod decides that his place is on the barricades . . . He is in the middle of the demonstrators . . . He guides the journalists and questions the head of police . . . Monod . . . discovers the lyrical character of the great hours of History, when rebellion only defends justice. This evening his name was on everyone's lips. He is the savior! He is the one to incarnate the triumphant science in the midst of the wounded, the oppressed, the misunderstood.[89]

Since the students' revolt was part of a larger political crisis, Monod's activism reflected a wider interest in politics per se, not just in academic reforms or science policy. In this regard, he differed from his former mentor Andre Lwoff, who being almost a decade older assumed the stance of a baffled bystander, as if the May 1968 events could only be *deja vu* for his generation, which had its formative moments in the interwar period. Student slogans such as "*Vive la révolution surrealiste*" ap-

[88] Debre, *Jacques Monod* (cit. n. 30).

[89] Ibid., p. 284 (author's translation). The French original is: "En France, au debut de mai 1968, les risques de survenue d'un soulevèment étudiant sont considerables . . . Daniel Cohn-Bendit, à la tête des étudiants de Nanterre, débarque dans la cour de la Sorbonne, d'où il est delogé par les CRS, contrairement à la regle qui veut qu'une université ne puisse être investié par les forces de police. La situation des lors ne peu que dégénerer et les manifestations de rue se multiplient, en même temps que les arrestations . . . Jacques Monod se rend à l'Assemblée Nationale à la tête d'une délégation de professeurs de la faculté des sciences. Il insiste pour que l'on porte un message urgent à Alain Peyerfitte, lui demandant de rendre publique l'intention du gouvernement d'amnistier les étudiants frondeurs . . . Monod decide que sa place est sur les barricades . . . Monod est au milieu des manifestants. Il commande aux journalistes et, saisissant les micros, interpelle le préfet de police . . . Monod . . . retrouve instinctivement le lyrisme des grandes heures de l'Histoire, quand seule la révolte prend la defense de la justice. Ce soir-la, son nom est sur toutes les levres. C'est lui le sauveur! C'est lui qui incarne la science triomphante au chevet des blessés, des opprimés, des incompris . . ."

peared to his generation as a pale imitation of the original surrealist revolution they had witnessed in Paris in the 1920s. Indeed, as his contemporary Raymond Aron (1905–1983), an astute observer of the political arena and a professor of sociology, commented (evidently with his own and the honoree's generation in mind), it was difficult to distinguish the real revolution from the unreal spectacle:

> Between 15 May and 30 May, national life was effectively paralyzed, most factories were not operating, workers were occupying their factories. . . . [T]he strike even reached the public services. . . . [T]here was, beyond the strikes and student speechifying, a genuinely political crisis, some deputies of the majority attacked the president, others attacked the prime minister. . . . [I]n the universities and even in the *lycées,* professors entered into more or less passionate conflicts with one another, some following or even leading the angry students, others raising the dam against the wave of demagoguery, utopianism, or dreaming, inflated by the illusion of living in historic times . . . 1968: an intellectual revolt of us against structure, of Sartre against Levi-Strauss . . . of "praxis" against institutions, of leftism against the Communist party . . . The ideological language of May, from students to workers, cut across party programs. It revived and popularized themes that could easily be found in books of cultural criticism . . . *One Dimensional Man* by Herbert Marcuse contained most of the themes that stirred the students: commercial society, forced consumption that was indispensible for the industrial apparatus, pollution, social repression, waste confronted with poverty, and so on. . . . [S]tudents and workers, simultaneously and separately, in a state of revolt; confrontations between police and demonstrators day after day with no shots fired on either side. . . . [F]inally, aside from the battles, there was often an atmosphere of gaiety, as though it were a holiday.[90]

Once the revolt ended, Monod continued to participate in university reforms together with Laurent Schwartz and Pierre Bourdieu, as members of a committee charged with following the recommendations of the policy meeting at Caen. Nine months later, in February 1969, Monod gave a series of lectures at Pomona College in southern California which sublimated the recent experience of social disorder and crisis into a vision of nature and of molecular evolution as a combination of chance or randomness with necessity or order. Nevertheless, in tune with the times, the concept of chance—with its connotation of freedom or escape from the structural determinism that reigned in the preceding decade—achieved rhetorical primacy. Billed as a natural philosophy of the universe in a world without causality, Monod's *Le hasard et la nécessité,* published in April 1970, became a bestseller that articulated the counterculture's brand of cosmic alienation and aggressive ethical despair.

Since Monod's emphasis on chance in relation to a new ethics of scientific objectivity directly challenged other teleological systems as sources of ethics (especially Marxism and Catholicism, the two main pillars of the French intellectual establishment), Monod's book eventually emerged as a media event. It further propelled Monod, whose name had become a household word as a result of his role in the May 1968 events, into the center of a large philosophical controversy, once the major

[90] Raymond Aron, *Memories* (Cambridge, Mass.: Harvard University Press, 1990); French edition as *Memoires:50 ans de réflexion politique* (Paris: Juillard, 1983, 1990), quotation on pp. 322–5 of the American edition. See also Luc Ferry and Alain Renault, *French Philosophy in the 1960s* (New York: State University of New York Press, 1988; French edition as *La Pensée 68* (Paris: Gallimard, 1985, 1988); Edgar Morin et al., *Mai 68: La brèche* (Paris: Complexe, 1988); Patrick Rambaud, *Les aventures de Mai* (Paris: Grasset, 1998).

Marxist structuralist philosopher of the 1960s, Louis Althusser, mocked his book as the "spontaneous philosophy of a scientist."[91] Nevertheless, Monod's bold and provocative bestseller appealed to the demobilized masses of younger readers as a work of scientific popularization that resonated with the ideologies of liberation that so animated them (and had animated Monod, just two years earlier, though he had since come to terms with the fact that society could not be reformed overnight).

By contrast, the French honoree's *L'ordre biologique,* published in 1969 as an updated translation of his *Biological Order,* a series of lectures delivered at M.I.T. in 1961 and published in English in 1962, remained in both title and overall message a eulogy to balance and order, whether in nature or society. This sober and balanced message had no appeal for the still-excited and not quite demobilized intellectual masses of the late 1960s, captive as they remained of the brief, quasi-revolutionary experience of May 1968 and its lasting legacy of antiestablishment sentiment.[92]

The profoundly different trajectories of the books by Monod and Lwoff in 1969 and 1970, coupled with Monod's activism during the May 1968 events and his subsequent appointment as director of the Institut Pasteur in 1971, created a complex situation. Their longstanding (exactly four decades, 1931–1971), asymmetrical relationship as patron and protégé was suddenly inverted, with Monod emerging as an uncontested figure of power and fame while the honoree was in "exile" at a less prestigious research institute. This inversion explains Monod's strange conduct, which ranged from delaying the French collective memory's publication until after his own book was published, ignoring the foreign coeditor, and vetoing the title favored by other authors. Last, but not least, Monod's commemorative essay was devoid of any insight into the honoree's intriguing career, reducing Lwoff to a mere genealogical bridge between Monod and Pasteur, but it abounded in presentations of his own career.

Monod's rise as a literary celebrity in 1970, on the coattails of molecular biology and the counterculture movement, parallels Watson's experience in 1968 when *The Double Helix* stirred a scandal once Harvard University Press refused to publish it following litigation threats by Crick and Wilkins. This book, too, emerged as a bestseller, as its exposé of the behind-the-scenes world of scientific discovery held a special appeal for the young, rebellious generation of the late 1960s. The book's

[91] Althusser, *Philosophie et philosophie spontanée* (cit. n. 59).

[92] See, for example, this quotation taken from p. 8 of the preface to the French edition of Andre Lwoff, *L'ordre biologique* (Paris: Fayard, 1969): "In France, in the course of this month of May 1968, a certain type of order was disturbed. A storm released the repressors [term from molecular biology for a small molecule which can prevent transcription of the DNA if it interacts with the promoter site on the DNA that regulates transcription], the operating genes lost control of the operons [coordinated system of structural and regulatory genes]. New molecules wished to take the place of old ones and contested the system of regulation. Of all this, unintended events had resulted, interesting and to say it all, quite remarkable.

Apparently, there is nothing in common between a molecular society and a human society . . . And then, the interactions which govern the molecular and cellular order resemble the phenomena which ensure the functioning of human society: the molecules and the men [generic for both men and women] are subject equally to tough constraints. Finally, the revolting molecules and the parasitic molecules have their equivalents in human societies." (*L'Ordre* was initially published in English in 1962 by M.I.T. Press since it contained the Compton Lectures given by Lwoff at M.I.T. in 1960. The French edition updated the science but deleted the footnotes, thus making it sound more like a manifesto than a scholarly book, especially since by 1969 the contents were not only no longer new but had lost the power of surprise that they had in 1962 when first published. See also Andre Lwoff, "The Prophage and I," in Cairns, Stent, and Watson, *Phage and Molecular Biology* (cit. n. 10), pp. 103–12.

heroes, who "take all," were members of a younger generation who achieved their scientific success because they departed from the outdated orthodoxies of their elders (who missed the most important discovery of the time in part because they ran the scientific world by gentlemen's agreement, rather than by cold-war style competition).

The literary success of Watson in the United States in 1968 and of Monod in France in 1970 solidified their personas as symbols of the molecular revolution in biology and as public representatives of the new discipline. Both soon became directors of leading institutions of molecular biology in their respective countries. However, attaining these positions also represented the culmination of several years of institutional entrepreneurship on their part. This activity was more enabled than constrained by the volatile political context of the 1960s, and it would bear directly on their role as organizers of commemorative occasions for honorees who must be sent to posterity so that they could present themselves rather quickly as new leaders—not only de facto, but also de jure—of a generation in search of novel idols who were willing and able to reinvent their own past, and by derivation, the past of the discipline they symbolized.

Finally, the commemorative texts assembled by the organizers-turned-editors began circulating detached from their initial commemorative occasions, preserving the occasions yet emerging as collective memories that became incorporated into the historical consciousness of molecular biology for future generations of scientists and other innocent readers.

## V. CONCLUSIONS: FROM COMMEMORATIONS TO COLLECTIVE MEMORY: FROM COLLECTIVE MEMORY TO COMPARATIVE HISTORY

This saga of the assembly of the first American and French collective memories, as well as the story of the relationship between them, identified major analytical constituents of the commemorative process and its product, the collective memory, while raising two major questions for future research: a) How the commemorative activity, encompassing symbolic texts, social pretexts, and political contexts, eventually becomes a collective memory shared by generations of scientists and even metascientists who were not there when the commemorations were staged from various timely mixtures of scientific, social, and political contingencies, but who came to accept it as an "actual past", even as a historical consciousness;[93] and b) How collective memories (especially the two examined here, which are part of a wider set of such publications in molecular biology) may inform the writing of a comparative and global history of this field, i.e., one that takes into account a wider range of collective memories and countermemories while transcending the national, disciplinary, or genealogical agendas that constrained the first American and French collective memories.

In the following section, the process of consolidating the three constituents of the commemorative projects (the political contexts, social pretexts, and memorializing texts) into decontextualized, quasi-objective, sentimental, cohesive, utopian, collective memories is summarized by highlighting the asymmetry in the deletion of some agendas and the encoding for lasting retention of others.

[93] For these concepts see Knapp, "Collective memory and the actual past" (cit. n. 77) and Funkenstein, "Collective memory and historical consciousness" (cit. n. 77).

### *From Commemorative Occasions to Collective Memories: The Selective Encoding of Genealogical, Disciplinary, and Institutional Pretexts*

This essay argued that the first two commemorative occasions in molecular biology in the United States and France served as pretexts for: a) assembling and circulating the collective memories of leading groups, while reflecting the agendas of new scientific leaders with regard to the public image of the discipline; b) representing molecular biology's origins via a founder figure selected for genealogical reasons of self-legitimation; and c) conquering a key institution from which power and authority could be exerted over the new discipline. In turn, those agendas, termed social pretexts, were shown to derive from a political context composed of drastic scientific change, international science policy, and domestic cultural politics.

Furthermore, the contingencies constituting the commemorative process were shown to range from the initial idea in the mind of a would-be organizer, whose better interface with the political context of the time made him grasp prior to others the key role that a certain past may play in scientific affairs; to his practical capacity to secure immediate allies who together could mobilize three dozen or so scientists, obtain their memories, and assemble and circulate them in the form of a collective memory.

The collective memory or quasi-objective interpretation of a certain past, by virtue of being the first such use of the past, could not only reconfigure a present that had been moving fast in new directions and leaving behind those whose best work was in the past, but would require that contender groups respond to the challenge of monopolizing the past. Indeed, latecomer responses could only be defensive, since the element of surprise was fully monopolized by the first who went "up for the grabs." The pursuit of a multidimensional agenda—genealogical, disciplinary, institutional—in a complex, commemorative format that attenuated the political edge of the enterprise with an uplifting cultural-historical patina invariably concluded with a symbolic transaction. The initiator-organizer's act of offering the published, leather-boxed volume of memories to its honoree became the high point of a ceremony which included the presence of dignitaries, oratorical speeches, banquets, receptions, and book signings. Moreover, it represented a subtle change of guard in which the honoree was accorded symbolic power and a place in posterity in exchange for vacating the leadership slot for the organizer.

The long commemorative process (two to three and a half years) can be seen as a prelude to or investment into the short (usually day-long) ceremonial outcome, which is the conversion of a new leader into a legitimate one by virtue of an elaborate collective and public demonstration of proper genealogical descent from a communal symbol (the honoree), whose premature transformation into a symbol and forced departure for posterity underlie the commemorative effort.

Since the majority of the scientists who agreed to publicly share their memories (in the belief that they were simply honoring the birthday of a senior, often admired, colleague) could not have known of the pursuit of hidden agendas by the organizers qua presumptive heirs, it follows that the symbolic texts of the collective memory were constructed via a complex process of negotiation among the organizers. In the same vein, the ratification of the individual agendas of the new leaders depended on their ability to align the many ordinary memories of various contributors through social networking and rhetorical devices embedded in the organizational and editorial processes. The process of ratification included:

1. The composition of *totemic memories* that survey the discipline and pronounce selectively on its past as reflected in the honoree's career and as refracted through the organizers' agendas;

2. The exclusion of *taboo memories* that might invalidate the totemic memories, such as the memories of scientific rivals, women scientists, and others who might supply countermemories; and

3. The inclusion of *visual memories* that condense the master memories in more memorable forms.

The genealogical and disciplinary agendas of the new leaders can be recovered, at least in part, from the published commemorative texts through a complex method of deconstruction. Still, those agendas (as well as the institutional agenda, which was invariably deleted from the published memories) had to be corroborated, clarified, and amplified by archival research that captured the behind-the-scenes interaction between the organizers qua editors of each commemorative occasion, as well as their relationship to the honoree and to various disciples who were not privy to the pragmatic triple agenda (genealogical, disciplinary, and institutional) that guided the lofty commemorative effort.

The question persists as to what extent the collective memory, which has so many illustrious contributors, is indeed a collective memory as opposed to being a social or collective ratification of the totemic or singular memory of a dominant figure, usually the heir apparent. How were the multiple contingencies and subjectivities of the commemorative process transformed into historical necessities, or into a collective memory accepted by and large as a reflection of the actual past? To fully answer these questions, further research is needed into the reception and actual use of these collective memories by scientists and metascientists.

Detailed research into the assembly process preceding the publication of each collective memory shows that, despite considerable control of the commemorative enterprise by the organizers (especially the heir apparent, who supplied the genealogical, disciplinary, and institutional agendas), the text of the collective memory contains sufficient information to deconstruct those agendas and recover missing, excluded, or taboo memories. This was particularly so in the French case, since the organizers did not act in unison and hence did not have the same degree of control as that mustered by the business alliance of the American organizers.

The implications of the published text and of the archival and oral resources for uncovering various countermemories (as well as the existence, side by side, of several collective memories, e.g., the first American and French collective memories) for the historiography of molecular biology pertain to transforming such memorialization efforts into resources for a global and comparative history. This can be done once they are deconstructed for what they are, namely, partisan representations of the past designed to secrete personalized genealogies, disciplinary programs, and institutional futures, from the viewpoint of heirs apparent. This is the only type of history that avoids reducing the history of an integrative discipline such as molecular biology to a single honoree (or unstructured amalgam of leading figures), a single "mother discipline," or a single "father institution," as each collective memory invariably does. This essay concludes with some suggestions for using collective memories, especially the two examined here, for writing a comparative history of molecular biology that benefits from the historical authenticity embedded in the genealogical, disciplinary, and institutional choices of each collective memory.

### *From Collective Memory to Comparative History*

While the American and the French collective memories became the core of historical and national consciousness of the discipline's past in their respective countries (much as a previously examined British case study did in the United Kingdom); this study aimed to uncover new historiographical vistas through a comparative, as well as more global, outlook. This section outlines further ways to move from this study of collective memories to a global and comparative history of molecular biology.

The very first genealogical choices in molecular biology were traced to honorees born in the first decade of this century—Max Delbrück (1906–1981) and Andre Lwoff (1902–1994). Notwithstanding important differences that dominate the American and French collective memories, most notably the revolutionary versus evolutionary models of scientific change, the honorees' career patterns illuminate the historicity of the molecular revolution in twentieth-century biology, a revolution of enormous ramifications for our time and for the coming century. A generational pattern emerges if we extend this observation to honorees of other collective memories in molecular biology, for example to the American Linus Pauling (1901–1994), the British J. Desmond Bernal (1901–1971) and Dorothy C. Hodgkin (1910–1994), and the French Jacques Monod (1910–1976). These honorees turned future leaders (whether social or intellectual) not only had their formative careers in the interwar period, especially in the 1930s, but often did major work related to the rise of molecular biology either in the 1930s or during World War II.

Following the rapid rise in support for science after World War II, these six honorees had the opportunity to inspire many younger scientists who flocked to their stimulating research programs. With the exception of Bernal, to whom the four 1962 British Nobelists expressed profound indebtedness (he directly recruited two and inspired all four), all the honorees won Nobel Prizes in the 1960s. Thus, they emerged as symbolic choices for collective memories (the two Americans were honored in the mid- and late-1960s, the two French in the 1970s, and the two British in the 1980s).[94]

This pattern supports the generalization that the century-long molecularization of biology can be best understood as a punctuated equilibrium sustained by three postwar settlements that favored, in each case, a different science as the cutting edge of this molecularization: namely biochemistry (as a biological spinoff from chemistry, the queen science in World War I); molecular biology (as a biological spinoff from physics, the dominant science of World War II); and biotechnology (as a biological spinoff of computer science, which reigned during the cold war).[95]

Beyond the genealogical level, extending the study of the first American and French collective memories into a comparative disciplinary history enriches and corrects the current historiography of molecular biology, which remains confined, to a significant level, to different disciplinary origins in different countries. For example,

[94] See Abir-Am, "Entre mémoire et histoire" (cit. n. 3).
[95] See Abir-Am, "The Molecularization of Biology" (cit. n. 13).

while the title of the first American collective memory encodes a claim that the origins of molecular biology were to be found in genetics, phage stands as a prop for genetics in order to defuse the provocation of a too-transparent attempt to appropriate the glory of a multidisciplinary field, for only one of its constituent disciplines, especially one whose classical phase was both remote from and antagonistic to anything molecular. In a similar vein, "microbes" in the title of the French collective memory stood as a prop not only for microbiology but also for an integrative, microbial-physiological approach that focused on the unity of basic life phenomena, rather than on the microbes' better-known association with infectious diseases.

Though neither the American nor the French collective memories chose to make explicit their respective disciplinary manifestoes, coding them instead under two categories of microorganisms, the key role of such disciplinary manifestoes can be retrieved by looking at the thematic scope of each collective memory. In contrast to the American collective memory, which argued that molecular biology originated in the transformation of forms of genetics (radiation, phage, bacterial, and molecular), the French collective memory suggested that molecular biology originated through a synthetic, integrative approach (physiological, biochemical, and genetic) to microorganisms.

The question persists as to how these profoundly different versions can be reconciled for a comparative history, especially since there was a certain degree of overlapping memories across the American and the French groups and considering the fact that Lwoff's induction of lysogeny in 1950, directly challenged the foundations of the American research program. Plausibly, a comparative history has to pay greater attention to the temporal succession of dominant disciplinary inputs into the integrative synthesis underlying molecular biology, so that different disciplines may play a role at different times. This divergence of the disciplinary manifestoes can be clarified by looking at the institutional agenda of each collective memory, since the manifestoes derived in part from linkage between the institutional and the disciplinary agendas.

Indeed, both the American and the French collective memories had institutional agendas built into the rationale for the commemorative occasions that led to their publication. While the American collective memory used the twenty-first anniversary of the first phage summer course that the honoree initiated at the Cold Spring Harbor Laboratory as its rationale, the French collective memory used the fiftieth anniversary of the French honoree's arrival at the Institut Pasteur. How central were these institutional agendas and for whom?

Indeed, institutional agendas, especially Watson's conquest of the directorship of CSHL and Monod's of the directorship of the Institut Pasteur, were the driving forces behind the organizers' initiatives. In a way, the disciplinary agenda of each collective memory was linked to the historical mission of each institution, since CSHL was established as a laboratory of genetics and eugenics in 1890, while the Institut Pasteur was established in 1888 as an integrative or multi-disciplinary laboratory addressing the then newly discovered microbial world.

Each heir apparent thus signalled his compatibility as would-be director while linking the historical legacy of his targeted institution to the latest scientific chic, even though molecular biology as a multidisciplinary synthesis extended well beyond the initial public missions of both institutions. Therefore, extending the comparative analysis to include institutions other than those linked to the personal aspirations of early disciplinary spokesmen is needed, in order to detach the complex institutional dimension of molecular biology from its first collective memories.

To conclude, major historiographical questions raised by this essay, e.g., the r/evolutionary nature of the rise of molecular biology; the historicity of its origins over a quarter century or half a century; the "paradoxical" spirit of the Phage Group versus the *gai-savoir* of the French group; the pretentious impulse of imitating the theoretical revolution in quantum physics as embodied in the Copenhagen spirit versus the culturally decadent or progress-challenging notion of evolutionary regression in lower life forms; and the institutional loci and disciplinary images of molecular biology remain to be expanded to several other groups that produced collective memories. Possible subjects include the American group of structural chemists, the British groups of X-ray crystallographers, the French group in cellular biochemistry; and the relationship between them. For only a "poetics and politics" of the international scientific "*vagabondage*" of genealogies, models, experiments, institutions, policies, and social movements in this century may capture the brave, new world of molecular biology.[96]

[96] On the social epistemology of *vagabondage* see the Fourth Graduate Students Conference on "Vagabondage: The Poetics and Politics of Movement," UC-Berkeley, 13 March 1996; see also the keynote address there by Pnina G. Abir-Am, "From Vienna to New York via Berlin and Paris: The Multicultural Poetics and Institutional Politics in the DNA Saga of Erwin Chargaff (1905–)."

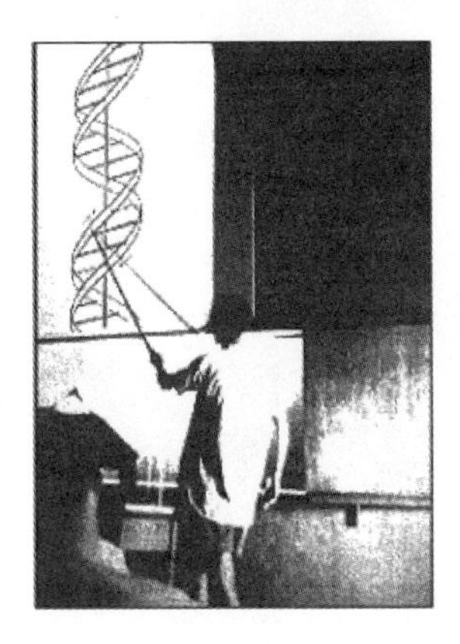

*Left to right, top to bottom: Delbrück and Luria, 1941; Delbrück and Luria, 1953; Watson; Delbrück (Courtesy of the Archives, California Institute of Technology); Szilard and Hershey, 1951; Throop Hall, California Institute of Technology (Courtesy of the Archives, California Institute of Technology); and Cold Spring Harbor. For further discussion of these images, see pp. 340–341.*

*Left to right, top to bottom: Lwoff, 1928; Lwoff, 1930s; Lwoff, 1950; Lwoff, 1959; Lwoff, late 1960s; caricature of Lwoff by Chatton; Chatton; a coffee break; Lwoff and Trouzet; Ephrussi, Faure-Fremiet, Millot, and Lwoff; another coffee break; a close one; Ephrussi and Lwoff. For further discussion of these images, see pp. 341–342.*

# Notes on Contributors

**Pnina G. Abir-Am** is the author of numerous essays on the history of molecular biology and has received the HSS History of Women in Science Prize. She is editor of *La Mise en memoire de la science: Pour une Ethnographie historique des rites commemoratifs* (Paris: Editions des Archives Contemporaines, 1998) and coeditor of *Creative Couples in the Sciences* (1996) and *Uneasy Careers and Intimate Lives, Women in Science, 1789–1979* (1987, 1989) both from Rutgers University Press. Currently a director of research at CNRS, Paris, she has taught history of science in Canada, Israel, and the United States, most recently at the University of California, Berkeley, where she is an NSF visiting associate professor. Abir-Am is completing a book titled *Research Schools of Molecular Biology in the United Kingdom, the United States, and France: Comparing Transnational Innovation in Twentieth-Century Science* and is collaborating on a new project: "The Logic of Life at Large Scale: A History of 'Big Science' in Biology."

**Daniela Barberis** is a Ph.D. candidate in the Conceptual Foundations of Science at the University of Chicago. Her dissertation is titled "Durkheim and Neo-Kantian Philosophy in France in the Early Twentieth Century." She has also worked on French psychology, especially on Ribot and Charcot, and on the topic of hysteria and hypnosis.

**Liliane Beaulieu** teaches at Collège du Vieux-Montréal in Québec and is a member of the Université de Montréal's Centre de recherches mathématiques. She has written on various aspects of the collective life and work of the Bourbaki group of mathematicians, a long-term project in which she is currently engaged.

**Mara Beller** is the Barbara Druss Dibner Professor in History and Philosophy of Science at Hebrew University of Jerusalem. Her book *Quantum Dialogue—The Making of a Revolution* was published by the University of Chicago Press in 1999. Her research interests include communication in science, sociology of knowledge, and the history of twentieth-century physics.

**Clark A. Elliott** served for a number of years as associate curator of the Harvard University Archives and most recently was librarian of the Burndy Library at the Dibner Institute for the History of Science and Technology in Cambridge, Massachusetts. He has published several reference works on American science and, with Margaret W. Rossiter, coedited *Science at Harvard University: Historical Perspectives* (1992).

**Owen Gingerich** is Professor of Astronomy and the History of Science at the Harvard-Smithsonian Center for Astrophysics in Cambridge, Massachusetts. His research has concentrated on the dissemination of the Copernican cosmology, and for many years he has worked on *An Annotated Census of Copernicus's* De Revolutionibus *(Nuremberg, 1543 and Basel, 1566).*

**Stanley Goldberg** was a longtime consultant to both the National Museum of American History and the National Air and Space Museum of the Smithsonian Institution. Earlier in his career, he had held teaching posts at Antioch College, the University of Zambia, and Hampshire College. He wrote *Understanding Relativity: Origins and Impact of a Scientific Revolution* (1984) and many other pieces addressing the history of modern physics and the interplay among science, culture, and politics. At the time of his death in 1996, he was working on a biography of General Leslie Groves, the commanding officer of the Manhattan Project. Robert Standish Norris has agreed to bring this project to completion as a book to be published by Steerforth Press.

**Joy Harvey** holds a doctorate from Harvard University in History of Science. She has taught at Harvard, Skidmore College, Sarah Lawrence College, and Virginia Tech. For a number of years she served as associate editor of the Darwin Correspondence Project, Cambridge University Library. Her book on Clémence Royer, *Almost a Man of Genius,* was published by Rutgers University Press in 1997. Currently, she is editing a two-volume *Dictionary of Women Scientists* with Marilyn Ogilvie for Routledge Press that is scheduled to appear in 2000.

**George E. Haddad** is a graduate student in the Department of the History of Medicine and Science at Yale University and a senior resident in the Department of Internal Medicine at Yale–New Haven Hospital. His dissertation examines storytelling and memory in professional medical culture. He plans to pursue a dual career as a clinician and medical historian.

**Dieter Hoffmann** is a research scholar at the Max Planck Institute for the History of Science and *Privatdozent* at the Humboldt University in Berlin. His work is focused on biographies and institutions in the history of modern physics; since 1989 he has increasingly dealt with the history of science and technology in the GDR. He is the author of numerous pieces on the history of modern physics, including his edition of the Farm Hall transcripts, *Operation Epsilon* (Berlin, 1993) and, most recently, *Science under Socialism: East Germany in Comparative Perspective* (Berlin, 1997; Cambridge, Mass., 1999), coedited with Kristie Macrakis.

**Charles S. Maier** is the author of *Dissolution: The Crisis of Communism and the End of East Germany* (1997); *The Unmasterable Past: History, Holocaust, and the German National Identity* (1988); and *Recasting Bourgeois Europe* (1975, 1988). He has also edited over half a dozen collections of essays, including *In Search of Stability* (1987). He is currently collaborating on a world history of the twentieth century. In February 1999 he received the Commander's Cross of the Federal Republic of Germany for his work in German history and his service as Director of the Minda de Gunzburg Center for European Studies at Harvard University since 1993.

**Dominique Pestre** was trained as both a physicist and a historian. He has published on cultural and political histories of physics, on the history of CERN, and on perceptions and definitions of science. He is currently working on the military and its relation to scientific practices.

**Robert W. Seidel** is the ERA Professor of the History of Technology and director of the Charles Babbage Institute at the University of Minnesota. He is the author or coauthor of two books on the history of the Atomic Energy Commission national laboratories and former director of the Bradbury Science Museum in Los Alamos, New Mexico.

**Christiane Sinding** is a senior researcher at the Institut National de la Santé et de la Recherche Médicale (INSERM). She was trained as a pediatrician and a historian of science. She is author of *Le clinicien et le chercheur: Des maladies de carence à le médecine moléculaire* (1991) and of several articles on the history of endocrinology. She is currently working on the history of diabetes mellitus.

**Vassiliki Betty Smocovitis** is Associate Professor in the Department of History at the University of Florida in Gainesville, Florida. Her research interests center on the history of evolutionary biology. She is the author of a book titled *Unifying Biology: The Evolutionary Synthesis and Evolutionary Biology* published in 1996 by Princeton University Press.

# Index